AF574894

PRACTICAL UROLOGIC CYTOPATHOLOGY

Practical Urologic Cytopathology

RICARDO H. BARDALES, M.D.

OXFORD
UNIVERSITY PRESS
2002

OXFORD
UNIVERSITY PRESS

Oxford New York
Athens Auckland Bangkok Bogotá Buenos Aires Cape Town
Chennai Dar es Salaam Delhi Florence Hong Kong Istanbul Karachi
Kolkata Kuala Lumpur Madrid Melbourne Mexico City Mumbai Nairobi
Paris São Paulo Shanghai Singapore Taipei Tokyo Toronto Warsaw

and associated companies in
Berlin Ibadan

Published by Oxford University Press, Inc.
198 Madison Avenue, New York, New York, 10016
http://www.oup-usa.org

Library of Congress Cataloging-in-Publication Data
Bardales, Ricardo H.
Practical urologic cytopathology / by Ricardo H. Bardales.
p. ; cm. Includes bibliographical references and index.
ISBN 0-19-513495-8
1. Urinary organs—Cytopathology.
[DNLM: 1. Urinalysis—methods. 2. Urologic Diseases—urine.
3. Biopsy, Needle—methods. 4. Urine—cytology. 5. Urologic
Neoplasms—diagnosis. QY 185 B245p 2001] I. Title.
RC900 .9 .B37 2001
616.6′07—dc21 2001036170

2 4 6 8 9 7 5 3 1

Printed in Hong Kong
on acid-free paper

To my parents, Ricardo and Walde,
and to Angela, Angie, and Ricky,
my role models
and forever companions. . . .
I owe you so much!

PREFACE

It is a common practice that *urine sediment* evaluation is conducted to help the study of "medical" or non-neoplastic conditions affecting the urinary tract. In contrast, *urine cytology* evaluation is performed primarily to search a neoplastic or surgically amenable process. The study of the urine sediment pertains to the microscopic findings of cellular and non-cellular elements and it is performed essentially in non-stained urine samples. In contrast, the cytological evaluation of urine requires the application of the Papanicolaou's stain. Both studies are conducted in different sections of the pathology laboratory with the clinical laboratory evaluating the urine sediment and the cytology laboratory the urine cytology. Informative urine sediment is of particular value to the nephrologist in the management of renal disorders. Likewise, an accurate urine cytology interpretation is invaluable for the urologist. Definitive therapy is based on these results.

Although the two laboratories seem to be divergent, they are complementary and share common findings that ultimately benefit the patient. For example, urolithiasis may be identified in the cytology laboratory but the definitive characterization of the crystals is made in the clinical laboratory. Similarly, the presence of blood or cell clusters, detected in the urine sediment, in the clinical laboratory should trigger the investigation of papillary urothelial carcinoma, which is often made by examination of urine cytology in the cytology laboratory.

As methods of urine collection, cytopreparation, and staining improved in the last decades, urine cytology is being used more frequently as a screening and diagnostic tool for the detection of neoplastic and non-neoplastic processes affecting the urinary tree. Description of cellular and non-cellular elements present in a urine cytology sample is the focus of this book and emphasis in the differential diagnosis and comparison of the cytologic features in both voided and instrumented urine specimens is frequently emphasized. The study of the urinary sediment as seen in a urine analysis performed in the clinical laboratory, is not the subject of this book. However, cellular and non-cellular components of clinical importance that need to be recognized and mentioned in a cytology report are covered.

This book is divided in 10 chapters organized in a conventional sequence. After a brief historical review in Chapter 1, the next four chapters describe normal, inflammatory, and infectious processes that affect the urinary tract. Chapters 6 and 7 review the literature in detail and provide practical guidelines for the cytologic interpretation of primary and secondary neoplasms of the urinary tract. The value of urine cytology is analyzed in Chapter 8. The closing chapters review the use and value of special methods applied to urine cytology specimens and the fine needle aspiration of urothelial neoplasms.

In order to facilitate the reading of this book, each disease entity is approached using the following format: definition and clinical presentation, cytomorphologic features, and differential cytologic diagnosis. Key concepts that include diagnostic clues for each entity are listed in boxes. Because the primary audience for this work will be individuals familiar with cytopathology, I limited the scope to cytology leaving the histopathology for exceptional cases and only if they conveyed a particular point. The aim is to cover in a concise manner the most salient definitional, clinical, and mainly cytologic features of the disease processes most frequently encountered in the urinary tract. Cytology can be very well described in prose, however, the art of cytopathology rests in the use of photomicrographs to illustrate the cytologic details. This book is extensively illustrated with approximately 300 color photomicrographs. I hope that this format provides a clear understanding of urologic cytology and guides the reader through this journey.

This book is primarily intended for those who interpret urine cytology specimens including general pathologists, cytopathologists, cytopathology fellows, pathology residents, cytotechnologists, and cytotechnology students. Our experience tells us that many of these individuals are frustrated with the difficulty and complexity of interpreting urine cytology specimens. I hope this book helps to solve these difficulties.

Minneapolis, Minnesota R.H.B.

ACKNOWLEDGMENTS

Without the contribution and support of several individuals I would not have been able to write this book. I owe my eternal gratitude to Drs. Leopold G. Koss, Norwin H. Becker, and Klaus Schreiber who introduced and guided me in the fascinating world of tissues and cells.

Dr. Michael W. Stanley, an outstanding teacher and coworker and better friend, played a vital role in the conception and realization of this book and I must thank him for his constant support and highly intelligent suggestions.

Angela, my wife, gave me the understanding, encouragement, and support that have permitted me to accomplish this and other important goals in life.

Last but not least, several persons with contagious enthusiasm and eagerness have contributed in the completion of this book. Drs. Luis E. De Las Casas and Perkins Mukunyadzi wrote two fundamental chapters, "Diagnostic Value of Urine Cytology Examination" and "Biomarkers and Special Techniques for Detection of Urothelial Cancer," respectively. Dr. Cynthia C. Nast of the Renal Pathology Service at Cedars-Sinai Medical Center, Dr. Linda K. Green of the Pathology Service at Baylor College of Medicine, Dr. Terry Oberly of the Department of Pathology at the University of Wisconsin in Madison, Dr. Andrew C. Henke of the Department of Pathology at the University of Iowa, Dr. Hassan N. Tawfik of the Department of Pathology at The National Cancer Institute and Cairo University, and Karen Ringsrud of the Department of Pathology at the University of Minnesota have given suggestions for the text and provided photographic material of outstanding quality. Elaine M. Challacombe, Curator in the Owen H. Wangensteen Historical Library of Biology and Medicine at the University of Minnesota, Shirah Winicur, Library Assistant, and the Service de la Communication of the "Musee du Louvre" have been invaluable in providing me with excellent material for the chapter on the history of urine cytopathology.

My appreciation to Oxford University Press editing and production departments in particular: Nancy J. Wolitzer, Managing Editor Medicine; Lauren M. Enck, Senior Editor Clinical Medicine; Edith Barry, Associate Editor; and Joanna Rulf, Editorial Assistant. This list would be incomplete without acknowledging Elisabeth Lanzl of the University of Chicago who patiently reviewed the grammar and syntax of this book and suggested pertinent changes.

CONTENTS

PRACTICAL UROLOGIC CYTOPATHOLOGY

1

FROM URINE CONTEMPLATION TO URINE EVALUATION: HISTORICAL OVERVIEW

A wealth of information about urinary examination is available within the history of urology. Symptoms and signs of disorders of the urinary tract, as common and remote as the origins of humanity, led to the development of urology and, later, nephrology. Therefore, urology was one of the first branches, if not the first branch, of medicine to emerge from the mists of antiquity and ignorance in an attempt to alleviate these disorders (Wershub, 1970).

Lack of knowledge about the etiology of diseases in antiquity led humans to ascribe and adopt superstitious beliefs to explain and cure their ailments. Urine was believed to be a soul substance and was used to rub the human body to protect the individual against evil spirits. Rain was believed to be the urine of a deity, and it was thought that exposure to rain would be soul-healing. Urine was used for bathing to beautify the skin, and because of its ammonium salts, urine was used as a mouthwash to prevent caries. There are indications that urine was used for macerating tobacco leaves for cigars, for flavoring cheese, and by bakers as a leavening agent. Some of these practices may still be prevalent among native people (Wershub, 1970).

URINARY TRACT DISORDERS IN ANTIQUITY

Two thousand years before the common era, Mesopotamians and Egyptians described urolithiasis. Schistosomiasis and hematuria as well as urinary incontinence were described in papyri from ancient Egypt (Fig. 1–1A,B). Paleopathologic studies have shown evidence of schistosomiasis in renal tubules and ureters of Egyptian mummies dating to around 2000 BCE (Ruffer, 1921).

Prescription for urologic problems was common practice in Egyptian and Assyrian medicine (Thompson, 1934). As in Babylonian texts, there are no descriptions in Egyptian papyri of therapy based on observations of the characteristics of urine. In contrast, such therapy was common practice in Byzantine and Western medieval medicine, where it persisted until the Renaissance, when therapy based on urine observation was disproved.

EARLY DEVELOPMENTS IN THE STUDY OF THE ANATOMY AND PHYSIOLOGY OF THE URINARY TRACT

Greek medicine is based in part on 2000 years of Babylonian and Egyptian observations. During a particularly creative period (500 BCE to 500 CE), the name of Hippocrates dominated all medical thought, and medicine became an art, a science, and a profession.

Hippocrates separated medicine from philosophy, magic, and superstition. He devoted a book to the study of urine, and described its physical characteristics and its clinical and prognostic implications. He be-

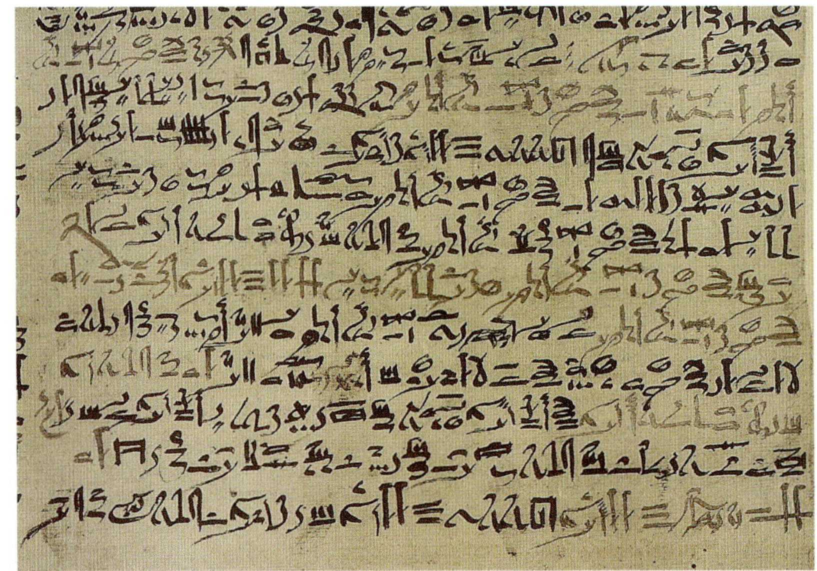

A

B

Figure 1–1A,B. *The Edwin Smith Surgical Papyrus.* Egyptians described diseases that affected the urogenital tract. This surgical case is that of a man who had a neck fracture that resulted in quadriplegia and urine incontinence. The translation of lines 16 and 17 (in red) is as follows: "One having a dislocation in a vertebra of his neck while he is unconscious of his two legs and two arms, and his urine dribbles. An ailment not to be treated." Reproduced from The Edwin Smith Surgical Papyrus. Courtesy of the Owen H. Wangensteen Historical Library of Biology and Medicine, University of Minnesota, Twin Cities Campus, Minneapolis, MN.

lieved that painless hematuria without fever did not indicate any serious problem (Wershub, 1970). He recognized at least three conditions of the urinary tract that could be evaluated by the nature and appearance of urinary outflow, namely, colic due to stones, suppuration, hematuria, and probably tuberculosis or malignant disease.

Galen (130–200 CE), a famous Roman physician and one of the most remarkable figures in medical history, enriched the concept of renal physiology based on animal experimentation. He attempted to explain the physiology of kidneys and the formation of urine. In addition, he understood the relationship of urolithiasis to the kidney, ureter, and bladder (Wershub, 1970).

After the fall of Rome, inquiries into the anatomy and physiology of the human body decreased notably. This period in history, known as "a thousand years of darkness," made minimal contributions to the advancement of medicine and, in particular, to the disorders of the urinary system.

URINE CONTEMPLATION

Long before the true significance of urine was assayed and evaluated, it played an important role in the evolution of medicine. It is probable that ancient humans were awed by the status and meaning of the appearance of urine (Wershub, 1970) (Figs. 1–2, 1–3).

Avicenna, the famous Persian physician and philosopher (980–1037), strongly believed in the physical characteristics of urine (color, density, sediment, and odor) as indicators of patient strength, the evolution and prognosis of illness, and the presence of

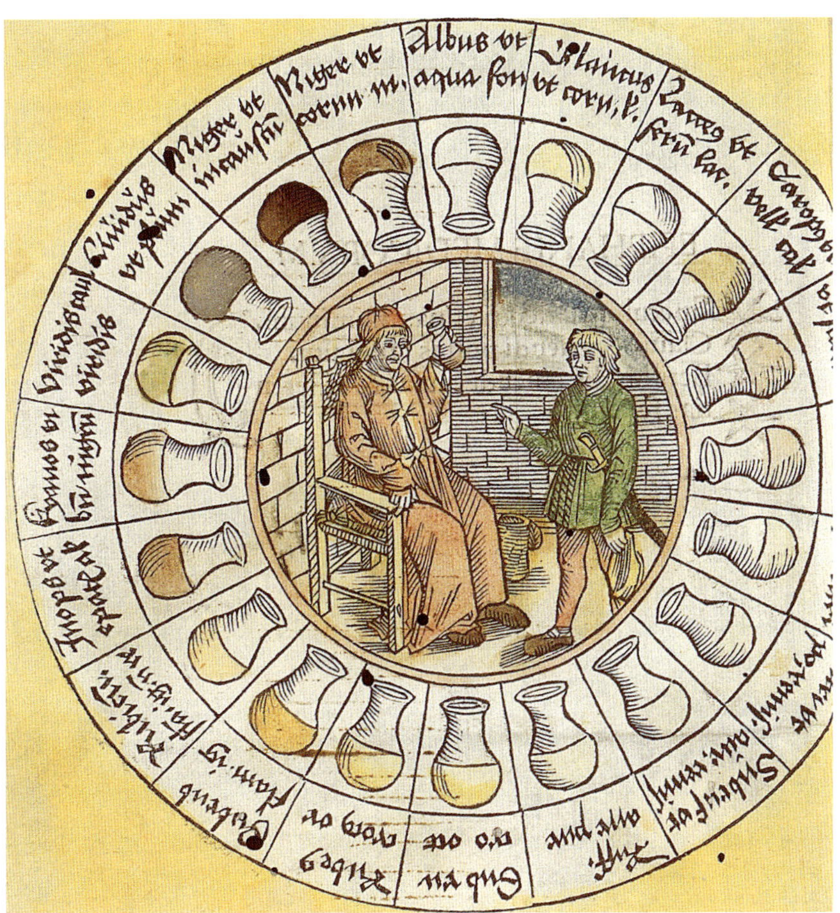

Figure 1–2. *Uroscopy wheel.* Visual examination of urine by a physician is depicted in manuscripts from as early as the fourteenth century. The treatment for an illness was based on comparison of the patient's urine with that shown in a uroscopy chart. This illustration appears in "Epiphanie medicorum," written by Ulrich Pinder, published in 1506. Courtesy of the Owen H. Wangensteen Historical Library of Biology and Medicine, University of Minnesota, Twin Cities Campus, Minneapolis, MN.

Figure 1–3. *"The Dropsical Woman."* Painted on wood by Gerard Dou (1613–1675) in 1663. This painting depicts a physician examining the urine of a woman with anasarca, presumably the result of a nephrotic syndrome. (Reproduced with permission from the Musée du Louvre)

liver abscess, among other conditions. He assessed the presence of calculus, abscess, or tumor by judging the consistency of the urine sediment. A bladder tumor was suspected when the urine had a dreadful odor (Wershub, 1970).

Visual examination of urine by a physician, an apothecary, and even a charlatan to diagnose disease is depicted in manuscripts dating as far back as the early fourteenth century. In the Middle Ages and even later, the urine flask became the symbol of the medical profession. Observation of the characteristics of urine contained in the flask was the main diagnostic method available to the physician, who supplemented this analyses by taking the pulse and studying the appearance of the facies, tongue, and excrement. Uroscopy, defined as "the art of making a diagnosis and prescribing therapy solely on the evidence afforded by looking at the urine contained in a specially made flask," was considered serious and almost scientific in the Middle Ages (Wershub, 1970).

Uroscopy, which was an art up to the sixteenth century and was practiced for more than 400 years, made little scientific progress. Interest in urine stimulated the search for the golden elixir by the alchemist, followed by the chemist, and finally by the scientist and the establishment of modern urine analysis.

FROM URINE CONTEMPLATION TO URINE ANALYSIS

Alchemy was a mixture of empirical observations combined with astrology, mysticism, and religious philosophy. Until nearly the end of the seventeenth century, diseases were diagnosed by visual examination of urine (uroscopy).

The earliest observations of urine with the aid of the microscope occurred in the first half of the seventeenth century. Using rudimentary microscopes, Nicolas-Claude Fabri de Peiresc was probably the first to observe lithiasis in urine samples in 1630 (Fogazzi and Cameron, 1996). A few decades later, Robert Hooke described urine crystals in his *Micrographia* (Hooke, 1665) (Fig. 1–4A,B). Notable advances in the histology and physiology of the kidney were made, but urinary microscopy was not a subject of great interest at the end of the seventeenth century.

The early years of the eighteenth century saw the advent of the scientific era and the application of chemistry to the examination of urine. Johann Baptista Van Helmont (1579–1644), who lived near Brussels, inaugurated the quantitative study of urine. He accurately determined the specific gravity of urine and noted the relationship between water ingestion and urine concentration (Leicester, 1956). Lorenzo Bellini (1643–1704), a famous Italian physician, and Hermann Boerhaave (1668–1738), a remarkable Dutch physician, made further advances in the chemistry of urine. Bellini skillfully used the microscope to study the structure of the kidneys, but he did not employ urinary microscopy (Fogazzi and Cameron, 1996). In addition, he wrote a classical description of the gross anatomy of the kidney (Wershub, 1970).

Progress in urine microscopy in the eighteenth century was minimal and limited to the examination of crystals, similar to the discoveries reported in the previous century. With the advent of better microscopes, the application of microscopy to all major medical fields, including the urinary tract, exploded in the nineteenth century. Nevertheless, the emphasis was on descriptions of crystals and casts, not of cells.

The quantitative and qualitative chemical evaluation of urine in the study of diseases advanced considerably in the nineteenth century. One of the earliest textbooks devoted to scientific urine analysis was written by George Owen Rees in 1836, which focused on the analysis of blood and urine in health and disease (Wershub, 1970). Pierre Rayer and Eugene N. Vigla introduced urine microscopy into everyday clinical practice around 1835 (Richet, 1991). In 1861 Lionel Beale published his well-known book on urine, which stressed the importance of microscopic examination. This microscopic evaluation of the unstained urine sediment, or *urinalysis,* to search for casts, crystals, and different cell types, is still performed in chemistry laboratories today.

In the first part of the nineteenth century, books and atlases on urine microscopy were concerned largely with crystals, with only a poor understanding of other components of the urinary sediment. In the second half of the century, technological advances in microscope design permitted the description of urinary changes in a wide spectrum of disorders.

Progress in the evaluation of urine sediment by nephrologists continued in the twentieth century. Phase contrast microscopy, developed in the early 1950s, was considered to be the state of the art for urine microscopy. Advances in clinical urine cytology were yet to come.

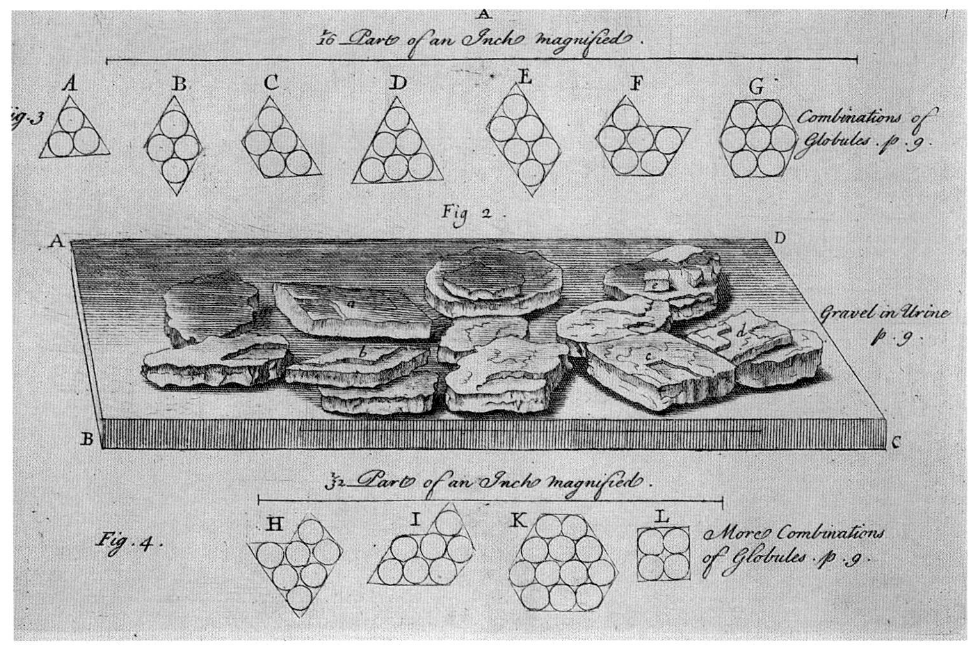

A

Figure 1–4. *Micrographia.* One of the first microscopic evaluations of urine crystals (*A*, Fig. 2, Gravel in Urine) was made by Dr. R. Hooke in 1665 and reproduced in his Micrographia Restaurata, published in 1745 (*B*). Courtesy of the Owen H. Wangensteen Historical Library of Biology and Medicine, University of Minnesota, Twin Cities Campus, Minneapolis, MN.

FROM URINE ANALYSIS TO URINE CYTOLOGY

Despite the invention of the microscope in the late sixteenth century, because of its limitations, evidence of its use for evaluating human cells in the next two centuries was sporadic. With the exception of the work of Abraham Vater (Vater, 1747), who used the microscope to evaluate the epithelial cells of human skin, before the nineteenth century there was virtually no mention of microscopic evaluation of cellular elements in urine samples.

Alfred Donne, in France in 1844, and Hermann Lebert, in Germany in 1845, independently published the first cytology atlases (Hajdu, 1977). Lebert, a pathologist, collected specimens for cytologic examination from effusions, respiratory tract secretions, and urine in addition to numerous examples of imprint cytology of neoplastic and nonneoplastic conditions (Lebert, 1845) (Fig. 1–5A,B). Shortly thereafter, in 1856, W. D. Lambl published one of the first documents on the subject of diagnostic urine cytology (Lambl, 1856) (Fig. 1–6). He described eight cases of carcinoma of the bladder detected from urine specimens and stressed the contribution of microscopic evaluation of urine in the diagnosis of this carcinoma (Koss, 1996; Long and Cohen, 1992). Several other reports on that subject were also written in the second half of the nineteenth century (Beale, 1864; Dickinson, 1869; Ferguson, 1893; Sanders, 1864).

Although Albert Daiber emphasized the role of urine cytology in clinical practice in his atlas on urine sediment published in 1896 (Daiber, 1896), the microscopic study of stained urine sediment was not introduced until the twentieth century. Except for a statement on the use of microscopic evaluation of stained histology sections of urine sediment for diagnosis of bladder cancer, made by Frank Ferguson at a meeting of the New York Pathological Society in 1892 (Koss, 1996), there are no records on the cytologic evaluation of cancers of the urinary tract until 1945, when Papanicolaou and Marshall published their findings in the journal *Science* (Papanicolaou and Marshall, 1945).

COPPER-PLATES

OF

Dr. *HOOKE*'s Wonderful Diſcoveries

BY THE

MICROSCOPE,

Reprinted and fully Explained:

Whereby the moſt Valuable PARTICULARS in that Celebrated AUTHOR's

MICROGRAPHIA

Are brought together in a narrow Compaſs;

AND

Intermixed, occaſionally, with many Entertaining and Inſtructive DISCOVERIES and OBSERVATIONS in NATURAL HISTORY.

Rerum Natura nuſquam magis quàm in minimis tota eſt.
PLIN. Hiſt. Nat. Lib. XI. C. 2.

LONDON:

Printed for and Sold by JOHN BOWLES, Printſeller at the *Black Horſe* in *Cornhill.* Sold alſo by R. DODSLEY, in *Pallmall,* and JOHN CUFF, Optician, in *Fleetſtreet.*

MDCCXLV.

B

Figure 1–4. *(Continued)*

Of historical interest is Dr. L. G. Koss's personal communication on the diagnosis of bladder carcinoma in situ made by Dr. Fred Stewart on a urine cytology preparation in 1942. The patient was Dr. James Ewing, Chief of Pathology at Memorial Hospital in New York City, who died of metastatic urethelial carcinoma a few months later (Koss, 1980).

In the last four decades of the twentieth century, substantial contributions to the epidemiology and cyto- and histomorphology, the role of cytology, and the natural history of bladder cancer originated from Dr. Koss's laboratories at Memorial Hospital and Montefiore Medical Center in New York.

At the turn of the twenty-first century, cytologic evaluation of urine sediment still is considered useful and difficult by the cytopathologist, controversial and difficult by the general surgical pathologist, and of limited use by most urologists.

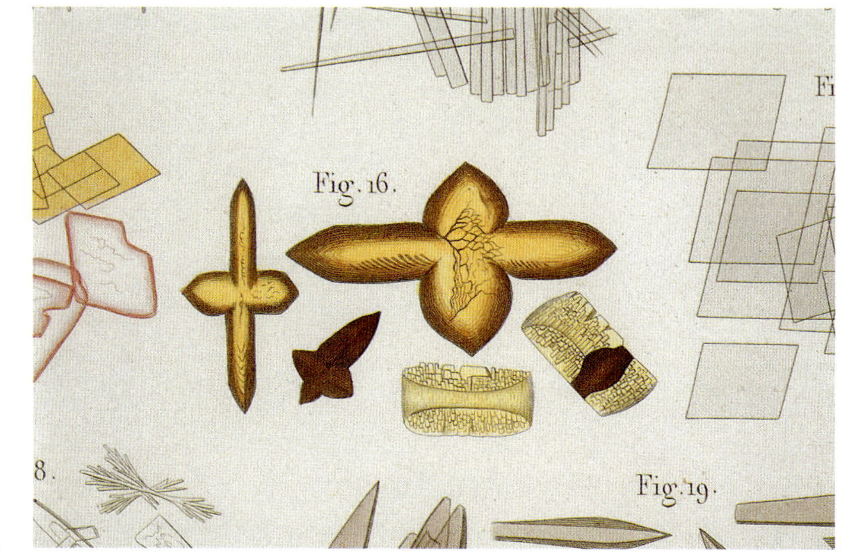

A

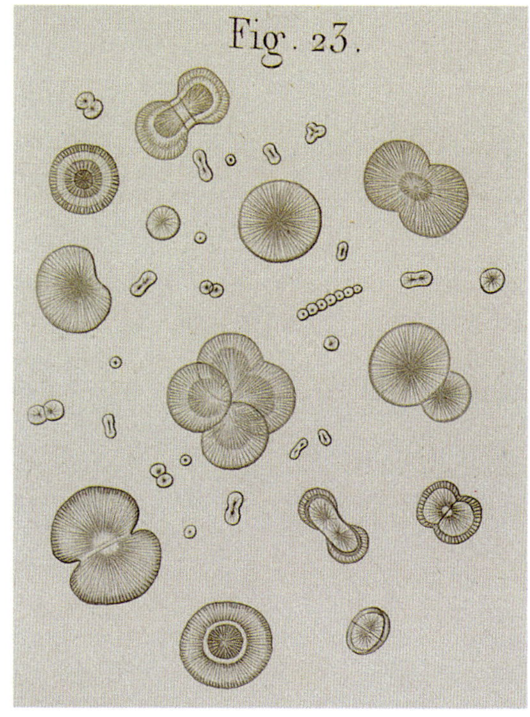

B

Figure 1–5. *Urine crystals.* Microscopic appearance of uric acid (*A*, Fig. 16) and calcium oxalate (*B*, Fig. 23) crystals, as illustrated by H. Lebert in his "Traite D'Anatomie Pathologique," published in 1861. Courtesy of the Owen H. Wangensteen Historical Library of Biology and Medicine, University of Minnesota, Twin Cities Campus, Minneapolis, MN.

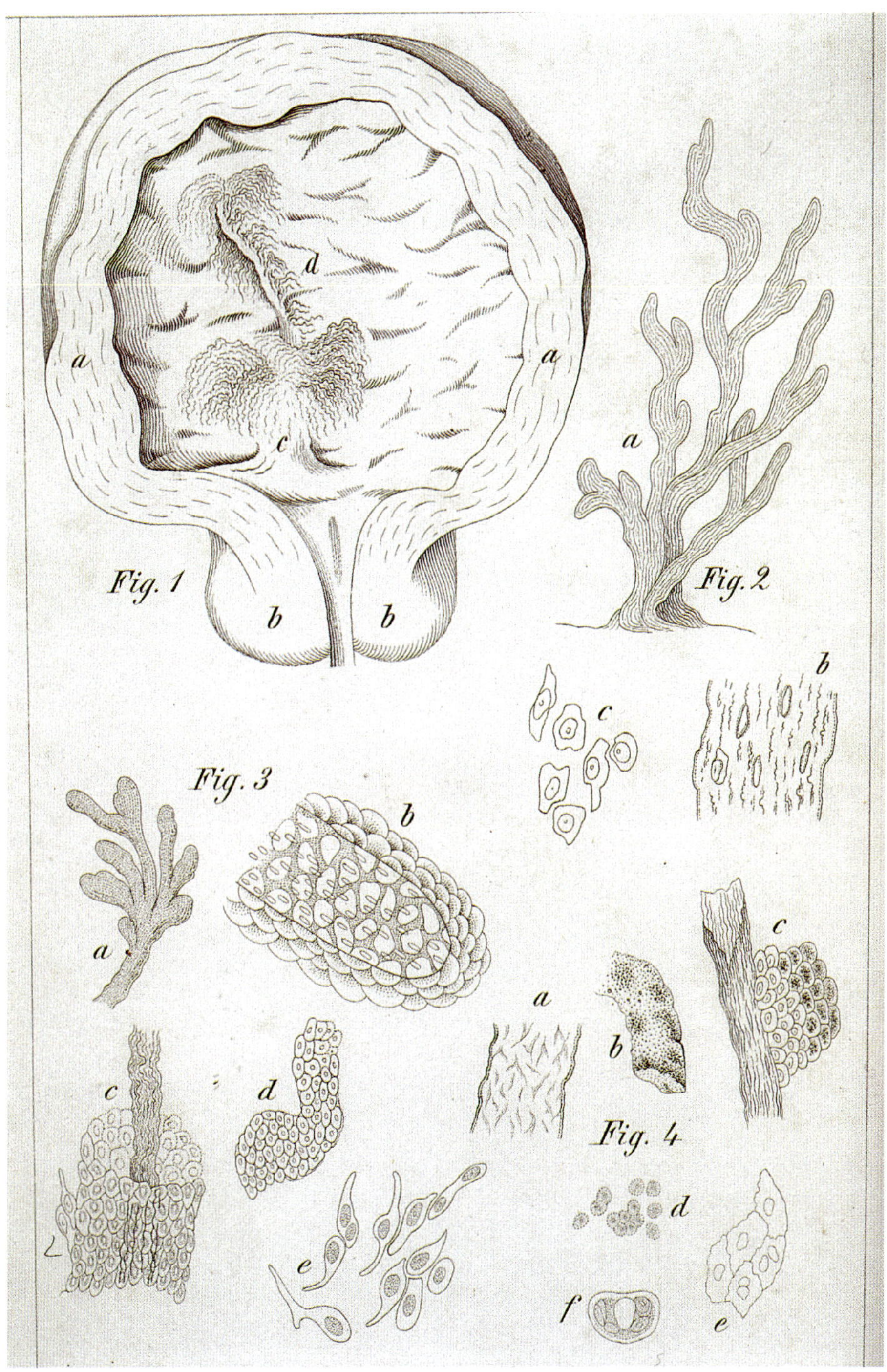

Figure 1–6. *Bladder cancer.* In 1856, Dr. W. D. Lambl published his monograph in *Vierteljahrschrift fur die Praktische Heilkunde* describing the gross appearance, histology, and cytology of bladder cancer. Figures 1d, 2a, and 3a,b show a low-grade papillary urothelial neoplasm of the bladder. Figures 2c and 3e illustrate urothelial cells with cytoplasmic elongation, round/oval nuclei, and evenly distributed chromatin characteristic of low-grade papillary tumors. Courtesy of the Owen H. Wangensteen Historical Library of Biology and Medicine, University of Minnesota, Twin Cities Campus, Minneapolis, MN.

As in any field in cytology, prompt preparation of the specimen after collection, adequate staining, and skillful interpretation are of transcendental importance for rendering a diagnosis worthy of trust for patient management. The advent of special techniques applied to urine cytology will enhance diagnostic accuracy, particularly for well-differentiated urothelial neoplasms, the main limitation of urine cytology in the diagnosis of tumors of the urinary tract.

REFERENCES

Beale, L. S. (1864). *The Microscope and Its Application to Clinical Medicine.* Highley, London.

Daiber, A. (1896). *Mikroskopie der Harnsedimente.* J. F. Bergmann, Wiesbaden.

Dickinson, W. H. (1869). Portions of cancerous growth passed by the urethra. *Trans Pathol Soc Lond* 20: 233–7.

Ferguson, F. (1893). The diagnosis of tumors of the bladder by microscopic examinations. *Proc NY Pathol Soc* 71–3.

Fogazzi, G. B., Cameron, J. S. (1996). Urinary microscopy from the seventeenth century to the present day. *Kidney Int* 50: 1058–68.

Hajdu, S. I. (1977). Cytology from antiquity to Papanicolaou. *Acta Cytol* 21: 668–676.

Hooke, R. (1665). *Micrographia: Or Some Physiological Descriptions of Minute Bodies Made by Magnifying Glasses with Observations and Inquiries Thereupon.* Martin and Allestry, London.

Koss, L. G. (1980). On the history of cytology. *Acta Cytol* 24: 475–7.

Koss, L. G. (1996). Historical note. In *Diagnostic Cytology of the Urinary Tract with Histopathologic and Clinical Correlations* (L. G. Koss, ed.), pp. xvii–xxi. Lippincott-Raven, Philadelphia.

Lambl, W. D. (1856). Ueber Harnblasenkrebs. Ein Beitrag zur mikroskopischen Diagnostik am Krankenbette. *Prager Viertelj Schr Heilkunde* 49: 1–15.

Lebert, H. (1845). *Physiologie Pathologique ou Recherches Cliniques, Experimentales et Microscopiques.* J. B. Bailliere, Paris.

Leicester, M. H. (1956). *The Historical Background of Chemistry.* Wiley, New York.

Long, S. R., Cohen, M. B. (1992). Classics in cytology. V: William Sanders and early urinary tract cytology. *Diagn Cytopathol* 8: 135–6.

Papanicolaou, G. N., Marshall, V. F. (1945). Urine sediment smears as a diagnostic procedure in cancers of the urinary tract. *Science* 101: 519–21.

Richet, G. (1991). From Bright's disease to modern nephrology: Pierre Rayer's innovative method for clinical investigation. *Kidney Int* 39: 787–92.

Ruffer, M. A. (ed.). (1921). *Studies of Paleopathology of Egypt.* University of Chicago Press, Chicago.

Sanders, W. K. (1864). Cancer of the bladder. Fragments forming urethral plugs discharged in the urine. Concentric colloid bodies. *Edinburgh Med J* 10: 273–4.

Thompson, S. (1934). Assyrian prescription for diseases of the urine. *Babyloniara* 14: 108–9.

Vater, A. (1747). A remarkable cutaneous disorder. In *The Philosophical Transactions, Abridged* (Vol. IX, p. 117). C & R Baldwin, London.

Wershub, L. P. (1970). *Urology. From Antiquity to the 20th Century.* Warren H. Green, St Louis, MO.

2

NONUROTHELIAL ELEMENTS PRESENT IN URINE

NONUROTHELIAL elements present in urinary tract specimens can be arbitrarily classified as cellular and noncellular (Table 2–1).

Because the presence of some of these elements may have significant clinical implications, their cytologic evaluation is as important as that of urothelial cells and should be mentioned in the urine cytology report.

CELLULAR ELEMENTS

Squamous Cells

Squamous cells are common findings, particularly in voided urine of women. They originate from a normal urothelial variant of the vaginal type located in the trigone of the urinary bladder but are rarely present in men (Wiener et al., 1979). This type of squamous epithelium lacks keratinization and contains abundant intracytoplasmic glycogen. This epithelium is histologically similar to cervical epithelium and probably follows the same cyclic hormonal pattern (Reuter, 1997). Although this epithelial variant is known as *squamous metaplasia*, there is no evidence that inflammation is associated with these changes (Box 2–1).

Cytology. The cytologic features are identical to those of nonkeratinized squamous cells seen in a normal cervico-vaginal cytologic smear in women or in the distal urethra in men (Fig. 2–1). The distinction may not be of clinical significance. Keratinization with hyperkeratosis or parakeratosis is a clinically significant change and should be investigated so that underlying leukoplakia or squamous carcinoma of the bladder can be ruled out (Koss, 1996).

Red Blood Cells

The presence of blood in a urine specimen (hematuria) may be gross or microscopic and symptomatic or asymptomatic. Although the presence of gross symptomatic hematuria indicates subjacent renal or urologic pathology, asymptomatic microscopic hematuria may not always be significant, and further workup may not reveal an obvious source of bleeding.

A basic problem in evaluating hematuria is the large number of diagnostic possibilities. Isolated hematuria, without proteinuria or casts, may be seen in neoplasms, lithiasis, or infection in any part of the urinary tract (urologic hematuria). In contrast, nephronal hematuria always indicates significant damage to the nephron, that is, glomerulonephritis, tubulointerstitial damage, or vasculitis, and is usually accompanied by red blood cell casts and proteinuria. Few primary renal diseases cause isolated hematuria without casts, for example, IgA nephropathy. Glomerular disease is the main cause of hematuria in children (Bonfante et al., 1996). In adults, the most common lesions are, in decreasing order, prostatic benign hypertrophy, urinary tract infection, urolithiasis, and neoplasms (Miguel-Gomara Perello et al., 1993).

Data indicate that the incidence of underlying urologic cancers in patients with asymptomatic microscopic hematuria ranges from 1% to 13%, and that up to 10% of patients with bladder cancer have asymptomatic microscopic hematuria and positive

Table 2–1. Nonurothelial Elements

Cellular	*Noncellular*
Squamous cells	Crystals
Red blood cells	Casts*
Leukocytes	Corpora amylacea
Renal tubular cells	Lubricant
Seminal vesicle cells and other cells	Artifacts

*Including cellular casts.

Box 2–1.

Squamous cells in urine are more common in women than in men.

Squamous cells are nonkeratinized and are identical to those seen in cervical Pap smears.

Box 2–2.

Occasional red blood cells may be seen in normal individuals.

Urologic hematuria is more common in the elderly.

Nephronal hematuria is more common in the young.

Abnormal red blood cell morphology and red blood cell casts indicate nephronal hematuria.

Intact red blood cells indicate hematuria of extrarenal origin.

urine cytology (Hattori et al., 1990; Mohr et al., 1986). Therefore, it appears that cytologic examination of voided urine should be the first test in patients with asymptomatic microscopic hematuria. However, screening of urine for microscopic hematuria as an indicator of urologic cancer remains controversial (Hiatt and Ordonez, 1994). Follow-up of patients with asymptomatic microscopic hematuria shows that the diagnostic yield of renal, prostatic, or urothelial cancer ranges from 0% to 5.2% of cases after 3 to 20 years (Howard and Golin, 1991; Murakami et al., 1990). The decision to pursue an exhaustive investigation will depend on the individual clinical setting and on the results of basic blood and urine analysis (Miguel-Gomara Perello et al., 1993) (Box 2–2).

Cytology. The cytology report must document the presence, number, and mor-

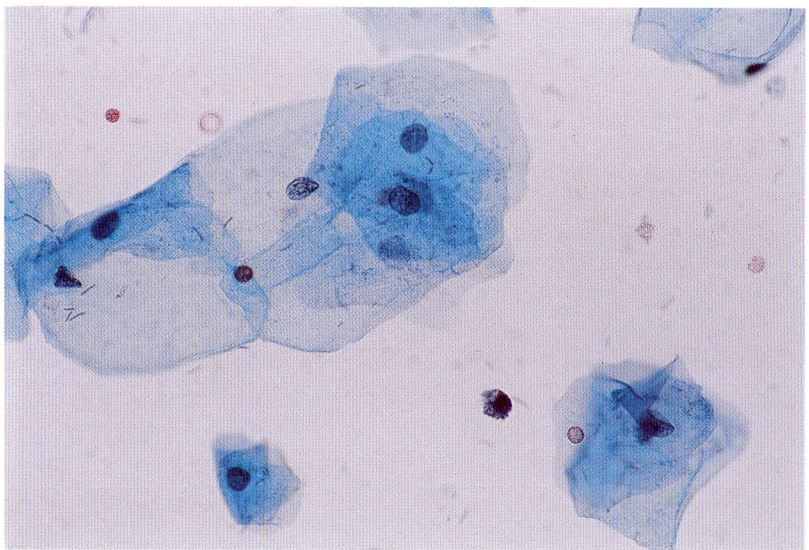

Figure 2–1. *Benign squamous cells.* A single degenerated urothelial cell and rare red blood cells are also present in this voided urine specimen from a healthy woman. (Papanicolaou stain, ×600)

phology of the red blood cells, as well as the presence of leading indicators of the origin of the hematuria. Intermittent hematuria correlates with the presence of serious urologic disorders, including bladder cancer (Messing et al., 1987).

Occasional red blood cells (one to eight per high-power field) are present in voided urine specimens from patients with minor urologic disorders and from normal individuals (Mohr et al., 1987). In contrast, significant isolated asymptomatic microscopic hematuria (more than eight red blood cells per high-power field) has a higher correlation than occasional red blood cells with the detection of urologic cancers in urine sediment from unselected patients (Froom et al., 1986; Mohr et al., 1987).

The morphology of red blood cells is helpful in suggesting the source of hematuria. Red blood cells of abnormal shape associated with red blood cell casts and/or proteinuria strongly suggest a nonneoplastic primary renal parenchymal cause (Fracchia et al., 1995; Marcussen et al., 1992). Similarly, red blood cells with a peripheral double ring and an empty center suggest a renal origin, in contrast to those that fail to show the empty center, suggesting an extrarenal source (Fig. 2–2) (Koss, 1996). Intact red blood cells suggest hematuria of extrarenal origin.

The advantages of cytocentrifuged urine cytology over membrane-filter or phase-microscopic techniques include improved sensitivity and specificity in the identification of elements of the urine sediment to detect renal parenchymal disease (Eggensperger et al., 1988). Urine cytology is also highly accurate in the detection of high-grade papillary urothelial carcinoma and carcinoma in situ in patients with asymptomatic microscopic hematuria (Carson et al., 1979; Fracchia et al., 1995).

Although cytology can suggest the source of hematuria, other methods permit a less subjective interpretation of the subtle abnormalities present in the red blood cells. Recently, automated comparative measurement of the mean corpuscular volume of red blood cells in urine and peripheral blood has been suggested to be a highly accurate method for differentiating nephronal from urologic hematuria (Angulo et al., 1999).

Neutrophils

Urinary neutrophils are not site- or disease-specific for the urinary tract. They may be associated with various disorders of the urinary tract or represent contamination from an external genital tract infection. Few urinary neutrophils are a common finding, particularly in women. They usually originate in

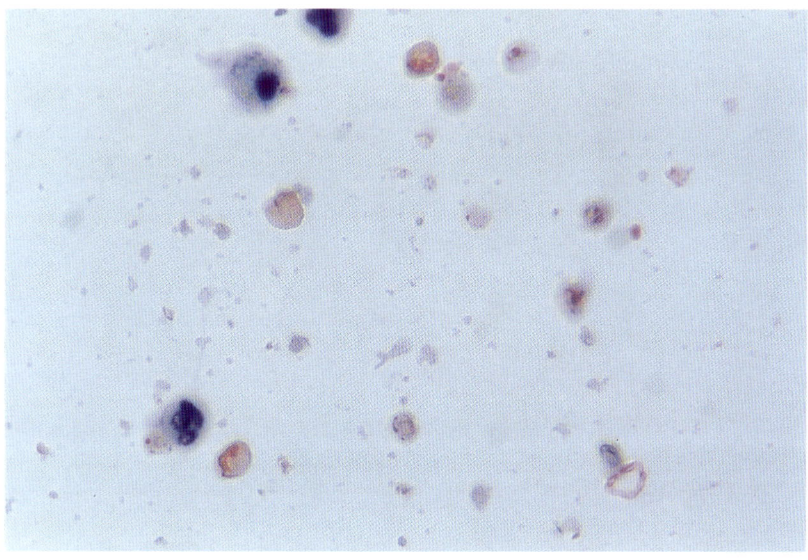

Figure 2–2. *Nephronal hematuria.* Markedly degenerated red blood cells show anisocytosis, ring-shaped features, and cytolysis. Voided urine from a patient with a renal transplant. (Papanicolaou stain, ×1000)

Box 2–3.

Few neutrophils are a common finding in women and uncircumcised men.

Neutrophils and leukocyte casts indicate a disease of renal origin, for example, pyelonephritis, interstitial nephritis, or lupus nephritis.

Large numbers of neutrophils indicate acute bacterial infection.

Box 2–4.

Eosinophiluria may be seen in hypereosinophilia, acute interstitial nephritis, and noninvasive bladder cancer treated with mitomycin C.

Eosinophils have a bilobed appearance, with inconspicuous or absent cytoplasmic granules on Papanicolaou stain.

the external genitalia and have no clinical significance.

An increased number of leukocytes in the urine, mainly neutrophils, is seen in almost all renal and urinary tract diseases (Lindqvist and Wahlin, 1975). The concomitant presence of leukocyte casts or mixed leukocyte-epithelial cell casts indicates the bacterial or nonbacterial disease of renal origin. Acute pyelonephritis, interstitial nephritis, and nonbacterial renal diseases such as lupus nephritis and glomerulonephritis exhibit moderate numbers of urinary neutrophils accompanied by leukocyte casts. Hematuria and red blood cell casts are also found in lupus nephritis. Inflammatory or bacterial diseases of the urinary bladder, urethra, prostate, or prepuce in uncircumcised men may also cause neutrophils to be present in the urine (Lindqvist and Wahlin, 1975).

Large numbers and/or clumps of neutrophils (more than 50 per high-power field) in a urine specimen strongly suggest an acute bacterial infection, including tuberculosis of the kidney or urinary tract or a ruptured abscess.

An increased number of urinary neutrophils may be seen as part of an inflammatory response or an associated bacterial infection in urolithiasis and urothelial cancer. Likewise, neutrophils are commonly present in the necrotic background of invasive malignancies (Box 2–3).

Eosinophils

The main characteristic of eosinophils in Papanicolaou-stained cytologic preparations is their bilobed nuclei. Eosinophilic granules are inconspicuous.

Eosinophiluria may be observed in a wide range of primary renal or secondary diseases associated with hypereosinophilia. Some of these processes include allergic responses, drug-induced reactions, eosinophilic helminthic infections, collagen vascular diseases, cholesterol embolization, essential hypereosinophilic syndrome, and malignancy. Prolonged eosinophilia produces multiorgan damage (Giudicelli et al., 1998).

Eosinophiluria may be present in acute interstitial nephritis caused by drugs, predominantly nonsteroidal anti-inflammatory agents. Eosinophiluria in cases of acute renal failure favors an acute allergic reaction over other causes of acute renal failure, but a definitive diagnosis requires renal biopsy (Hadrick et al., 1996; Linton et al., 1980). The sensitivity of eosinophiluria in the diagnosis of acute interstitial nephritis is 60% (Sutton, 1986).

Eosinophiluria in interstitial cystitis may be of allergic significance and may reflect eosinophilic infiltration of the bladder wall (Yamada et al., 1992). However, it may not be possible to make this diagnosis by means of urine cytology alone, because eosinophiluria may not be present in all cases and urinary neutrophils may predominate in some (Garcia-Ureta et al., 1978). Acute cystitis with abundant eosinophiluria is present in 60% of patients with noninvasive bladder cancer treated with mitomycin C (Gelabert Mas et al., 1991) (Box 2–4).

Lymphocytes and Plasma Cells

Urine cytology is more accurate than phase contrast microscopy in evaluation of the lymphocyte count and morphology (Corey et al., 1997; Eggensperger et al., 1988). A few

Box 2–5.

Few lymphocytes are a common finding.

Lymphocyturia exceeding 20% suggests acute cellular rejection in renal transplantation.

Lymphocytes found in association with plasma cells and histiocytes suggest lupus nephritis or a chronic infectious process.

Lymphocytes of large size and abnormal cytomorphology may indicate malignant lymphoma.

small lymphocytes are commonly found in urine specimens and usually lack clinical significance. However, finding plasma cells in the urine sediment is a significant finding and should be investigated.

Lymphocyturia can be seen in acute cellular allograft rejection, lupus nephritis, tuberculosis, and other infectious or inflammatory responses. A cytodiagnosis of renal acute cellular allograft rejection is suggested in the presence of lymphocyturia of 20% or more (Corey et al., 1997). The presence of more than 30% lymphocytes, plasma cells, and histiocytes correlates with lupus nephritis (Lindqvist and Wahlin, 1975).

More recently, the presence of lymphocyte markers in urine samples has been correlated with diverse disease states such as IgA nephropathy (Hotta et al., 1993), and with the immune response in patients with noninvasive bladder cancers treated with recombinant interleukin II (Nouri et al., 1994).

The presence of lymphoid cells with abnormal cytomorphology warrants further investigation so that malignant lymphoma involving the kidney or the urinary tract can be ruled out (Cheson et al., 1981, 1984). In such instances, the use of Romanovsky stains in air-dried cytospin smears allows better characterization of the nuclear characteristics. Special techniques can be applied to urine samples in support of such a diagnosis (Yam and Janckila, 1985; Zakowski et al., 1996) (Box 2–5).

Cytology. Benign small lymphocytes measure approximately 10 μm in diameter and have a round nucleus, a scant cytoplasmic rim, smooth or slightly irregular nuclear contours, dense, uniform chromatin, and no nucleoli. Their size should not be greater than that of a segmented leukocyte or the nucleus of an intermedi-

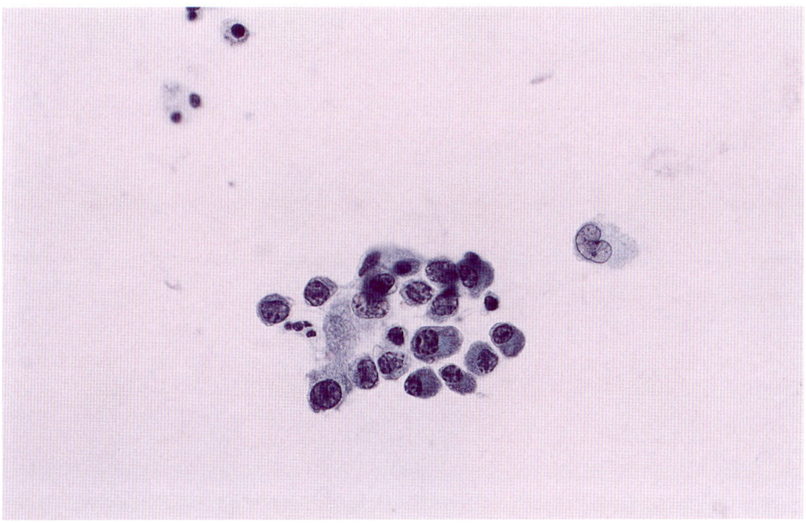

Figure 2–3. *Benign lymphocytes, plasma cells, and a single histiocyte.* Note the scant cytoplasm of lymphocytes, the dense cytoplasm with perinuclear clearing, and the clock-face chromatin pattern of the plasma cells. The bean-shaped nucleus with a thin nuclear contour and powdery chromatin is evident in the isolated histiocyte. Instrumented urine obtained from a patient with a history of intravesical BCG immunotherapy for noninvasive urothelial carcinoma. (Papanicolaou stain, ×1000)

Box 2–6.

Histiocytes are absent in normal urine.

They may be single or form clusters, that is, granulomas.

Histiocytes have variable size, clear cytoplasm, a bean-shaped nucleus, pale chromatin, and no nucleolus.

They signify the presence of a granulomatous process of the urinary system, that is, infectious, noninfectious, or intravesical immunotherapy with BCG.

ate squamous cell (Fig. 2–3). The characteristic eccentric nucleus with coarse peripheral chromatin, perinuclear clearing, and absent nucleolus in plasma cells may be difficult to identify in Papanicolaou-stained specimens (Fig. 2–3).

Histiocytes

Histiocytes may be found singly or forming loosely cohesive granulomas. After exclusion of the induction of a granulomatous response in the bladder wall by intravesical immunotherapy for noninvasive bladder cancer with bacillus Calmette-Guerin (BCG), a granulomatous infectious process is usually the explanation for the presence of histiocytes in urine (Bhan et al., 1998).

Cytology. The presence of epithelioid histiocytes in urine is abnormal. Their size varies, but most range between 20 and 40 μm. They exhibit a characteristic bean-shaped nucleus with smooth contours, pale chromatin, and no nucleolus. Their cytoplasm is thin, clear, or foamy, with an ill-defined cytoplasmic membrane. Multinucleated histiocytes may be present (Figs. 2–3, 2–4; Box 2–6).

Macrophages

Macrophages (CD68+/25F9+ cells) are absent in urine samples from normal individuals. Their presence is not specific but it implies a subjacent pathologic process, particularly renal parenchymal disease.

Macrophages are present in greatest number in patients who have highly progressive diseases such as rapidly progressive glomerulonephritis or diffuse proliferative lupus

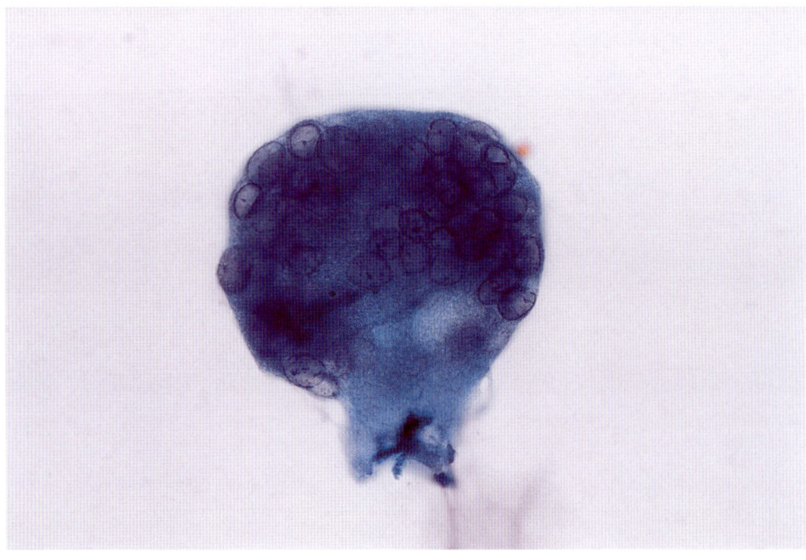

Figure 2–4. *Multinucleated histiocyte.* Note the numerous peripherally aggregated nuclei with powdery chromatin and inconspicuous nucleoli. Instrumented urine obtained from a patient with a history of intravesical BCG immunotherapy for noninvasive urothelial carcinoma. (Papanicolaou stain, ×1000)

Box 2–7.

Macrophages are absent in normal urine.

Their presence implies subjacent pathology, usually renal parenchymal disease.

They are usually seen singly and occasionally in loose clusters.

Macrophages have a single round nucleus, powdery chromatin, and cytoplasmic debris.

nephropathy. They are less abundant in patients with moderately progressive diseases (IgA nephropathy and Alport's syndrome) and virtually absent in patients with nonprogressive diseases (minimal change nephrotic syndrome and urolithiasis) (Hotta et al., 1993, 1998). Hemosiderin-laden macrophages may be present in urine samples from patients with acute allograft renal transplant rejection.

Cytology. Macrophages usually present singly, but occasionally they are arranged in loosely cohesive clusters. Their size varies considerably, but most range between 20 and 40 μm, and their cytoplasm may contain phagocytic debris or hemosiderin. Their nucleus is usually single, central or excentric, round or bean-shaped, and has a powdery chromatin with a small nucleolus (Fig. 2–5; Box 2–7).

Differential diagnosis. The round cell shape and clear, lacy cytoplasm permit the distinction of macrophages from superficial urothelial cells, which commonly have a biconvex shape and dense cytoplasm. Macrophages may be confused with renal tubular cells, although the latter are usually smaller, with pyknotic nuclei and granular cytoplasm. Distinction of macrophages from the exfoliated cells of clear cell renal cell carcinoma may be difficult. In such cases, the presence of cells with a red macronucleolus and radiographic evidence of a renal mass supports the diagnosis of renal cell carcinoma.

Renal Tubular Cells

It is important to recognize renal tubular cells in a urine specimen because their presence almost always signifies renal parenchymal diseases, such as renal allograft rejection, acute tubular necrosis, renal infarct, or

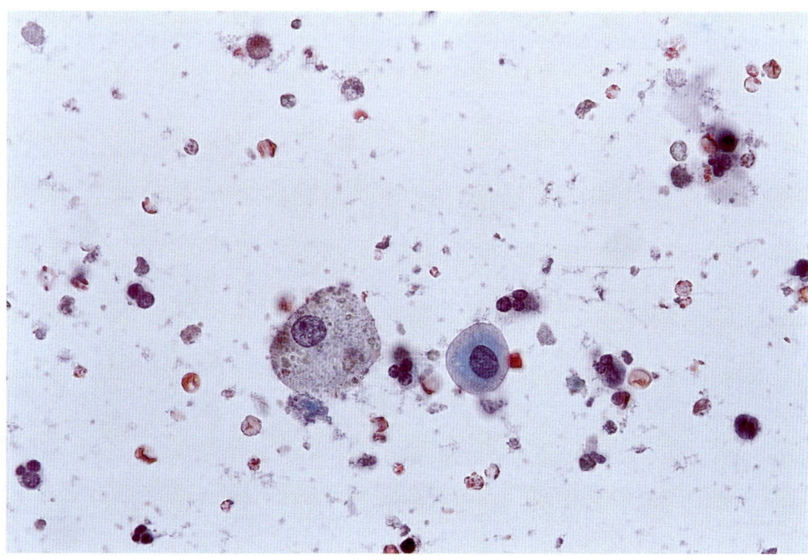

Figure 2–5. *Macrophage.* Brown intracytoplasmic granules of hemosiderin are present. One benign umbrella cell, degenerated segmented leukocytes, and lysed red blood cells are also noted. Voided urine from a patient with a renal transplant. (Papanicolaou stain, ×600)

Box 2–8.

Few renal tubular cells may be seen in normal urine.

Numerous renal tubular cells indicate renal parenchymal disease.

Collecting duct fragments indicate severe tubular injury.

Renal tubular cells are associated with pathologic casts, and with white and red blood cells.

papillary necrosis (Eggensperger et al., 1988, 1989). However, a few single degenerated renal tubular cells may be seen in normal urine samples.

Renal tubular cells line the proximal and distal convoluted tubules of the renal cortex and the collecting ducts of the renal medulla. The cytomorphology of these cells can be assessed accurately by use of the Papanicolaou stain, which permits their distinction from other mononuclear cells (Schumann et al., 1977). The cytologic features of renal tubular cells are summarized in Table 2–2.

The precise differentiation between proximal and distal convoluted renal tubular cells is difficult in cytologic smears (Fig. 2–6). Similarly, distal convoluted renal tubular cells may be difficult to differentiate from collecting ductal cells and the latter from basal/intermediate urothelial cells. The precise distinction between proximal and distal renal tubular cells may not be critical for patient management.

In addition to single renal tubular cells, the presence of renal epithelial fragments of collecting-duct origin in clusters of three or more cells indicates a more severe form of renal tubular injury. Exfoliated renal epithelial fragments may be identified by their attachment to renal casts, as well as by their cylindrical configuration and honeycomb arrangement. The exfoliated epithelium may contain spindle or elongated cells (Schumann et al., 1981).

The presence of renal tubular cells in urine should lead the pathologist to search for other cytologic evidence of renal damage such as pathologic casts, inflammatory cells, and red blood cells. The cytology report should address the presence of these cells, both singly and in fragments, as well as all other cellular and noncellular elements of clinical significance (Box 2–8).

Seminal Vesicle Cells

In urine cytology, high-grade malignant cells, seminal vesicle cells, and urothelial cells infected with human polyomavirus exhibit similar characteristics. The recognition of such characteristics may be difficult even for the experienced microscopist.

Seminal vesicle cells can be suspected if spermatozoa are found in the urine specimen. Although their presence in the urine is uncommon, cellular characteristics permit their distinction (Droese et al., 1976).

Cytology. The most common and distinctive features of seminal vesicle cells are yellowish-brown cytoplasmic pigment granules and large, irregular, hyperchromatic nuclei. Intranuclear cytoplasmic inclusions, nuclear folding, and prominent nucleoli are observed in about one-third of cases (Fig. 2–7A–D) (Koivuniemi and Tyrkko, 1976).

Seminal vesicle cells can cause abnormal DNA ploidy measurements, such as aneu-

Table 2–2. Cytologic Features of Renal Tubular Cells

	Cortical	*Medullary*
Size	15–60 μm	15–60 μm
Shape	Oval/round	Cuboidal/columnar
Cell arrangement	Single	Single/honeycomb
Cell borders	Indistinct	Distinct
Cytoplasm	Granular	Finely granular
Nucleus	Pyknotic	Not pyknotic

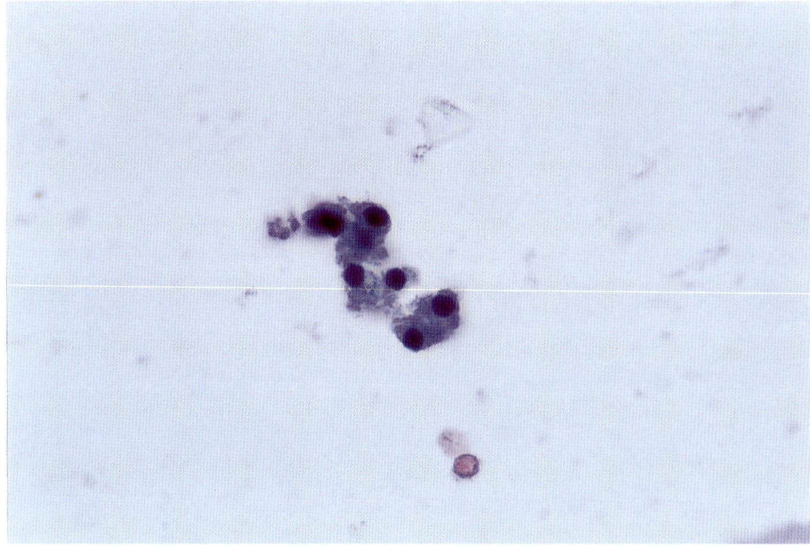

Figure 2–6. *Renal tubular cells.* Aggregated proximal renal tubular cells show pyknotic nuclei and granular cytoplasm. Numerous granular and epithelial casts were seen in other fields of this voided urine specimen from a patient with nephronal hematuria. (Papanicolaou stain, ×1000)

ploidy and tetraploidy. This may be of particular importance in the evaluation of urine samples from patients with high-grade bladder cancer by the use of this parameter. Morphologic criteria remain vital for an accurate cytologic diagnosis (Wojcik et al., 1999) (Box 2–9).

Prostate Cells

Like seminal vesicle cells, prostate cells may occasionally be found in a urine sample, frequently accompanied by variable numbers of spermatozoa and macrophages, particularly after prostatic massage. Corpora amylacea may be identified.

Prostate cells usually form small sheets of cuboidal or elongated cells arranged in monolayers. The cells have distinct borders, pale and scant cytoplasm, a round nucleus, powdery chromatin, and a small chromocenter (Fig. 2–8). Honeycombing is usually observed when the focus of the microscope is adjusted.

The presence of nucleoli and acinus formation may indicate prostatic adenocarcinoma, as discussed in Chapter 7.

Box 2–9.

Distinction of seminal vesicle cells from high-grade carcinoma or the polyomavirus cytopathic effect may be difficult.

Seminal vesicle cells may be associated with spermatozoa.

These cells are large and exhibit yellowish-brown cytoplasmic pigment, large, irregular, hyperchromatic nuclei, and prominent nucleoli.

They have an abnormal DNA ploidy value.

NONCELLULAR ELEMENTS

Noncellular elements are usually incidental findings in urine specimens. These elements are usually sought by microscopic evaluation of unstained urine sediment smears in the clinical laboratory (e.g., urinalysis), but evaluation of Papanicolaou-stained cytologic preparations is equally if not more informative than the urinalysis. Correlation of cytologic findings with those of urinalysis is necessary in questionable cases, and the final interpretation needs to be supported by appropriate clinical information.

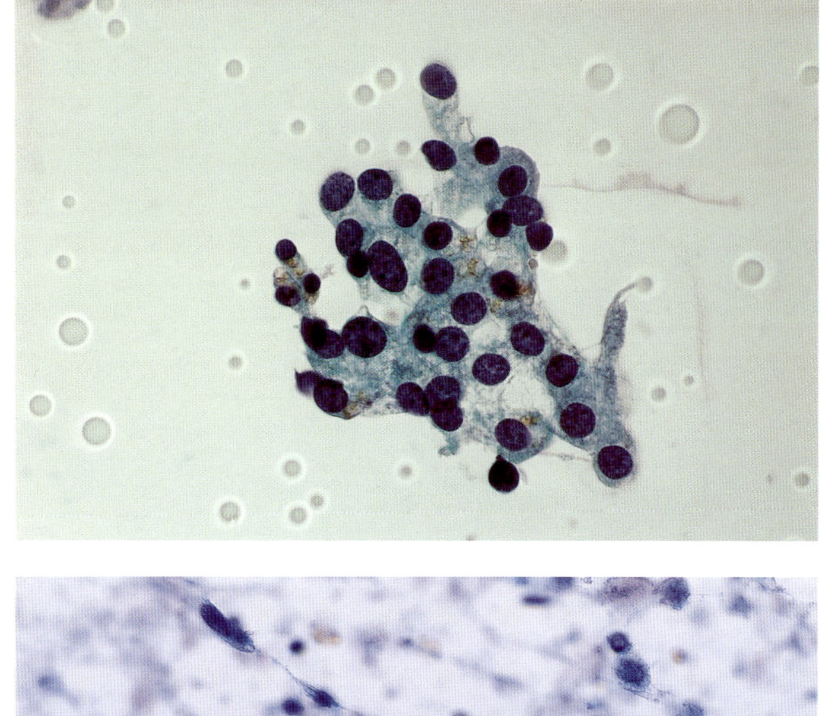

Figure 2–7. *Seminal vesicle cells.* Various forms of seminal vesicle cells exhibiting variable amounts of the characteristic cytoplasmic yellow pigment are illustrated. (*A*) A sheet of round or oval, uniform cells with hyperchromatic nuclei resembles urothelial carcinoma in situ. (*B*) A single cell with a large, homogeneous nucleus resembles cells in the human polyomavirus cytopathic effect. (*C,D*) Cells with large nuclei with a nuclear cytoplasmic invagination and marked pleomorphism and coarse chromatin resemble those of malignant melanoma or a poorly differentiated malignancy. Smears from fine-needle aspiration biopsy of the prostate (*A*) and from a surgically resected seminal vesicle (*B–D*). (Papanicolaou stain, ×600)

Urinary Crystals

Crystals in the urine are of limited clinical significance, particularly if they are found in urine standing at room temperature. Most crystals disappear when urine is heated to 37°C, and clinical correlation is necessary when they remain at such a temperature (Bradley and Schumann, 1984).

In general, normal crystals may be present in acidic, neutral, or alkaline urine, in contrast to abnormal crystals, which are usually

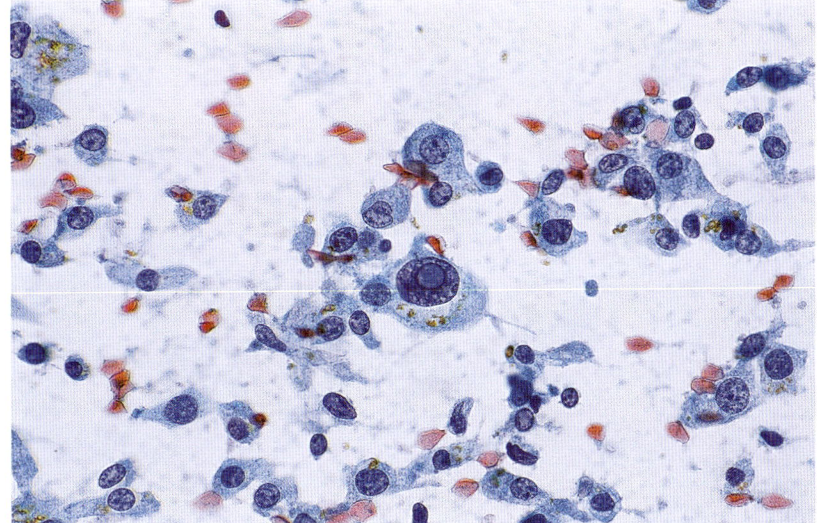
C

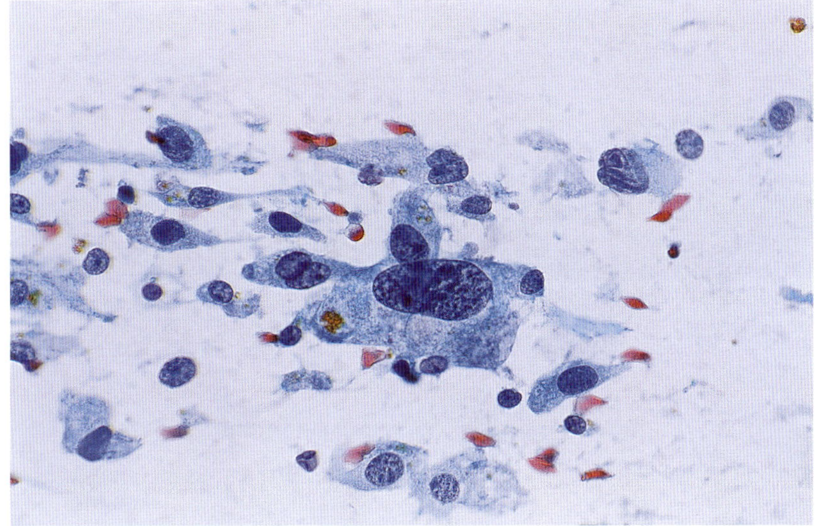
D

Figure 2–7. *(Continued)*

found in acidic or neutral urine but only exceptionally in alkaline urine, that is, indinavir and acyclovir (K. M. Ringsrud, University of Minnesota, personal observation; Thompson, 1981). Evaluation of the pH and other physical and chemical characteristics of the urine, which is helpful for proper crystal identification, is performed in the clinical laboratory, not the cytology laboratory, and the cytopathologist rarely knows the results at the time of urine evaluation.

Crystals may be found in the urine of both healthy individuals and patients with urolithiasis, toxic renal damage, chronic renal failure, or other systemic processes (Geyer, 1993). Phosphates, urates, and oxalates are particularly common and occur in normal urine. The recognition of less common, clinically significant urinary crystals is based on the experience gained in recognizing more common, clinically nonsignificant crystals. The microscopic appearance and clinical significance of urinary crystals are summarized in Table 2–3.

Amorphous urates and *amorphous phosphates* crystallize when urine has been refrigerated or left at room temperature. They usually have no clinical significance. *Triple phosphate*

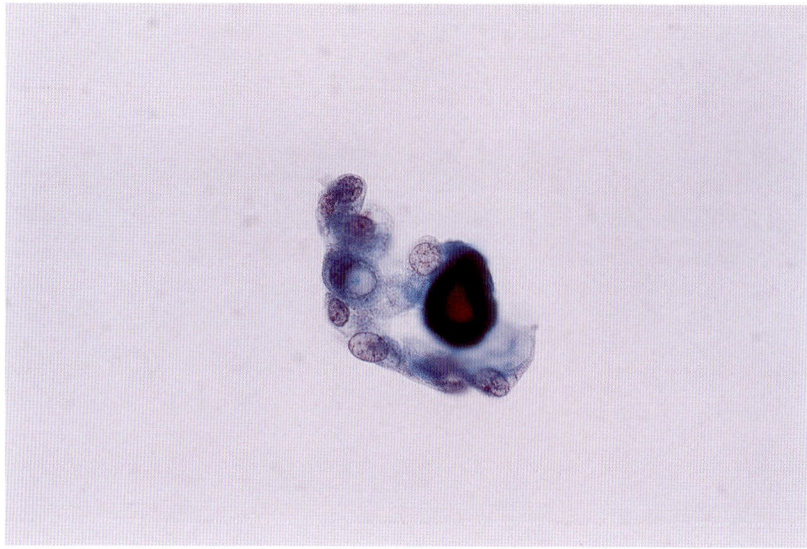

Figure 2–8. *Prostate cells.* Epithelial cells show thin cytoplasm, a round nucleus, a uniform thin nuclear contour, and finely granular chromatin. These cells are presumably of prostatic origin and surround a fragment of the corpora amylacea. Instrumented urine. (Papanicolaou stain, ×600)

(struvite) crystals may not be clinically significant but may be present in urine from patients with urease-positive, gram-negative (*Proteus, Klebsiella,* and *Pseudomonas* spp.) urinary tract infections that cause an alkaline urine with an excess of urinary ammonium (Fig. 2–9). *Amorphous phosphates* or *calcium carbonate* crystals generally appear in the urine after ingestion of a heavy meal or large quantities of vegetables, respectively (Thompson, 1981).

The most common crystals found in urinary sediment are those of *uric acid* or its salt, amorphous urate. These crystals form in almost any urine specimen that stays at room temperature, and they can mimic the shape of many other crystals (Fig. 2–10A,B). Uric acid crystals are considered pathologic only when they are present in fresh voided urine, indicating increased purine catabolism (Thompson, 1981). Therefore, caution in their interpretation is recommended, particularly when there is no clinical evidence of a metabolic imbalance or urolithiasis. Formation of uric acid stones requires uric acid excretion, low urine volume, and acidic urine that permits the crystallization of undissociated uric acid. Gout, strenuous exercise with dehydration, and gastrointestinal disorders cause abnormally acidic urine (Rivers et al., 2000) (Box 2–10).

The presence and amount of *calcium oxalate* crystals in the urine of healthy individuals are directly related to the consumption of foods rich in oxalic acid (apples, oranges, tomatoes, carbonated water) (Thompson,

Box 2–10.

Crystals have clinical significance only if they are found in freshly voided urine.

Not uncommonly, crystals are found in urine standing at room temperature.

Phosphates, urates, and oxalates are common in normal urine.

Caution in the interpretation of uric acid crystals in the absence of a metabolic disorder, systemic disease, or urolithiasis is recommended.

Calcium oxalate crystals may be pathologic or associated with consumption of foods rich in oxalic acid, such as apples, tomatoes, or carbonated water.

Drug-related crystalluria should be suspected in the presence of unusual crystals.

Table 2–3. Urinary Crystals

Type	*Appearance*	*Clinical Significance*
Amorphous urates	Clumps of yellow-brown granules	None
Calcium carbonate	Colorless small granules, spheres, and dumbbell-shaped	None
Amorphous phosphate	Clumps of colorless granules	None
Triple phosphate (magnesium ammonium phosphate or struvite)	Coffin lid, flat fern, wedge-shaped	Little, if any
Ammonium urates	Yellow-brown, round, with twisted protuberances	Little, if any
Uric acid	Various shapes and colors. Needle, hexagonal, rosette, rhombic, "lemon" shapes. Colorless, yellow, brown	Pathologic only if found in freshly voided urine. Chemotherapy, fever, leukemia, lymphoma, gout, etc.
Calcium oxalate	Octahedral, dumbbell, ovoid	May be pathologic. Chronic renal failure, Crohn's disease, ethylene glycol and methoxyfluorane toxicity
Cystine	Colorless, flat hexagonal plates	Pathologic. Cystinuria
Tyrosine	Sheaves of colorless fine needles	Pathologic. Liver disease
Leucine	Yellow, spherical, with concentric laminations and striations	Pathologic. Liver disease
Bilirrubin	Yellow granules, needles, rhombic plates	Pathologic. Hepatobiliary or hemolytic disorders
Drugs	Different shapes, depending on the drug (sulfa, antacids, indinavir, etc.)	May indicate high dosage or overdosage

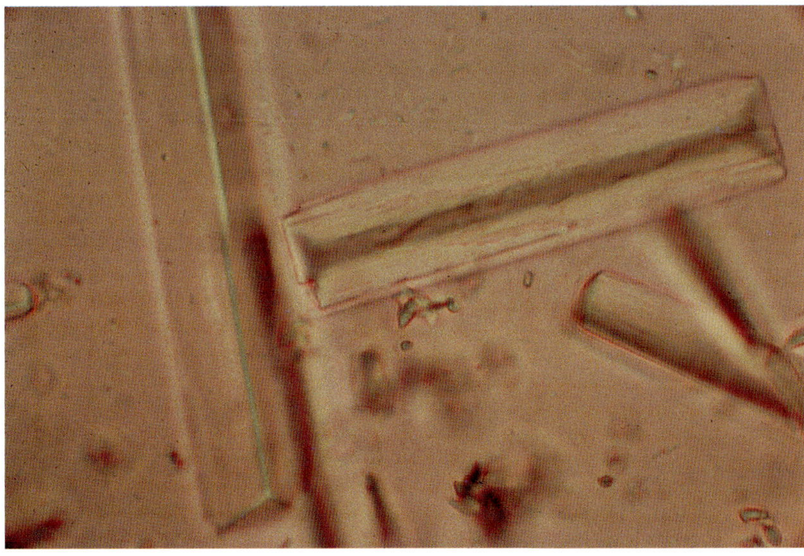

Figure 2–9. *Triple phosphate crystals.* These are large, common "coffin-lid" forms. Such crystals, also referred to as *struvite*, have limited clinical significance. (Unstained, ×440). Courtesy of Karen M. Ringsrud, BS, MT(ASCP), Assistant Professor, Department of Pathology, University of Minnesota, Minneapolis, MN.

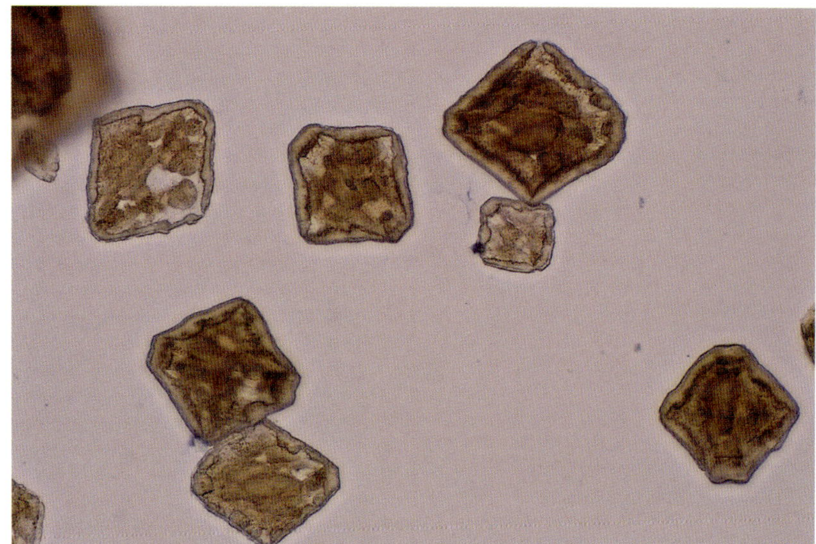

A

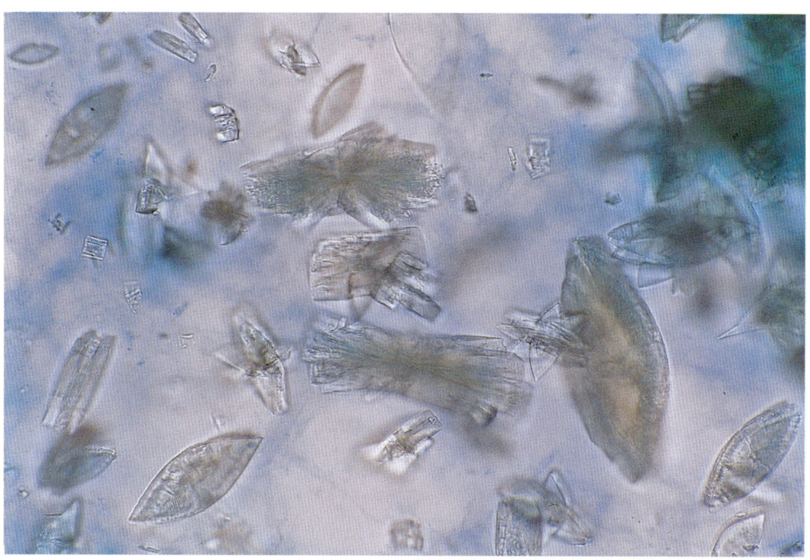

B

Figure 2–10. *Uric acid crystals.* These are the most common crystals seen in urine specimens that remain at room temperature. Uric acid crystals may have various shapes, including rhombic plates (*A*), pointed lemon, and rare forms (*B*) usually with a yellow or red-brown color. (Papanicolaou stain, ×600)

1981). Overabsorption of dietary oxalate with oxaluria, as seen in fat malabsorption secondary to jejunoileal bypass in the treatment of obesity, ileal resection for Crohn's disease, and chronic pancreatic disease, is the most common cause of calcium oxalate stones in such patients. Other disorders accompanied by calcium oxalate stones include idiopathic hypercalciuria, primary hyperparathyroidism, hypercalcemia of nonparathyroid origin (malignancy, granulomatous disease, hyperthyroidism, pheochromocytoma, immobilization, thiazide diuretic use), and hyperuricosuria (Balaji and Menon, 1997). The incidence of calcium oxalate stone formation is also increased in patients with hyperuricosuria, most commonly associated with excess dietary purine ingestion. These patients develop not only uric acid stones but also calcium oxalate stones (Rivers et al., 2000). In conclusion, hyperoxaluria from any cause can produce tubulointerstitial nephropathy and nephrolithiasis (Fig. 2–11A,B).

Cystine crystals are seen in urine from patients with the hereditary metabolic recessive disorder cystinuria, which is characterized by a transepithelial transport defect in the intestine and kidneys. Most heterozygous individuals have crystalluria in the absence of stones, in contrast to most homozygous persons, who are stone formers (Balaji and Menon, 1997). In any case, the presence of these crystals in the urine is pathologic and needs to be mentioned in the cytology report (Fig. 2–12). The diagnosis of cystinuria should be considered when the characteristic crystals are identified in the first urine voided in the morning. The diagnosis is confirmed by measurement of urinary cystine levels (Rivers et al., 2000).

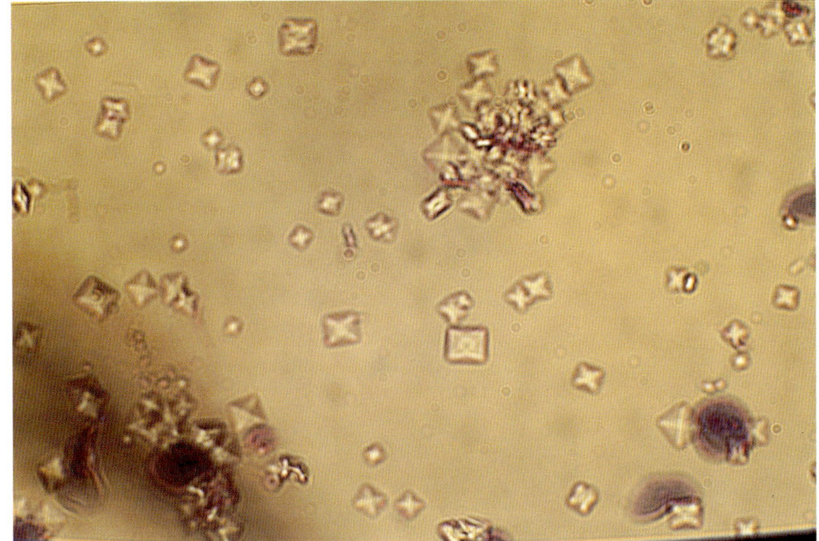

A

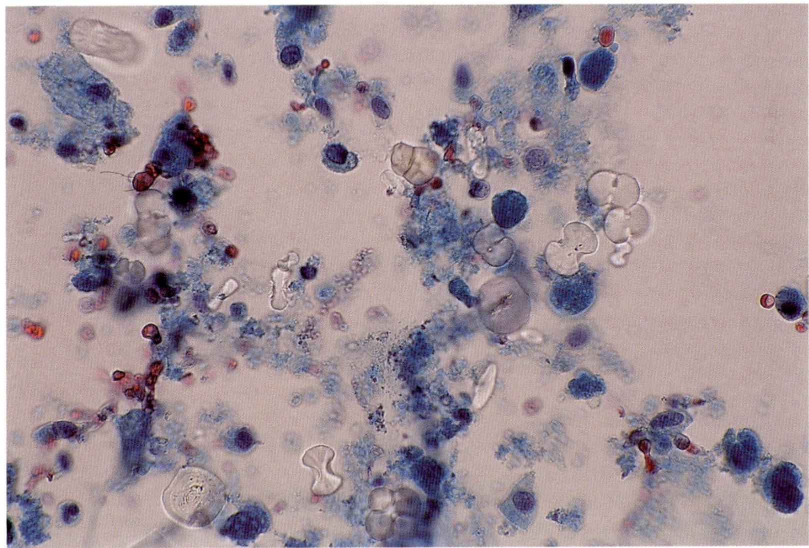

B

Figure 2–11. *Calcium oxalate crystals.* These are the most common crystals seen in normal urine. They are also the most common constituents of most urinary stones. (*A*) They are colorless and have an octahedron shape. (*B*) The less common ovoid and dumbbell shapes are smaller and may be mistaken for red blood cells. (Papanicolaou stain, ×600)

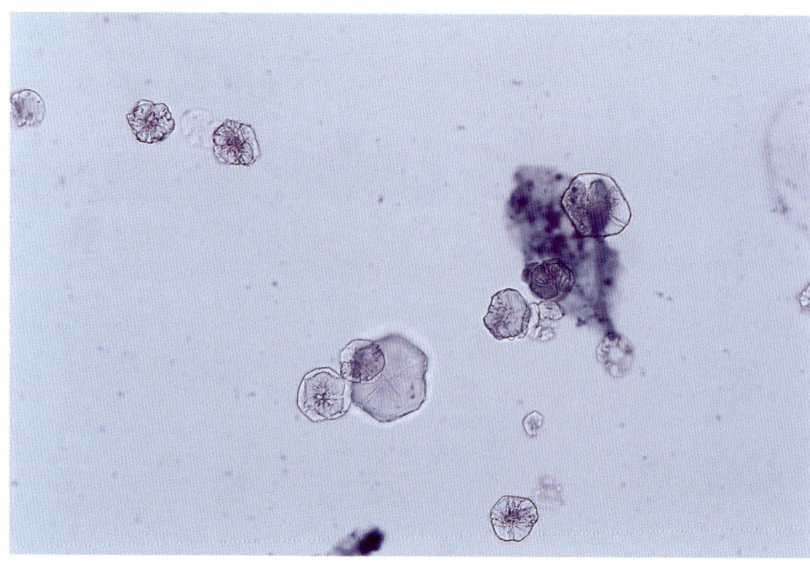

Figure 2–12. *Cystine crystals.* These are probably the most important of the abnormal crystals of metabolic origin. They are thin, colorless, hexagonal, and refractile with laminations. (Wright stain, ×600). Courtesy of Karen M. Ringsrud, BS, MT(ASCP), Assistant Professor, Department of Pathology, University of Minnesota, Minneapolis, MN.

Tyrosine and *leucine* crystals are rare in urine specimens. They are occasionally seen in the urine of patients with systemic diseases associated with severe hepatocyte degeneration and necrosis. *Bilirubin* crystals may occasionally be present in urine and indicate hematopoietic, biliary, or liver disease with severe jaundice.

Urinary crystals with *unusual shapes* may be related to the administration of therapeutic drugs or radiologic contrast substances. This finding should prompt clinical correlation because of the potential association with lithiasis. Laxative abuse, carbonic anhydrase inhibitors such as acetazolamide and methazolamide, triamterene, sulfa drugs (Fig. 2–13), silicates, indinavir, and herbal remedies may be associated with nephrolithiasis (Rivers et al., 2000). Chronic ingestion of herbal teas containing ephedrine or cough medications can result in calculi that contain ephedrine (Drach, 2000). The protease inhibitor indinavir, administered to patients with acquired immune deficiency syndrome (AIDS), causes urinary excretion of crystals, leading to stone formation in some cases. Indinavir crystals have a "bow-tie" or fan-like appearance in cytologic preparations and "stars" of flat, thin crystals in urinalysis (Fig. 2–14) (Filie et al., 2000; Schwartz et al., 1999). Approximately 1% of calculi have drug-related components (Daudon et al., 1983).

Casts

Casts are cylinders formed in the renal tubules. They are composed of an amorphous, colorless, and semitransparent core called *Tamm-Horsfall protein.* Pigment, debris, and different cell types are added to this core during movement from the renal tubules to the renal pelvis. The microscopic appearance, classification, and pathologic significance of casts are a reflection of the cells—for example, red blood cells, renal tubular cells, inflammatory cells, and debris—added during this migration. The characteristics of the different types of casts are summarized in Table 2–4.

The finding of *red blood cell casts* is very significant and is indicative of nephronal hematuria (Fig. 2–15). These casts are the result of glomerular capillary basement membrane disruption and, less commonly, of tubular basement membrane disruption. Clinical conditions associated with the presence of these casts include glomeru-

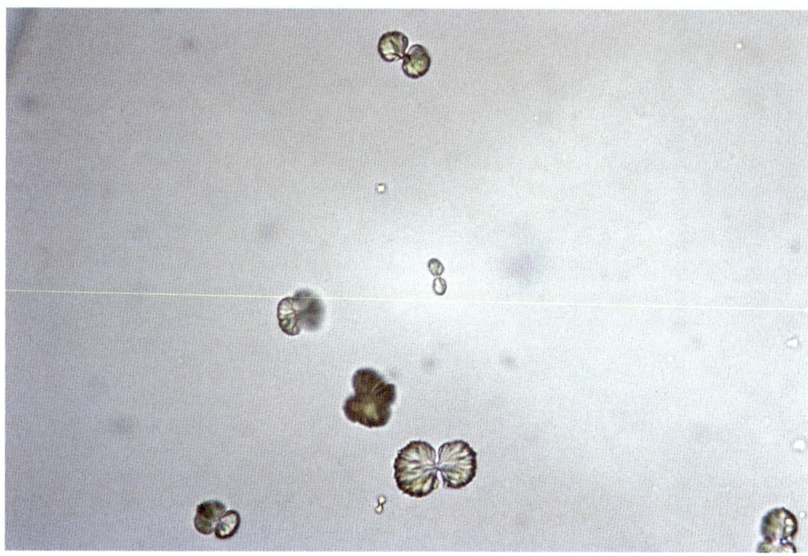

Figure 2–13. *Sulfa crystal.* Crystals of sulfamethoxazole can be present in urine specimens after a high dose of Bactrim or Septra is administered. (Unstained, ×320). Courtesy of Karen M. Ringsrud, BS, MT(ASCP), Assistant Professor, Department of Pathology, University of Minnesota, Minneapolis, MN.

lonephritis, lupus nephritis, vasculitis with glomerular damage, and, less commonly, tubulointerstitial damage. Other noninflammatory glomerular conditions are renal infarction, and severe acute pyelonephritis. When red blood cell degeneration occurs, the cast may appear granular, with pigmented granules adhering to the Tamm-Horsfall protein core (pigmented granular casts) (Fig. 2–16) (Bradley and Schumann, 1984; Haber, 1981) (Box 2–11).

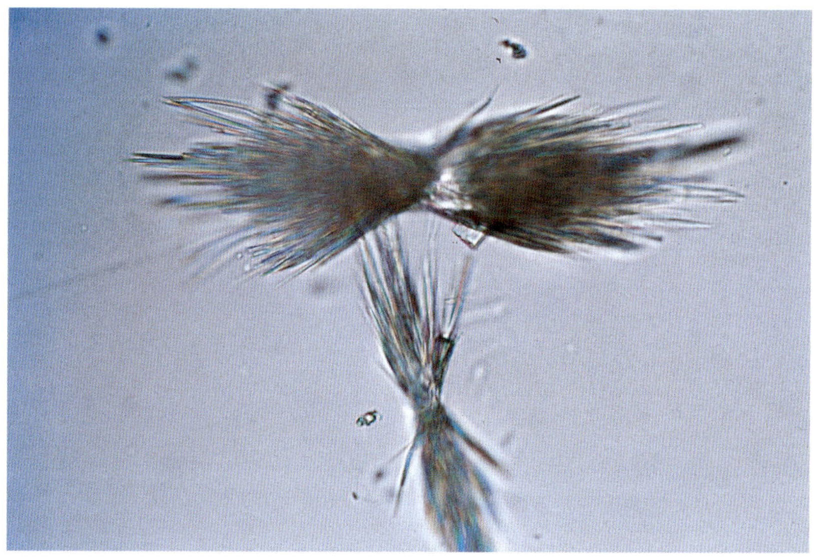

Figure 2–14. *Indinavir sulfate crystal.* Thin indinavir crystals have a “bow-tie” or fan-like appearance. (Unstained, 320). Courtesy of Karen M. Ringsrud, BS, MT(ASCP), Assistant Professor, Department of Pathology, University of Minnesota, Minneapolis, MN.

Box 2–11.

Casts are composed of a semitransparent core (Tamm-Horsfall protein).

Cellular casts are best identified in Papanicolaou-stained smears.

The clinical significance of granular and hyaline casts depends on finding other pathologic casts.

Detection of casts in the first morning specimen of urine predicts the presence or absence of renal disease and correlates with pathologic changes seen in the renal biopsy specimen.

White blood cell casts are most often formed by migration of inflammatory cells through the tubular wall that surrounds the Tamm-Horsfall protein core. Their presence is often an indication of acute pyelonephritis, and they may be associated with bacteriuria and casts of bacteria, usually in the absence of hematuria. In contrast, when the acute inflammatory component affects primarily the glomeruli, the urine often contains red blood cell casts and other casts and lacks bacteriuria. As a result of tubular stasis, transit time, and cell degeneration, white cell casts may form total or partial nonpigmented granular casts. The final interpretation requires clinical correlation (Bradley and Schumann, 1984; Haber, 1981). White

Table 2–4. Cytologic Evaluation of Urinary Casts

Type	*Appearance*	*Clinical Significance*
Red blood cell	Red blood cells covering more than one-third of the cast; hyaline core may show reddish-brown granules	Glomerular damage
White blood cell	Predominantly neutrophils	Tubulointerstitial disease
Epithelial cell	Sheets of tubular cells or single tubular cells with different degrees of degeneration	Tubular damage
Mixed cell	Red, white, and epithelial cells with different degrees of degeneration	Tubulointerstitial disease
Waxy	Smooth, thick, and opaque, with surface notches and crevices, sharp edges, and "broken-off ends"	Chronic renal disease
Granular	Fine or coarse granules	May be pathologic
Hyaline	Transparent and colorless	May be pathologic

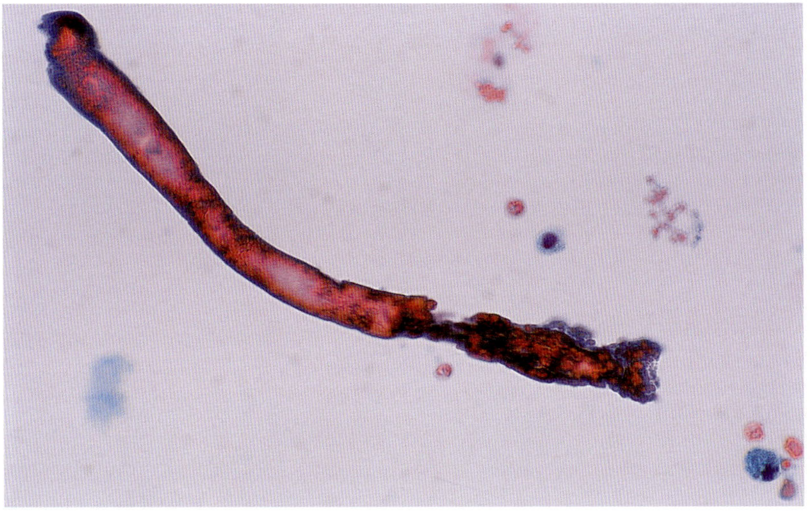

Figure 2–15. *Red blood cell cast.* Partially degenerated red blood cells are embedded in the Tamm-Horsfall protein core. These casts indicate glomerular and/or tubular basement membrane damage. (Papanicolaou stain, ×600)

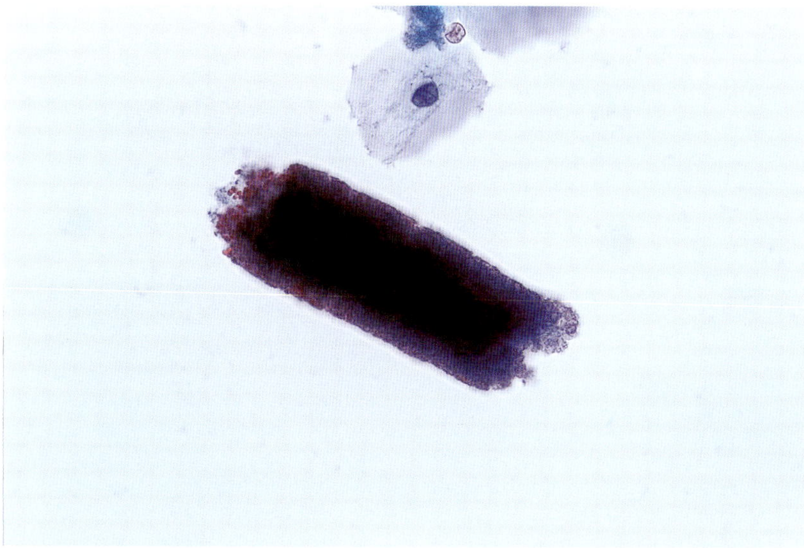

Figure 2–16. *Red blood cell/granular cast.* Marked degeneration of red blood results in a predominantly granular cast with a red hue in the center of the core. (Papanicolaou stain, ×600)

blood cell casts can also be seen in other disorders affecting the tubulointerstitial region, such as lupus nephritis and, occasionally, renal transplant acute cellular rejection.

Renal tubular epithelial cell casts are best identified in Papanicolaou-stained cytologic specimens. In contrast to phase contrast microscopy, this method permits a more accurate distinction of red cell casts from white cell casts. These casts are formed when sloughed sheets or single tubular cells adhere to the Tamm-Horsfall protein core resulting in an epithelial cast composed of parallel rows of cells or randomly arranged cells (Fig. 2–17). These subtle differences can be seen best in well-preserved specimens. In the

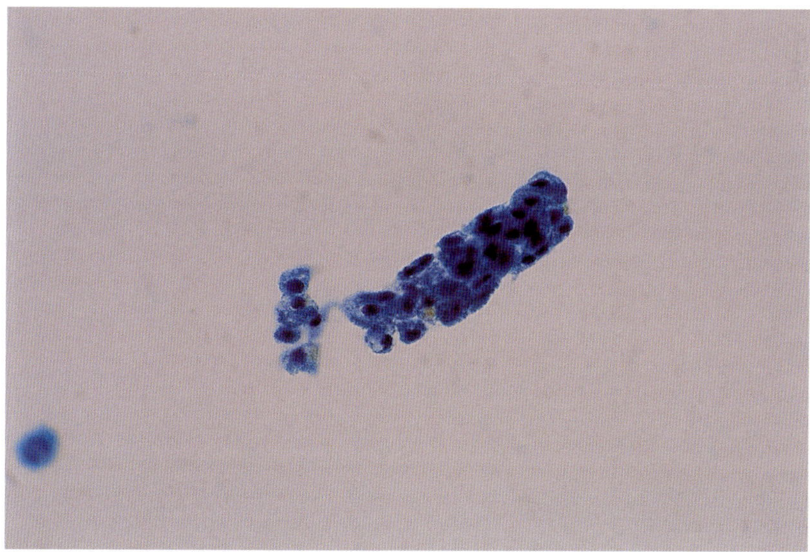

Figure 2–17. *Epithelial cell cast.* These casts are formed by a parallel arrangement of degenerated cells with granular cytoplasm and pyknotic nuclei. They indicate a serious injury of the renal tubular epithelium. (Papanicolaou stain, ×600)

presence of cellular degeneration, the distinction of epithelial from white cell casts may be difficult and ultimately, like white cell casts, they may become nonpigmented granular casts (Fig. 2–18). Epithelial cell casts are present in urine from patients with serious acute and subacute renal diseases, most commonly renal allograft rejection, acute tubular necrosis, renal infarct, papillary necrosis, renal viral infection, ethylene glycol or heavy metal poisoning, and salicylate intoxication (Eggensperger et al., 1988, 1989).

Mixed cell casts are composed of renal tubular cells, white blood cells, and red blood cells and usually reflect tubulointerstitial disease. The cellular component may undergo degeneration, and the cellular cast may have different degrees of a granular component. In some instances, the cellular component cannot be determined with certainty, and some inferences may be drawn from the cells present in the background (Bradley and Schumann, 1984).

Waxy casts are present in urine specimens from patients with long-lasting and progressive chronic renal disease. They are formed presumably in dilated and atrophic tubules, with delayed transit time of the luminal contents sufficient to produce molding and hardness (Fig. 2–19). Waxy casts are narrow or broad and have a smooth, thick, opaque (waxy) appearance with sharp edges and "broken-off ends." They are brittle, and usually have surface notches and crevices (Haber, 1981). Waxy casts are broad in cases of advanced renal insufficiency or chronic allograft nephropathy. They may be narrow in cases of acute cellular rejection.

Granular casts, in contrast to the previously mentioned types of casts, may not be pathologic. They may be present in the urine of healthy individuals after heavy exercise; the formation of these casts under such circumstances is not fully understood. The texture of the constituent granules may vary from fine to coarse, but this has no bearing on the clinical significance of the casts. Because granular casts may represent degenerated cellular casts, the clinical significance of granular casts depends on the finding of other pathologic casts in the urine specimen (Fig. 2–20) (Bradley and Schumann, 1984).

Hyaline casts can be seen in the urine of healthy individuals after strenuous exercise. As with granular casts, the clinical significance of this finding depends on the presence of cellular casts in the urine specimen. Characteristically, these casts are of variable length, transparent and colorless, and have tapered ends (Fig. 2–21). Essentially, they are composed mostly of the fibrillar Tamm-

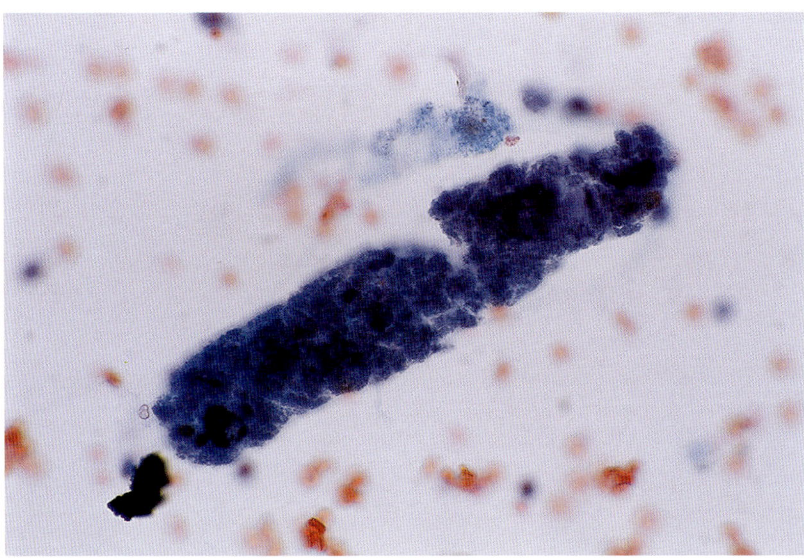

Figure 2–18. *Epithelial cell/granular cast.* Degeneration of epithelial cells results in a mixed epithelial and granular cast. (Papanicolaou stain, ×600)

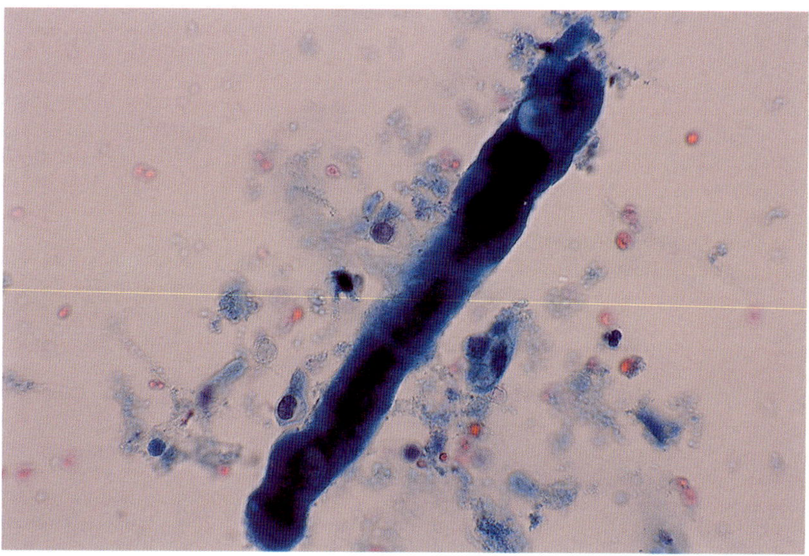

Figure 2–19. *Waxy cast.* A thick, smooth, glassy appearance is characteristic of these casts. Voided urine of a patient with chronic renal insufficiency. (Papanicolaou stain, ×600)

Horsfall protein core (Cohen, 1981). The filaments comprising the casts may be confused with amyloid because of ultrastructural similarities, especially when observed in urinalysis samples (Cohen, 1981).

Urine cytodiagnosis performed by examination of Papanicolaou-stained cytocentrifuge preparations of urine sediment for the identification of casts has been proved valuable in the evaluation of patients with renal parenchymal disease, allograft dysfunction in renal transplant patients, and acute renal failure of diverse etiology (Eggensperger et al., 1988, 1989; Marcussen et al., 1995; Schumann et al., 1981). The finding of casts in the first morning specimen of urine is highly pre-

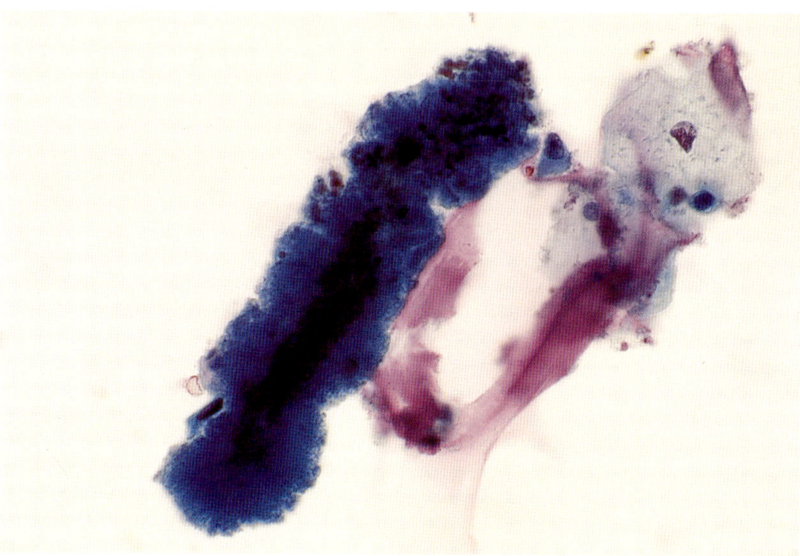

Figure 2–20. *Granular cast.* A coarsely granular cast was present in the voided urine smear of a patient with chronic renal insufficiency. (Papanicolaou stain, ×600)

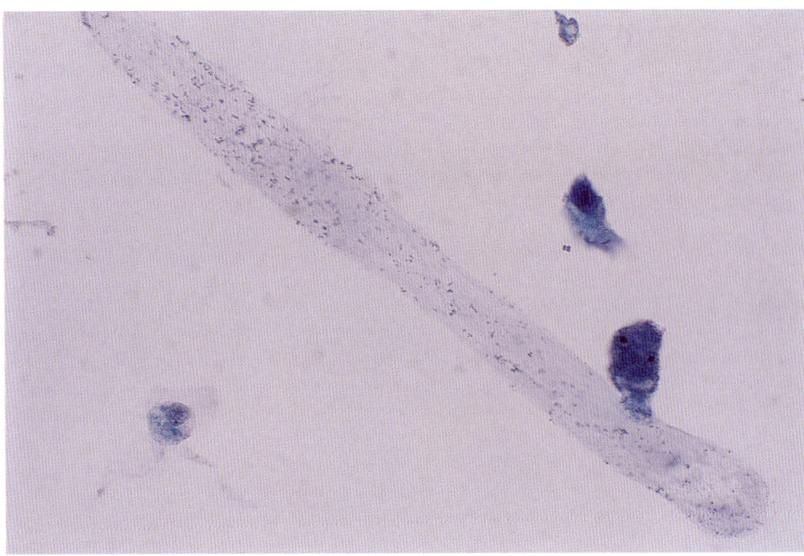

Figure 2–21. *Hyaline cast.* A thin, colorless, transparent cylinder with a round, tapered end is characteristic of this cast, composed almost exclusively of the fibrillary Tamm-Horsfall protein core. Voided urine specimen. (Papanicolaou stain, ×600)

dictive of the presence of renal disease. If no casts are detected in two early-morning specimens, the likelihood of finding significant pathologic changes in a renal biopsy sample is low (Gyory et al., 1984).

Corpora Amylacea

Prostatic glandular tissue in men is associated with varying forms of acellular structures, including corpora amylacea (Fig. 2–8). These lamellated, eosinophilic concretions are infrequently present in urine specimens, although they may be present in urine after ejaculation or after a prostatic massage. In these cases, they are accompanied by spermatozoa and, rarely, by seminal vesicle cells. On electron microscopy, corpora amylacea are found to be composed of bundles of fibrils and occasional interspersed electron-dense areas (Drachenberg and Papadimitriou, 1996). It has been proposed that prostatic corpora amylacea represent localized amyloidosis of beta-2 microglobulin origin (Cross et al., 1992).

Bacteria

The presence of bacteria in a urine specimen is not an uncommon finding. Bacteria are particularly prominent when the urine has not been refrigerated and transportation to the cytology laboratory has been delayed for several hours. The lack of neutrophils and the predominance of bacterial organisms strongly suggest bacterial overgrowth. Bacterial organisms in the presence of neutrophils may be significant only in the presence of bacterial phagocytosis by neutrophils. The presence of bacterial casts is indicative of bacterial pyelonephritis or renal infection.

Artifacts and Contaminants

Artifacts and contaminants that may be found in a urine specimen include lubricant, talc powder, cotton and wood fibers, fecal matter, pollen grains, and hair, among others (Box 2–12).

Cytologically, *lubricant* appears as a well-delineated, flat, opaque, amorphous fragment. It usually has a grayish-blue color and frayed, spiculated, or filiform borders (Fig. 2–22). Rare small bubbles may be admixed occasionally. The identification of lubricant in a urine specimen indicates the use of an instrument for obtaining the urine sample. As a result, urothelial cellular changes related to instrumentation must be considered

Box 2–12.

The presence of lubricant indicates recent instrumentation or instrumentally obtained urine.

Lubricant appears grayish-blue, flat, and opaque, with sharp edges that are frequently scalloped or frayed.

Fecal matter shows partly digested vegetable cells and muscle fibers, and it may indicate fecal contamination or an enterovesical fistula.

The Maltese cross formation is typical of talcum powder.

Pollen grains should be distinguished from human cells and worm ova, that is, *Schistosoma* and *Enterobius vermicularis.* The latter represents perineal contamination.

in the differential diagnosis of low-grade papillary urothelial carcinoma.

Distinction of lubricant from a mucinous background of mucin-producing adenocarcinomas may be difficult. Helpful clues include evaluation of the interface between the background and the substance in question. The interface is inconspicuous, with imperceptible blending in a mucinous background, and is sharp and well defined when lubricant is present. In addition, a mucinous background is diffuse and at times difficult to see, whereas lubricant is localized and confined to small clumps.

Fecal matter in urine cytology is characterized by a "dirty" background, partly digested vegetable cells with a cell wall, and degenerated muscle fibers (Fig. 2–23). This usually indicates fecal contamination in infants and in debilitated, bedridden, and incontinent patients, but it may also indicate an enterovesical fistula.

Cornstarch and *talc* are the most common contaminants seen in urine cytology. Cornstarch, used to line surgical and laboratory protective gloves, has replaced talc. Whether they originate from dusting powder applied to the perineal area of infants or bedridden patients or from surgical gloves used by the operator to obtain the urine specimen, the semitransparency, birefringency, and identification of the typical Maltese cross with polarized light are helpful features for its recognition. Starch granules are usually round, with a central dimple or slit (Fig. 2–24).

Cotton, hair, and other fibers, usually the result of contamination at the bedside or dur-

Figure 2–22. *Lubricant.* Clumps of dense basophilic material with sharp edges are characteristic of lubricant. (Papanicoalou stain, ×600)

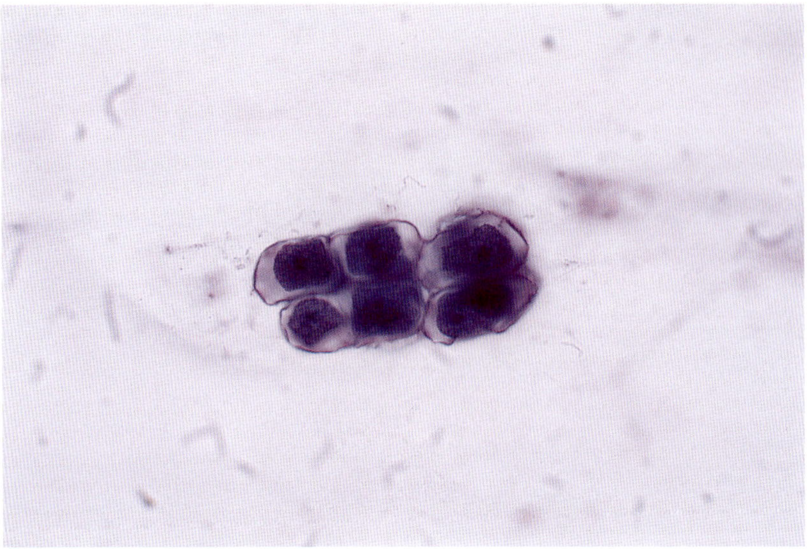

Figure 2–23. *Fecal matter.* A thick cell wall, a cytoplasmic membrane, and granular cytoplasm are seen in this cluster of partially degenerated vegetable cells. Catheterized urine from a bedridden woman. (Papanicolaou stain, ×600)

ing urine preparation in the laboratory, are easily recognizable. Occasionally they may be mistaken for hyaline or granular casts, although they are polarizable and longer.

Pollen grains have wide variations in size, shape, and texture. Their size is helpful in distinguishing them from human cells. Additional features include the presence of a cell wall and a large central nucleus (Fig. 2–25).

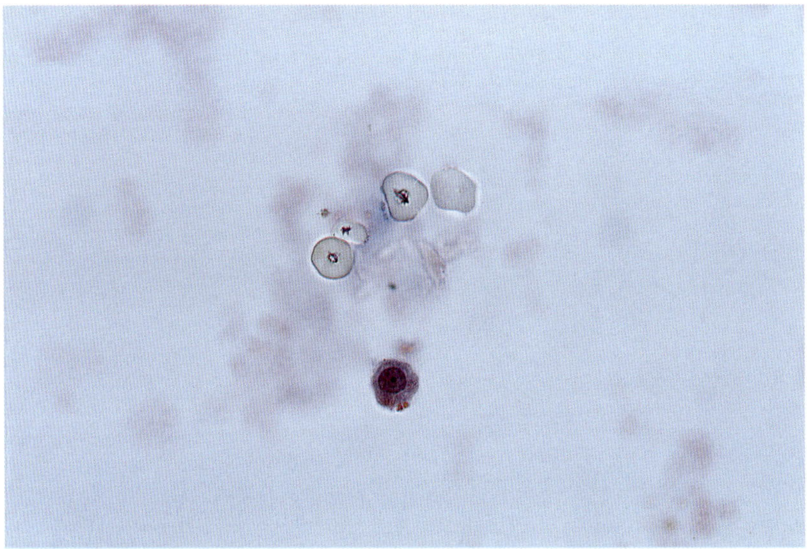

Figure 2–24. *Starch granules.* Colorless, semitransparent, round granules with a central dimple or slit are characteristic. The ill-defined Maltese cross is more evident under polarized light. (Papanicolaou stain, ×600)

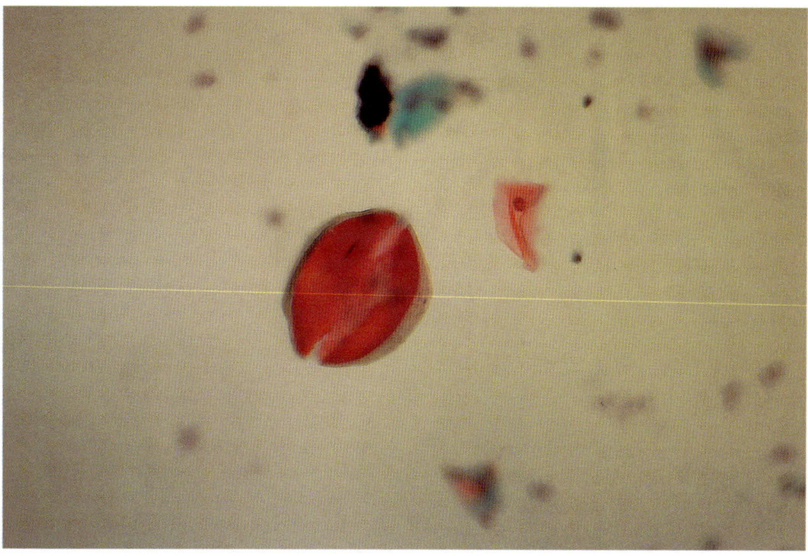

Figure 2–25. *Pollen grain.* A pollen grain resembling a pinworm ovum was identified in this voided urine specimen. (Papanicolaou stain, ×200)

CYTOLOGY REPORT OF THE URINARY PROFILE

Cytodiagnostic urine sediment analysis using Papanicolaou-stained smears has proved to be reliable and highly sensitive in detecting urine sediment abnormalities indicative of renal parenchymal disease (Eggensperger et al., 1988, 1989). This diagnostic tool is underutilized by most nephrologists.

In addition to the assessment of urothelial cells, the report should include all elements of clinical importance, namely, erythrocytes, leukocytes, tubular cells, casts, crystals, and microorganisms. These findings must be interpreted and, whenever possible, combined into urinary profiles (e.g., the nephritic sediment, the nephrotic sediment).

The *nephritic sediment* contains erythrocytes and red blood cell casts. The *nephrotic sediment* is composed of granular and waxy casts. These findings, combined with other laboratory tests, pathologic findings, and clinical data, allow the definition and management of urinary tract diseases (Fogazzi et al., 1998).

A urine sample that contains red blood cells, cellular casts, waxy casts, red blood cells casts, and granular casts indicates a progressive smoldering process affecting different parts of the nephron and has been called *telescoped sediment.* This abnormality is found in chronic glomerulonephritis with active glomerulitis including lupus nephritis and subacute bacterial endocarditis, among other diseases (Box 2–13).

Box 2–13.

The nephritic sediment is composed of red blood cells and red blood cell casts and correlates with acute glomerular inflammation, that is, acute glomerulonephritis.

The nephrotic sediment is composed of granular and waxy casts and correlates with significant proteinuria and chronic glomerular basement membrane damage.

Telescoped sediment is composed of various casts and correlates with simultaneous glomerular, tubular, and interstitial damage.

REFERENCES

Angulo, J. C., Lopez-Rubio, M., Guil, M., Herrero, B., Burgaleta, C., Sanchez-Chapado, M. (1999). The value of comparative volumetric analysis of urinary and blood erythrocytes to localize the source of hematuria. *J Urol* 162: 119–26.

Balaji, K. C., Menon, M. (1997). Mechanism of stone formation. In *The Urologic Clinics of North America* (M. I. Resnick, ed.), Vol. 24, pp. 1–11. W. B. Saunders, Philadelphia.

Bhan, R., Pisharodi, L. R., Gudlaugsson, E., Bedrossian, C. (1998). Cytological, histological, and clinical correlations in intravesical Bacillus Calmette-Guerin immunotherapy. *Ann Diagn Pathol* 2: 55–60.

Bonfante, L., D'Angelo, A., Favaro, S., Giacomini, A., Normanno, M., Calo, L., Antonello, A., Bordin, V., Vianello, D., Meani, A., Borsatti, A. (1996). 7-year-follow up (1987–1994) of a population with asymptomatic persistent microhematuria and normal renal function. *Minerva Med* 87: 525–9.

Bradley, M., Schumann, G. B. (1984). Examination of urine. In *Clinical Diagnosis and Management* (J. B. Henry, ed.), 17th ed., pp. 380–457. W. B. Saunders, Philadelphia.

Carson, C. C. D., Segura, J. W., Greene, L. F. (1979). Clinical importance of microhematuria. *JAMA* 241: 149–50.

Cheson, B. D., Johnston, J. L., del Junco, G., Kjeldsberg, C. R. (1981). Cytologic evidence for disseminated immunoblastic lymphoma. *Am J Clin Pathol* 75: 621–5.

Cheson, B. D., Schumann, G. B., Johnston, J. L. (1984). Urinary cytodiagnosis of renal involvement in disseminated histiocytic lymphoma. *Acta Cytol* 28: 148–52.

Cohen, A. H. (1981). Morphology of renal tubular hyaline casts. *Lab Invest* 44: 280–7.

Corey, H. E., Alfonso, F., Hamele-Bena, D., Greenstein, S. M., Schechner, R., Tellis, V., Geva, P., Koss, L. G. (1997). Urine cytology and the diagnosis of renal allograft rejection. I. Studies using conventional staining. *Acta Cytol* 41: 1732–41.

Cross, P. A., Bartley, C. J., McClure, J. (1992). Amyloid in prostatic corpora amylacea. *J Clin Pathol* 45: 894–7.

Daudon, M., Protat, M. F., Reveillaud, R. J. (1983). Detection and diagnosis of drug induced lithiasis. *Ann Biol Clin* 41: 239–49.

Drach, G. W. (2000). Secondary and miscellaneous urolithiasis. Medications, urinary diversions, and foreign bodies. *Urol Clin North Am* 27: 269–73.

Drachenberg, C. B., Papadimitriou, J. C. (1996). Prostatic corpora amylacea and crystalloids: similarities and differences on ultrastructural and histochemical studies. *J Submicrosc Cytol Pathol* 28: 141–50.

Droese, M., Voeth, C., Konetzke, C. (1976). The role of cells originating from seminal vesicles in aspiration biopsy smears of the prostate (author's transl). *Urologe [A]* 15: 18–20.

Eggensperger, D. L., King, C., Gaudette, L. E., Robinson, W. M., O'Dowd, G. J. (1989). Cytodiagnostic urinalysis. Three years experience with a new laboratory test. *Am J Clin Pathol* 91: 202–6.

Eggensperger, D., Schweitzer, S., Ferriol, E., O'Dowd, G., Light, J. A. (1988). The utility of cytodiagnostic urinalysis for monitoring renal allograft injury. A clinicopathological analysis of 87 patients and over 1,000 urine specimens. *Am J Nephrol* 8: 27–34.

Filie, A., Wilder, A., Brosky, K., Kopp, J., Arora, K., Abati, A. (2000). Urinary cytology associated with human BK virus and indinavir in HIV positive patients. A study of 153 cases. *Mod Pathol* 13: 45A (# 246).

Fogazzi, G. B., Grignani, S., Colucci, P. (1998). Urinary microscopy as seen by nephrologists. *Clin Chem Lab Med* 36: 919–24.

Fracchia, J. A., Motta, J., Miller, L. S., Armenakas, N. A., Schumann, G. B., Greenberg, R. A. (1995). Evaluation of asymptomatic microhematuria. *Urology* 46: 484–9.

Froom, P., Gross, M., Froom, J., Caine, Y., Margaliot, S., Benbassat, J. (1986). Factors associated with microhematuria in asymptomatic young men. *Clin Chem* 32: 2013–15.

Garcia-Ureta, E., Llorente Apat, P., Martinez-Penuela, J. M., Sarmiento Gomez, C. (1978). Cytology of urine in eosinophilic cystitis. Apropos of 19 cases. *Arch Esp Urol* 31: 337–50.

Gelabert Mas, A., Vesa Llanes, J., Corominas, J., Arango Toro, O., Bielsa Gali, O., Llado Carbonell, C. (1991). Eosinophilic cystitis as a special form of response to mitomycin C. Analysis and comments on our cases. *Arch Esp Urol* 44: 929–32.

Geyer, S. J. (1993). Urinalysis and urinary sediment in patients with renal disease. *Clin Lab Med* 13: 13–20.

Giudicelli, C. P., Didelot, F., Duvic, C., Desrame, J., Herody, M., Nedelec, G. (1998). Eosinophilia and renal pathology. *Med Trop (Mars)* 58: 477–81.

Gyory, A. Z., Hadfield, C., Lauer, C. S. (1984). Value of urine microscopy in predicting histological changes in the kidney: double blind comparison. *Br Med J (Clin Res Ed)* 288: 819–22.

Haber, M. H. (1981). Urinary casts. In *Urinary Sediment: A Textbook Atlas* (M. H. Haber, ed.), pp. 55–93. American Society of Clinical Pathology, Chicago.

Hadrick, M. K., Vaden, S. L., Geoly, F. J., Cullen, J. M., Douglass, J. P. (1996). Acute tubulointerstitial nephritis with eosinophiluria in a dog. *J Vet Intern Med* 10: 45–7.

Hattori, R., Matsuura, O., Takeuchi, N., Hashimoto, J., Ohshima, S., Ono, Y., Kinukawa, T., Miyake, K. (1990). Clinical importance of microhematuria as an initial sign of bladder tumor. *Nippon Hinyokika Gakkai Zasshi—Jpn J Urol* 81: 414–19.

Hiatt, R. A., Ordonez, J. D. (1994). Dipstick urinalysis screening, asymptomatic microhematuria, and subsequent urological cancers in a population-based sample [published erratum appears in Cancer Epidemiol Biomarkers Prev 1994 Sep;3(6):523]. *Cancer Epidemiol, Biomarkers Prevention* 3: 439–43.

Hotta, O., Kitamura, H., Taguma, Y. (1998). Detection of mature macrophages in urinary sediments: clinical significance in predicting progressive renal disease. *Ren Fail* 20: 413–18.

Hotta, O., Taguma, Y., Ooyama, M., Yusa, N., Nagura, H. (1993). Analysis of CD14+ cells and CD56+ cells in urine using flow cytometry: a useful tool for monitoring disease activity of IgA nephropathy. *Clin Nephrol* 39: 289–94.

Howard, R. S., Golin, A. L. (1991). Long-term followup of asymptomatic microhematuria. *J Urol* 145: 335–6.

Koivuniemi, A., Tyrkko, J. (1976). Seminal vesicle epithelium in fine-needle aspiration biopsies of the prostate as a pitfall in the cytologic diagnosis of carcinoma. *Acta Cytol* 20: 116–19.

Koss, L. G. (1996). The cellular and acellular components of the urinary sediment. In *Diagnostic Cytology of the Urinary Tract* (L. G. Koss, ed.), pp. 17–37. Lippincott-Raven, Philadelphia.

Lindqvist, B., Wahlin, A. (1975). Differential count of urinary leucocytes and renal epithelial cells by phase contrast microscopy. *Acta Med Scand* 198: 505–9.

Linton, A. L., Clark, W. F., Driedger, A. A., Turnbull, D. I., Lindsay, R. M. (1980). Acute interstitial nephritis due to drugs: review of the literature with a report of nine cases. *Ann Intern Med* 93: 735–41.

Marcussen, N., Schumann, J., Campbell, P., Kjellstrand, C. (1995). Cytodiagnostic urinalysis is very useful in the differential diagnosis of acute renal failure and can predict the severity. *Ren Fail* 17: 721–9.

Marcussen, N., Schumann, J. L., Schumann, G. B., Parmar, M., Kjellstrand, C. (1992). Analysis of cytodiagnostic urinalysis findings in 77 patients with concurrent renal biopsies. *Am J Kidney Dis* 20: 618–28.

Messing, E. M., Young, T. B., Hunt, V. B., Emoto, S. E., Wehbie, J. M. (1987). The significance of asymptomatic microhematuria in men 50 or more years old: findings of a home screening study using urinary dipsticks. *J Urol* 137: 919–22.

Miguel-Gomara Perello, J., Orfila Timoner, J., Riera Mari, V. (1993). Asymptomatic microhematuria in the adult. *Anal Med Intern* 10: 403–8.

Mohr, D. N., Offord, K. P., Melton, L. J. D. (1987). Isolated asymptomatic microhematuria: a cross-sectional analysis of test-positive and test-negative patients. *J Gen Intern Med* 2: 318–24.

Mohr, D. N., Offord, K. P., Owen, R. A., Melton, L. J. D. (1986). Asymptomatic microhematuria and urologic disease. A population-based study. *JAMA* 256: 224–9.

Murakami, S., Igarashi, T., Hara, S., Shimazaki, J. (1990). Strategies for asymptomatic microscopic hematuria: a prospective study of 1,034 patients. *J Urol* 144: 99–101.

Nouri, A. M., Hyde, R., Oliver, R. T. (1994). Clinical and immunological effect of intravesical interleukin-2 on superficial bladder cancer. *Cancer Immunol Immunother* 39: 68–70.

Reuter, V. E. (1997). Urinary bladder, ureter, and renal pelvis. In *Histology for Pathologists* (S. S. Sternberg, ed.), 2nd ed., pp. 835–47. Lippincott-Raven, Philadelphia.

Rivers, K., Shetty, S., Menon, M. (2000). When and how to evaluate a patient with nephrolithiasis. *Urol Clin North Am* 27: 203–13.

Schumann, G. B., Burleson, R. L., Henry, J. B., Jones, D. B. (1977). Urinary cytodiagnosis of acute renal allograft rejection using the cytocentrifuge. *Am J Clin Pathol* 67: 134–40.

Schumann, G. B., Johnston, J. L., Weiss, M. A. (1981). Renal epithelial fragments in urine sediment. *Acta Cytol* 25: 147–52.

Schwartz, B. F., Schenkman, N., Armenakas, N. A., Stoller, M. L. (1999). Imaging characteristics of indinavir calculi. *J Urol* 161: 1085–7.

Sutton, J. M. (1986). Urinary eosinophils. *Arch Intern Med* 146: 2243–4.

Thompson, A. (1981). Crystalluria. In *Urinary Sediment: A Textbook Atlas* (M. H. Haber, ed.), pp. 33–54. American Society of Clinical Pathology, Chicago.

Wiener, D. P., Koss, L. G., Sablay, B., Freed, S. Z. (1979). The prevalence and significance of Brunn's nests, cystitis cystica and squamous metaplasia in normal bladders. *J Urol* 122: 317–21.

Wojcik, E. M., Bassler, T. J., Jr., Orozco, R. (1999). DNA ploidy in seminal vesicle cells. A potential diagnostic pitfall in urine cytology. *Anal Quant Cytol Histol* 21: 29–34.

Yam, L. T., Janckila, A. J. (1985). Immunocytochemical diagnosis of lymphoma from urine sediment. *Acta Cytol* 29: 827–32.

Yamada, T., Murayama, T., Taguchi, H. (1992). The clinical significance of eosinophils in urine. *Hinyokika Kiyo* 38: 173–6.

Zakowski, M. F., Feiner, H., Finfer, M., Thomas, P., Wollner, N., Filippa, D. A. (1996). Cytology of extranodal Ki-1 anaplastic large cell lymphoma. *Diagn Cytopathol* 14: 155–61.

3

CONSTITUENTS OF URINARY TRACT SPECIMENS IN THE ABSENCE OF DISEASE

A few important considerations in the handling of urinary tract specimens are the following:

- Fresh urine specimens should be promptly transported to the cytology laboratory for immediate processing.
- Refrigeration of the specimen is strongly recommended when immediate transport to the laboratory is not possible.
- An alternative and more frequently used method is to fix the urine with an equal volume of 50% or 70% ethyl alcohol, preferably with added 2% Carbowax (polyethylene glycol). This procedure can be performed at the bedside or in the laboratory when delay in specimen transport or processing is expected.
- Conventional cytospin smears should be prepared immediately after the specimen is received. Few studies comparing conventional cytospin preparation and newer monolayer technologies, that is, Cytec (Marlborough, MA) and Autocyte (Burlington, NC), have been conducted, and experience in the cytologic interpretation of urinary specimens prepared with these monolayer technologies is limited (Luthra et al., 1999). However, the significant reduction in the number of inadequate samples and better cell morphology preservation appear to be advantageous and should be weighed against the higher cost of these new processing techniques.
- The use of coated glass slides is recommended for adequate adherence of the cellular and noncellular elements to the slide surface after centrifugation. The use of totally frosted glass slides is discouraged.

VOIDED URINE

Voiding of urine is the most gentle, noninvasive, and easiest method of obtaining a urine specimen. Important for successful cytologic interpretation is thorough cleaning of the urinary meatus, a task more difficult to accomplish in women than in men because of the more complex external anatomy of the urogenital organs.

The second morning midstream urine specimen is most adequate. The first morning urine specimen contains the largest number of cellular elements, but its interpretation may be limited by the cellular degeneration present. Serial collections of similar samples during three consecutive days provide a high yield for the diagnosis of urothelial carcinoma, particularly the high-grade form (Koss et al., 1985). Voided urine cytology is also a good adjunct for assessing the suitability of flexible cystoscopy in the follow-up of patients with urothelial carcinoma (Baltaci et al., 1996; Loh et al., 1996).

In experienced hands, voided urine cytology may be as accurate as and comparable to DNA flow cytometry in predicting bladder cancer recurrence (Gregoire et al., 1997) (Box 3–1).

Box 3–1.

The second morning midstream voided urine specimen is most adequate.

Smear shows low cellularity, mostly large umbrella and squamous cells, and few single basal-intermediate urothelial cells; however, an occasional cluster may be identified.

Umbrella cells have a biconvex shape with a central nucleus and powdery chromatin.

Basal-intermediate urothelial cells are cuboidal or columnar and show a round central nucleus with regular borders and powdery chromatin.

Cytology. As a result of spontaneous nontraumatic urothelial exfoliation, voided urine specimens from normal individuals are usually of low cellularity and are composed almost entirely of large urothelial and squamous cells. Few degenerated renal tubular cells, red blood cells, segmented leukocytes, or lymphocytes may be present in adequately collected and well-preserved samples (see Chapter 2).

The urothelial cells originate almost exclusively from the superficial (luminal) urothelial field of umbrella cells in contact with the urine. Squamous cells may be the result of either contamination with cells from the external genitalia or squamous metaplasia of the bladder trigone, which is seen more commonly in women than in men (Fig. 3–1).

The most common cells found in voided urine are umbrella cells with a single nucleus, although rare multinucleated forms may be seen. Umbrella cells commonly exhibit biconvex cytoplasmic outlines, with one border slightly flatter than the other, and show a round central nucleus with powdery and regular chromatin, small chromocenters, and a delicate, thin cytoplasm (Fig. 3–2). Their size is comparable to that of squamous cells, although larger sizes are not uncommonly found.

The larger multinucleated umbrella cells may show anisonucleosis, but other nuclear features and their overall biconvex cellular appearance are similar to those of single umbrella cells. The cytoplasm may be dense, thin, or vacuolated, with well-defined scalloped but smooth cytoplasmic borders (Fig. 3–3).

Normal cuboidal or columnar urothelial cells from intermediate or basal layers are

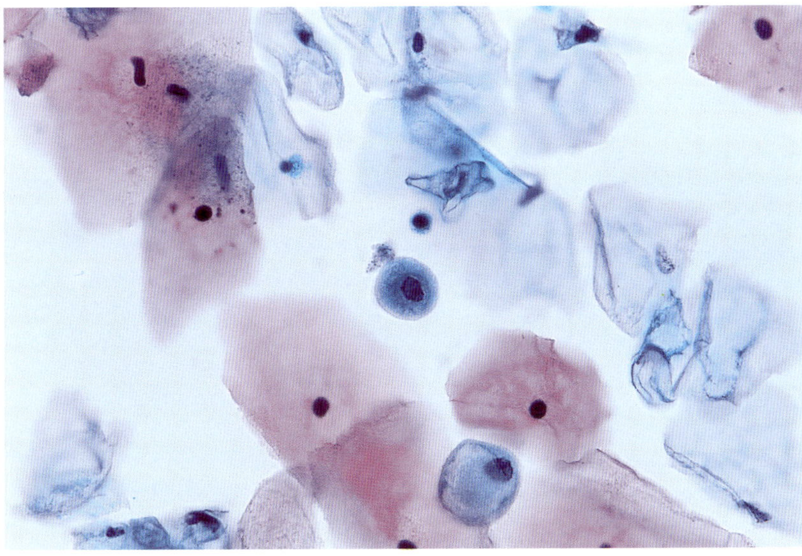

Figure 3–1. *Voided urine.* Rare benign urothelial cells and numerous squamous cells are present in this sample obtained from a healthy 26-year-old woman. Nuclear degeneration is present in the urothelial cells. Thin layer preparation. (Papanicolaou stain, ×600)

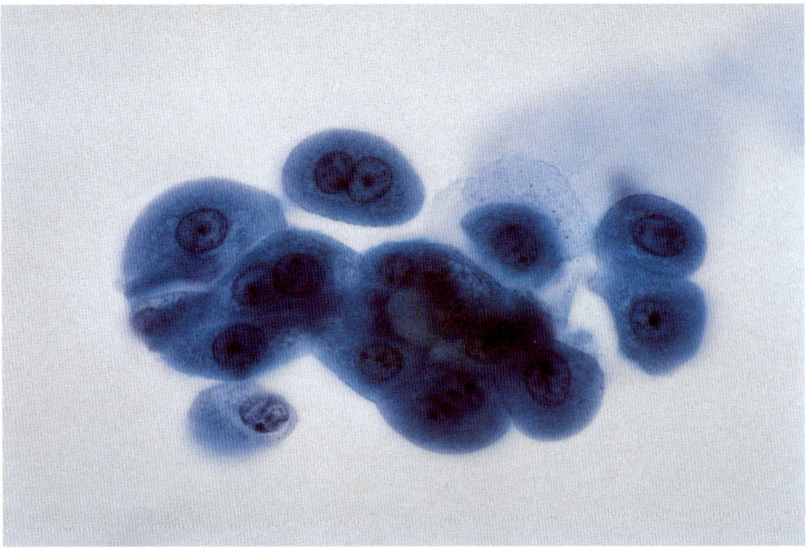

Figure 3–2. *Normal umbrella cells.* Biconvex shape, dense cytoplasm, a low nuclear-cytoplasmic ratio, and round nuclei are characteristic features. Bladder washing. (Papanicolaou stain, ×600). Courtesy of Dr. Michael W. Stanley, Chair, Department of Pathology, Hennepin County Medical Center. Minneapolis, MN.

small and may be present singly or in small numbers. The nuclear-cytoplasmic ratio is low, the cytoplasm is thin, the nuclear contour is round, smooth, and thin, and the chromatin is powdery and regularly distributed (Fig. 3–4). A rare cluster of such cells may be identified in voided urine specimens. Careful examination of cell morphology at the edge of the cluster will reveal characteristics similar to those of single cells. Odd cell shapes may be indicative of urothelial pathology. Numerous clusters of benign-

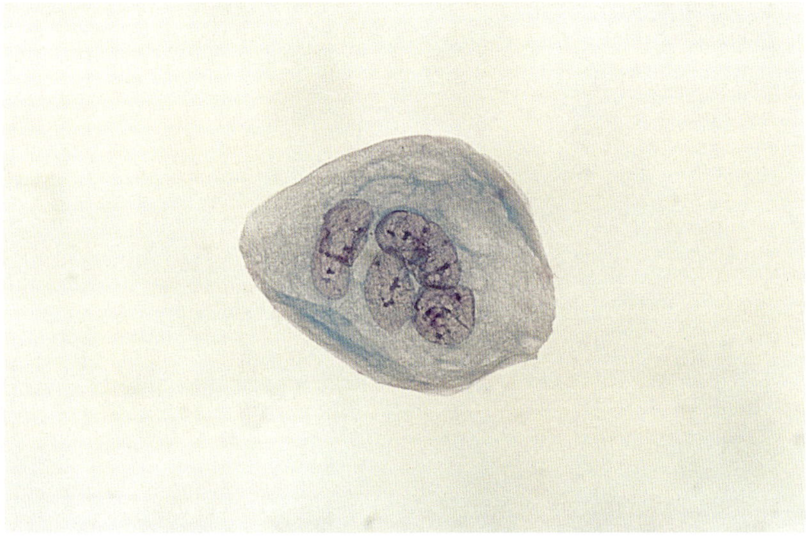

Figure 3–3. *Normal multinucleated umbrella cell.* Thin cytoplasm, bi-convex shape, and round/oval nuclei with powdery chromatin and multiple chromocenters are noted. Bladder washing. (Papanicolaou stain, ×600)

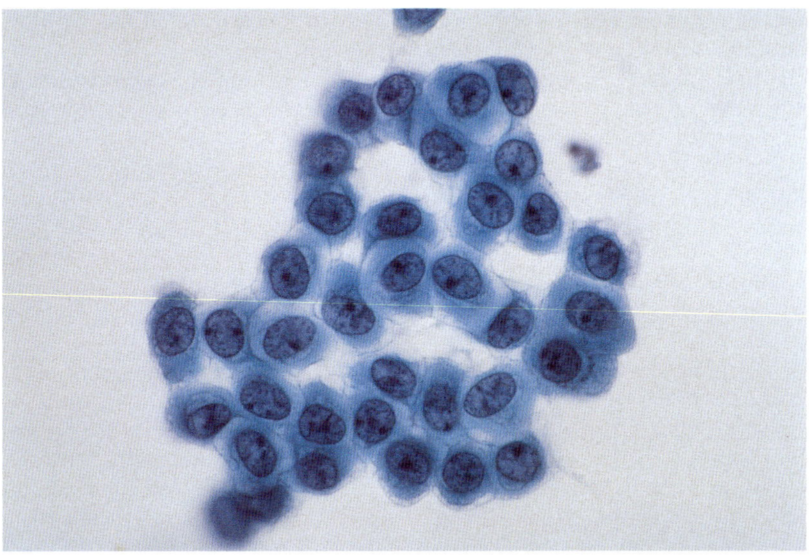

Figure 3–4. *Normal basal-intermediate urothelial cells.* Cuboidal cells with homogeneous dense cytoplasm, a low nuclear-cytoplasmic ratio, and uniform central round/oval nuclei with regular nuclear contours and small nucleoli are observed. Bladder washing. (Papanicolaou stain, ×600). Courtesy of Dr. Michael W. Stanley, Chair, Department of Pathology, Hennepin County Medical Center, Minneapolis, MN.

appearing urothelial cells raise the possibility of urolithiasis, low-grade papillary urothelial carcinoma, or a recent catheterization. Such a finding is unusual in voided urine specimens, and its presence should be stressed and further investigation suggested.

Because the turnover rate of urothelial maturation is more than 300 days, normal mitotic figures are exceedingly rare in the absence of reactive or neoplastic urothelial pathology.

Differential diagnosis. Umbrella cells are always benign, although they may coexist with urothelial neoplasms. Parabasal squamous cells must be distinguished from umbrella cells. In contrast to the latter, parabasal cells are round and lack a biconvex shape. Their distinction may lack diagnostic significance because both types of cells are benign. More significant is the correct distinction of umbrella cells from uni- or multinucleated macrophages or histiocytes, which are present in infectious or inflammatory processes.

The evaluation of columnar cells in urinary specimens should include tumoral entities of urothelial and nonurothelial origin. Clusters of columnar cells with uniform and oval nuclei with powdery chromatin may be seen in cystitis cystica and rarely in benign prostatic epithelial polyps. Admixed goblet cells may be present in both entities (Schnadig et al., 2000).

Distinction of clusters of basal and intermediate cells and granulomas from patients followed up for urothelial carcinoma requires awareness of and familiarity with these changes (Mack and Frick, 1994). Further details are presented in Chapter 4. Presenting clinical information and discussing the findings with the treating physician are always helpful so that an erroneous cytologic interpretation can be avoided.

URINE OBTAINED BY INSTRUMENTATION

Evaluation of the bladder and upper portions of the urinary system requires cystoscopic guidance; however, noncystoscopic catheterization of the urinary bladder may be used to obtain urine or to perform bladder washes/barbotages.

Cystoscopy of the upper tract (ureter, renal pelvis, and calices) may allow systematic inspection of these areas with a flexible ureteropyeloscope under general or epidural anesthesia. This procedure is relatively safe and particularly useful in the evaluation and diagnosis of space-occupying lesions and upper-tract hematuria of unknown etiology (Yazaki et al., 1999).

Urethral tumors may best be evaluated by voided urethral cytology and urethroscopy with washes or brushings (Pfister, 1976). However, brush cytology is a specific and more sensitive sampling method than cytology of urine obtained by irrigation or catheterization in detecting urothelial carcinoma of the upper urinary tract. As with urinary cytology in general, the technique is less effective in diagnosing low-grade than high-grade tumors (Dodd et al., 1997).

These diagnostic techniques have enhanced the ability to detect and accurately stage urothelial carcinoma. As discussed in later chapters, urine cytology is a sensitive method for detecting occult high-grade tumors in both the upper and lower urinary tracts (Grossman, 1986; van der Poel et al., 1997).

Washes must be performed with a sterile isotonic solution such as Ringer's lactate or saline solution. The specimen must be transported immediately to the laboratory. Refrigeration or fixation is strongly recommended when delay in transportation or processing is expected. Cell transfer from a brush to a glass slide must be performed promptly at the bedside to avoid air-drying. Immediate immersion of the brush in a cytology fixative is an alternative, particularly to make smears by thin-layer preparation methods.

Because of the high cellular yield, bladder washes are commonly used to obtain cells for ancillary tests such as DNA ploidy analysis by flow cytometry or image analysis (Amberson and Laino, 1993; Hermansen et al., 1988; Murphy et al., 1986; Slaton et al., 1997; Wiener et al., 1999), karyometric analysis (van der Poel et al., 1996), G-actin, p300 tumor antigen evaluated by quantitative fluorescence image analysis (Hemstreet et al., 1999), fluorescence in situ hybridization (Sauter et al., 1995), and telomerase activity (Dalbagni et al., 1997). These tests are objective, reproducible, and highly specific, and when used in conjunction with urinary cytology, they are valuable in detecting recurrences in the follow-up of patients with bladder cancer. DNA ploidy analysis has been proposed as being most suitable for monitoring high-risk populations such as industrial workers or others exposed to carcinogens, in addition to persons with a history of urothelial tumors (Melamed, 1990) (Box 3–2).

Cytology. Under normal conditions, the urinary tract lumen is a virtual space except for the bladder, which harbors a minimal volume of residual urine. Therefore, the introduction of an instrument dislodges the urothelium and yields a highly cellular urine specimen composed primarily of a mixture of umbrella, cuboidal, and columnar basal and intermediate, squamous, and red blood cells. In general, the more distally the cystoscope travels, the higher the cellularity of the specimen. Hence, ureteral and renal pelvis specimens are usually the most cellular (Fig. 3–5).

Low-power microscopic examination shows cohesive cell aggregates forming balls or pseudopapillary clusters, although a mixture of large and small clusters as well as triplets, pairs, or single cells in smaller numbers may

Box 3–2.

Washings yield large numbers of umbrella cells.

Brushings yield large numbers of basal and intermediate cells.

Preservation and cellular detail are better than those seen in voided urine.

Renal tubular cells and red blood cells are variably present.

Smears show numerous urothelial cell clusters and single urothelial cells.

Urothelial cells have thin or vacuolated cytoplasm, a low nuclear-cytoplasmic ratio, a round nucleus with regular borders, and powdery chromatin.

A similar smear pattern is seen in urine specimens from patients with urolithiasis and low-grade papillary urothelial neoplasms.

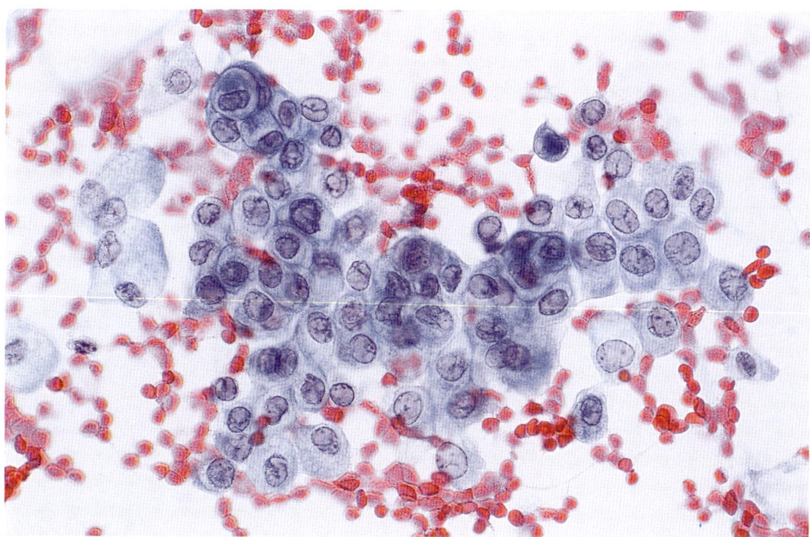

Figure 3–5. *Ureteral washing.* Numerous urothelial cells, single and in small clusters, are commonly seen in ureteral samples from nonneoplastic processes. Cells feature vacuolated cytoplasm, a low nuclear-cytoplasmic ratio, and eccentric round nuclei. Specimen obtained from the opposite ureter at the time of ureteral washing for urothelial carcinoma. (Papanicolaou stain, ×600)

be observed (Fig. 3–6A–D). The predominance of a particular cell type depends largely on the sampling device used and the topographic site sampled. Umbrella cells are more numerous in ureteral specimens, and are often associated with urothelial columnar cells. Similar cell aggregates are also found in bladder samples (Fig. 3–7). Few or no squamous cells are present in bladder, renal pelvis, and ureteral samples. In contrast,

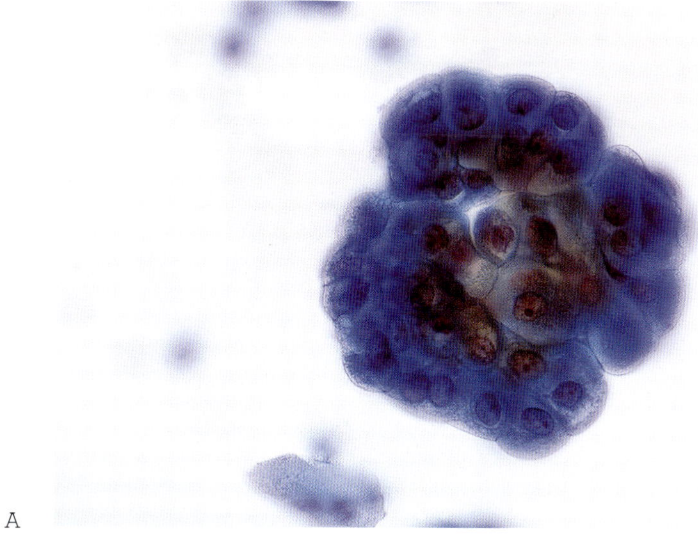

Figure 3–6. *Instrumentation effect.* (*A*) Cohesive, round, tri-dimensional aggregate of superficial urothelial umbrella cells with prominent nucleoli are commonly seen in the instrumentation effect. (*B,C*) Pseudopapillary clusters of basal-intermediate cells and umbrella cells with vacuolated cytoplasm are also present. (*D*) Less frequently, cells exhibit lack of cohesion, vacuolated cytoplasm, and prominent nucleoli. Bladder (*A,B*) and ureteral (*C.D*) washings. (Papanicolaou stain, ×600)

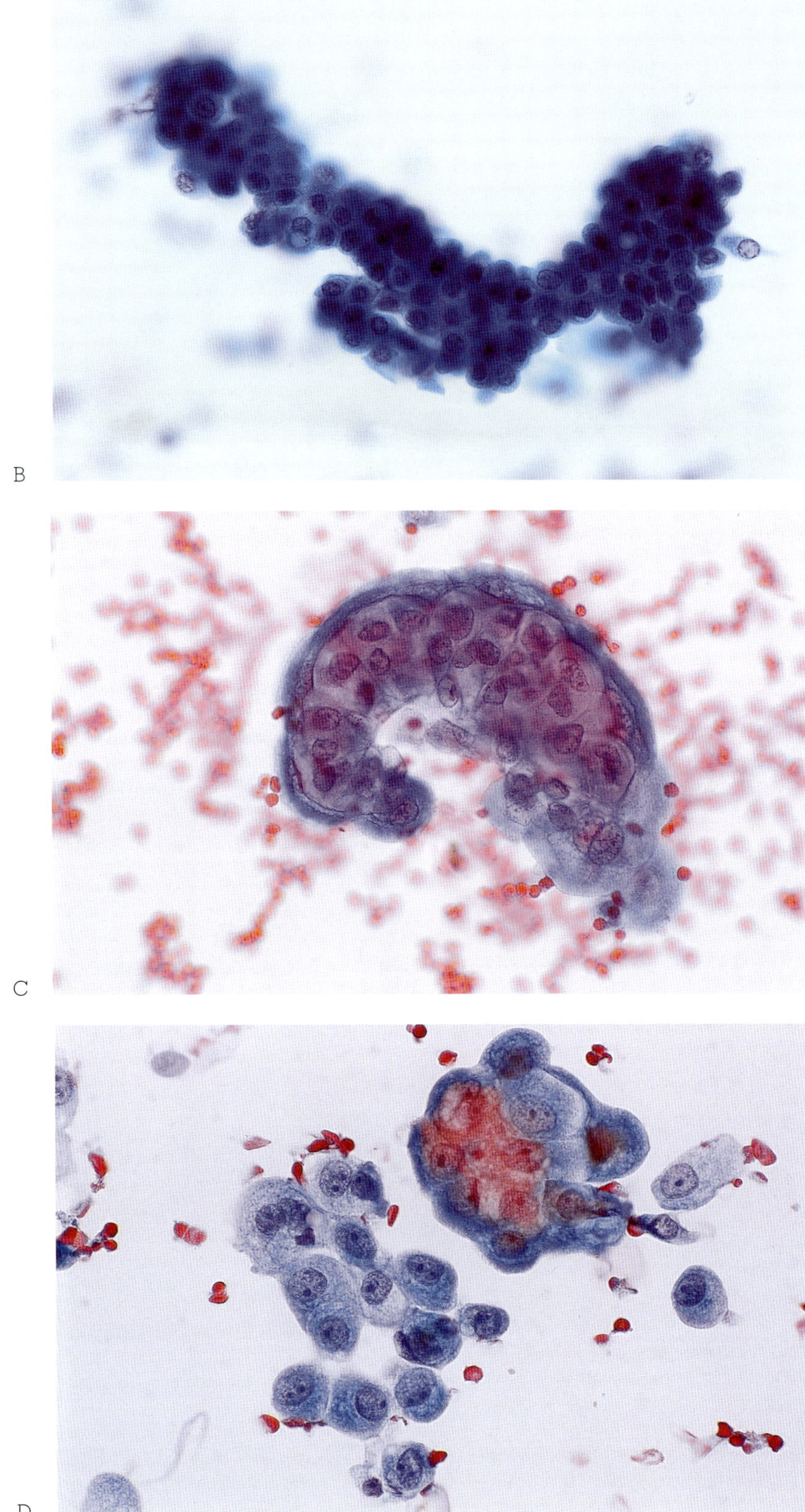

Figure 3–6. *(Continued)*

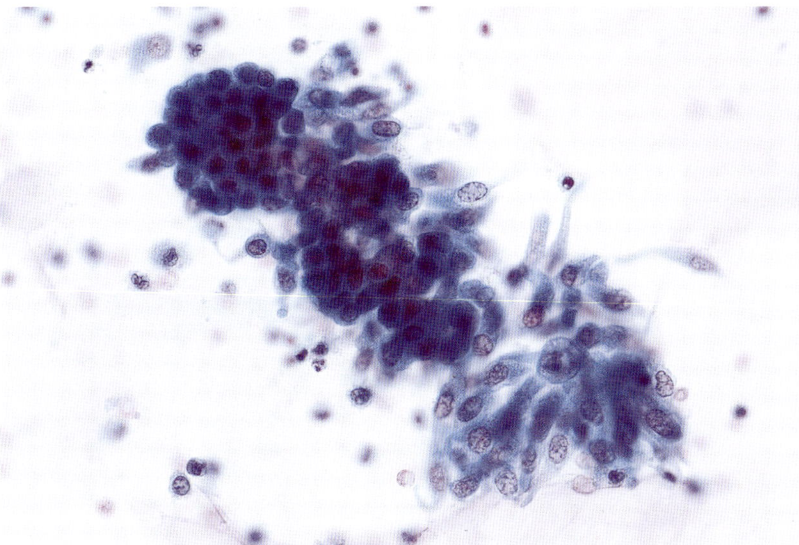

Figure 3–7. *Bladder washing.* Cellular specimen showing benign cuboidal and columnar cells of basal-intermediate cell layers. (Papanicolaou stain, ×600)

umbrella and squamous cells predominate in urethral specimens (Fig. 3–8). Washes yield clusters containing more umbrella cells and fewer basal and intermediate cells. Although there is a high yield of umbrella cells, because of the more abrasive nature of the procedure, brushings yield more basal and intermediate cell clusters than do washes. In both instances, urothelial cells of superficial and deeper layers may be found in the same clusters.

The cellular detail is optimal in instrumented specimens when these specimens are handled appropriately, and the cytologic characteristics are similar to those described for normal urothelial cells in voided speci-

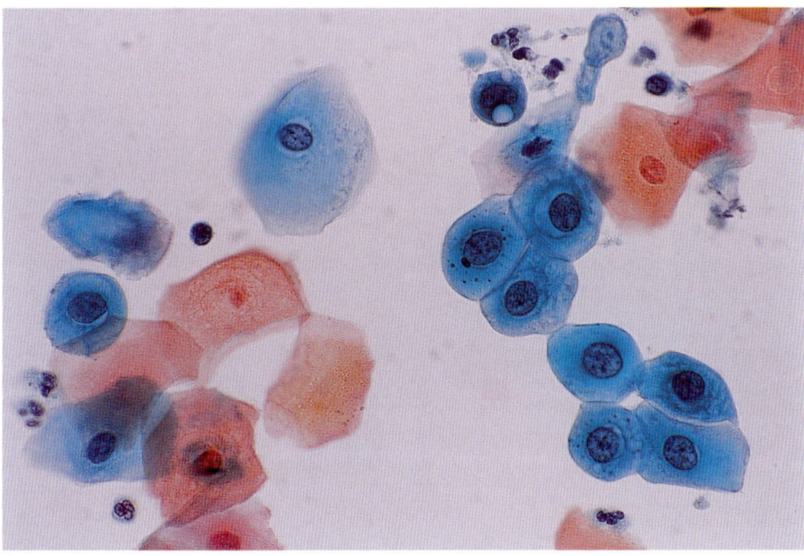

Figure 3–8. *Urethral washing.* A mixture of umbrella and squamous cells characterizes urethral samples from patients with nonneoplastic urethral disorders. (Papanicolaou stain, ×600)

mens. Briefly, urothelial cells have thin or vacuolated cytoplasm, a low nuclear-cytoplasmic ratio, a round nucleus with regular borders, and powdery chromatin.

Renal tubular cells may be identified, but inflammatory cells are variably present in these specimens. Intact red blood cells are more numerous in brushings than in washings because of the inherent traumatic injury of the brush.

The value of cytologic evaluation of ureteral or renal pelvis washes is greatly enhanced when bilateral specimens are submitted. Comparison of the cell morphology from both sides facilitates the detection of subtle but helpful cytologic abnormalities.

Differential diagnosis. Careful evaluation of nuclear and nucleolar characteristics permits an accurate distinction between a reactive instrumentation effect and high-grade urothelial carcinoma.

In some cases, washings or brushings allow an accurate diagnosis of low-grade urothelial carcinoma, particularly in the presence of high cell discohesiveness and subtle but recognizable nuclear abnormalities. However, in most cases, the difficulty of diagnosing low-grade urothelial carcinoma is compounded by the high cellularity and the cell clusters so characteristic of instrumented urine specimens. Because low-grade urothelial carcinoma may display an overlying umbrella cell layer, the absence or presence of umbrella cells in the cell clusters of brushings or washings is meaningless.

The abrasive effect of urolithiasis on the urothelium induces exfoliation of urothelial cell clusters that are indistinguishable from the cell clusters seen in washings and brushings and in low-grade urothelial carcinoma. Further discussion on the cytology of these entities is presented in Chapters 5 and 6.

URINE OBTAINED THROUGH A STENT

The development of the double-J ureteral stent has been a significant advance in the management of ureteral obstruction. The stent, inserted endoscopically or surgically, is used to divert the urine internally into the urinary bladder. One end of the stent is placed in the renal pelvis, and the other remains in the bladder so that the urine flows freely through it (Finney, 1978).

The double-J ureteral stent is frequently used for palliation of symptoms in cases of obstructive ureteropathy caused by urolithiasis. This procedure may delay or avoid surgery, particularly in cases of friable ureteral stones. The palliative use of the stent, when no other treatment is possible for malignant ureteral obstruction, avoids the need for ureterostomy and offers the patient a comfortable quality of life (Giannakopoulos et al., 1995). This stent may also be placed at the time of reparative surgery for ureteral trauma, remaining in place for four to eight weeks to allow complete ureteral healing (Cormio, 1995) (Box 3–3).

Box 3–3.

Double-J ureteral stent is used to divert urine internally from the renal pelvis to the bladder in cases of ureteral obstruction.

The smear pattern is similar to that of instrumented urine, with marked cellular degeneration and squamous metaplasia.

Reactive urothelial and squamous cells are present in addition to cellular and noncellular elements associated with the underlying obstructive process.

Cytology. Urine cytology specimens obtained through a stent share the characteristics of urine obtained with other instrumentation methods. Smears show reactive urothelial cells in addition to the cellular or noncellular elements associated with the underlying obstructive process. Crystals will be seen easily in urolithiasis, the most common cause of intrinsic obstruction. High-grade carcinoma will readily be identified by the presence of biphasic benign and malignant cell populations. Low-grade urothelial carcinoma will require careful evaluation of cellular details, comparison with urothelial cells from the contralateral side, and clinical–radiologic correlation.

When the stent is in place for a long period, the constant rubbing effect of the tip on the renal pelvis urothelium produces

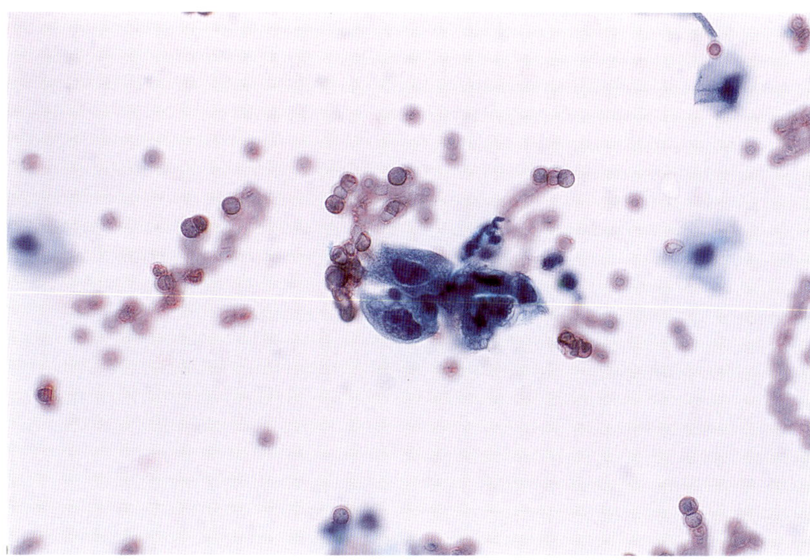

Figure 3–9. *Double-J ureteral stent effect.* A cluster of degenerated benign umbrella cells has a preserved biconvex shape. Vacuolated cytoplasm and dark, degenerated nuclei are also noted. These changes should not be interpreted as atypical or malignant. (Papanicolaou stain, ×600)

marked cytologic changes that include squamous metaplasia, which is difficult to distinguish from the changes of urothelial or squamous carcinoma (Figs. 3–9 and 6–34). Similar changes may be identified in urine samples from patients with a neurogenic bladder and a permanent urinary bladder catheter (Fig. 3–10). These striking cell abnormalities, without a proper clinical setting, invite one to make an erroneous diagnosis.

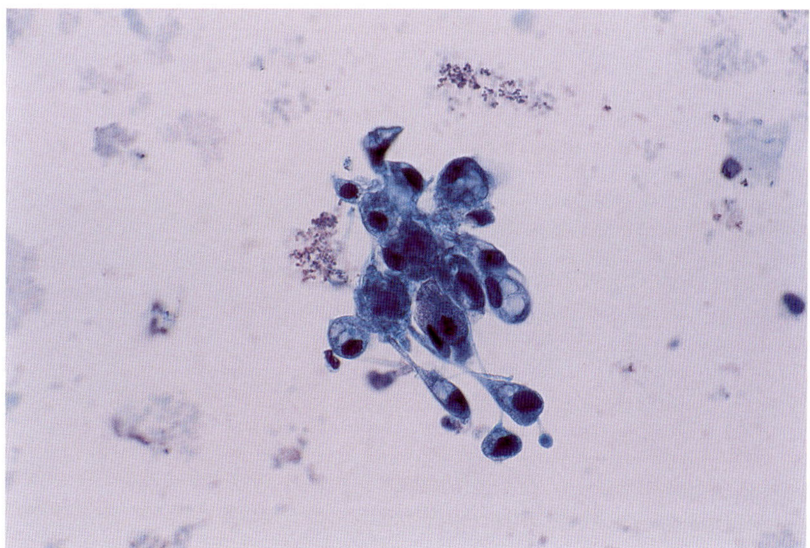

Figure 3–10. *Permanent bladder catheter effect.* Marked degeneration is seen in the urothelial cells. A preserved nuclear-cytoplasmic ratio, vacuolated cytoplasm, and a homogeneous, dark nucleus with smooth nuclear contours are evident. The urine sample was obtained from a paraplegic patient who had a Foley catheter inserted for 18 years. (Papanicolaou stain, ×600)

ILEAL-CONDUIT URINE

Of the various intestinal conduits, the ileal conduit remains the most commonly used form of urinary diversion, especially for elderly patients undergoing total cystectomy, particularly for bladder cancer. This procedure involves anastomosis of the ureters to an ileal loop and of the latter to the skin of the abdomen, where an ostomy bag receives the drained urine. A surgical variant includes anastomosis of the ileal pouch to the urethral stump in men, allowing the patient a fairly normal lifestyle after surgery (Skinner and Skinner, 1994). Complications include strictures, electrolyte imbalance, pyelonephritis with renal function impairment, lithiasis, and, less commonly, urothelial carcinoma (Azimuddin et al., 1999).

Histologic changes in the intestinal epithelium are those of a functional adaptation of the mucosa to the new environment. Progressive shortening of the villi, loss of microvilli, an increased number of goblet cells, and an abnormal pattern of mucin production occur during the first three years after surgery. These changes in the pattern of mucin production in the ileum appear to represent a transient defense mechanism of the mucosa exposed to urine. After three years, the mucosa progressively flattens and mucin production is reduced, and after four years the epithelium is flat and avillous, with a tendency toward stratification but without evidence of mucin secretion, dysplasia, or malignancy. Rare islands of normal epithelium are found even many years after surgery (Aragona et al., 1998; Parenti et al., 1999).

In contrast, significant histologic abnormalities have been found in the epithelium of other intestinal segments used for urinary diversion, such as a colon conduit, cecal reservoir, or ureterosigmoidostomy, after many years of urinary diversion. The incidence of benign and malignant intestinal neoplasms in these enteric segments is extremely high. Potential carcinogens include oxygen radicals produced by neutrophils, high levels of urinary *N*-nitrosamines, and increased epithelial cell proliferation at the urothelial-intestinal suture line (Azimuddin et al., 1999; Malone et al., 1997; Nurse and Mundy, 1989). In contrast to these enteric urinary diversions—in particular ureterosigmoidostomy, with a high incidence of intestinal and, less commonly, urothelial malignant neoplasms—the ileal conduit appears to have a low risk of malignancy. Although urothelial carcinoma may arise within an ileal loop conduit after cystectomy for urothelial carcinoma, it appears to be exceedingly rare (Carter and Gillatt, 1996; Inobe et al., 1999; Yamada et al., 1998). No case of adenocarcinoma of the ileum has been reported. However, it is possible that ileal conduits create the risk of tumorigenesis (Malone et al., 1997).

Thus, periodic urine cytology and close long-term follow-up of patients after urinary diversion with an ileal conduit are necessary. Hematuria and ureteral obstruction are the earliest clinical manifestations of malignancy, and urine cytology provides a rapid, accurate diagnosis in these cases (Shokeir et al., 1995). Patients with ureterosigmoidostomy should be followed closely by means of colonoscopy, urine cytology, and examination for occult blood in feces (Azimuddin et al., 1999) (Box 3–4).

Box 3–4.

The ileal conduit remains the most commonly used form of urinary diversion for patients with total cystectomy.

Patients with an ileal conduit have a low risk of developing malignancy.

Smears are cellular and show markedly degenerated, round intestinal cells with pyknotic nuclei, granular cytoplasm, and indistinct cytoplasmic borders.

Well-preserved fragments of intestinal epithelial cells are rare.

Red blood cells and inflammatory cells are absent.

Hematuria and obstruction are manifestations of malignancy.

Cytology. Because of the high turnover rate of intestinal epithelium cells, ileal-conduit urine specimens are markedly cellular and degenerated. These characteristics, in conjunction with the lack of cell cohesiveness, the granular background, and the lack of inflammation or red blood cells, permit instantaneous recognition (Fig. 3–11A).

The numerous single and degenerated cells are round/cuboidal, with indistinct cell borders, granular cytoplasm, and a pyknotic, eccentric nucleus. Round eosinophilic cytoplasmic inclusions are not uncommon. Less degenerated intestinal cells show ample vacuolated cytoplasm and a round, hyperchromatic nucleus. This cellular appearance is highly characteristic of ileal-conduit urine cytology (Malmgren et al., 1971).

Small, loosely cohesive clusters of degenerated cells are usually present, but recog-

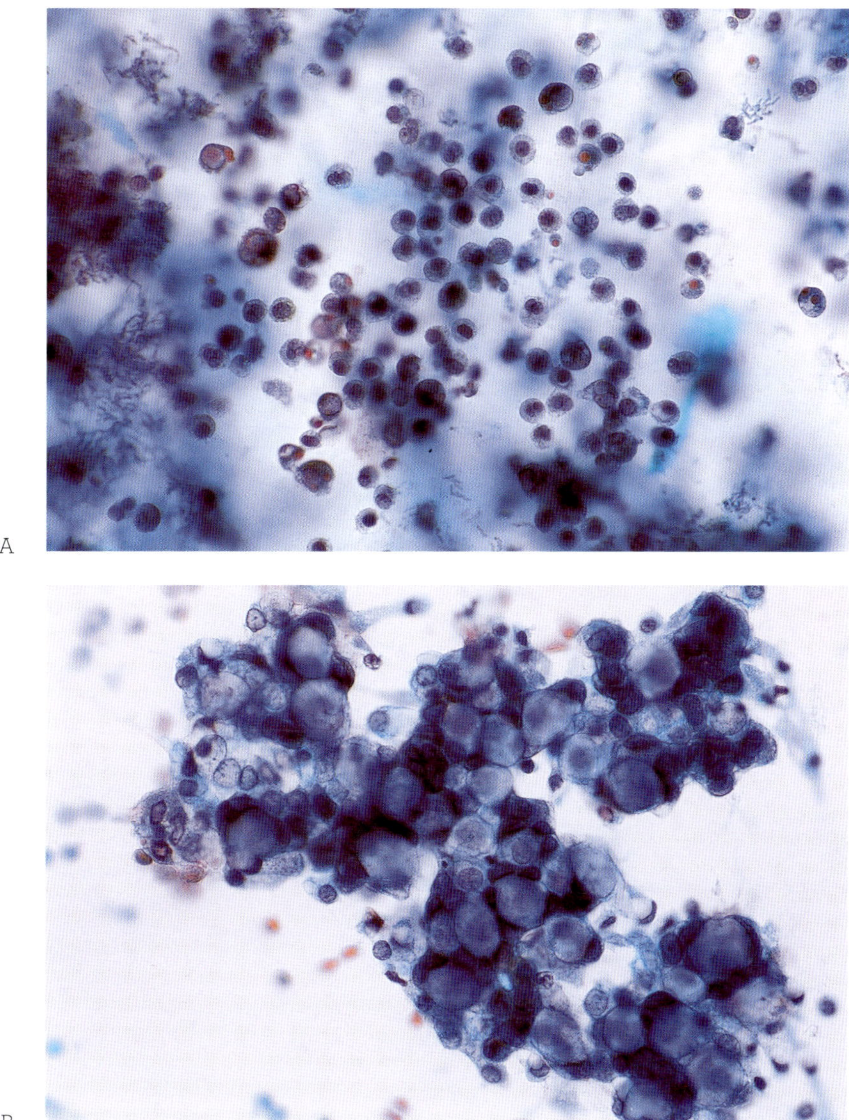

Figure 3–11. *Ileal conduit urine.* (*A*) Numerous round, markedly degenerated cells and a granular background are characteristic of ileal-conduit urine. (*B,C*) Clusters of intestinal epithelium with goblet cells and columnar cells with basally located nuclei may be identified. (*D,E*) Clusters of intestinal epithelium showing various degrees of nuclear degeneration and residual columnar cells are also present. These changes should not be interpreted as atypical or malignant.

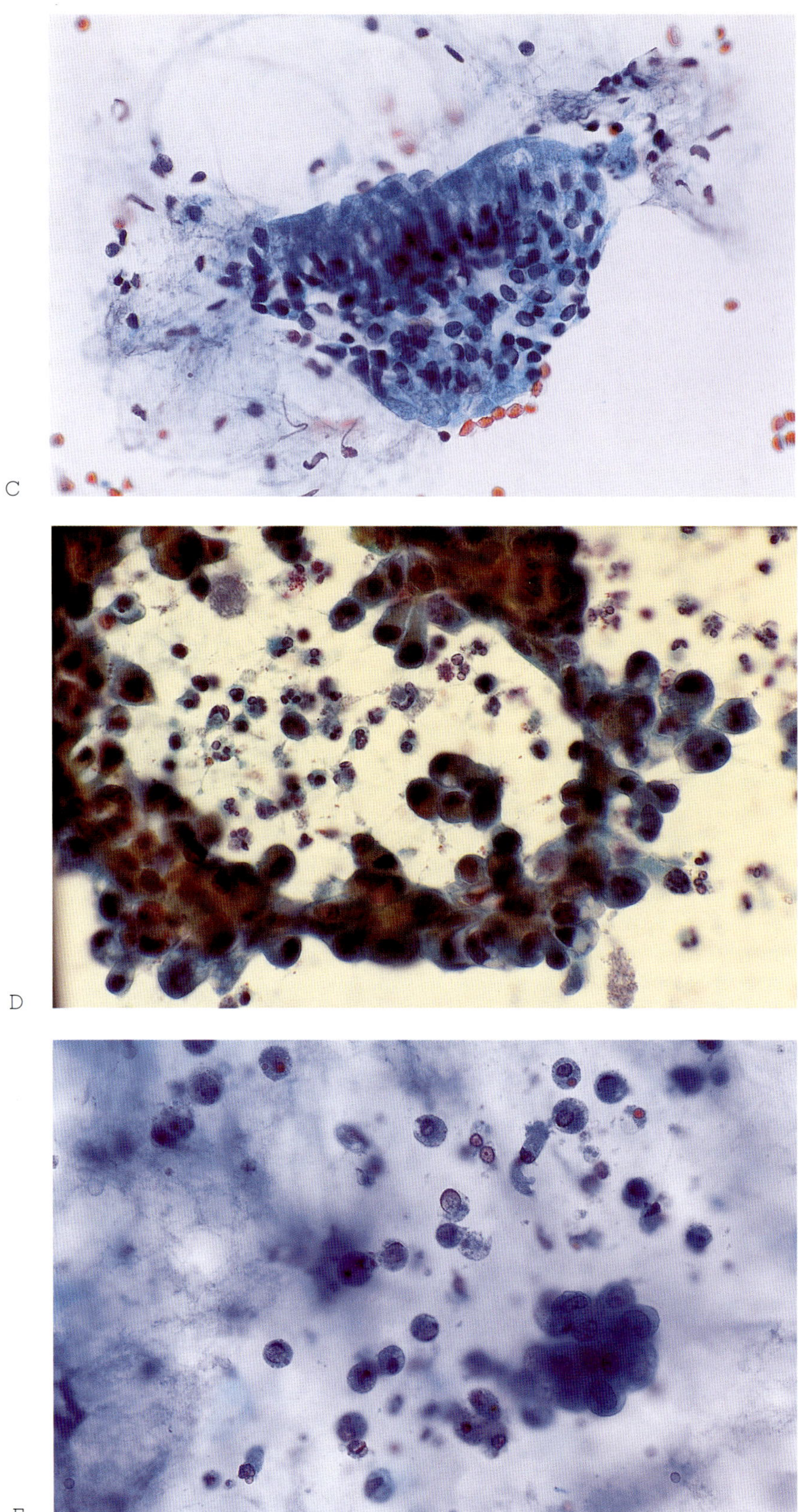

Figure 3–11. *(Continued)*

nizable, well-preserved, intact clusters of intestinal cells are rare. The latter are cohesive and composed of columnar cells commonly admixed with goblet cells (Fig. 3–11B). A honeycomb appearance or basally located, uniform, round nuclei, when viewed in profile, help to identify the intestinal origin of the cells (Fig. 3–11C). These cell clusters may be most apparent shortly after surgery. Less clustering and more degeneration are characteristic in samples obtained several years after surgery (Fig. 3–11D). A cytologic finding of nonintestinal cell clusters several years after surgery must trigger a careful evaluation to exclude recurrent urothelial carcinoma.

Cytologic evaluation of ileal-conduit urine from patients with cystectomy is accurate and identifies recurrent urothelial carcinoma even before it is clinically evident. Periodic cytologic examination of urine at three to six month intervals is considered advisable in these patients (Anagnostopoulou et al., 1995).

Differential diagnosis. A high index of suspicion and familiarity with the cytologic features are needed to identify the smear pattern of ileal-conduit urine in the absence of clinical information.

Extensive cellular degeneration and lack of cohesiveness are characteristics of invasive high-grade urothelial carcinomas. Well-preserved cells with distinct cytoplasmic outlines, anaplastic nuclei, and the tumor diathesis so characteristic of an invasive malignancy are absent in ileal-conduit urine.

When single and clustered, clearly malignant cells are identified, the diagnosis of carcinoma can be made with confidence. Close communication with the treating physician is always recommended when chemotherapy and radiation therapy effects are considered in the differential diagnosis.

The precise interpretation of clusters of cytologically bland cells may be challenging, particularly several years after ileal-conduit surgery. Because the intestinal epithelium occasionally may show cell clustering, the search for columnar and goblet cells in the clusters will point to an intestinal origin. The presence of cohesive clusters of round or cuboidal cells with clear cytoplasmic borders, dense cytoplasm, and subtle nuclear abnormalities may be suggestive of a low-grade urothelial tumor, and further investigation is needed. We must remember that small clusters of vacuolated cells admixed with columnar cells showing varying degrees of degeneration are not indicative of a papillary neoplasm but are commonly found in ileal-conduit cytology specimens (Fig. 3–11E).

REFERENCES

Amberson, J. B., Laino, J. P. (1993). Image cytometric deoxyribonucleic acid analysis of urine specimens as an adjunct to visual cytology in the detection of urothelial cell carcinoma. *J Urol* 149: 42–5.

Anagnostopoulou, I., Rammou-Kinia, R., Likourinas, M. (1995). Urine cytology evaluation in cases of uretero-ileal cutaneous diversion. *Cytopathology* 6: 268–72.

Aragona, F., De Caro, R., Parenti, A., Artibani, W., Bassi, P., Munari, P. F., Pagano, F. (1998). Structural and ultrastructural changes in ileal neobladder mucosa: a 7-year follow-up. *Br J Urol* 81: 55–61.

Azimuddin, K., Khubchandani, I. T., Stasik, J. J., Rosen, L., Riether, R. D. (1999). Neoplasia after ureterosigmoidostomy. *Dis Colon Rectum* 42: 1632–8.

Baltaci, S., Suzer, O., Ozer, G., Beduk, Y., Gogus, O. (1996). The efficacy of urinary cytology in the detection of recurrent bladder tumours. *Int Urol Nephrol* 28: 649–53.

Carter, A., Gillatt, D. A. (1996). Recurrent transitional cell carcinoma arising within an ileal conduit following cystectomy. A case report and review of the literature. *Eur Urol* 30: 519–20.

Cormio, L. (1995). Ureteric injuries. Clinical and experimental studies. *Scand J Urol Nephrol Suppl* 171: 1–66.

Dalbagni, G., Han, W., Zhang, Z. F., Cordon-Cardo, C., Saigo, P., Fair, W. R., Herr, H., Kim, N., Moore, M. A. (1997). Evaluation of the telomeric repeat amplification protocol (TRAP) assay for telomerase as a diagnostic modality in recurrent bladder cancer. *Clin Cancer Res* 3: 1593–8.

Dodd, L. G., Johnston, W. W., Robertson, C. N., Layfield, L. J. (1997). Endoscopic brush cytology of the upper urinary tract. Evaluation of its efficacy and potential limitations in diagnosis. *Acta Cytol* 41: 377–84.

Finney, R. P. (1978). Experience with new double J ureteral catheter stent. *J Urol* 120: 678–81.

Giannakopoulos, X., Gartzios, A., Giannakis, D., Tsamboulas, K., Kotoulas, K. (1995). Internal urinary diversion in pelvic cancers and quality of life. Value of double "J" endoprosthesis. *J Urol* 101: 221–7.

Gregoire, M., Fradet, Y., Meyer, F., Tetu, B., Bois, R., Bedard, G., Charrois, R., Naud, A. (1997). Diagnostic accuracy of urinary cytology, and deoxyribonucleic acid flow cytometry and cytology on bladder washings during followup for bladder tumors. *J Urol* 157: 1660–4.

Grossman, H. B. (1986). Diagnosis and treatment of bladder carcinoma. *Compr Ther* 12: 42–7.

Hemstreet, G. P., 3rd, Rao, J., Hurst, R. E., Bonner, R. B., Mellott, J. E., Rooker, G. M. (1999). Biomarkers in monitoring for efficacy of immunotherapy and chemoprevention of bladder cancer with dimethylsulfoxide. *Cancer Detect Prev* 23: 163–71.

Hermansen, D. K., Reuter, V. E., Whitmore, W. F., Jr., Fair, W. R., Melamed, M. R. (1988). Flow cytometry and cytology as response indicators to M-VAC (methotrexate, vinblastine, doxorubicin and cisplatin). *J Urol* 140: 1394–6.

Inobe, T., Kanda, K., Murakami, Y., Tsuji, M., Tamura, M., Kagawa, S. (1999). Recurrent bladder adenocarcinoma in an ileal conduit stoma: a case report. *Int J Urol* 6: 467–70.

Koss, L. G., Deitch, D., Ramanathan, R., Sherman, A. B. (1985). Diagnostic value of cytology of voided urine. *Acta Cytol* 29: 810–16.

Loh, C. S., Spedding, A. V., Ashworth, M. T., Kenyon, W. E., Desmond, A. D. (1996). The value of exfoliative urine cytology in combination with flexible cystoscopy in the diagnosis of recurrent transitional cell carcinoma of the urinary bladder. *Br J Urol* 77: 655–8.

Luthra, U. K., Dey, P., George, J., Abdulla, M. A., Shaheen, A. A., Sheikh, Z. A., George, S. S. (1999). Comparison of ThinPrep and conventional preparations: urine cytology evaluation [letter]. *Diagn Cytopathol* 21: 364–6.

Mack, D., Frick, J. (1994). Diagnostic problems of urine cytology on initial follow-up after intravesical immunotherapy with Calmette-Guerin bacillus for superficial bladder cancer. *Urol Int* 52: 204–7.

Malmgren, R. A., Soloway, M. S., Chu, E. W., Del Vecchio, P. R., Ketcham, A. S. (1971). Cytology of ileal conduit urine. *Acta Cytol* 15: 506–9.

Malone, M. J., Izes, J. K., Hurley, L. J. (1997). Carcinogenesis. The fate of intestinal segments used in urinary reconstruction. *Urol Clin North Am* 24: 723–8.

Melamed, M. R. (1990). Flow cytometry detection and evaluation of bladder tumors. *J Occup Med* 32: 829–33.

Murphy, W. M., Emerson, L. D., Chandler, R. W., Moinuddin, S. M., Soloway, M. S. (1986). Flow cytometry versus urinary cytology in the evaluation of patients with bladder cancer. *J Urol* 136: 815–19.

Nurse, D. E., Mundy, A. R. (1989). Assessment of the malignant potential of cystoplasty. *Br J Urol* 64: 489–92.

Parenti, A., Aragona, F., Bortuzzo, G., De Caro, R., Pagano, F. (1999). Abnormal patterns of mucin secretion in ileal neobladder mucosa: evidence of preneoplastic lesion? *Eur Urol* 35: 98–101.

Pfister, R. R. (1976). Endoscopy and the detection of genitourinary carcinoma. *Cancer* 37: 471–4.

Sauter, G., Moch, H., Wagner, U., Novotna, H., Gasser, T. C., Mattarelli, G., Mihatsch, M. J., Waldman, F. M. (1995). Y chromosome loss detected by FISH in bladder cancer. *Cancer Genet Cytogenet* 82: 163–9.

Schnadig, V. J., Adesokan, A., Neal, D., Jr., Gatalica, Z. (2000). Urinary cytologic findings in patients with benign and malignant adenomatous polyps of the prostatic urethra. *Arch Pathol Lab Med* 124: 1047–52.

Shokeir, A. A., Shamaa, M., el-Mekresh, M. M., el-Baz, M., Ghoneim, M. A. (1995). Late malignancy in bowel segments exposed to urine without fecal stream. *Urology* 46: 657–61.

Skinner, E. C., Skinner, D. G. (1994). Management of bladder cancer. In *Geriatric Urology* (P. D. O'Donnell, ed.), pp. 157–65. Little, Brown, Boston.

Slaton, J. W., Dinney, C. P., Veltri, R. W., Miller, C. M., Liebert, M., O'Dowd, G. J., Grossman, H. B. (1997). Deoxyribonucleic acid ploidy enhances the cytological prediction of recurrent transitional cell carcinoma of the bladder. *J Urol* 158: 806–11.

van der Poel, H. G., Boon, M. E., van Stratum, P., Ooms, E. C., Wiener, H., Debruyne, F. M., Witjes, J. A., Schalken, J. A., Murphy, W. M. (1997). Conventional bladder wash cytology performed by four experts versus quantitative image analysis. *Mod Pathol* 10: 976–82.

van der Poel, H. G., Witjes, J. A., van Stratum, P., Boon, M. E., Debruyne, F. M., Schalken, J. A. (1996). Quanticyt: karyometric analysis of bladder washing for patients with superficial bladder cancer. *Urology* 48: 357–64.

Wiener, H. G., Remkes, G. W., Schatzl, G., Susani, M., Breitenecker, G. (1999). Quick-staining urinary cytology and bladder wash image analysis with an integrated risk classification: a worthwhile improvement in the follow-up of bladder cancer? *Cancer* 87: 263–9.

Yamada, Y., Fujisawa, M., Nakagawa, H., Tanaka, H., Gotoh, A., Gohji, K., Arakawa, S., Kamidono, S. (1998). Squamous cell carcinoma in an ileal conduit. *Int J Urol* 5: 613–14.

Yazaki, T., Kamiyama, Y., Tomomasa, H., Shimizu, H., Okano, Y., Iiyama, T., Iizumi, T., Umeda, T. (1999). Ureteropyeloscopy in the diagnosis of patients with upper tract hematuria: an initial clinical study. *Int J Urol* 6: 219–25.

4

INFECTIONS AND INFESTATIONS OF THE URINARY TRACT

UROLOGIC examination that includes urine cytology should be requested for all men with urinary tract infections; for women with recurrent infections, painless hematuria, or intrinsic obstruction of the urinary tract; and for other selected patients (Stamm and Turck, 1987).

Specific etiologic organisms such as parasites, viruses, and fungi, and less frequently bacteria, can be identified by means of cytology. However, the final diagnostic and therapeutic implications of the last two rest on culture and drug sensitivity test results. In addition, cytology helps to detect a coexisting neoplasm that masquerades as a urinary tract infection.

Infections of the urinary tract can be classified into two anatomic categories: upper (pyelonephritis) and lower (urethritis, cystitis, and prostatitis). With the exception of prostate infections, all other urinary tract infections and infestations will be covered in the following paragraphs.

KIDNEY

Disease processes can affect one or more of the four compartments of the kidney: glomeruli, tubules, interstitium, and vessels. The glomeruli and vessels are more frequently affected by degenerative and immunologic disorders than by infectious processes. Infections preferentially affect both the tubules and the interstitium, hence the name *tubulointerstitial diseases.*

Diseases in which tubulointerstitial injury appears to be the primary event can be classified according to the clinical findings into acute and chronic, each having a specific etiology. Causes of acute tubulointerstitial disease include infections, toxins, metabolic disorders, chemotherapy, radiation therapy, and transplant rejection, among others, infections being by far the most common. Chronic tubulointerstitial diseases include chronic granulomatous and nongranulomatous pyelonephritis (Cotran et al., 1999a). Associated glomerular and vascular damage can be present, but this damage either is mild or appears late in the course of the disease (Box 4–1).

Acute Pyelonephritis

Acute pyelonephritis is a suppurative inflammation of the renal parenchyma commonly caused by bacterial infection. Routes of infection include hematogenous dissemination from sepsis, or more frequently ascending from a lower urinary tract infection, commonly cystitis. The most frequent pathogenic organisms are coliform bacteria from the lower intestinal tract such as *Escherichia coli, Proteus, Enterobacter,* and *Klebsiella.* Bacteria, fungi, and viruses are common agents in hematogenous dissemination.

Patients with acute bacterial pyelonephritis complain of fever, flank pain, and malaise. When urethritis or cystitis coexists, dysuria, frequency, and urgency are also present. Although the disease is more common in women in the reproductive age group, it may be present in men or women of any age in association with urinary obstruction, instrumentation, vesicoureteral reflux, diabetes mellitus, or immunosuppression (Cotran et al., 1999a).

Histologically, an acute inflammatory exudate is seen initially within the medullary

Box 4–1.

The kidney has four compartments: glomerular, tubular, interstitial, and vascular.

Degenerative and immunologic disorders affect the glomeruli and vessels.

Infection is the most common form of acute tubular and interstitial disease.

Granulomatous and nongranulomatous pyelonephritis are common causes of chronic tubular and interstitial disease.

tubules and interstitium in acute bacterial pyelonephritis. Hematogenous dissemination of the infection initially affects the renal cortex, with microabscess formation. Complications of acute pyelonephritis are renal abscess, papillary necrosis, and pyonephrosis.

Fungal acute renal involvement is commonly seen in patients who are immunocompromised due to AIDS, cancer, or diabetes mellitus. Immunosuppression is usually associated with multiorgan involvement and systemic clinical manifestations. Diffuse renal parenchymal disease, single or multiple abscesses, and obstructive uropathy are types of upper urinary tract fungal involvement seen particularly with *Candida* and *Aspergillus* infections. Renal cryptococcal infection (pyelonephritis or abscesses) is a manifestation of systemic cryptococcosis seen in AIDS and renal transplant patients (Wise et al., 1999). Granulomas and abscesses are seen with renal coccidioidomycosis and histoplasmosis. Acute interstitial nephritis due to cytomegalovirus (CMV), herpes virus, and the BK strain of polyomavirus is also seen in immunocompromised patients (Fig. 4–1A–D) (Ljutic and Glavina, 1995; Nickeleit et al., 1999; Silbert et al., 1990). Unfortunately, no routine screening for human polyomavirus infection is available, so its frequency is probably underestimated. No specific therapy is currently available for this infection (Boubenider et al., 1999).

Cytology. As a result of the intratubular collection of neutrophils, white blood cell casts in urine samples are characteristic of acute bacterial pyelonephritis. However, the absence of white blood cell casts does not exclude this diagnosis in the appropriate clinical setting. Other cytologic findings are not specific. Neutrophils and reactive urothelial cells are present, although they are more numerous when cystitis or urethritis coexists. Red blood cells may be present in cases of severe bacterial infections.

Fungi and viruses must be searched for thoroughly, particularly in an immunocompromised host. Filamentous fungi may not be present in voided urine specimens, and the diagnosis may require direct ureteral or renal pelvis catheterization. However, the cytodiagnosis of renal mucormycosis or renal coccidioidomycosis has been made by means of urine examination (Florentine et al., 1997; Kuntze et al., 1988). Fungal casts may be a diagnostic clue to renal candidiasis (Argyle, 1985).

Intranuclear inclusions are seen in the renal tubular cells in CMV, polyomavirus, and adenovirus infections. The morphologic characteristics of infectious agents that affect the kidney are similar to those present in other organs and are described in the section on the urinary bladder (Box 4–2).

Differential diagnosis. In the absence of acute inflammation, the identification of

Box 4–2.

Acute pyelonephritis is frequently caused by coliform bacteria ascending from a lower urinary tract infection.

Predisposing factors include urinary obstruction, instrumentation, vesicoureteral reflux, diabetes mellitus, and immunosuppression.

Fungal or viral pyelonephritis is frequently seen in immunocompromised patients.

Fungal casts or a viral cytopathic effect may be identified in urine cytology smears.

White blood cell casts, neutrophils, red blood cells, and reactive urothelial cells are present in cases of bacterial pyelonephritis.

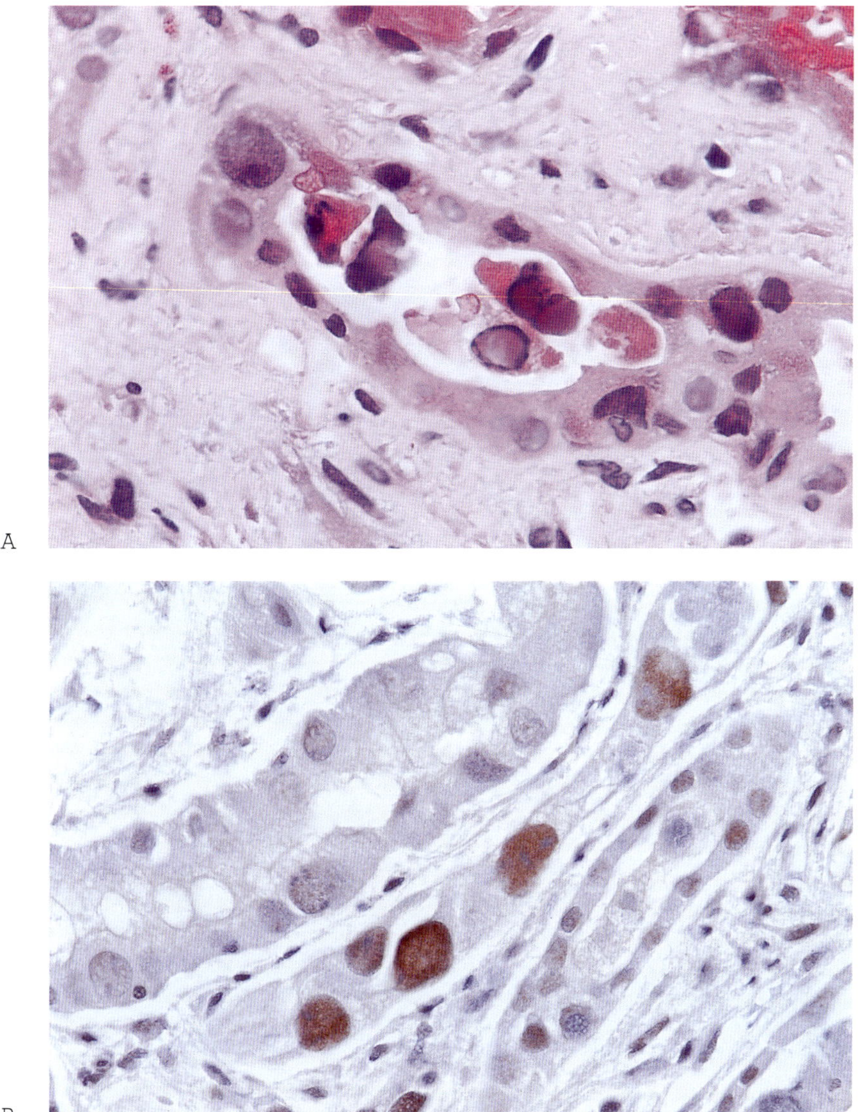

Figure 4–1. *Tubulointerstitial nephritis due to polyomavirus and cytomegalovirus infection.* (*A*) Polyomavirus infection shows dense intranuclear inclusions that occupy almost the entire nuclear area of the renal tubular cells. No perinuclear halo is present. (*B*) Immunoperoxidase stain for polyomavirus is positive. (*C*) Ultrathin sections of the infected cells examined by electron microscopy show loose crystalline clusters and free polyomavirus particles measuring 40 to 45 nm in the nucleus. Note the small, irregular clumps of chromatin at the periphery of the nucleus. (*D*) Cytomegalovirus-infected large renal tubular cell with a large, eosinophilic intranuclear inclusion and a perinuclear halo are present together with interstitial chronic inflammation. (*E*) Ultrathin sections show the intranuclear inclusion to be composed of viral particles measuring 100 to 200 nm, with a round core and a double membrane, embedded within a dense reticular matrix. Renal biopsy specimens obtained from renal allograft recipients with a clinical suspicion of acute cellular rejection. (A,D, Hematoxylin and eosin stain, ×600). (Uranyl acetate stain, ×16,000 and inset ×30,000 (C); ×10,200 and inset ×38,000 (E)). Figures C–E are courtesy of Dr. Terry D. Oberly, Professor and Vice-Chair, Department of Pathology and Laboratory Medicine, University of Wisconsin-Madison School of Medicine, Madison, WI.

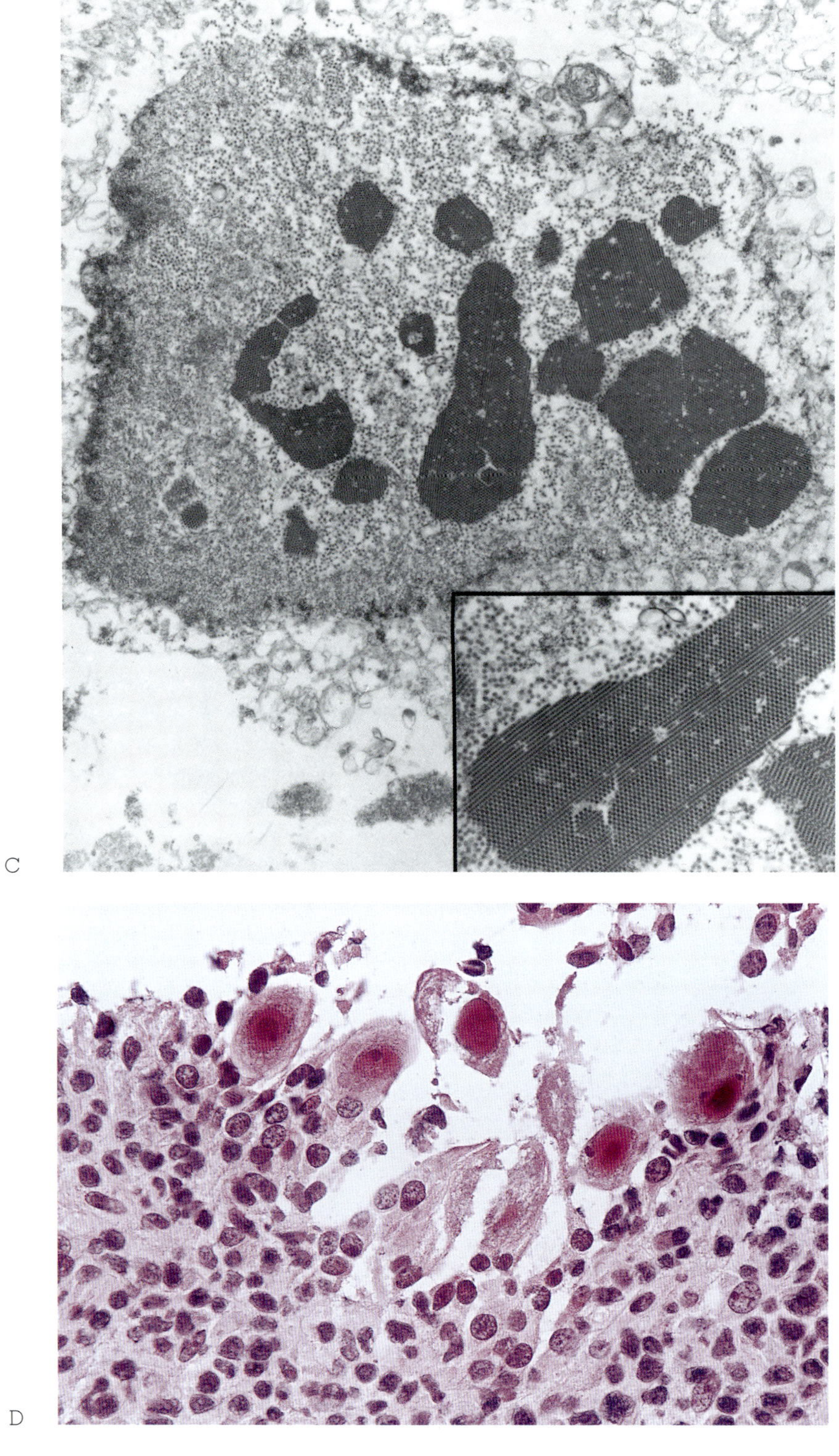

Figure 4–1. (*Continued*)

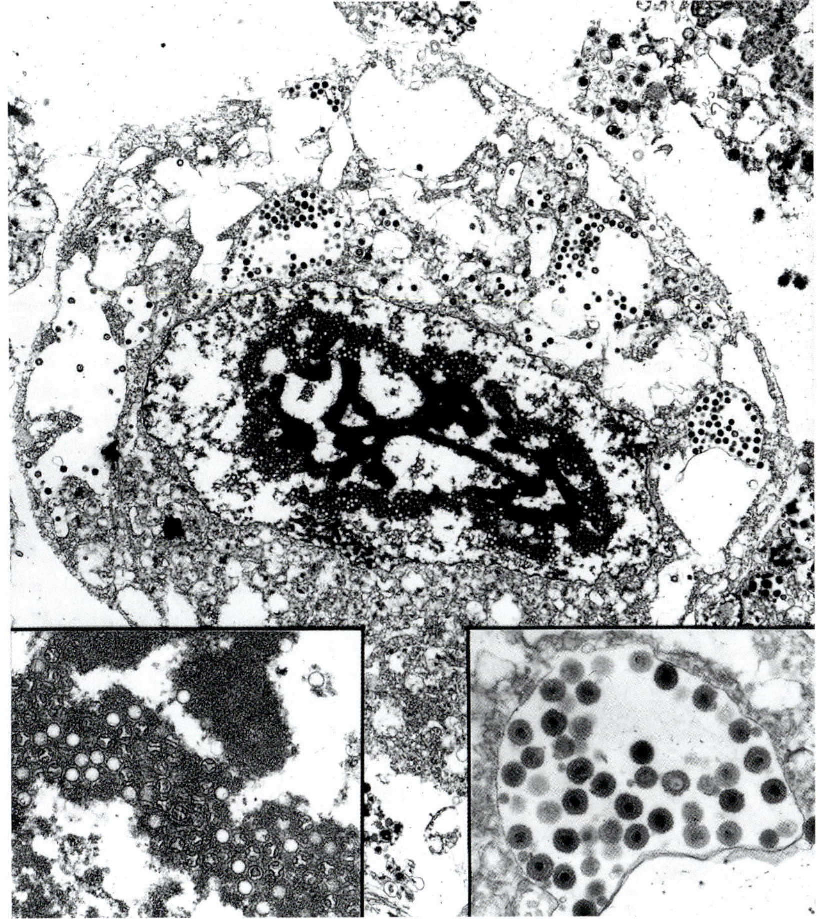

Figure 4–1. (*Continued*)

bacteria or budding yeast in the smear is usually associated with overgrowth and not with active infection.

Renal Papillary Necrosis

Papillary necrosis is a complication of acute pyelonephritis, most commonly found in diabetics and in the presence of urinary tract obstruction or analgesic abuse (Griffin et al., 1995). This is a process of coagulative necrosis of the renal medulla at the tips of the papillae, with a limited acute inflammatory response in the damaged tissue. The inflammatory response is usually restricted to the interface between the necrotic tissue and the kidney parenchyma (Cotran et al., 1999a).

The clinical picture may vary from a severe, life-threatening disease to a more chronic process causing insidious renal failure and polyuria due to inability to concentrate urine (isostenuric state).

Cytology. Small fragments of degenerated tissue may be identified, accompanied by granular background debris. Hematuria is common. The acute inflammatory response that is present is variable and occa-

> **Box 4–3.**
>
> Renal papillary necrosis is a complication of acute pyelonephritis in diabetics.
>
> Smears show fragments of degenerated tissue, hematuria, and granular debris.
>
> The acute inflammatory response is minimal.

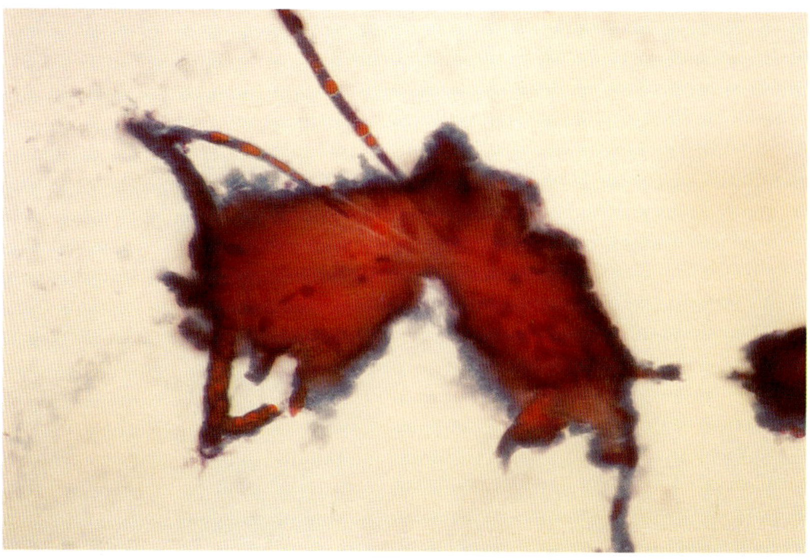

Figure 4–2. *Renal papillary necrosis.* Fragment of necrotic renal papilla with capillaries identified in a urine sample obtained from a diabetic patient with acute pyelonephritis and isostenuria. (Papanicolaou stain, ×600)

sionally is minimal (Fig. 4–2). The findings must be correlated with the clinical scenario (Box 4–3).

Chronic Pyelonephritis

Chronic tubulointerstitial inflammation and renal scarring with damage of the pelvicaliceal region, characteristic of chronic pyelonephritis, accounts for 10% to 20% of patients in transplant or dialysis units (Cotran et al., 1999a). Chronic pyelonephritis develops as a result of a chronic obstructive process with bouts of bacterial infection or chronic nonobstructive (reflux) nephropathy, a congenital disorder that occurs early in childhood and is by far the most common form of chronic pyelonephritis.

Histologic findings include tubular atrophy and dilatation, interstitial fibrosis, and interstitial chronic inflammation with occasional intra- and extratubular neutrophilic inflammation, particularly during bouts of acute infection. The characteristic "thyroidization" of the tubules results from dilatation and accumulation of eosinophilic casts within the lumen.

Box 4–4.

Chronic nonobstructive (reflux) nephropathy is the most common form of chronic pyelonephritis.

The renal parenchyma shows tubular atrophy and dilatation, interstitial fibrosis, and chronic inflammation.

Smears show a predominant mononuclear cell inflammation and occasionally broad waxy casts, hyaline casts, and granular casts.

Cytology. Urine cytology findings are nonspecific. The inflammation that is present is usually sparse and composed predominantly of mononuclear cells. Neutrophils predominate in periods of acute infection, and occasional white blood cell casts may be found. The background is granular, indicative of tissue damage, with broad waxy casts and occasional hyaline or fine granular casts. Coarse granular casts may be present when proteinuria coexists (Fig. 4–3; Box 4–4).

Xanthogranulomatous Pyelonephritis

Xanthogranulomatous pyelonephritis (XGP) is a well-recognized but rare type of chronic

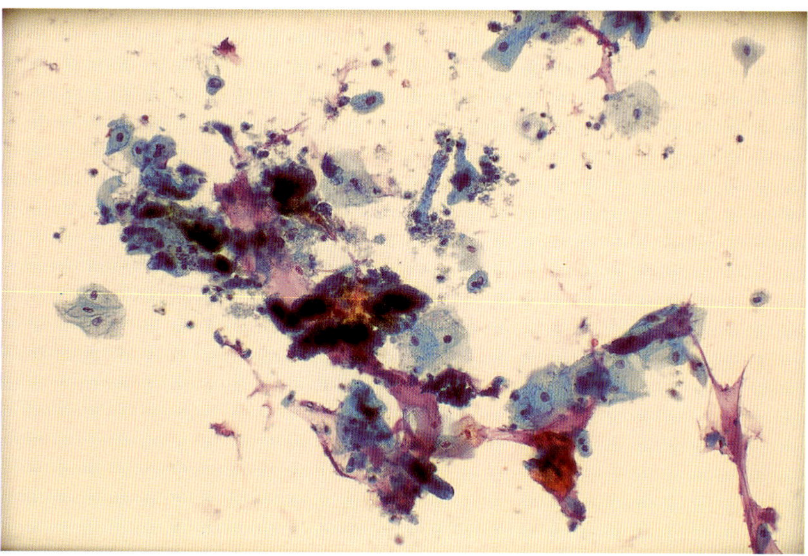

Figure 4–3. *Chronic pyelonephritis.* Urine sample from a patient with chronic pyelonephritis shows numerous casts and amorphous debris together with a few squamous cells. (Papanicolaou stain, ×200)

pyelonephritis. Classically, it occurs in middle-aged women, but it may occur at any age and may have a variable clinical presentation ranging from lack of renal symptoms to repetitive bouts of pyuria or bacteriuria or chronic renal failure (Gregg et al., 1999). *Escherichia coli* (67%) and *Proteus mirabilis* (26%) are the most common organisms isolated from voided urine, kidney, and blood (Chuang et al., 1992).

A staghorn calculus associated with a nonfunctional renal mass strongly suggests the diagnosis of XGP. Aside from this classical presentation, the preoperative diagnosis of XGP may be difficult to sustain both clinically and radiographically, particularly in the absence of lithiasis and in the presence of an extensive renal mass with extrarenal involvement mimicking a neoplasm. The use of fine-needle aspiration biopsy to distinguish between XGP and a neoplasm is recommended in both children and adults. Because nephrectomy is always necessary, the distinction is often made postoperatively (Goodman et al., 1998). However, a precise preoperative pathologic diagnosis of XGP, renal cell carcinoma, urothelial carcinoma, or another malignancy is needed to determine the mode of surgical exploration.

Coalescent masses of lobulated yellow tissue involving the renal parenchyma, with ulceration and necrosis of the perirenal fat and occasional extension to adjacent organs, are seen on gross examination (Anonymous, 1995). Numerous foamy histiocytes forming granulomas with occasional plasma cells and lymphocytes are present histologically. Differentiation between the foamy histiocytes of XGP and clear cells from renal cell carcinoma may be difficult on frozen and even permanent sections, particularly when clinical findings indicate an extensive renal mass in-

Box 4–5.

Xanthogranulomatous pyelonephritis is a rare form of chronic pyelonephritis associated with coliform bacteria and pyuria.

A nonfunctional renal mass and a staghorn calculus suggest XGP.

Smears show intense lymphoplasmacytic inflammation, foamy and multinucleated histiocytes, and a hemorrhagic background. Squamous metaplastic cells may be numerous.

Periodic acid–Schiff (PAS)-negative intracytoplasmic inclusions may be identified.

volving adjacent tissues. Keratin and CD68 immunohistochemical stains that separate epithelial cells from histiocytes are helpful.

Review of the literature shows the association between urologic tumors and XGP to be 4% (27 tumors in 676 cases of XGP up to 1988) (Ballesteros et al., 1988). The significance of this association is unkown.

Cytology. Rare reports on the urine cytology of XGP have been published. Some claim an accuracy rate of 80% when serial voided urine and ureteral washings are examined (Ballesteros et al., 1980, 1983). Characteristically, the urine shows an intense inflammatory and hemorrhagic background, making cellular identification difficult. Lymphocytes and plasma cells are also present, along with scattered histiocytes showing abundant foamy cytoplasm and an eccentric nucleus. Occasional giant multinucleated histiocytes may be identified. Basophilic intracytoplasmic PAS-negative inclusions have been reported (Bosch-Princep et al., 1998). The constant long-term abrasive effect of staghorn stones in the renal pelvis yields squamous metaplastic cells in urine samples (Beyer-Boon et al., 1978) (Box 4–5).

Differential diagnosis. The differential diagnosis of XGP in urine includes a nonspecific inflammatory response, inflammatory diseases with giant cells such as tuberculosis, and entities with histiocytes such as malacoplakia (Bosch-Princep et al., 1998). Occasionally, clear cell renal cell carcinoma involving the renal pelvis may exfoliate malignant cells in urine. In these cases, cytologic atypia, mitosis, and a prominent nucleolus help to distinguish XGP from clear cell renal cell carcinoma. However, well-differentiated neoplasms may be difficult to distinguish, and the final diagnosis requires fine-needle aspiration biopsy or nephrectomy. Pathologic entities that exfoliate clear cells in urine specimens are detailed in Table 7–2 of Chapter 7.

Renal Tuberculosis

Renal tuberculosis is an uncommon infectious complication seen after renal transplantation and other states of immunosuppression (Lenk et al., 1988). The kidney is involved via hematogenous dissemination from a pulmonary focus; however, most patients do not have pulmonary disease at the time of renal involvement.

The initial renal involvement occurs in the cortex and remains quiescent for a variable period of time. It becomes active in the medulla, causing a coalescent and destructive caseating granulomatous and cavitary process. This process produces papillary necrosis, with subsequent seeding of the tuberculous infection in the ureters and bladder (Murphy et al., 1982). At this point, the symptoms are isostenuria, dysuria, and hematuria.

Finding of Langhans-type giant histiocytes, epithelioid histiocytes, and necrotic debris on fine-needle aspiration cytology with bacteriologic culture of the aspirated material or saline rinses of the fine-needle specimen can be used to support the diagnosis of renal tuberculosis (Baniel et al., 1991) (Box 4–6).

Box 4–6.

Renal tuberculosis is the result of hematogenous dissemination from a pulmonary focus.

Renal tuberculosis causes caseating granulomas in the renal medulla and papillary necrosis.

Smears show epithelioid histiocytes, granulomas, Langhans-type giant cells, and amorphous debris.

Keratinized, anucleated squamous cells (leukoplakia) may be present in chronic or inactive tuberculosis.

Cytology. As in bladder tuberculosis, the presence of epithelioid histiocytes, granulomas, multinucleated Langhans-type giant cells, and amorphous debris in urine cytology supports the diagnosis of tuberculosis. However, the frequent coexistence of ureteral or bladder tuberculosis with identical cell characteristics makes it difficult to ascertain the exact origin of the process, and final diagnosis requires radiologic correlation.

Urinary exfoliation of numerous anucleated squamous cells or leukoplakia, as a result of squamous metaplasia of the urothelium, may be found in inactive renal tuberculosis (Armora Mani et al., 1992). The association of renal tuberculosis with leukoplakia of the renal pelvis was quite common in the era before chemotherapy of tuberculosis, but apparently it has not been seen since effective drugs became available (Byrd et al., 1976).

Differential diagnosis. The differential diagnosis includes XGP, renal cell carcinoma, and malacoplakia. These entities are listed in Table 7–2 of Chapter 7. In addition, a granulomatous response secondary to local immunotherapy with BCG in patients with superficial bladder cancer must be excluded before a diagnosis of tuberculosis is rendered.

RENAL PELVIS AND URETERS

The renal pelvis and ureters, which collect and transport the urine from the kidneys to the bladder, are affected by developmental, inflammatory, and neoplastic conditions but rarely by primary infectious processes. When present, infections are usually the result of a contiguous process, primarily from the kidneys or bladder or from involvement with a systemic process (Braf and Smith, 1977; Egawa et al., 1994; Hugosson and Olsen, 1986; Saad and Hanafy, 1974; Wehrheim, 1973).

A case of xanthogranulomatous ureteritis has been described, with clinical and pathologic findings similar to those of XGP. Patients may have a nonspecific pyuria and gram-negative bacteriuria (Remigio et al., 1975).

Rare occurrences of CMV ureteritis in immunosuppressed patients have been reported (Moudgil et al., 1997; Mueller et al., 1995).

URINARY BLADDER

The most common cause of cystitis is bacterial, frequently caused by gram-negative bacilli of intestinal origin. Less common causes include tuberculosis and fungal, parasitic, and viral infections.

Bacterial Infection

Bacterial cystitis affects predominantly women in the reproductive age group. The clinical symptoms include suprapubic pain, frequency, urgency, dysuria, and turbid, malodorous, and occasionally bloody urine. The most common cause is an ascending bacterial infection that gained access to the bladder through the urethra.

Anatomic factors related to the proximity of the urethra to the perineum in women are believed to result in bacterial colonization of the introitus and distal urethra by gram-negative enteric bacteria, the first step in the pathogenesis of bacterial cystitis. The most common pathogenic organism is *E. coli*, followed by *Proteus*, *Enterobacter*, and *Klebsiella* (Cotran et al., 1999b) (Box 4–7).

Cytology. Cytologic findings of bacterial cystitis are nonspecific. Voided urine specimens show variable numbers of segmented leukocytes and intact red blood cells in a nonnecrotic background. The number of urothelial cells is increased, and they exhibit reactive changes consisting of cell enlargement, enlarged and eccentric nuclei, and prominent nucleoli. The cytoplasm may be vacuolated and contain neutrophils and the nuclei may be hyperchromatic, but the chromatin distribution is uniform and the nuclear contours are smooth. Cells are arranged in sheets or small clusters or remain single, and the nuclear-cytoplasmic ratio is normal

Box 4–7.

Bacterial cystitis is commonly caused by gram-negative enteric bacilli.

Smears show neutrophils, red blood cells, and reactive urothelial cells.

Cellular changes may be more pronounced in chronic than in acute cystitis.

The association with white blood cell casts indicates acute pyelonephritis.

or slightly increased (Fig. 4–4A–D). Degenerated cells, debris, and eosinophils may be present in subacute or chronic cystitis. Reactive cellular changes may be more pronounced in the latter (Fig. 4–5A–C).

A search for white blood cell casts indicative of pyelonephritis must be conducted in all urine samples with leukocyturia. Association of bacterial cystitis with pyelonephritis is not uncommon.

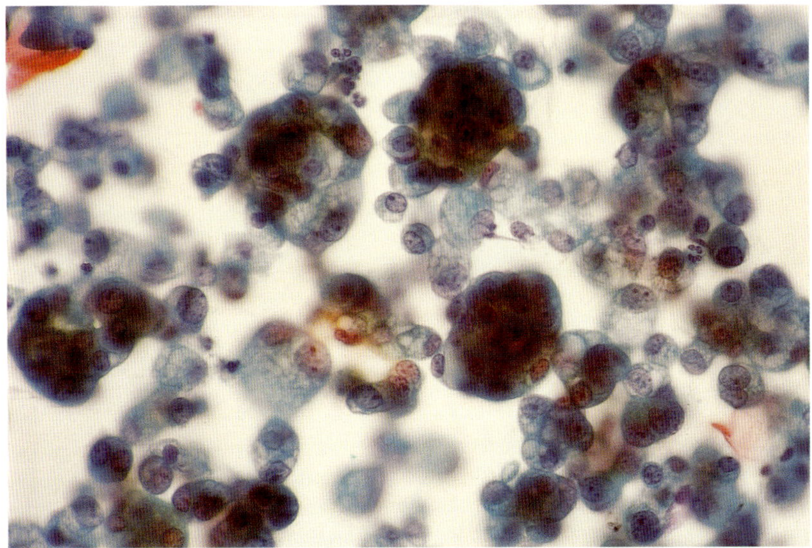

A

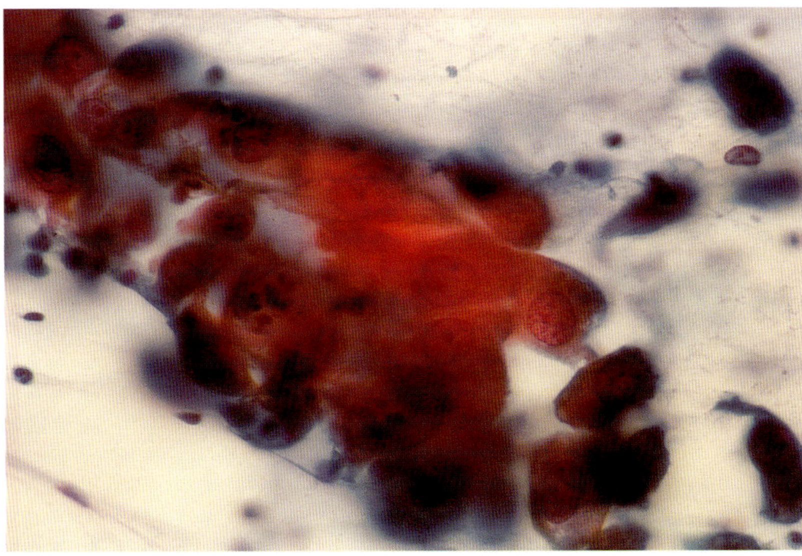

B

Figure 4–4. *Acute cystitis.* Various reactive urothelial cell changes secondary to acute inflammation are present. (*A*) Prominent changes are lack of cell cohesion showing small clusters and single cells with vacuolated cytoplasm, eccentric bland nuclei, and a few segmented leukocytes. (*B*) A sheet of large cells with intracytoplasmic neutrophils, round, bland, large nuclei, and prominent eosinophilic, round nucleoli. (*C*) A cohesive sheet of basal-intermediate cells shows dense homogeneous cytoplasm and hyperchromatic nuclei with granular chromatin. Features that prevent the diagnosis of atypia or malignancy include nuclear roundness, a regular chromatin distribution, a preserved nuclear-cytoplasmic ratio, and the presence of acute inflammation without necrosis. (*D*) A cohesive sheet of reparative cells is admixed with neutrophils. Bladder washings. (Papanicolaou stain, ×600)

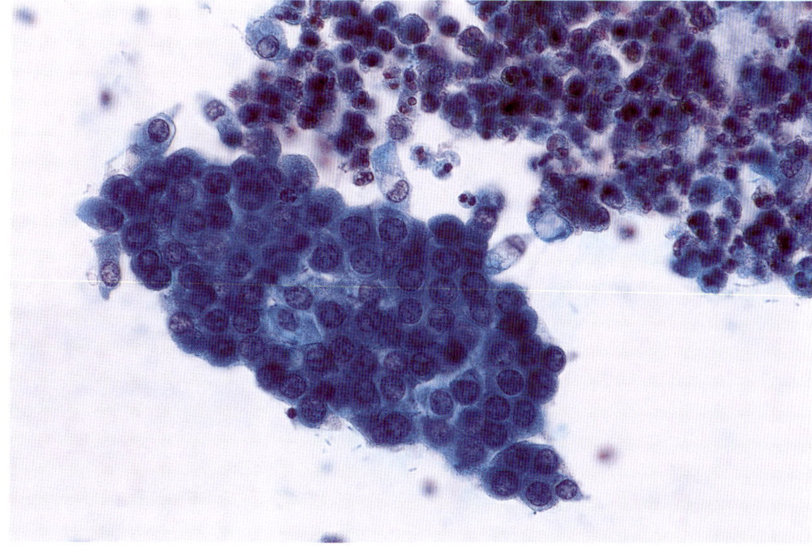

C

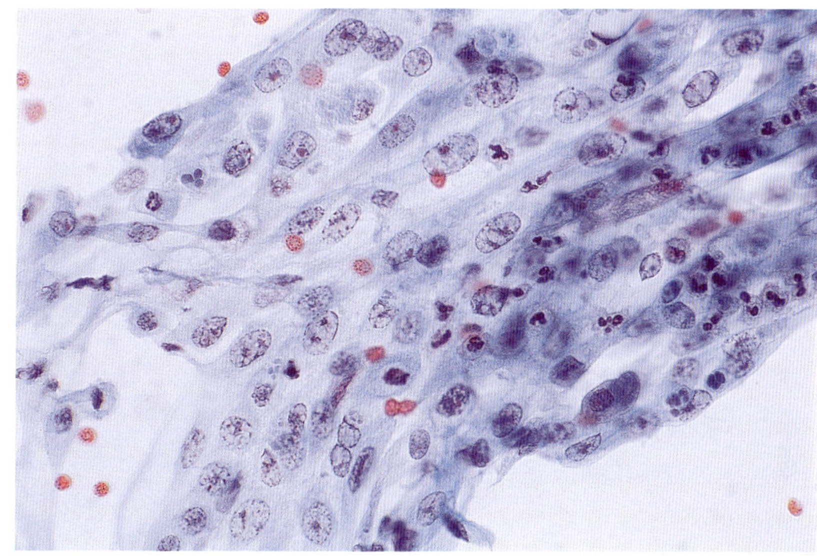

D

Figure 4–4. (*Continued*)

TUBERCULOSIS

In contrast to bacterial cystitis, which is frequently due to an ascending infection, in most cases of tuberculous cystitis the route of infection is descending from and almost always secondary to renal tuberculosis. Tuberculosis may be related to the socioeconomic status of the population in certain geographic areas of the world or to the immunologic status of the individual. Sterile leukocyturia (pyuria) may be a manifestation of urinary tuberculosis in immunosuppressed patients (Gomez et al., 1998).

The infection is usually caused by *Mycobacterium tuberculosis*, although other species may be seen, particularly *Mycobacterium avium intracellulare* as part of hematogenous dissemination in AIDS patients.

Cytology. The triad of epithelioid granulomas, multinucleated Langhans-type giant cells, and a granular necrotic background, characteristic of tuberculous cystitis, is rarely present in urine samples. However, the diagnosis can be suspected in the presence of epithelioid and/or Langhans giant cells,

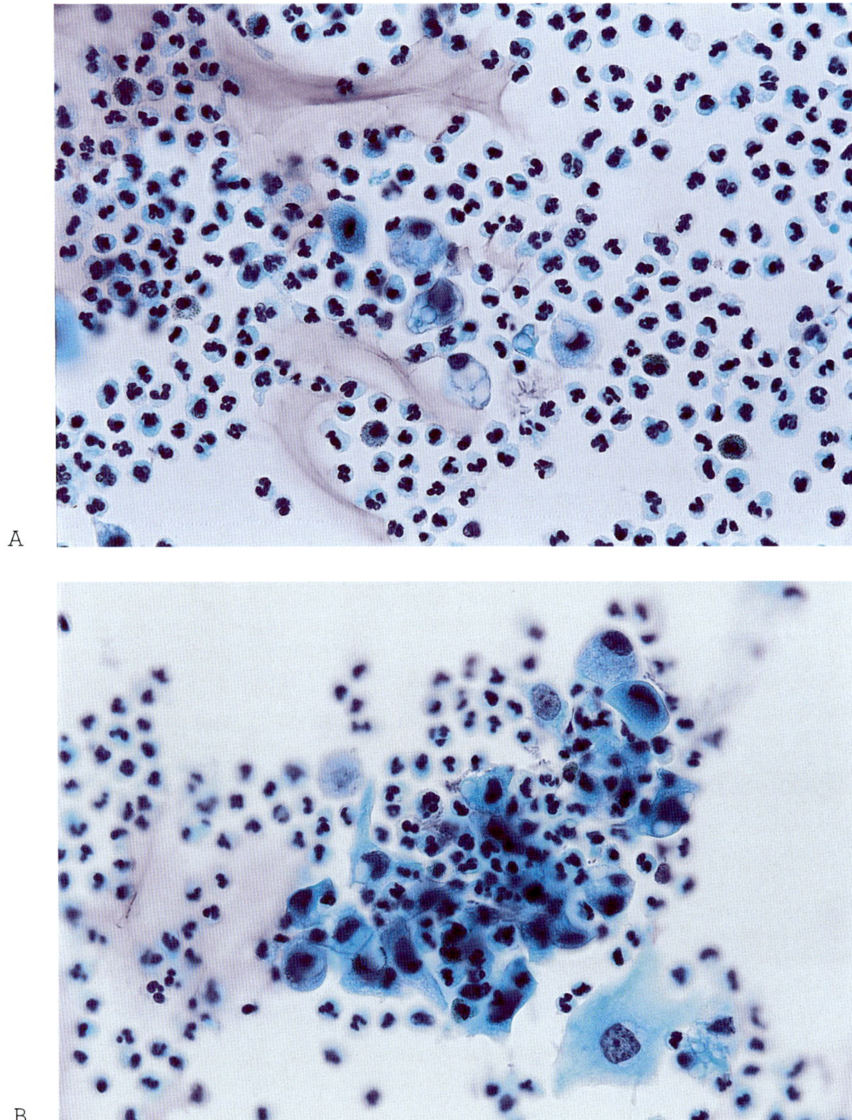

Figure 4–5. *Subacute and chronic cystitis.* The smear background shows neutrophils and eosinophils. (*A,B*) The cellular changes include cytoplasmic vacuolization, hyperchromatic nuclei, and minimal irregularity of the nuclear contours. (*C*) Cells with elongated cytoplasmic processes, enlarged nuclei, and lack of cohesion together with lymphoid cells may be observed in the urine of patients with chronic urinary tract infection. Voided urine, thin-layer preparation (*A,B*) and ureteral washing (*C*). (Papanicolaou stain, ×600)

which are found in up to 85% of cases (Piscioli et al., 1985). The cytologic findings reflect the coalescent and necrotizing nature of the frequently ulcerated tuberculoid granulomas formed in the lamina propria. Lymphoplasmacytic inflammation of variable degree is present.

Epithelioid granulomas can be recognized by the loose aggregation of histiocytes with thin or finely vacuolated cytoplasm, indistinct cytoplasmic borders almost forming a syncitium, oval or bean-shaped nuclei with delicate nuclear contours, fine chromatin, and a small chromocenter. The characteristic peripheral-cell palisading is rarely seen, and the presence of cells with frayed borders and wispy cytoplasm is more common. Langhans-type giant cells are characterized by a

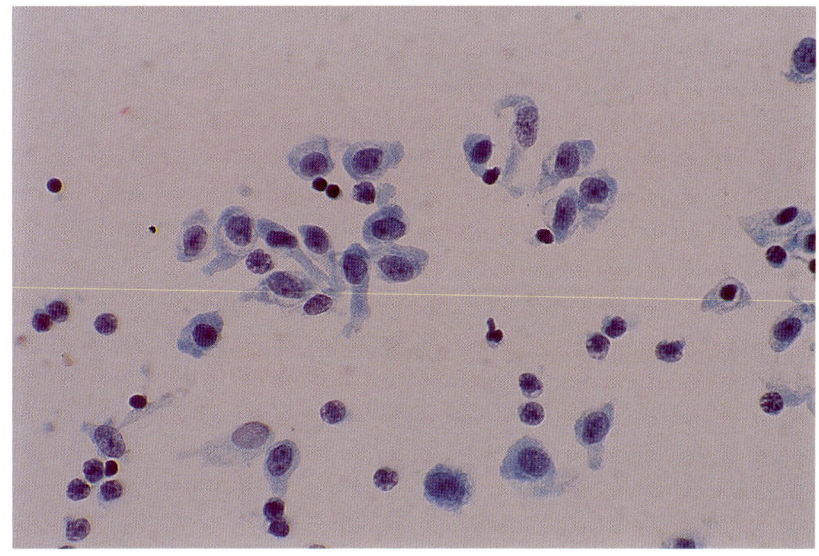

Figure 4–5. *(Continued)*

horseshoe-like eccentric aggregate of nuclei; however, most cells show a more random nuclear distribution.

Varying degrees of urothelial cell reactivity are present. Mild urothelial cell changes are noted in most cases. However, severe urothelial atypia has been identified in rare cases (Piscioli et al., 1985) (Box 4–8).

Differential diagnosis. Because tuberculosis of the bladder may cause atypical cells to be present in the urine, this lesion has to be differentiated from urothelial carcinoma (Piscioli et al., 1985). A biphasic population of epithelioid granulomas and atypical urothelial cells makes granulomatous cystitis the diagnosis of choice. The absence of clearly malignant nuclear features will exclude a high-grade invasive malignancy in a background of necrosis. Tissue confirmation is required in questionable cases.

It is important to exclude other possibilities before a diagnosis of tuberculous cystitis is rendered. Epithelioid granulomas may be associated with intravesical BCG immunotherapy for urothelial carcinoma, and occasionally necrobiotic granulomas develop after transurethral surgery with cauterization (Spagnolo and Waring, 1986).

> **Box 4–8.**
>
> Bladder tuberculosis occurs secondary to renal tuberculosis.
>
> Smears show epithelioid granulomas, Langhans-type giant cells, granular debris, and variable numbers of lymphocytes and plasma cells.
>
> Reactive urothelial cells are present. Marked cellular changes may be identified.

Fungal Infection

Fungal cystitis (Box 4–9) is rare and occurs particularly in diabetics and hosts with compromised immune function. Other predisposing factors include prior treatment with broad-spectrum antibiotics and corticosteroids, bladder outlet obstruction, and the use of indwelling catheters (Greene, 1992). The clinical picture includes severe urgency, frequency, and nocturia with sterile pyuria and microhematuria (Rohner and Tuliszewski, 1980).

Cystitis caused by *Candida albicans*, the most common etiologic agent, has been associated with anaerobic bacterial infections of the bladder or with formation of a "fungus ball" leading to obstructive uropathy,

Box 4–9.

Cytologic diagnosis of fungal cystitis should be suggested only in the presence of acute inflammation and tissue damage.

Candida albicans is the most common agent causing fungal cystitis. Smears show pseudohyphal forms and budding yeasts.

Septated hyphae are seen in *Aspergillus* spp., *Zygomycetes, Psedoallescheria boydii,* and *Fusarium* spp. The final diagnosis needs culture correlation.

Broad-based budding yeasts are seen in *Blastomyces* cystitis.

Cryptococcus has a mucin-positive capsule.

Urothelial cells show reactive changes.

particularly in diabetics (Comiter et al., 1996; Sultana et al., 1998). Other, less common causes of fungal cystitis include *Aspergillus, Mucor, Blastomyces, Histoplasma, Cryptococcus,* and *Coccidioides,* which are also detected in urine samples, particularly in immunosuppressed hosts (Eickenberg et al., 1975; Florentine et al., 1997; Plunkett et al., 1981; Torrington et al., 1979; Wise and Silver, 1993).

Cytology. *Candida* is the most common fungal agent present in urine and the most common contaminant. In the absence of chronic debilitating disease or vascular disease, candiduria may represent colonization or low-grade infection that resolves once the underlying disorder is controlled. Persistent candiduria, which is most common in critically ill patients, may represent a disseminated or more extensive (not only bladder) urinary tract infection (Wise et al., 1999). The presence of *Candida* organisms in the urine on examination should be mentioned in the cytology report, but the diagnosis of urinary tract candidiasis based on urine cytology examination should be suggested only in the presence of acute inflammation, tissue damage, and reactive cellular changes (Fig. 4–6). The final diagnosis must correlate the results of fungal cultures with clinical data.

Pseudohyphal forms and budding yeasts are characteristic of *Candida* spp. Broad-base budding yeast is diagnostic of blastomycosis (Fig. 4–7). An intracellular location is characteristic of histoplasmosis (Fig. 4–8). Large,

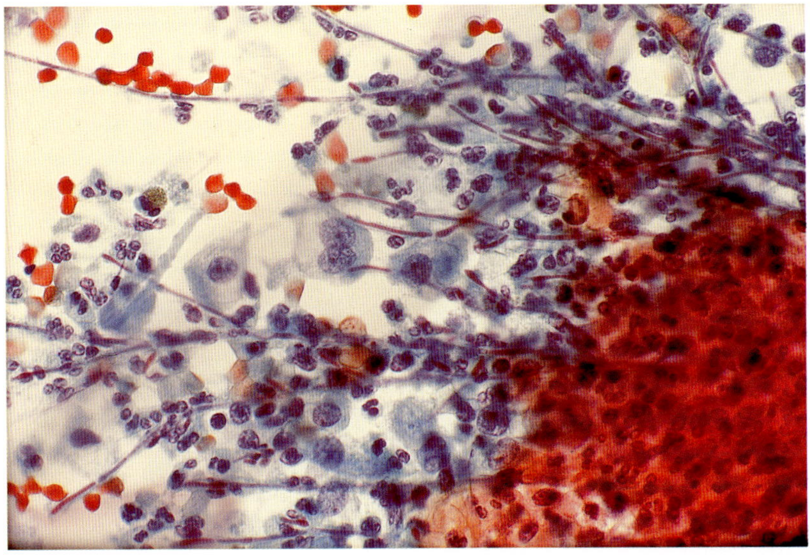

Figure 4–6. *Candida cystitis.* Pseudohyphae, marked acute inflammation, and reactive urothelial cells are present. Bladder washing. (Papanicolaou stain, ×600)

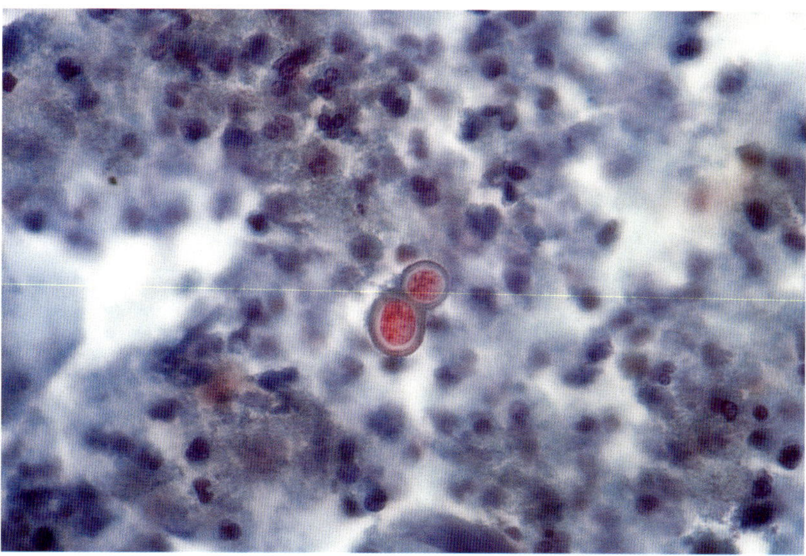

Figure 4–7. *Blastomycosis.* Diagnostic broad-based budding yeast is present in a background of acute inflammation. The patient died of systemic blastomycosis. Bladder washing. (Papanicolaou stain, ×600)

clear capsules surrounding yeast forms are characteristic of *Cryptococcus* as well. Further morphologic characterization can be obtained with the use of special stains. The intracellular location of *Histoplasma* is best seen with the use of PAS or silver stain. Mucicarmine stain or india ink preparations are useful in the diagnosis of *Cryptococcus* (Fig. 4–9A,B).

The diagnosis of *Aspergillus* needs culture confirmation because of the morphologic similarity with other nonpigmented hyphae such as *Rhizopus* spp. and other zygomycetes, *Pseudalescheria boydii,* and *Fusarium* spp.

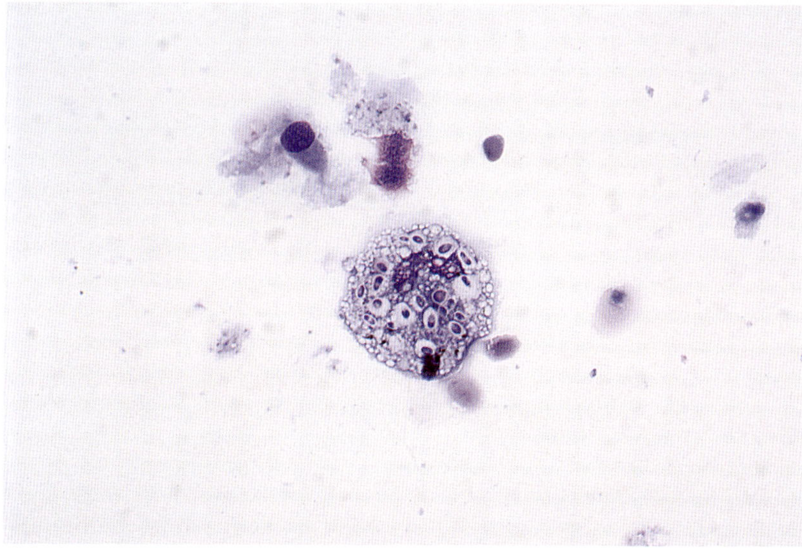

Figure 4–8. *Histoplasmosis.* Macrophage with intracellular yeasts is present in the urine sample of a patient with epididymitis. Bladder washing. (Papanicolaou stain, ×600)

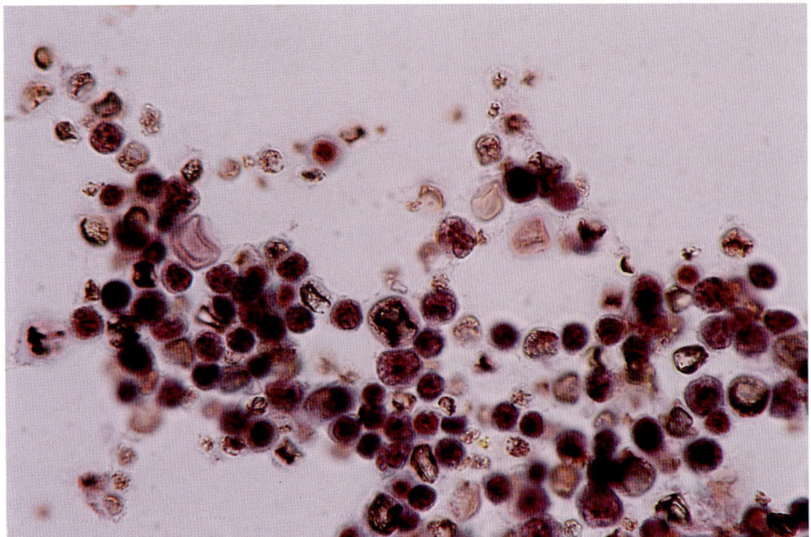

A

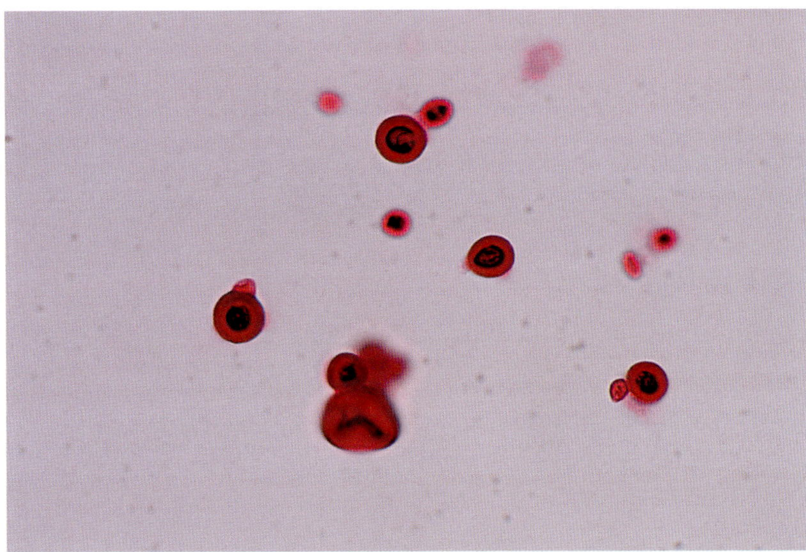

B

Figure 4–9. *Cryptococcosis.* Semitransparent fungal organisms with double contours are seen together with marked inflammation (*A*) (Papanicolaou stain, ×600). The mucoid capsule stains red with mucin stain (*B*) (mucicarmine stain. ×600). Bladder washing. This human immunodeficiency virus (HIV)-positive patient died of systemic cryptococcosis.

Briefly, *Rhizopus* has wide hyphae with irregular contours and infrequent septation. The other genera are thinner, with parallel contours and frequent septation. Branching at a 45° angle is characteristic of *Aspergillus* spp.; however, this feature may be difficult to evaluate because of the tissue damage and cell degeneration that are almost invariably present in urine samples from patients with these angioinvasive forms of fungal infections (Fig. 4–10). The presence of filamentous fungi must be stated in the cytology report, but the specific diagnosis should not be attempted on the basis of cytology only.

The accompanying background includes lysed blood, tissue damage, acute inflammation, and extensive degeneration. Urothelial cells show reactive cellular changes and degeneration similar those described for bacterial infection.

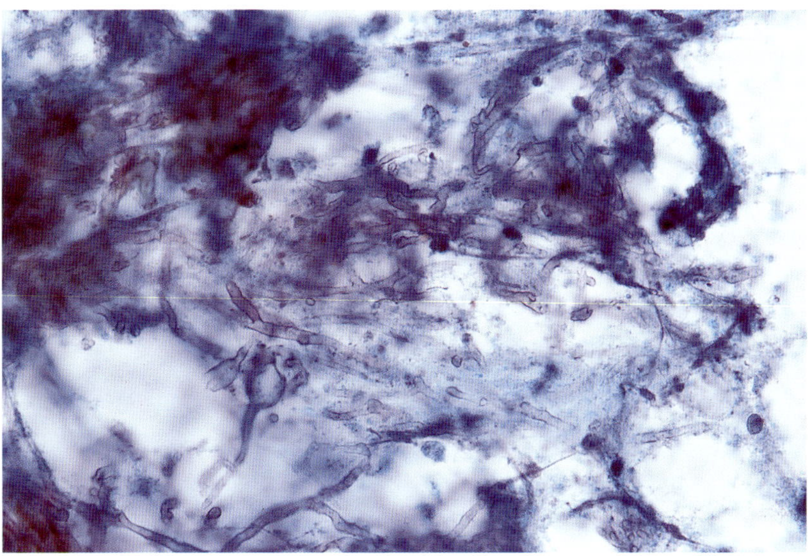

Figure 4–10. *Aspergillosis.* Fungal hyphae with septation and 45° angle branching are characteristic of this fungus. However, the final diagnosis requires colony identification in the urine culture. Bladder washing. (Papanicolaou stain, ×600)

Differential diagnosis. Because of degeneration, blood, inflammation, and tissue damage, the low-power appearance of the smear resembles that of invasive carcinoma. Once again, the discrepancy between the degree of tissue damage and the lack of malignant cells should initiate the search for pathogenic organisms.

Actinomycosis

Localized actinomycosis (Box 4–10) of the urinary bladder may be present in association with intrauterine devices or develop primarily in immunosuppressed patients. A hypogastric abdominal mass is frequently found in this rare form of localized actinomycosis, and the preoperative clinical diagnosis is almost always that of a malignant neoplasm (Guermazi et al., 1996; Villani et al., 1987).

Cytology. Papanicolaou stain best shows the characteristics of *Actinomyces,* similar to those present in cervicovaginal smears. The spidery aggregates of *Actinomyces* (sulfur granules) have a dark center and fuzzy edges composed of numerous filamentous organisms radiating from the center. In well-prepared stains, the cytomorphology is characteristic, particularly if accompanied by a dense, acute inflammatory reaction and evidence of tissue damage. Urothelial cells show reactive inflammatory changes.

Differential diagnosis. Before a diagnosis is made, actinomycosis needs to be distinguished from aggregates of other filamentous organisms such as *Candida, Aspergillus, Nocardia,* and so on, and from urine contamination with clusters of lubri-

Box 4–10.

Bladder actinomycosis is a rare form of localized actinomycosis.

Smears show sulfur granules with a dense basophilic center and spidery edges.

Marked acute inflammation, tissue damage, and reactive urothelial cells are also present.

Sulfur granules should be distinguished from other filamentous organisms, particularly *Nocardia.*

Table 4–1. Most Common Forms of Viral Cystitis: Cytologic Features

Virus Group	*Cytology*	*References*
Herpes-varicella		
Herpes simplex	Large cells. Multinucleation with molding. Ground-glass nuclei. Nuclear inclusion.	Bossen and Johnston (1975), Masukawa et al. (1972)
Cytomegalovirus	Large cells. Large nuclear inclusion with perinuclear halo. Eosinophilic ytoplasmic cinclusions. No ground-glass nuclei.	Cederqvist and Johansson (1967), Iwa et al. (1984), Kline and Craighead (1967), Schumann et al. (1977)
Adenovirus	Large intranuclear basophilic or amphophilic, homogeneous inclusion that almost completely fills the nuclear area without a perinuclear clear halo. Single or multiple small intranuclear eosinophilic inclusions with ill-defined borders and peri-inclusion nuclear degeneration/clearing. Multinucleated cells with cytoplasmic eosinophilic inclusions.	Belshe and Mufson (1974), Lipsey and Bolande (1967)
Papovavirus		
Human papilloma virus	Koilocytes.	Cecchini et al. (1988) Fralick et al. (1994), Fralick et al. (1986)
Polyomavirus	Large basophilic, homogeneous nuclear inclusion with a thin perinuclear space. Beaded nuclear contours.	Boon et al. (1989), Coleman (1975), Kahan et al. (1980)
	Large, degenerated nuclear inclusion with a clumped, coarse appearance.	Braza et al. (1989), Itoh et al. (1998), Kupper et al. (1993), Minassian et al. (1994)

cant, *Lactobacillus vaginalis,* or synthetic fibers.

Viral Infection

Viral cystitis is rare (Box 4–11). However, virus particles representing polyomavirus, papillomavirus, and adenovirus type II, in decreasing frequency, have been detected in urine specimens from healthy pregnant women by the use of negative-staining electron microscopy (Lecatsas and Boes, 1980). Similar studies have shown that polyomavirus, adenovirus, and herpes virus are also present in immunosuppressed.patients. No RNA viruses or oncogenic RNA viruses (retroviridae) have been detected (Lecatsas and van Wyk, 1978).

Herpex simplex type 2, CMV, and polyomavirus are the DNA viruses that most frequently affect the urinary tract. Characteristics of the viral cytopathic effects in urine cytology samples are listed in Table 4–1.

1. *Herpes virus type 2* (genitalis) has been isolated from human bladder urine (Masukawa et al., 1972). Hemorrhagic cystitis due to herpes simplex virus may develop in isolation or in the setting of disseminated herpetic infection in immunocompetent and immunosuppressed hosts and is frequently associated with genital herpetic lesions (Fig. 4–11) (DeHertogh and Brettman, 1988; McClanahan et al., 1994).

2. *Herpes zoster* is an infection caused by the varicella virus. The inflammatory reaction

Box 4–11.

Hemorrhagic cystitis due to herpes simplex virus is frequently associated with genital herpes.

Cytomegalovirus cystitis usually occurs in the setting of immunosuppression.

Bladder or urethral condyloma often occurs in association with aggressive anogenital condylomas. Koilocytes may be present in urine smears.

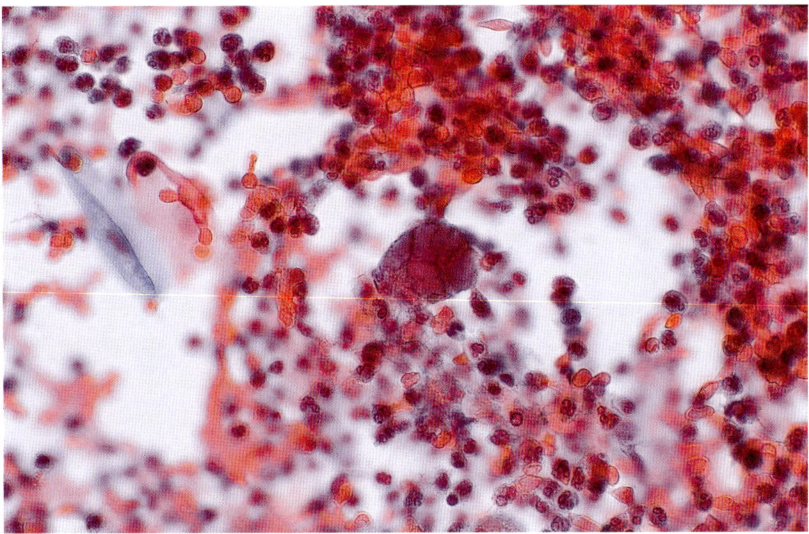

Figure 4–11. *Herpes virus infection.* Cell enlargement, multinucleation with nuclear molding, and a "ground-glass" appearance along with marked inflammation are present in the voided urine of a renal allograft recipient receiving immunosuppressive therapy. (Papanicolaou stain, ×600)

can involve the spinal cord and anterior-horn cells, causing neurologic disorders including urologic alterations such as acute urinary retention, urinary incontinence secondary to flaccid detrusor paralysis, and most commonly a cystitis-like syndrome (Broseta et al., 1993). Recovery is usually complete (Jellinek and Tulloch, 1976). Identification of this virus in urine specimens has not been reported.

3. *Cytomegalovirus*-induced hemorrhagic cystitis can occur after bone marrow transplantation (Spach et al., 1993), with AIDS (Zurlo et al., 1993), and with other causes of immunosuppression (Fig. 4–12). Disseminated CMV infection has been found in

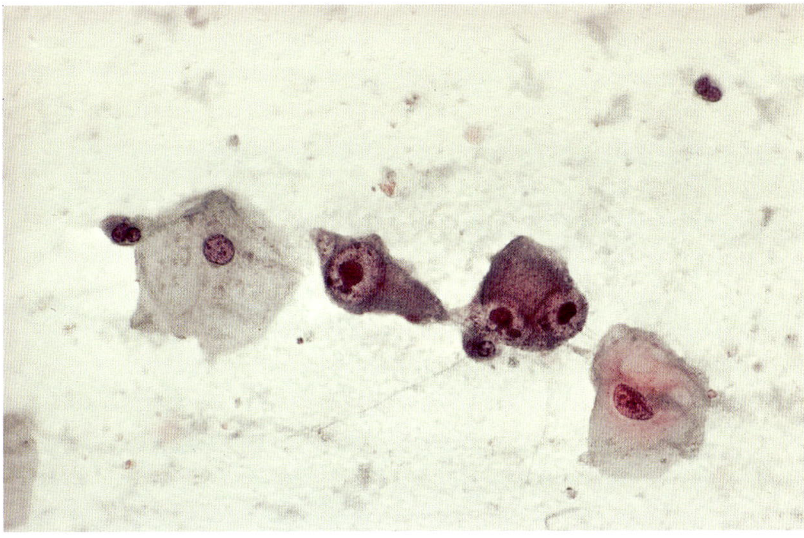

Figure 4–12. *Cytomegalovirus infection.* Large cells, presumably of renal tubular origin, with an eosinophilic nuclear inclusion and a perinuclear halo, are present in the voided urine of a patient with systemic CMV infection. (Papanicolaou stain, ×400)

52.4% of autopsies of patients with adult T-cell leukemia and involving the urinary bladder in 27.3% (Nagaoka et al., 1997). Urine culture has a poor diagnostic and predictive value in predicting CMV disease in transplant and AIDS patients (Falagas et al., 1997; Zurlo et al., 1993). Future studies should examine the value of alternative markers, such as CMV antigenemia (Gondo et al., 1994) or the detection of viral DNA by polymerase chain reaction (PCR) (Olive et al., 1989) for predicting CMV disease.

4. *Papilloma virus.* Bladder condyloma is more common in women than in men, usually develops in the setting of immunosuppression, and frequently results from locally aggressive anogenital condylomas extending into the urethra and bladder (Benoit et al., 1988; Wiedemann et al., 1995). Likewise, although rare, human papilloma virus (HPV) infection of the urinary bladder may result in solitary or widespread bladder condylomatosis (Ginsberg et al., 1989; Keating et al., 1985; van Poppel et al., 1986).

Human papilloma virus types 6 and 11 have been found in bladder condylomas (Benoit et al., 1988; Del Mistro et al., 1988; Iwasawa et al., 1992). Only rare cases of progression of condyloma to verrucous carcinoma in the urinary bladder and/or renal pelvis have been described (Batta et al., 1990; Bishop et al., 1998; Bruske et al., 1997).

Human papilloma virus infection and its pathogenetic role in chronic recurrent urethritis and cystitis, as well as in urothelial carcinoma, are controversial (Agliano et al., 1994a; Aynaud et al., 1998). Some suggest that HPV 16 and 18 may carry the risk of development of malignancy in the urinary tract because they occur in the anogenital regions (Agliano et al., 1994b). Others suggest that HPV types play a minor role in the development of urothelial carcinoma of the bladder in the general population, although they can act as oncogenic agents particularly in immunosuppressed patients (Noel et al., 1994).

Cytology. As in urethral condylomas, the yield of urine cytology is low in comparison with that of brush cytology. However, koilocytes may be detected in urine samples (Cecchini et al., 1988; Fralick et al., 1994; Nahhas et al., 1986). Koilocytes are characterized by a wrinkled, hyperchromatic nucleus, a perinuclear cavity with thick outlines, and dense cytoplasm (Shirai et al., 1988). Before the diagnosis of bladder condyloma is made, condyloma of the urethra and external genitalia should be excluded.

Condyloma is recognized histologically by the HPV cytopathic effect, which permits its differentiation from urothelial papilloma and carcinoma.

5. *Adenovirus type II.* Adenoviruses (Box 4–12) are excreted in the urine by approximately 11% of immunosuppressed patients (Lecatsas and van Wyk, 1978) and may not play a major pathogenetic role in interstitial cystitis (Hukkanen et al., 1996). However, necrotizing tubulointerstitial nephritis (Ito et al., 1991) and hemorrhagic cystitis associated with adenovirus type II are known causes of morbidity in immunosuppressed patients, particularly children (Echavarria et al., 1999).

Two types of intranuclear inclusions can be identified in the infected cells of these patients. One type, a large amphophilic or basophilic inclusion that occupies almost the entire nuclear area and is almost identical to that seen in human polyomavirus infection, gives the "smudged" appearance to the infected cell. The other type of inclusion is eosinophilic, small, and may be multiple, and, in contrast to the Cowdry A herpetic inclusion, it has ill-defined borders. The infected cells are slightly enlarged. Finding both types of inclusion in the urine sample

Box 4–12.

Adenovirus type II may cause acute tubulointerstitial disease and hemorrhagic cystitis in immunosuppressed patients, particularly children.

Two types of intranuclear inclusions are identified: a single large, basophilic inclusion occupying almost the entire nuclear area and a small, eosinophilic inclusion with a peri-inclusion nuclear clearing.

Multinucleated cells with eosinophilic cytoplasmic inclusions may be seen.

strongly favors the diagnosis of adenovirus (Fig. 4–13A–C). In cases of measles, epithelial tubular and transitional multinucleated cells containing characteristic eosinophilic cytoplasmic inclusions can be detected by light microscopy in Papanicolaou-stained smears of urinary sediment or by indirect immunofluorescence (Belshe and Mufson, 1974; Lipsey and Bolande, 1967). Currently, PCR is considered a valuable test for detect-

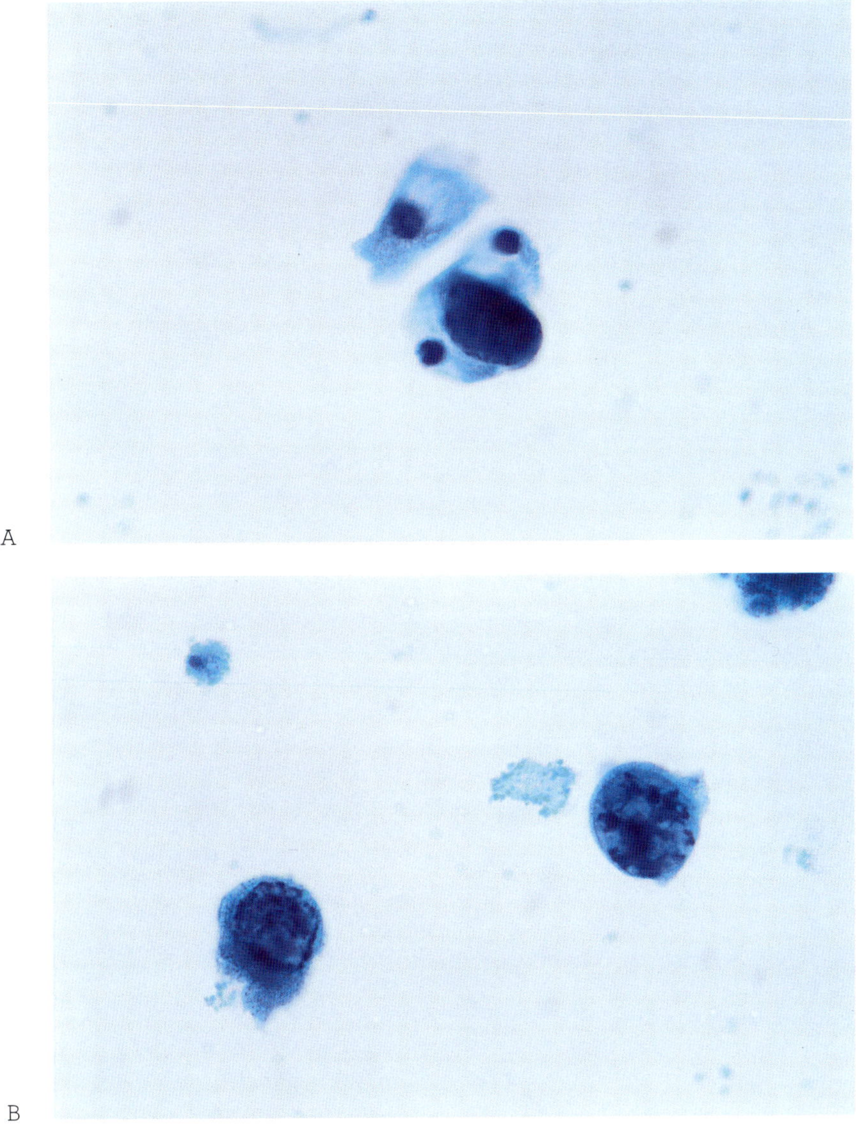

Figure 4–13. *Adenovirus infection.* Two types of intranuclear inclusions are seen. (*A*) A large, homogeneous, amphophilic inclusion occupies almost the entire nuclear area and resembles the human polyomavirus effect. (*B*) Small, eosinophilic inclusions have ill-defined borders with peri-inclusion nuclear clearing. (*C*) Ultrathin sections show the intranuclear inclusion to be composed of hexagonal viral particles measuring 60 to 90 nm in diameter, arranged in a lattice-like pattern. A perinuclear halo is evident. The patient had a systemic adenovirus infection. On voided urine cytology, this finding was initially diagnosed as a human polyomavirus effect. (A,B, Papanicolaou stain, ×1000). (Uranyl acetate stain, ×15,000 and inset ×64,000 (C). Courtesy of Dr. Terry D. Oberly, Professor and Vice-Chair, Department of Pathology and Laboratory Medicine, University of Wisconsin-Madison School of Medicine, Madison, WI.

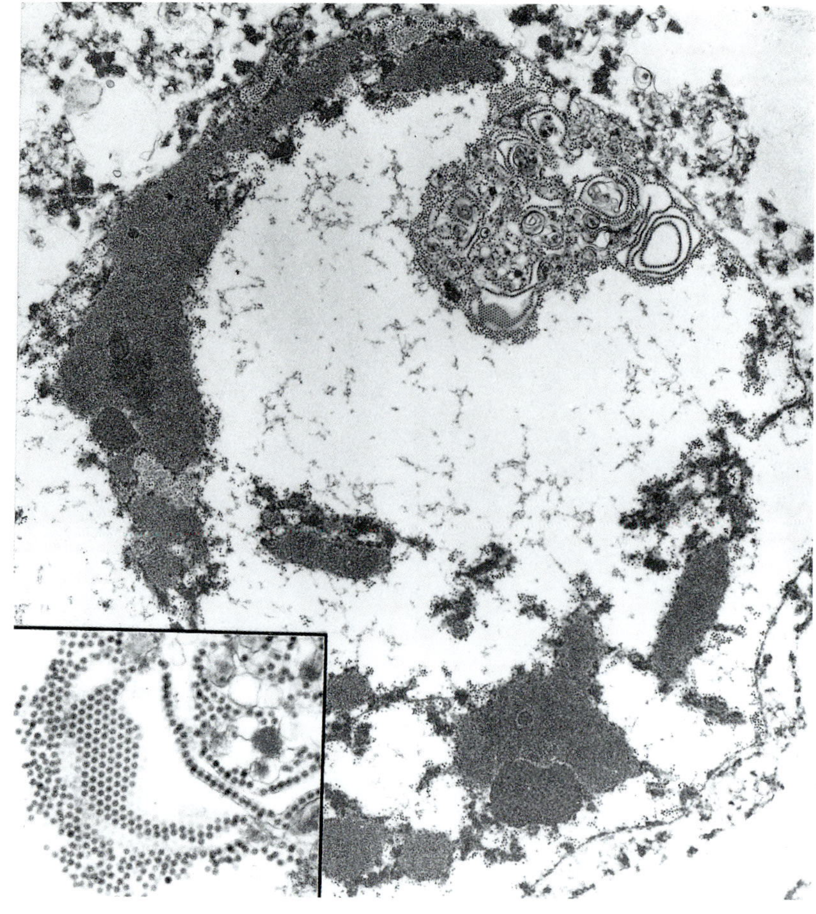

C

Figure 4–13. (*Continued*)

> **Box 4–13.**
>
> Polyomavirus is acquired early in life and remains quiescent, only to reappear often in the setting of immunosuppression.
>
> Polyomavirus cystitis often causes hematuria with few and vague symptoms.
>
> Viruria may be present months after the symptoms subside.
>
> Polyomavirus may cause tubulointerstitial disease in renal allograft recipients.
>
> The nuclear cytopathic effect of polyomavirus must be distinguished from urothelial carcinoma in situ.

ing this virus in urine and blood samples (Henderson et al., 1998).

6. *Polyomavirus* DNA (viruria) is found in the urine of 4% of urologic patients, is found twice as frequently after the seventh decade of life, and is proportional to the severity of immunosuppression (Kitamura et al., 1990; Shah et al., 1997). The virus is acquired early in life and remains quiescent thereafter, only to reappear often in the setting of immunosuppression (Apperley et al., 1987; Masuda et al., 1998; Tani et al., 1983) (Box 4–13).

Rarely, polyomavirus causes clinically evident tubulointerstitial disease in renal allograft recipients after reactivation of latent virus in the renal epithelium. In these cases, urine cytology detects numerous cells with intranuclear inclusions, but the final diagnosis is made by renal biopsy. However, preliminary data suggest that urine cytology is more sensitive for the diagnosis (Drachenberg et al., 1999; Nickeleit et al., 1999).

Polyomavirus infection may be seen in both children and adults; the viral cytopathic effect is detected in urothelial cells and renal tubular cells (Braza et al., 1989). Children usually show the classic large, homogeneous, basophilic intranuclear inclusions, and most adults yield degenerated intranuclear inclusion–bearing cells with coarse chromatin. These features lack clinical significance. However, the presence of cells with a typical cytopathic effect in adults may be indicative of marked immunosuppression (Itoh et al., 1998).

The infection may be subclinical, correlated with hematuria with or without cystitis, or systemic (Bogdanovic et al., 1996; Cottler-Fox et al., 1989; Tani et al., 1983; Vogeli et al., 1999). A polyomavirus cytopathic effect may be detected in urine for several months after the symptoms subside (Masuda et al., 1998).

Cytology is the most inexpensive, most reliable, and highly accurate test for detecting the viral cytopathic effect of polyomavirus (Fig. 4–14A–D) (Braza et al., 1989; Masuda et al., 1998). In questionable cases, enzyme-linked immunosorbent assays of urinary supernatants, DNA hybridization assays of urothelial cells, PCR techniques, immunohistochemistry, immunofluorescence, or electron microscopy may be used (Akura et al., 1988; Arthur et al., 1986; Bogdanovic et al., 1996; Hogan et al., 1980).

Differential Diagnosis. Human polyomavirus-infected cells in the urinary sediment are characterized by large, homogeneous, basophilic nuclear inclusions that may mimic the nuclear changes of urothelial carcinoma (Boon et al., 1989; Koss et al., 1984). Furthermore, urothelial cells infected with polyomavirus may have an aneuploid or a hyperdiploid DNA content with mildly elevated proliferative activity (Koss et al., 1984; Wojcik et al., 1997). Meticulous evaluation of the nuclear characteristics is helpful in making the correct diagnosis. However, when the nuclear inclusion is degenerated and coarse, the distinction from high-grade urothelial carcinoma, particularly in situ, will require further investigation. The problem of differential diagnosis may be compounded in the rare instances when polyomavirus and carcinoma in situ coexist. Distinction from the cytopathic effect of herpes infection is based on the nuclear characteristics and multinucleation. Cytomegalovirus can be distinguished by the larger size of the infected cells, the more pronounced perinuclear halo, and eosinophilic nuclear and cytoplasmic inclusions. Adenovirus may have a large, homogeneous intranuclear basophilic inclusion closely resembling that of polyomavirus; however, cells with single and multiple small, irregular intranuclear inclusions and nuclear clearing are commonly found in the same sample in cases of adenovirus.

In summary, most viral infections affecting the urinary tract are detected accurately by urine exfoliative cytology. This is particularly valuable when a viral etiology can be demonstrated in the setting of immunosuppression, including renal allograft transplant patients who have hemorrhagic cystitis, acute tubular necrosis, or tubulointerstitial disease. Sequentially repeated examinations allow a prompt diagnosis, which is valuable in guiding the proper clinical management (Nickeleit et al., 2000; Stella et al., 1983).

Schistosomiasis

Schistosomiasis is endemic in Africa and Mediterranean Asia. Rare cases have been reported in the United States, usually in immigrants from these countries (Bock and Neal, 1995). Freshwater snails, the intermediate hosts, transmit the disease. The 1 to 2 cm trematode *Schistosoma haematobium* and its eggs reside in the definitive host, the human being, and cause hematuria or chronic cystitis with dysuria and frequency. This type of infection is often indolent, but it can have significant long-term sequelae. Occasionally, *Schistosoma mansoni* eggs, which localize preferentially in the intestine and liver and are excreted through the stools, may be found in the urine (Cheever et al., 1975).

The infectious larvae (cercaria) present in water penetrate the human skin, travel though the superficial vascular network, and lodge in the venous system of the lower urinary tract, where they develop into adult worms. These worms produce several hundred eggs a day that are deposited in the small vessels, eventually migrate through the bladder wall into the lumen, and are ex-

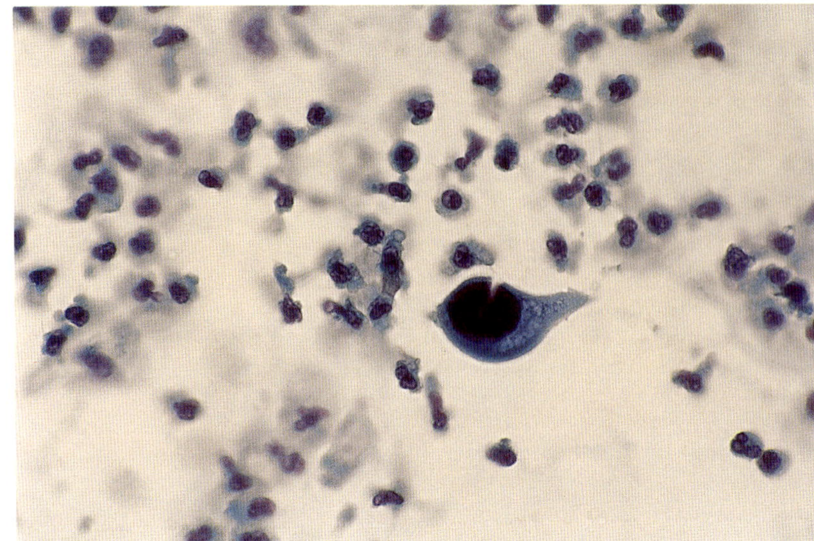

A

B

Figure 4–14. *Polyomavirus infection.* Various forms of the human polyoma virus cytopathic effect are illustrated. (*A*) Cell exhibits a dense, homogeneous, basophilic intranuclear inclusion that fills the area entirely and causes nuclear cracking. The cytoplasm is dense and granular. (*B*) Cell with vacuolated cytoplasm bearing a basophilic, homogeneous inclusion that partially fills the nuclear area, leaving a thin zone of clearing that surrounds the inclusion. The patient had received systemic chemotherapy for acute leukemia. (*C*) Beading of the nuclear membrane is prominent in the larger cell; a clumped and degenerated inclusion is present in the smaller cell. (*D*) Threads and small, round clumps of degenerated chromatin are noted in these cells, which show nonspecific cytoplasmic eosinophilic inclusions. Bladder washings. (Papanicolaou stain, ×1000)

creted with the urine. The eggs are released into the water, where the encased miracidium is released and infects the snails to complete the cycle (Samuelson, 1999).

The severity and frequency of the sequelae of urinary schistosomiasis (hydroureter, hydronephrosis, bladder ulcer, and polyposis) and of its complications (bacterial urinary tract infection, renal failure, and urothelial cancers) depend on the intensity and duration of infection (Smith and Christie, 1986). An association between bladder schistosomiasis and the development of squamous cell carcinoma of the urinary

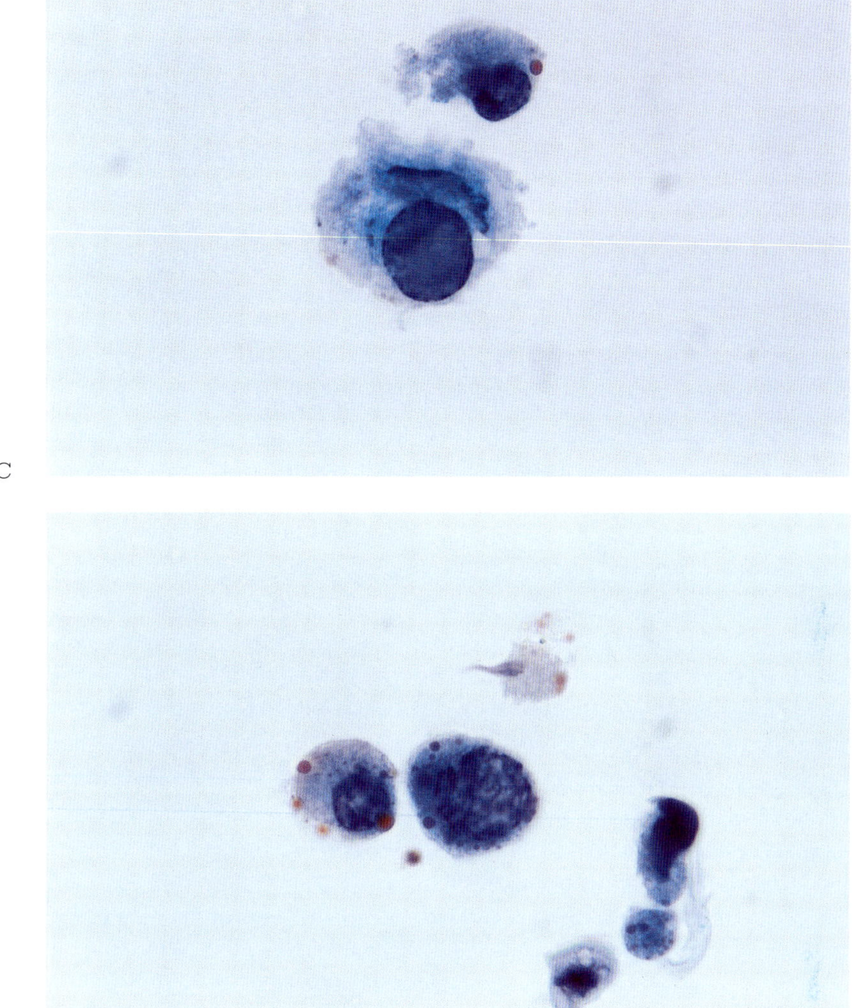

Figure 4–13. *(Continued)*

bladder has been postulated; however, the specific etiologic relationship and pathogenesis have not yet been defined (Godwin and Hanash, 1984).

Cystoscopic examination of the bladder shows a granular mucosa with ulcers and polyps, particularly in late stages of the disease (Smith et al., 1977). Tissue biopsy reveals granulomas, fibrosis, and calcified eggs (Smith and von Lichtenberg, 1976). Histologic changes in the urothelium include squamous and intestinal metaplasia (Smith and Christie, 1986).

Detection of proteinuria and hematuria associated with *S. haematobium* infection has been used as a screening test for selecting schistosoma egg–positive individuals (Mott et al., 1983). The degree of eosinophiluria has also been found to correlate with active infestation (Eltoum et al., 1992). However, definitive evidence of active disease is provided by detection of schistosoma eggs by microscopic examination (Bosompem et al., 1996; Hassan et al., 1994). The number of eggs excreted in the urine is directly related to the burden of infestation (Box 4–14).

Cytology. *Schistosoma haematobium* eggs are oval, measure between 100 and 150 μm, and have a characteristic small but distinct

> **Box 4–14.**
>
> Schistosomiasis is endemic in Africa and Mediterranean Asia and rare in the United States.
>
> *Schistosoma haematobium* causes hematuria and chronic cystitis in the definitive host, the human being.
>
> The adult worms produce hundreds of eggs, many of which are excreted in urine following diurnal periodicity.
>
> Complications of urinary schistosomiasis include hydroureter, hydronephrosis, renal failure, and squamous cancer of the bladder.

terminal spine (Fig. 4–15A,B). *Schistosoma mansoni* has a lateral spine. Cytologic smears demonstrate viable, degenerated, or calcific eggs and empty shells. The egg is bigger than a squamous cell. It is not uncommon to encounter urothelial cells with squamous metaplasia and varying degrees of keratinization and nuclear atypia.

Urinary egg excretion has diurnal periodicity, and it may be necessary to obtain 24 hour urine specimens or urine voided between 10 A.M. and 2 P.M. even in cases of active infection (Anonymous, 1994) (Box 4–15).

> **Box 4–15.**
>
> Hematuria and eosinophiluria correlate with active schistosomal infestation.
>
> *Schistosoma haematobium* eggs may be calcified and degenerated. Viable eggs are oval, measure 100 to 150 μm, and have a terminal spine. *Schistosoma mansoni* has a lateral spine.
>
> Urothelial cells may show squamous metaplasia and keratinization.
>
> A search for squamous cells with anaplastic nuclei is recommended to exclude squamous carcinoma, a not uncommon complication of schistosomiasis.

TRICHOMONIASIS

Trichomoniasis of the urinary tract is rare (Sokol and Min, 1972). This sexually transmitted infestation presents as urethritis and/or cystitis characterized by dysuria, frequency, and sterile pyuria. Men are commonly asymptomatic carriers. However, when the infestation is clinically evident in men, it may remain undetected for a period of time because of the low level of clinical suspicion and the low diagnostic accuracy of urinalysis in the detection of the organism.

In women, when present, urinary tract trichomoniasis is frequently associated with genital trichomoniasis. Although a Pap smear is routinely obtained in the presence of any gynecologic infection, urine cytology is not frequently ordered for evaluation of urinary tract infections. Therefore, the disease may remain underrecognized, especially in men.

Cytologic findings in urine specimens are similar to those of cervicovaginal trichomoniasis. Therefore, the cytologic diagnosis of bladder trichomoniasis in women should be reserved for those cases supported by clinical evidence. The cytologic diagnosis in men is less controversial.

Cytology. Smears show marked acute inflammation with clumps of aggregated neutrophils, squamous metaplastic cells, and few urothelial cells (Niewiadomski et al., 1998). Hematuria is common, but necrosis is absent.

Trichomonas is a light gray, pear-shaped protozoan that exhibits a small, dark, slit-like, eccentric, diagnostically helpful nucleus. The average size of the organism ranges between 15 and 50 μm, and it frequently shows cytoplasmic eosinophilic granules. In contrast to fresh urine wet-mount preparations, on Papanicolaou-stained cytology smears the flagellum is not readily apparent. Trichomonads may be found surrounding the squamous cells, upon which numerous neutrophils aggregate and form the "cannonball" so characteristic of trichomoniasis in cytologic preparations (Fig. 4–16; Box 4–16).

Differential diagnosis. *Trichomonas* in urine must be distinguished from superficial

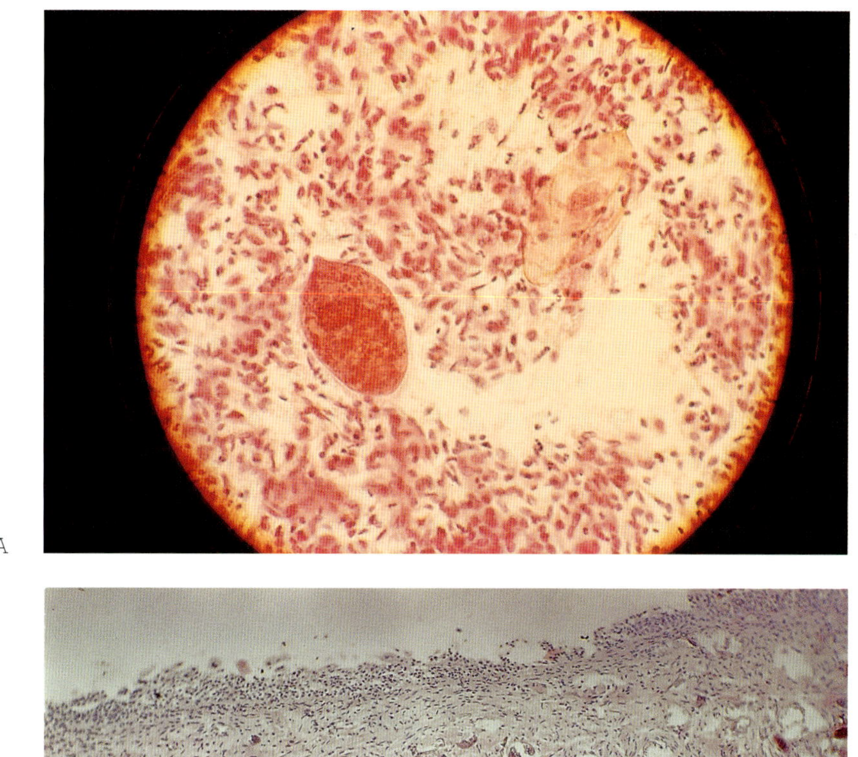

A

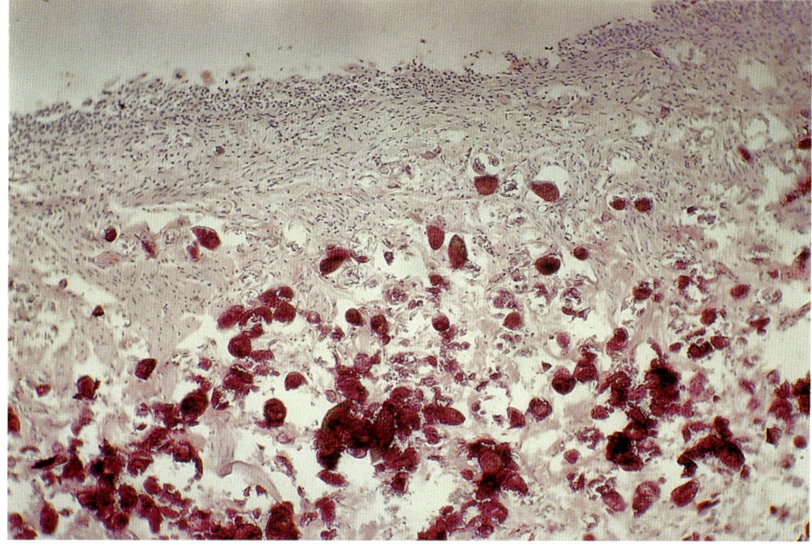

B

Figure 4–15. *Schistosomiasis.* (*A*) Egg of *Schistosoma haematobium* shows the characteristic terminal spine and a miracidium inside. There are abundant exfoliated benign urothelial cells and some inflammatory cells. (*B*) Heavy deposition and calcification of bilharzia ova in the lamina propria of the urinary bladder are present. Atrophy and loss of surface urothelium give the bladder the gross appearance of a sandy patch. (A, Papanicolaou stain, ×600; B, Hematoxylin and eosin stain, ×100). Courtesy of Dr. Hassan Nabil Tawfik, Professor and Director of Anatomic and Surgical Pathology, National Cancer Institute, Cairo University, Cairo, Egypt.

urothelial cells and renal tubular cells. Identification of the nuclear characteristics is helpful.

Other Parasites

Enterobius vermicularis (pinworm) is a helminth that primarily infests the cecum and vermiform appendix, particularly in children. Occasionally, pinworm ova are detected in urine specimens (Fig. 4–17). Gravid female worms migrate to the anus and perianal skin, where they deposit their ova.

Toxoplasma can be present as a dark, crescent-shaped organism in the cytoplasm

Box 4–16.

Trichomoniasis of the urinary tract is associated with genital trichomoniasis.

The urine cytologic diagnosis of bladder trichomoniasis in women needs to be supported by clinical evidence.

Smears show acute inflammation, clumps of neutrophils, red blood cells, and reactive urothelial cells.

Trichomonas is a pear-shaped protozoan with a dark, slit-like, eccentric nucleus.

The flagellum is not readily apparent in cytologic preparations.

of macrophages in urine specimens obtained from immunosuppressed patients.

Various other helminth ova can be identified occasionally in urine specimens contaminated with feces of infested patients. Occasionally, larvae or adult mites, particularly *Sarcoptes scabiei*, appear in the urine sediment after falling into the urine from the skin and perineal areas during urine collection.

URETHRA

Urethritis

Acute urethritis is classified into gonococcal and nongonococcal types. The latter is commonly caused by gram-negative bacilli, particularly *E. coli*, and is associated with cystitis in women and prostatitis in men. *Chlamydia* and *Ureaplasma* are common causes of nongonococcal urethritis, particularly in men. Other rare pathogens implicated include fungi and *Trichomonas*, as well as primary infection with herpes simplex virus (Schwartz and Hooton, 1998).

Gram stain and culture of urethral material, urine sediment, and urine cultures are used for the diagnosis of gonococcal and chlamydial infections (Desai and Robson, 1982). The diagnosis of nongonococcal, nonchlamydial urethritis is supported by the presence of more than 4 polymorphonuclear leukocytes per field (×1000) in a Gram-stained urethral specimen or more than 15 polymorphonuclear leukocytes in all five fields (×400) in the first voided urine

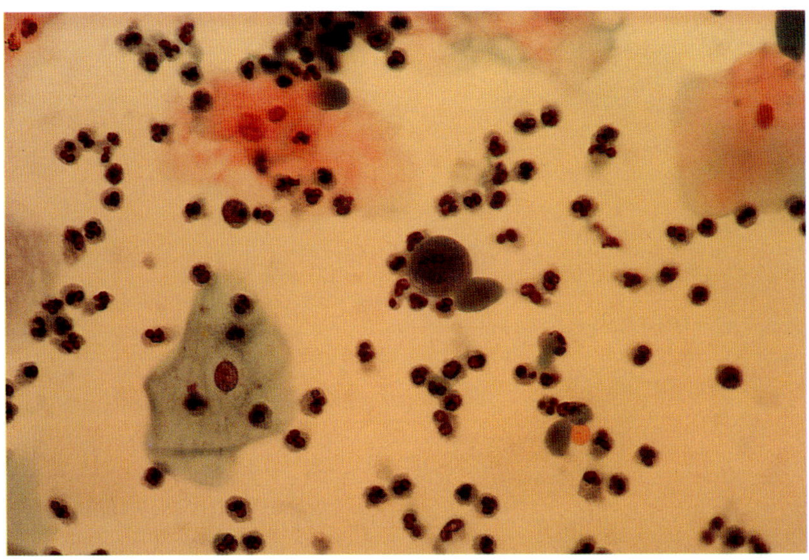

Figure 4–16. *Trichomoniasis.* Scattered trichomonads, a few reactive urothelial cells, and acute inflammatory cells are present in the voided urine of a woman with vaginal trichomoniasis and acute cystitis. (Papanicolaou stain, ×600)

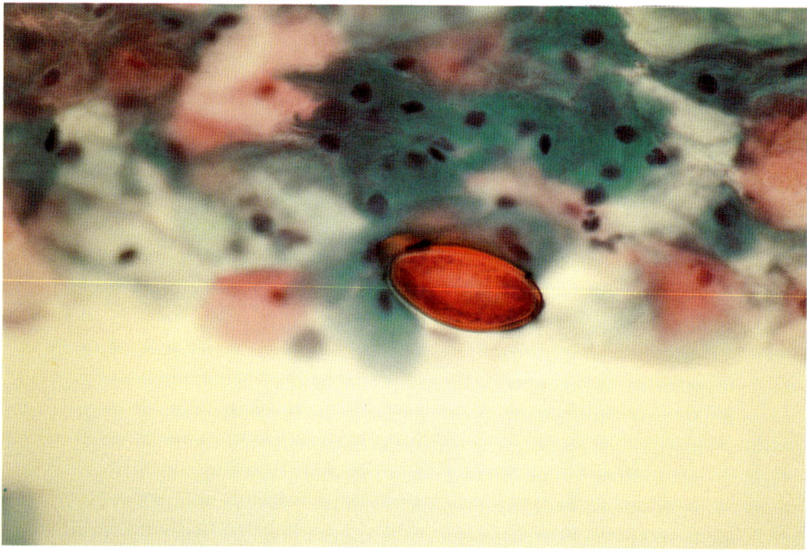

Figure 4–17. *Enterovius vermicularis.* Egg showing a flattened side and internal contents. Numerous squamous cells are also noted in this voided urine obtained from an adolescent girl with cystitis. (Papanicolaou stain, ×400)

specimen (Bowie, 1978). Diagnosis, especially of *Chlamydia trachomatis,* has been improved by the use of amplification assays for the detection of DNA or rRNA. These methods have high sensitivity, especially for specimens with a small number of infectious agents. In addition, they enable the use of voided urine as a noninvasive type of specimen (Stary, 1998) (Box 4–17).

Box 4–17.

Urine cytology is a test rarely requested for patients with urethritis.

The diagnosis of gonococcal urethritis is made by urine culture, examination of the urine sediment, and Gram stain with culture of urethral material.

Urethral smears from patients with gonococcal urethritis show reactive urothelial cells, numerous neutrophils, and numerous diplococci, some of them intracellular.

Lymphocytes and tingible-body macrophages in a urethral smear suggest chlamydial urethritis. Crescent-shaped elementary bodies of *Chlamydia* in the smears can be seen with the use of Romanovsky stains.

Cytology. The clinical manifestations of itching, burning, and urinary frequency with urethral discharge (yellow in gonococcal urethritis, watery in nongonococcal urethritis) are so characteristic that urine cytology is rarely requested.

Minimal cytologic changes in urothelial and squamous cells are seen in urethritis. Urethral smears from gonococcal urethritis show numerous acute inflammatory cells, epithelial cells, and numerous diplococci best seen under oil immersion. The presence of neutrophils is not specific because they are found in other types of urethritis and cystitis as well.

However, the presence of lymphocytes accompanied by tingible-body macrophages in a urethral smear, as in cases of follicular cervicitis in Pap smears, may suggest the diagnosis of chlamydial urethritis. Similar to follicular cervicitis of the uterine cervix in women with genital chlamydial infections, subepithelial urethral infiltration of lymphocytes is found histologically in cases of chlamydial urethritis (Woolf, 1998). Lymphocytes may also be present in voided urine (Shahmanesh et al., 1996).

Box 4–18.

Urethral condylomas are usually associated with condylomas of the external genitalia.

Human papilloma virus types 6, 11, 16, and 18 are common, with high congruence with the types present in sexual partners.

Condylomas may progress to a high-grade squamous lesion or squamous carcinoma.

Occasionally, koilocytes may be present in urine samples.

On cytology, one may detect the crescent-shaped elementary bodies of *Chlamydia* in urethral smears, particularly with the use of Romanovsky stains (Woolf, 1998). The detection yield of chlamydia in urethral smears can be improved by the use of fluorescent antibodies.

The cytology of viral, fungal, and parasitic urethritis is similar to that in the urinary bladder.

CONDYLOMA

Condyloma, a sexually transmitted viral infection involving the urethra, is commonly associated with condylomas of the external genitalia, although approximately 5% of patients with cytologic evidence of urethral condyloma do not have genital lesions (Kotoulas et al., 1996). Condyloma usually is present initially in the external genitalia of men, that is, the glans penis or shaft, or in the perineum or external genitalia of women and may involve the urethral orifice secondarily. A case of systemic mucosal involvement including the urethra and urinary bladder has been described (Bishop et al., 1998). The most common types of human papilloma virus found in these lesions are 6, 11, 16, and 18, with a high grade of congruence with those present in sexual partner(s) (Del Mistro et al., 1987). The course of the disease is usually indolent, although progression to high-grade squamous intraepithelial lesions and squamous cell carcinoma is known (Box 4–18).

Histologically, condyloma of the urethra is identical to condylomas arising in other mucosal epithelia. The pathognomonic feature is the presence of koilocytes in these lesions.

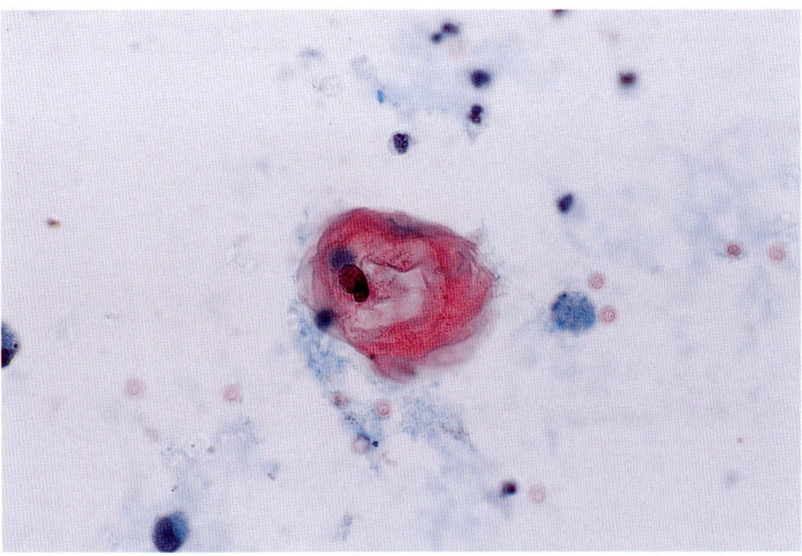

Figure 4–18. *Urethral condyloma.* Squamous cells with features highly suggestive of the human papilloma virus effect were seen in the voided urine of a woman with a urethral condyloma. (Papanicolaou stain, ×600)

Spontaneous cell exfoliation of urethral condyloma is uncommon, but occasionally koilocytes are detected in a voided urine sample (Fralick et al., 1994). A more effective method of cytologic detection is urethroscopy with urethral brushing, particularly for detection of cytologic abnormalities within the urethra (Kotoulas et al., 1996). Polymerase chain reaction has been used with a high degree of accuracy for identifying the HPV DNA in urine samples (Iwasawa et al., 1997; Melchers et al., 1989). The HPV types detected in urine are identical to those detected in urethral condylomas (Iwasawa et al., 1997). Assessment of HPV types by immunocytochemistry in brush cytologic samples of the distal portion of the urethra has been described (Cardamakis et al., 1997).

Cytology. The cytomorphologic features of urethral condyloma are identical to those present in koilocytes of cervicovaginal cytologic specimens. Briefly, squamous cells are somewhat oval, with eccentric and wrinkled, hyperchromatic nuclei and a well-defined cytoplasmic cavity with thick borders containing amorphous debris (Fig. 4–18).

PROSTATE

Infection of the prostate, that is, prostatitis, can be classified into acute and chronic and bacterial and nonbacterial. The diagnosis of prostatitis requires bacteriologic cultures from urine specimens obtained before and after prostatic massage, as well as urethral smears prepared from the prostatic secretion after prostatic massage (Krieger et al., 2000; Ludwig et al., 2000). Rarely, these specimens are submitted for cytologic evaluation. However, urine cytology examination may be requested so that a malignant process can be excluded, particularly in elderly men with associated lower urinary tract symptoms.

The most common etiologic agents causing prostatitis are bacteria similar to those causing lower urinary tract infections (Meares, 1987). Rare causes of prostatitis include nonspecific and specific granulomatous processes such as tuberculous, fungal, parasitic, and viral processes (Corachan et al., 1994; de la Rosette et al., 1993a).

Cytology. Urine samples show mild to moderate inflammatory urothelial cell changes, which are seen in 20% of patients with prostatitis syndrome. The presence of severe urothelial abnormalities merits further investigation to rule out urothelial carcinoma, which is present in up to 50% of cases. Cytologic findings are nonspecific in most cases. However, occasionally the etiologic agent may be evident in urine samples (Corachan et al., 1994). Leukocyturia and hematuria are not uncommon in prostatitis (de la Rosette et al., 1993b) (Box 4–19).

Box 4–19.

Urine cytology is a test rarely requested for patients with prostatitis. The diagnosis is made by bacteriologic cultures of pre- and postprostatic massage urine specimens.

Smears show reactive urothelial cells, leukocyturia, and hematuria.

Cytologic changes are nonspecific in most cases.

REFERENCES

Agliano, A. M., Gazzaniga, P., Cervigni, M., Gradilone, A., Napolitano, M., Pastore, L. I., Manzari, V., Frati, L., Vecchione, A. (1994a). Detection of human papillomavirus type 16 DNA sequences in paraffin-embedded tissues from the female urinary tract. *Urol Int* 52: 208–12.

Agliano, A. M., Gradilone, A., Gazzaniga, P., Napolitano, M., Vercillo, R., Albonici, L., Naso, G., Manzari, V., Frati, L., Vecchione, A. (1994b). High frequency of human papillomavirus detection in urinary bladder cancer. *Urol Int* 53: 125–9.

Akura, K., Hatakenaka, M., Kawai, K., Takenaka, M., Kato, K. (1988). Use of immunocytochemistry on urinary sediments for the rapid identification of human polyomavirus infection. A case report. *Acta Cytol* 32: 247–51.

Anonymous. (1994). Case records of the Massachusetts General Hospital. Weekly clinicopathological exercises. Case 1—1994. A 27-year-old woman with secondary infertility and a bladder mass [clinical conference]. *N Engl J Med* 330: 51–7.

Anonymous. (1995). Case records of the Massachusetts General Hospital. Weekly clinicopathological exercises. Case 2—1995. A 71-year-old man with masses in the pancreas, presacral region, and left kidney [clinical conference]. *N Engl J Med* 332: 174–9.

Apperley, J. F., Rice, S. J., Bishop, J. A., Chia, Y. C., Krausz, T., Gardner, S. D., Goldman, J. M. (1987). Late-onset hemorrhagic cystitis associated with urinary excretion of polyomaviruses after bone marrow transplantation. *Transplantation* 43: 108–12.

Argyle, C. (1985). The identification of fungal casts. A new method for diagnosing visceral candidiasis. *Clin Lab Med* 5: 331–54.

Armora Mani, J., Munoz Segui, J., Perez Cespedes, M., Aguilo Lucia, F., Torrecilla Ortiz, C., Serrallach Mila, N. (1992). Leukoplakia of the upper urinary tract. *Arch Espanoles Urol* 45: 65–7.

Arthur, R. R., Shah, K. V., Baust, S. J., Santos, G. W., Saral, R. (1986). Association of BK viruria with hemorrhagic cystitis in recipients of bone marrow transplants. *N Engl J Med* 315: 230–4.

Aynaud, O., Tranbaloc, P., Orth, G. (1998). Lack of evidence for a role of human papillomaviruses in transitional cell carcinoma of the bladder [see comments]. *J Urol* 159: 86–9; discussion 90.

Ballesteros, J. J., Faus, R., Gironella, J. (1980). Preoperative diagnosis of renal xanthogranulomatosis by serial urinary cytology: preliminary report. *J Urol* 124: 9–11.

Ballesteros, J. J., Munne, A., Rivalta, T., Palma, C., Manzano, I. (1988). Association of renal xanthogranuloma and urological neoplasia. *Eur Urol* 15: 306–10.

Ballesteros, J. J., Serrano, S., Faus, J., Gironella, J. (1983). Urinary cytology in the diagnosis of renal xanthogranulomatosis. *Eur Urol* 9: 343–5.

Baniel, J., Manning, A., Leiman, G. (1991). Fine needle cytodiagnosis of renal tuberculosis. *J Urol* 146: 689–91.

Batta, A. G., Engen, D. E., Reiman, H. M., Winkelmann, R. K. (1990). Intravesical condyloma acuminatum with progression to verrucous carcinoma. *Urology* 36: 457–64.

Belshe, R. B., Mufson, M. A. (1974). Identification by immunofluorescence of adenoviral antigen in exfoliated bladder epithelial cells from patients with acute hemorrhagic cystitis. *Proc Soc Exp Biol Med* 146: 754–8.

Benoit, G., Orth, G., Vieillefond, A., Sinico, M., Charpentier, B., Jardin, A., Fries, D. (1988). Presence of papilloma virus type 11 in condyloma acuminatum of bladder in female renal transplant recipient. *Urology* 32: 343–4.

Beyer-Boon, M. E., Cuypers, L. H., de Voogt, H. J., Brussee, J. A. (1978). Cytological changes due to urinary calculi: a consideration of the relationship between calculi and the development of urothelial carcinoma. *Br J Urol* 50: 81–9.

Bishop, J. W., Emanuel, J. M., Sims, K. L. (1998). Disseminated mucosal papilloma/condyloma secondary to human papillomavirus. *Am J Surg Pathol* 22: 1291–5.

Bock, B., Neal, P. M. (1995). Urinary schistosomiasis in Nebraska. *Nebraska Med J* 80: 264–7.

Bogdanovic, G., Ljungman, P., Wang, F., Dalianis, T. (1996). Presence of human polyomavirus DNA in the peripheral circulation of bone marrow transplant patients with and without hemorrhagic cystitis. *Bone Marrow Transplant* 17: 573–6.

Boon, M. E., van Keep, J. P., Kok, L. P. (1989). Polyomavirus infection versus high-grade bladder carcinoma. The importance of cytologic and comparative morphometric studies of plastic-embedded voided urine sediments. *Acta Cytol* 33: 887–93.

Bosch-Princep, R., Salvado-Usach, M. T., Martinez-Gonzalez, S., Alvaro-Naranjo, T. (1998). Xanthogranulomatous pyelonephritis: urinary and fine needle aspiration cytology [letter]. *Acta Cytol* 42: 1062–4.

Bosompem, K. M., Ayi, I., Anyan, W. K., Nkrumah, F. K., Kojima, S. (1996). Limited field evaluation of a rapid monoclonal antibody-based dipstick assay for urinary schistosomiasis. *Hybridoma* 15: 443–7.

Bossen, E. H., Johnston, W. W. (1975). Exfoliative cytopathologic studies in organ transplantation IV. The cytologic diagnosis of herpesvirus in the urine of renal allograft recipients. *Acta Cytol* 19: 415–19.

Boubenider, S., Hiesse, C., Marchand, S., Hafi, A., Kriaa, F., Charpentier, B. (1999). Post-transplantation polyomavirus infections. *J Nephrol* 12: 24–9.

Bowie, W. R. (1978). Comparison of Gram stain and first-voided urine sediment in the diagnosis of urethritis. *Sex Transm Dis* 5: 39–42.

Braf, Z. F., Smith, M. J. (1977). Bilateral chronic ureteritis associated with *Serratia marcescens*. *Urology* 10: 44–6.

Braza, F., Johnson, K. A., Grigg, L. M., Lopez, V. F., Kavirajan, V. (1989). Human papovavirus in a routine urine specimen of a four-year-old boy. *Diagn Cytopathol* 5: 286–8.

Broseta, E., Osca, J. M., Morera, J., Martinez-Agullo, E., Jimenez-Cruz, J. F. (1993). Urological manifestations of herpes zoster. *Eur Urol* 24: 244–7.

Bruske, T., Loch, T., Thiemann, O., Wirth, B., Janig, U. (1997). Panurothelial condyloma acuminatum with development of squamous cell carcinoma of the bladder and renal pelvis. *J Urol* 157: 620–1.

Byrd, R. B., Viner, N. A., Omell, G. H., Trunk, G. (1976). Leukoplakia associated with renal tuberculosis in the chemotherapeutic era. *Br J Urol* 48: 377–81.

Cardamakis, E., Kotoulas, I. G., Relakis, K., Metalinos, K., Stathopoulos, E., Diamantis, A., Korantzis, A., Michopoulos, J., Mantouvalos, C., Tzingounis, V. (1997). Peoscopic diagnosis of flat condyloma and penile intraepithelial neoplasia. Clinical manifestation. *Gynecol Obstet Invest* 43: 255–60.

Cecchini, S., Cipparrone, I., Confortini, M., Scuderi, A., Meini, L., Piazzesi, G. (1988). Urethral cytology of Cytobrush specimens. A new technique for detecting subclinical human papillomavirus infection in men [see comments]. *Acta Cytol* 32: 314–17.

Cederqvist, L., Johansson, O. (1967). Cytological diagnosis of generalized cytomegalic inclusion disease. *Acta Paediatr Scand* 56: 127–31.

Cheever, A. W., Torky, A. H., Shirbiney, M. (1975). The relation of worm burden to passage of *Schistosoma haematobium* eggs in the urine of infected patients. *Am J Trop Med Hyg* 24: 284–8.

Chuang, C. K., Lai, M. K., Chang, P. L., Huang, M. H., Chu, S. H., Wu, C. J., Wu, H. R. (1992). Xanthogranulomatous pyelonephritis: experience in 36 cases. *J Urol* 147: 333–6.

Coleman, D. V. (1975). The cytodiagnosis of human polyomavirus infection. *Acta Cytol* 19: 93–6.

Comiter, C. V., McDonald, M., Minton, J., Yalla, S. V. (1996). Fungal bezoar and bladder rupture secondary to candida tropicalis. *Urology* 47: 439–41.

Corachan, M., Valls, M. E., Gascon, J., Almeda, J., Vilana, R. (1994). Hematospermia: a new etiology of clinical interest. *Am J Trop Med Hyg* 50: 580–4.

Cotran, R. S., Kumar, V., Collins, T. (1999a). Diseases affecting tubules and interstitium. In *Robbins Patho-*

logic Basis of Disease (R. S. Cotran, V. Kumar, T. Collins, eds.), 6th ed., pp. 968–81. W. B. Saunders, Philadelphia.

Cotran, R. S., Kumar, V., Collins, T. (1999b). Urinary bladder. In *Robbins Pathologic Basis of Disease* (R. S. Cotran, V. Kumar, T. Collins, eds.), 6th ed., pp. 1000–8. W. B. Saunders, Philadelphia.

Cottler-Fox, M., Lynch, M., Deeg, H. J., Koss, L. G. (1989). Human polyomavirus: lack of relationship of viruria to prolonged or severe hemorrhagic cystitis after bone marrow transplant. *Bone Marrow Transplant* 4: 279–82.

de la Rosette, J. J., Hubregtse, M. R., Meuleman, E. J., Stolk-Engelaar, M. V., Debruyne, F. M. (1993a). Diagnosis and treatment of 409 patients with prostatitis syndromes. *Urology* 41: 301–7.

de la Rosette, J. J., Hubregtse, M. R., Wiersma, A. M., Debruyne, F. M. (1993b). Value of urine cytology in screening patients with prostatitis syndromes. *Acta Cytol* 37: 710–12.

DeHertogh, D. A., Brettman, L. R. (1988). Hemorrhagic cystitis due to herpes simplex virus as a marker of disseminated herpes infection. *Am J Med* 84: 632–5.

Del Mistro, A., Braunstein, J. D., Halwer, M., Koss, L. G. (1987). Identification of human papillomavirus types in male urethral condylomata acuminata by in situ hybridization. *Hum Pathol* 18: 936–40.

Del Mistro, A., Koss, L. G., Braunstein, J., Bennett, B., Saccomano, G., Simons, K. M. (1988). Condylomata acuminata of the urinary bladder. Natural history, viral typing, and DNA content. *Am J Surg Pathol* 12: 205–15.

Desai, K., Robson, H. G. (1982). Comparison of the Gram-stained urethral smear and first-voided urine sediment in the diagnosis of nongonococcal urethritis. *Sex Transm Dis* 9: 21–5.

Drachenberg, C. B., Beskow, C. O., Cangro, C. B., Bourquin, P. M., Simsir, A., Fink, J., Weir, M. R., Klassen, D. K., Bartlett, S. T., Papadimitriou, J. C. (1999). Human polyoma virus in renal allograft biopsies: morphological findings and correlation with urine cytology. *Hum Pathol* 30: 970–7.

Echavarria, M. S., Ray, S. C., Ambinder, R., Dumler, J. S., Charache, P. (1999). PCR detection of adenovirus in a bone marrow transplant recipient: hemorrhagic cystitis as a presenting manifestation of disseminated disease. *J Clin Microbiol* 37: 686–9.

Egawa, S., Utsunomiya, T., Uchida, T., Mashimo, S., Koshiba, K. (1994). Emphysematous pyelonephritis, ureteritis, and cystitis in a diabetic patient. *Urol Int* 52: 176–8.

Eickenberg, H. U., Amin, M., Lich, R., Jr. (1975). Blastomycosis of the genitourinary tract. *J Urol* 113: 650–2.

Eltoum, I. A., Suliaman, S. M., Ismail, B. M., Ismail, A. I., Ali, M. M., Homeida, M. M. (1992). Evaluation of eosinophiluria in the diagnosis of schistosomiasis hematobium: a field-based study. *Am J Trop Med Hyg* 46: 732–6.

Falagas, M. E., Snydman, D. R., Ruthazer, R., Werner, B. G., Griffith, J. (1997). Surveillance cultures of blood, urine, and throat specimens are not valuable for predicting cytomegalovirus disease in liver transplant recipients. Boston Center for Liver Transplantation Cytomegalovirus Immune Globulin Study Group [see comments]. *Clin Infect Dis* 24: 824–9.

Florentine, B. D., Carriere, C., Abdul-Karim, F. W. (1997). Mucor pyelonephritis. Report of a case diagnosed by urine cytology, with diagnostic considerations in the workup of funguria. *Acta Cytol* 41: 1797–800.

Fralick, R. A., Malek, R. S., Goellner, J. R., Hyland, K. M. (1994). Urethroscopy and urethral cytology in men with external genital condyloma. *Urology* 43: 361–4.

Ginsberg, P. C., Williams, J. J., Klaus, R. L. (1989). Bilateral ureteral obstruction secondary to condylomata acuminata of the urinary bladder. *J Am Osteopath Assoc* 89: 69–72.

Godwin, J. T., Hanash, K. (1984). Pathology of bilharzial bladder cancer. *Progr Clin Biol Res* 162A: 95–143.

Gomez, E., Aguado, S., Baltar, J., Alvarez, R., Laures, A., Alvarez-Grande, J. (1998). Sterile leukocyturia as a manifestation of urinary tuberculosis in renal transplant patients [letter]. *Nephrol Dial Transplant* 13: 1610–1.

Gondo, H., Minematsu, T., Harada, M., Akashi, K., Hayashi, S., Taniguchi, S., Yamasaki, K., Shibuya, T., Takamatsu, Y., Teshima, T., et al. (1994). Cytomegalovirus (CMV) antigenaemia for rapid diagnosis and monitoring of CMV-associated disease after bone marrow transplantation [see comments]. *Br J Haematol* 86: 130–7.

Goodman, T. R., McHugh, K., Lindsell, D. R. (1998). Paediatric xanthogranulomatous pyelonephritis. *Int J Clin Pract* 52: 43–5.

Greene, M. H. (1992). Emphysematous cystitis due to *Clostridium perfringens* and *Candida albicans* in two patients with hematologic malignant conditions. *Cancer* 70: 2658–63.

Gregg, C. R., Rogers, T. E., Munford, R. S. (1999). Xanthogranulomatous pyelonephritis. *Curr Clin Top Infect Dis* 19: 287–304.

Griffin, M. D., Bergstralhn, E. J., Larson, T. S. (1995). Renal papillary necrosis—a sixteen-year clinical experience. *J Am Soc Nephrol* 6: 248–56.

Guermazi, A., de Kerviler, E., Welker, Y., Zagdanski, A. M., Desgrandchamps, F., Frija, J. (1996). Pseudotumoral vesical actinomycosis. *J Urol* 156: 2002–3.

Hassan, S. I., Talaat, M., el Attar, G. M. (1994). Evaluation of urinalysis reagent strips versus microscopical examination of urine for *Schistosoma haematobium*. *J Egypt Soc Parasitol* 24: 603–9.

Henderson, Y. C., Liu, T. J., Clayman, G. L. (1998). A simple and sensitive method for detecting adenovirus in serum and urine. *J Virol Methods* 71: 51–6.

Hogan, T. F., Padgett, B. L., Walker, D. L., Borden, E. C., McBain, J. A. (1980). Rapid detection and identification of JC virus and BK virus in human urine by using immunofluorescence microscopy. *J Clin Microbiol* 11: 178–83.

Hugosson, C. O., Olsen, P. (1986). Early ureteric changes in *Schistosoma haematobium* infection. *Clin Radiol* 37: 501–3.

Hukkanen, V., Haarala, M., Nurmi, M., Klemi, P., Kiilholma, P. (1996). Viruses and interstitial cystitis: adenovirus genomes cannot be demonstrated in urinary bladder biopsies. *Urol Res* 24: 235–8.

Ito, M., Hirabayashi, N., Uno, Y., Nakayama, A., Asai, J. (1991). Necrotizing tubulointerstitial nephritis associated with adenovirus infection. *Hum Pathol* 22: 1225–31.

Itoh, S., Irie, K., Nakamura, Y., Ohta, Y., Haratake, A., Morimatsu, M. (1998). Cytologic and genetic study of polyomavirus-infected or polyomavirus-activated cells in human urine. *Arch Pathol Lab Med* 122: 333–7.

Iwa, N., Masuda, K., Yutani, C., Katayama, Y., Ito, K., Ueda, S., Kato, S. (1984). Urinary cytology of cytomegalovirus infection associated with systemic lupus erythematosus [letter]. *Acta Cytol* 28: 775–6.

Iwasawa, A., Hiltunen-Back, E., Reunala, T., Nieminen, P., Paavonen, J. (1997). Human papillomavirus DNA in urine specimens of men with condyloma acuminatum. *Sex Transm Dis* 24: 165–8.

Iwasawa, A., Kumamoto, Y., Maruta, H., Fukushima, M., Tsukamoto, T., Fujinaga, K., Fujisawa, Y., Kodama, N. (1992). Presence of human papillomavirus 6/11 DNA in condyloma acuminatum of the urinary bladder. *Urol Int* 48: 235–8.

Jellinek, E. H., Tulloch, W. S. (1976). Herpes zoster with dysfunction of bladder and anus. *Lancet* 2: 1219–22.

Kahan, A. V., Coleman, D. V., Koss, L. G. (1980). Activation of human polyomavirus infection-detection by cytologic technics. *Am J Clin Pathol* 74: 326–32.

Keating, M. A., Young, R. H., Carr, C. P., Nikrui, N., Heney, N. M. (1985). Condyloma acuminatum of the bladder and ureter: case report and review of the literature. *J Urol* 133: 465–7.

Kitamura, T., Aso, Y., Kuniyoshi, N., Hara, K., Yogo, Y. (1990). High incidence of urinary JC virus excretion in nonimmunosuppressed older patients. *J Infect Dis* 161: 1128–33.

Kline, T. S., Craighead, J. E. (1967). Renal homotransplantation. The cytology of the urine sediment. *Am J Clin Pathol* 47: 802–6.

Koss, L. G., Sherman, A. B., Eppich, E. (1984). Image analysis and DNA content of urothelial cells infected with human polyomavirus. *Anal Quant Cytol* 6: 89–94.

Kotoulas, I. G., Cardamakis, E., Relakis, K., Diamantis, A., Papathanasiou, Z., Korantzis, A., Mantouvalos, C. (1996). Peoscopic diagnosis of flat condyloma and penile intraepithelial neoplasia. IV. Urethral reservoir. *Gynecol Obstet Invest* 41: 55–60.

Krieger, J. N., Jacobs, R., Ross, S. O. (2000). Detecting urethral and prostatic inflammation in patients with chronic prostatitis. *Urology* 55: 186–91; discussion 191–2.

Kuntze, J. R., Herman, M. H., Evans, S. G. (1988). Genitourinary coccidioidomycosis. *J Urol* 140: 370–4.

Kupper, T., Stoffels, U., Pawlita, M., Burrig, K. F., Pfitzer, P. (1993). Morphological changes in urothelial cells replicating human polyomavirus BK. *Cytopathology* 4: 361–8.

Lecatsas, G., Boes, E. G. (1980). Urinary virus excretion in pregnancy. *S Afr Med J* 57: 988–90.

Lecatsas, G., van Wyk, J. A. (1978). DNA viruses in urine after renal transplantation. *S Afr Med J* 53: 787–8.

Lenk, S., Oesterwitz, H., Scholz, D. (1988). Tuberculosis in cadaveric renal allograft recipients. Report of 4 cases and review of the literature. *Eur Urol* 14: 484–6.

Lipsey, A. I., Bolande, R. P. (1967). The exfoliative source of abnormal cells in urine sediment of patients with measles. *Am J Dis Child* 113: 677–82.

Ljutic, D., Glavina, M. (1995). Tubulointerstitial nephritis with uveitis syndrome following varicella zoster reactivation [letter]. *Nephron* 71: 485–6.

Ludwig, M., Schroeder-Printzen, I., Ludecke, G., Weidner, W. (2000). Comparison of expressed prostatic secretions with urine after prostatic massage—a means to diagnose chronic prostatitis/inflammatory chronic pelvic pain syndrome [in process citation]. *Urology* 55: 175–7.

Masuda, K., Akutagawa, K., Yutani, C., Kishita, H., Ishibashi-Ueda, H., Imakita, M. (1998). Persistent infection with human polyomavirus revealed by urinary cytology in a patient with heart transplantation. A case report. *Acta Cytol* 42: 803–6.

Masukawa, T., Garancis, J. C., Rytel, M. W., Mattingly, R. F. (1972). Herpes genitalis virus isolation from human bladder urine. *Acta Cytol* 16: 416–28.

McClanahan, C., Grimes, M. M., Callaghan, E., Stewart, J. (1994). Hemorrhagic cystitis associated with herpes simplex virus. *J Urol* 151: 152–3.

Meares, E. M., Jr. (1987). Acute and chronic prostatitis: diagnosis and treatment. *Infect Dis Clin North Am* 1: 855–73.

Melchers, W. J., Schift, R., Stolz, E., Lindeman, J., Quint, W. G. (1989). Human papillomavirus detection in urine samples from male patients by the polymerase chain reaction. *J Clin Microbiol* 27: 1711–14.

Minassian, H., Schinella, R., Reilly, J. C. (1994). Polyomavirus in the urine: follow-up study. *Diagn Cytopathol* 10: 209–11.

Mott, K. E., Dixon, H., Osei-Tutu, E., England, E. C. (1983). Relation between intensity of *Schistosoma haematobium* infection and clinical haematuria and proteinuria. *Lancet* 1: 1005–8.

Moudgil, A., Germain, B. M., Nast, C. C., Toyoda, M., Strauss, F. G., Jordan, S. C. (1997). Ureteritis and cholecystitis: two unusual manifestations of cytomegalovirus disease in renal transplant recipients. *Transplantation* 64: 1071–3.

Mueller, B. U., MacKay, K., Cheshire, L. B., Choyke, P. L., Kitchen, B., Widemann, B., Pizzo, P. A. (1995). Cytomegalovirus ureteritis as a cause of renal failure in a child infected with the human immunodeficiency virus. *Clin Infect Dis* 20: 1040–3.

Murphy, D. M., Fallon, B., Lane, V., O'Flynn, J. D. (1982). Tuberculous stricture of ureter. *Urology* 20: 382–4.

Nagaoka, H., Tokimatsu, I., Yamasaki, T., Nagai, H., Otsuka, E., Hashimoto, A., Goto, Y., Nasu, M., Kikuchi, H., Daa, T., Akizuki, S. (1997). [A pathological study of cytomegalovirus infections in autopsied cases with adult T-cell leukemia]. *Kansenshogaku Zasshi—J Jpn Assoc Infect Dis* 71: 222–8.

Nahhas, W. A., Marshall, M. L., Ponziani, J., Jagielo, J. A. (1986). Evaluation of urinary cytology of male sexual partners of women with cervical intraepithelial neoplasia and human papilloma virus infection. *Gynecol Oncol* 24: 279–85.

Nickeleit, V., Hirsch, H. H., Binet, I. F., Gudat, F., Prince, O., Dalquen, P., Thiel, G., Mihatsch, M. J. (1999). Polyomavirus infection of renal allograft recipients: from latent infection to manifest disease. *J Am Soc Nephrol* 10: 1080–9.

Nickeleit, V., Hirsch, H. H., Zeiler, M., Gudat, F., Prince, O., Thiel, G., Mihatsch, M. J. (2000). BK-virus nephropathy in renal transplants-tubular necrosis, MHC-class II expression and rejection in a puzzling game. *Nephrol Dial Transplant* 15: 324–32.

Niewiadomski, S., Florentine, B. D., Cobb, C. J. (1998). Cytologic identification of trichomonas vaginalis in urine from a male with long-standing sterile pyuria [letter]. *Acta Cytol* 42: 1060–1.

Noel, J. C., Thiry, L., Verhest, A., Deschepper, N., Peny, M. O., Sattar, A. A., Schulman, C. C., Haot, J. (1994).

Transitional cell carcinoma of the bladder: evaluation of the role of human papillomaviruses. *Urology* 44: 671–5.

Olive, D. M., al Mufti, S., Simsek, M., Fayez, H., al Nakib, W. (1989). Direct detection of human cytomegalovirus in urine specimens from renal transplant patients following polymerase chain reaction amplification. *J Med Virol* 29: 232–7.

Piscioli, F., Pusiol, T., Polla, E., Failoni, G., Luciani, L. (1985). Urinary cytology of tuberculosis of the bladder. *Acta Cytol* 29: 125–31.

Plunkett, J. M., Turner, B. I., Tallent, M. B., Johnson, H. K. (1981). Cryptococcal septicemia associated with urologic instrumentation in a renal allograft recipient. *J Urol* 125: 241–2.

Remigio, P., Ramos, C., Basirico, P. (1975). Xanthogranulomatous ureteritis. *J Urol* 114: 621–3.

Rohner, T. J., Jr., Tuliszewski, R. M. (1980). Fungal cystitis: awareness, diagnosis and treatment. *J Urol* 124: 142–4.

Saad, S. A., Hanafy, H. M. (1974). Bilharzial (schistosomal) ureteritis cystica. *Urology* 4: 261–6.

Samuelson, J. (1999). Schistosomiasis. In *Robbins Pathologic Basis of Disease* (R. S. Cotran, V. Kumar, T. Collins, eds.), 6th ed., pp. 396–7. W. B. Saunders, Philadelphia.

Schumann, G. B., Berring, S., Hill, R. B. (1977). Use of the cytocentrifuge for the detection of cytomegalovirus inclusions in the urine of renal allograft patients. A case roport. *Acta Cytol* 21: 168–72.

Schwartz, M. A., Hooton, T. M. (1998). Etiology of nongonococcal nonchlamydial urethritis. *Dermatol Clin* 16: 727–33, xi.

Shah, K. V., Daniel, R. W., Strickler, H. D., Goedert, J. J. (1997). Investigation of human urine for genomic sequences of the primate polyomaviruses simian virus 40, BK virus, and JC virus. *J Infect Dis* 176: 1618–21.

Shahmanesh, M., Pandit, P. G., Round, R. (1996). Urethral lymphocyte isolation in non-gonococcal urethritis. *Genitourin Med* 72: 362–4.

Shirai, T., Yamamoto, K., Adachi, T., Imaida, K., Masui, T., Ito, N. (1988). Condyloma acuminatum of the bladder in two autopsy cases. *Acta Pathol Japon* 38: 399–405.

Silbert, P. L., Matz, L. R., Christiansen, K., Saker, B. M., Richardson, M. (1990). Herpes simplex virus interstitial nephritis in a renal allograft [see comments]. *Clin Nephrol* 33: 264–8.

Smith, J. H., Christie, J. D. (1986). The pathobiology of *Schistosoma haematobium* infection in humans. *Hum Pathol* 17: 333–45.

Smith, J. H., Torky, H., Kelada, A. S., Farid, Z. (1977). Schistosomal polyposis of the urinary bladder. *Am J Trop Med Hyg* 26: 85–8.

Smith, J. H., von Lichtenberg, F. (1976). Tissue degradation of calcific *Schistosoma haematobium* eggs. *Am J Trop Med Hyg* 25: 595–601.

Sokol, A. B., Min, D. S. (1972). Trichomonas cystitis in a six-week-old infant—effective treatment with oral metronidazole. *J Indiana State Med Assoc* 65: 1084–6.

Spach, D. H., Bauwens, J. E., Myerson, D., Mustafa, M. M., Bowden, R. A. (1993). Cytomegalovirus-induced hemorrhagic cystitis following bone marrow transplantation. *Clin Infect Dis* 16: 142–4.

Spagnolo, D. V., Waring, P. M. (1986). Bladder granulomata after bladder surgery. *Am J Clin Pathol* 86: 430–7.

Stamm, W. E., Turck, M. (1987). Urinary tract infection, pyelonephritis, and related conditions. In *Harrison's Principles of Internal Medicine* (E. Braunwald, K. J. Isselbacher, R. G. Petersdorf, J. D. Wilson, J. B. Martin, A. S. Fauci, eds.), 11th ed., pp. 1189–95. McGraw-Hill, New York.

Stary, A. (1998). Urethritis. Diagnosis of nongonococcal urethritis. *Dermatol Clin* 16: 723–6, xi.

Stella, F., Alfani, D., Renna-Molaioni, E., Berloco, P., Stella, C., Famulari, A., Troccoli, R., Cortesini, R. (1983). Laboratory diagnosis in clinical renal transplants. II. Exfoliative urinary cytology in non-immunologic complications. *Quaderni Sclavo di Diagnostica Clinica e di Laboratorio* 19: 507–22.

Sultana, S. R., McNeill, S. A., Phillips, G., Byrne, D. J. (1998). Candidal urinary tract infection as a cause of pneumaturia. *J R Coll Surgeons Edinburgh* 43: 198–9.

Tani, E. M., Montenegro, M. R., Viana de Camargo, J. L. (1983). Urinary bladder washings in autopsies. A cytopathologic study of 63 cases. *Acta Cytol* 27: 128–32.

Torrington, K. G., Old, C. W., Urban, E. S., Carpenter, J. L. (1979). Transurethral passage of *Aspergillus* fungus balls in acute myelocytic leukemia. *South Med J* 72: 361–3.

van Poppel, H., Stessens, R., de Vos, R., van Damme, B. (1986). Isolated condyloma acuminatum of the bladder in a patient with multiple sclerosis: etiological and pathological considerations. *J Urol* 136: 1071–3.

Villani, U., Leoni, S., Pastorello, M., Casolari, E., Manferrari, F. (1987). Actinomycosis of bladder and intrauterine devices. *Br J Urol* 60: 463–4.

Vogeli, T. A., Peinemann, F., Burdach, S., Ackermann, R. (1999). Urological treatment and clinical course of BK polyomavirus-associated hemorrhagic cystitis in children after bone marrow transplantation. *Eur Urol* 36: 252–7.

Wehrheim, W. (1973). Crohn's disease simulating tuberculous ureteritis. *Int Urol Nephrol* 5: 359–62.

Wiedemann, A., Diekmann, W. P., Holtmann, G., Kracht, H. (1995). Report of a case with giant condyloma (Buschke-Lowenstein tumor) localized in the bladder. *J Urol* 153: 1222–4.

Wise, G. J., Silver, D. A. (1993). Fungal infections of the genitourinary system. *J Urol* 149: 1377–88.

Wise, G. J., Talluri, G. S., Marella, V. K. (1999). Fungal infections of the genitourinary system: manifestations, diagnosis, and treatment. *Urol Clin North Am* 26: 701–18, vii.

Wojcik, E. M., Miller, M. C., Wright, B. C., Veltri, R. W., O'Dowd, G. J. (1997). Comparative analysis of DNA content in polyomavirus-infected urothelial cells, urothelial dysplasia and high grade transitional cell carcinoma. *Anal Quant Cytol Histol* 19: 430–6.

Woolf, N. (1998). Disorders of the urethra. In *Pathology. Basic and Systemic* (N. Woolf, ed.), pp. 717–18. W. B. Saunders, London.

Zurlo, J. J., O'Neill, D., Polis, M. A., Manischewitz, J., Yarchoan, R., Baseler, M., Lane, H. C., Masur, H. (1993). Lack of clinical utility of cytomegalovirus blood and urine cultures in patients with HIV infection [see comments]. *Ann Intern Med* 118: 12–17.

5

NONINFECTIOUS INFLAMMATORY CONDITIONS OF THE URINARY TRACT

KIDNEY, RENAL PELVIS, AND URETER

TUBULOINTERSTITIAL NEPHROPATHY

TUBULOINTERSTITIAL nephropathy includes acute tubular necrosis, acute interstitial nephritis, and chronic interstitial nephritis. The first two are characterized by a rapid deterioration of renal function, with oliguria in many instances. The third usually follows a more insidious course, with gradual loss of renal function.

Histologically, acute tubular necrosis shows interstitial edema with minimal lymphoplasmacytic infiltration. In contrast, acute interstitial nephritis exhibits interstitial lymphocytes, plasma cells, histiocytes, and eosinophils. Sparing of the glomeruli is characteristic of these acute processes, and, despite their names, no acute inflammation is present within the tubules or in the interstitium. Chronic interstitial nephritis is characterized by interstitial fibrosis and tubular atrophy (Bonsib, 1997).

The causes of tubulointerstitial nephropathy include hypersensitivity, numerous drugs, chemicals, heavy metals, systemic diseases, sarcoidosis, bacterial and viral diseases, and transplant rejection, among others. Many of these etiologic agents may produce acute or chronic interstitial nephritis or acute tubular necrosis and commonly do not elicit a distinct pathologic picture (Bonsib, 1997). Therefore, a precise diagnosis and identification of the specific cause require amalgamating a meticulous history, an appropriate clinical setting, and laboratory data with renal biopsy findings.

Cytology. Cytologic findings in the acute tubulointerstitial processes are nonspecific. The urine shows red blood cells, but white cells including white blood cell casts, as well as red blood cell casts, are absent. Granular casts are present in acute tubular necrosis, and lymphoplasmacytic inflammation with eosinophiluria is found in acute interstitial nephritis. None of these findings is specific for acute interstitial nephritis or acute tubular necrosis. However, eosinophiluria may prove useful for distinguishing between them (Corwin and Haber, 1987). Eosinophils are normally absent from urine.

It may be difficult to separate urothelial cells from renal tubular cells. Immunocytochemistry and electron microscopy of urine sediment are useful for cell characterization and diagnosis of certain types of acute interstitial nephritis (Mandal, 1988; Segasothy et al., 1988) (Box 5–1).

Differential diagnosis. The finding of eosinophils in the urine has been suggested to be useful in supporting the diagnosis of acute interstitial nephritis (Corwin and Haber, 1987). However, more than one eosinophil per 100 cells has been associated with a variety of clinical conditions affecting the kidney or urinary tract, including rapidly progressive glomerulonephritis, acute prostatitis, and occasionally acute cystitis or acute glomerulonephritis (Nolan et al., 1986). The use of Hansel's stain improves

Box 5–1.

Tubulointerstitial nephropathy includes acute and chronic interstitial nephritis and acute tubular necrosis. The cytologic features are nonspecific.

Lymphoplasmacytic inflammation and eosinophils may be present in acute interstitial nephritis. However, neutrophils and white blood cell casts are absent.

Eosinophiluria should not be used as the sole criterion for the diagnosis of acute interstitial nephritis.

Granular casts are present in acute tubular necrosis.

the sensitivity and predictive value of eosinophiluria in the diagnosis of acute interstitial nephritis, but in the absence of renal biopsy or other clinical clues suggesting the diagnosis, eosinophiluria should not be used as the sole criterion for this diagnosis (Corwin et al., 1989; Nolan and Kelleher, 1988).

The presence of red blood cell casts indicates glomerulonephritis, and white blood cell casts indicate acute pyelonephritis.

Acute Renal Allograft Rejection

The clinical diagnosis of acute renal allograft rejection can often be predicted or confirmed on the basis of characteristic cytologic findings in cytocentrifuged and Papanicolaou-stained urine samples in correlation with renal biopsy and clinical findings.

The presence of more than 20 exfoliated renal collecting duct tubular cells per 10 high-power fields (hpf) has been found to be useful in predicting an acute rejection episode, and their persistence has prognostic importance (Eggensperger et al., 1988; Schumann et al., 1977). Based on this parameter, 64% of cytodiagnostic urinalyses precede the clinical diagnoses. (Eggensperger et al., 1988). Likewise, a cytodiagnosis of acute cellular rejection can be suggested when the lymphocyte count is 20% or 13 per 10 hpf, particularly in the presence of large numbers of renal tubular cells (Corey et al., 1997; Eggensperger et al., 1988).

However, activated lymphocytes may not be present in all cases of allograft rejection with acute renal failure. For this reason, a finding of red blood cells, lymphocytes, tubular cells, casts, mixed cell clusters, and nuclear changes in epithelial cells, including mitoses, should be mentioned in the cytology report of the urinary profile of renal transplant patients. Rejection can be strongly suggested in the presence of lymphocyturia and an increased number of renal tubular cells (Bossen and Johnston, 1977). Nevertheless, it is not possible to differentiate between acute allograft rejection and acute renal failure of other origins (Winkelmann et al., 1985). Clinical correlation is necessary (Box 5–2).

Sarcoidosis

Sarcoidosis has been associated with various renal manifestations, including calcemic nephropathy, granulomatous interstitial nephritis, tubular dysfunction, and glomerulonephritis. Granulomatous parenchymal infiltration is found in up to 40% of patients with sarcoidosis; noncaseating granulomas are commonly located in the cortex but may also be found in the medulla. In contrast to interstitial involvement, glomerular disease is rare and may yield hematuria, proteinuria, and red blood cell casts in the urine sedi-

Box 5–2.

The smear shows numerous lymphocytes and tubular cells, red blood cells, casts, and reactive urothelial cells.

Cytodiagnosis of acute renal allograft rejection is suggested strongly in the presence of lymphocyturia and increased numbers of renal tubular cells.

Without a renal biopsy, it is not possible to distinguish between acute cellular rejection and acute renal failure of other etiology.

Box 5–3.

Noncaseating granulomas are commonly located in the renal cortex.

Exfoliated histiocytes and granulomas are rare findings in urine specimens.

Red blood cell casts in urine specimens suggest sarcoidal glomerulopathy in patients with systemic sarcoidosis.

ment (Casella and Allon, 1993). Intraluminal localization with involvement of the renal pelvis and hydronephrosis is exceptional (Schillinger et al., 1999).

Cytology. In contrast to the finding in tuberculosis, the parenchymal location and nonnecrotizing nature of the granulomas make difficult the exfoliation of epithelioid histiocytes and multinucleated cells into the urine. The nonspecific presence of red blood cells and red blood cell casts suggests sarcoidal glomerulopathy in patients with generalized sarcoidosis. Fine-needle aspiration cytology may provide a more precise diagnosis (Box 5–3).

Ureteropyelitis Cystica and Glandularis

Along with von Brunn's nests, pyelitis glandularis and pyelitis cystica are the most common noninfectious and nonneoplastic lesions of the urinary tract. These lesions are identical to those present in the urinary bladder; details of their histopathology and cytology are given later in this chapter.

Squamous Metaplasia (Leukoplakia)

Keratinizing squamous metaplasia of the upper urinary tract is uncommon, and has been attributed to infection or mechanical injury to the urothelium and/or to genetic factors (Armora Mani et al., 1992; Zipkin et al., 1985). The morphologic and clinical term commonly used as a synonym is *leukoplakia.*

Leukoplakia can occur in association with renal tuberculosis and was most commonly seen in the prechemotherapeutic era (Byrd et al., 1976). Most important, leukoplakia is a premalignant disease, and patients with this diagnosis deserve careful and frequent follow-up (Benson et al., 1984).

Cytology. The presence of exfoliated plaques of anucleated squamous cells is characteristic of leukoplakia. Evidence of cytologic atypia should be sought carefully. The presence of rare single, anucleated squamous cells very likely represents contamination from the external genitalia, particularly in women (Box 5–4).

Glandular Metaplasia

Colonic-type goblet cells are a relatively common component of reactive urothelial lesions, but extensive intestinal metaplasia is rare. When the latter occurs, the likelihood of developing adenocarcinoma is high. Despite its innocuous histology, widespread intestinal metaplasia apparently identifies a disturbed urothelium that creates a high risk of cancer (Bullock et al., 1987; Ward, 1971). Extensive intestinal metaplasia has been seen in association with chronic irritation and/or infection of the upper urinary tract.

URINARY BLADDER

A classification of noninfectious bladder lesions is presented in Table 5–1.

Box 5–4.

Keratinizing squamous metaplasia is known as leukoplakia.

Leukoplakia is a premalignant condition leading to squamous carcinoma.

Smears show single and clustered keratinized, anucleated squamous cells.

Evidence of cytologic atypia should be sought carefully.

Table 5–1. Noninfectious Disorders of the Urinary Bladder

Inflammatory
Malacoplakia
Vesicoenteric fistula
Nonspecific cystitis: hemorrhagic, interstitial, eosinophilic, and follicular
Inflammatory pseudotumor
Epithelial proliferations
Cystitis glandularis
Cystitis cystica
Squamous metaplasia
Nephrogenic adenoma
Miscellaneous
Endometriosis
Endocervicosis
Mullerianosis

Malacoplakia

Malacoplakia is an uncommon granulomatous disease of unclear etiology that affects predominantly middle-aged women and involves primarily the urinary tract. It is more frequent in the urinary bladder and ureters than in the renal pelvis and kidney, where it is seen as a complication of recurrent pyelonephritis. It can cause fever, flank pain, and acute renal failure (al-Sulaiman et al., 1993; Long and Althausen, 1989; Tsung, 1982). Bilateral renal malacoplakia is a progressive, destructive, and occasionally fatal disease (Ho et al., 1979). Patients with bladder malacoplakia develop signs and symptoms of urinary tract infection, including microscopic or gross hematuria. Urine culture frequently isolates *Escherichia coli* (Long and Althausen, 1989).

Grossly, the lesions are soft, tan-yellow plaques, often with a central umbilication and a hyperemic base. Histologically, there are submucosal aggregates of large histiocytes (of von Hansemann). with eosinophilic and granular cytoplasm, containing basophilic inclusions with concentric lamination (Michaelis-Gutmann bodies) that are PAS-, calcium-, and iron-positive. Varying numbers of plasma cell are present.

The histiocytes appear to be due to abnormal phagocytosis, and the Michaelis-Gutmann bodies may represent mineralized fragments of bacteria, the result of impaired autophagolysosomal activity of the cell to kill phagocytosed bacteria (August et al., 1994; Curran, 1987). Michaelis-Gutmann bodies may not be found by light microscopy in all cases. However, ultrastructural studies have been useful for detecting early stages in the formation of these bodies and for distinguishing malacoplakia from other histiocytic interstitial renal processes (August et al., 1994).

Cytology. Histiocytes show an eosinophilic cytoplasm containing single or multiple Michaelis-Gutmann bodies. These bodies are usually basophilic, measure 5 to 10 μm in diameter, and have concentric laminations (Fig. 5–1). The background is granular, with necrotic debris and a few plasma cells. Urothelial cells show reactive changes. The cytologic findings of malacoplakia of the renal pelvis seen in catheterized ureteral urine are similar to those in the bladder (Tsung, 1982) (Box 5–5).

Differential diagnosis. Findings of macrophages or clear cells in urine should be included in the differential diagnosis of malacoplakia. See Table 7–2.

Vesicoenteric Fistula

Vesical fistulas are more common in men than in woman and are seen as complications of diverticulitis, colon cancer, inflammatory bowel disease, radiation therapy, and surgery or, more rarely, in infectious processes (Carson et al., 1978). A history of

Box 5–5.

Malacoplakia is a granulomatous disease of unknown etiology.

Patients have symptoms and signs or urinary tract infection.

Urine culture frequently shows *E. coli.*

Smears show histiocytes of von Hansemann with eosinophilic and granular cytoplasm containing Michaelis-Gutmann bodies (PAS-, iron-, and calcium positive).

Granular debris, plasma cells, and reactive urothelial cells are also present.

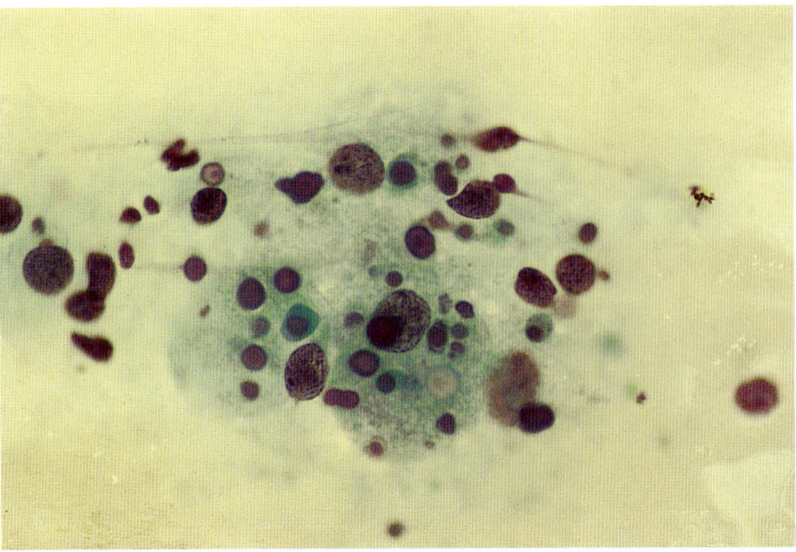

Figure 5–1. *Malacoplakia.* This bladder washing shows Michaelis-Gutman bodies within histiocytes. (Papanicolaou stain, ×600). Courtesy of Dr. Linda Green, Associate Professor of Pathology, Baylor College of Medicine, Houston, TX.

fever, an abdominal mass, or cystitis is common (Carson et al., 1978; Walser et al., 1999). In contrast, brown, turbid, odorous urine (fecaluria) or passing gas at the time of urination (pneumaturia) is uncommon, and not all patients with isolated pneumaturia have a vesicoenteric fistula (Synhaivsky and Malek, 1985).

Computed tomography (CT) scan and cystoscopy are helpful in establishing the diagnosis. Urine culture may show a mixed infection (Sandison and Jones, 1993).

Cytology. Fecal matter in urine can be identified by the presence of vegetable cells, fibers, degenerated striated muscle cells, and amorphous debris. Numerous bacilli are seen in the background. A few reactive urothelial cells may be present (Fig. 5–2; Box 5–6).

Box 5–6.

Fecaluria and pneumaturia are uncommon manifestations of vesicoenteric fistula. Fever, cystitis, and an abdominal mass are more common.

Smears show vegetable cells, muscle fibers, amorphous debris, and numerous bacilli.

Hemorrhagic Cystitis

Cyclophosphamide is the prototype agent that causes hemorrhagic cystitis with ninefold increased incidence of bladder cancer (Fairchild et al., 1979; Stillwell and Benson, 1988). Other cytotoxic drugs, radiotherapy, and infectious agents may be associated with hemorrhagic cystitis.

Cytology. This nonspecific form of cystitis does not have a distinct cytopathology, except for cellular changes associated with the causative factor, such as chemotherapy, radiation treatment, viral infection, and so on, which are covered in other chapters in this book. Urothelial cell groups have been described in these cases, with no specific significance (Goldstein et al., 1998).

Interstitial Cystitis (Hunner's Ulcer)

Interstitial cystitis is a chronic, possibly inflammatory bladder condition of unknown cause and pathogenesis that lacks specific clinical, cystoscopic, and histopathologic di-

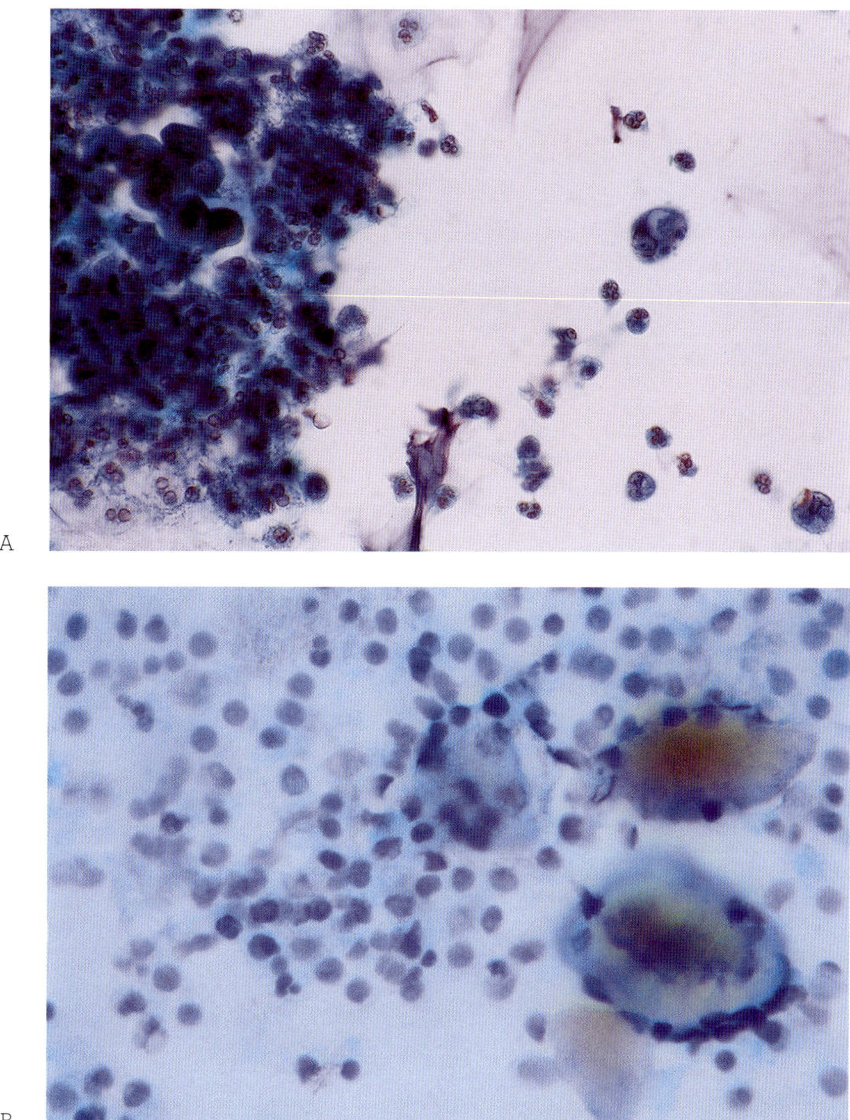

Figure 5–2. *Vesicoenteric fistula.* (*A*) Reactive urothelial cells and marked acute inflammatory reaction are seen in this bladder washing from a patient with a vesicoenteric fistula. (*B*) The voided urine from the same patient shows fragments of partly digested vegetable matter and degenerated acute inflammatory cells. (Papanicolaou stain, ×600). Courtesy of Dr. Linda Green, Associate Professor of Pathology, Baylor College of Medicine, Houston, TX.

agnostic criteria (Koziol, 1994). Although specific consensus criteria have been given for the diagnosis of interstitial cystitis, the final diagnosis rests on the exclusion of other bladder diseases (Hanno, 1994). The characteristic wedge-shaped Hunner's ulcer is not always present.

Changes found on bladder biopsy include superficial ulceration with extensive denudation of the urothelium, granulation tissue, and chronic inflammation. An exudate composed of fibrin admixed with red blood cells and inflammatory cells may be present on the surface of the denuded areas. Focal suburothelial hemorrhage may be seen in nonulcerated areas. Increased numbers of mast cells are found mainly in the muscularis propria (Johansson and Fall, 1994).

Cytology. As expected, urine cytology findings are consistent with the tissue findings but are nonetheless nonspecific. One cytologic study of 33 urine specimens from 14 patients with interstitial cystitis described an inflammatory smear pattern of mostly neutrophils, mononuclear cells, and, less frequently, eosinophils. Reactive urothelial cells, degenerated cells, and red blood cells are commonly found. No mast cells are identified by light microscopy or toluidine blue stain (Dodd and Tello, 1998) (Box 5–7).

Differential diagnosis. Other forms of cystitis, including eosinophilic cystitis, should be included in the differential diagnosis.

Eosinophilic Cystitis

Eosinophilic cystitis is rare but affects people of all ages. Its cause is unknown, but it may be seen in association with allergic diseases, hypereosinophilic syndrome, and bladder trauma, including transurethral or open bladder surgery. It may be idiopathic. The clinical presentation varies, but hematuria is a constant feature, with urinary frequency and dysuria being common (Okafo et al., 1985).

Histologically, there is edema of the lamina propria with a mixed inflammatory infiltrate, predominantly eosinophils (Young and Able, 1997).

Cytology. It is not possible to diagnose eosinophilic cystitis by means of cytologic study of the urine alone. Neutrophils may be the predominant inflammatory cell population, with only small numbers of eosinophils (Garcia-Ureta et al., 1978).

Box 5–7.

Smears show mixed acute and chronic inflammatory cells, reactive urothelial cells, red blood cells, and amorphous debris.

Tissue biopsy shows superficial ulceration, extensive denudation, and granulation tissue.

Follicular Cystitis

The presence of mucosal nodules containing lymphoid follicles within the lamina propria of the bladder characterizes follicular cystitis. This lesion may be identified in patients with urinary tract infection, particularly in prepubertal girls and in patients with bladder cancer (Sarma, 1970). These cystoscopically and microscopically detected changes are reversible and usually disappear after successful eradication of the urinary infection (Hansson et al., 1990).

Cytology. The location of the lymphoid follicles within the lamina propria prevents shedding of spontaneously exfoliated cells. However, it is possible that a post-bladder-biopsy urine sample may contain a pleomorphic population of lymphocytes and tingible-body macrophages, suggestive of follicular cystitis (Fig. 5–3). In contrast to follicular cervicitis, the association with *Chlamydia* remains to be elucidated.

Inflammatory Pseudotumor

Inflammatory pseudotumor is a rare pseudoneoplastic spindle cell proliferation unrelated to surgical bladder trauma. It occurs typically in young adults, more frequently in women (Young, 1997). Commonly, patients present clinically with gross hematuria. The pseudotumor is a polypoid intraluminal mass measuring several centimeters. Microscopically, the lesions show spindle myofibroblasts, myxoid stroma, and variable numbers of lymphocytes, plasma cells, and red blood cells. Besides immunoreactivity with vimentin and actin, myofibroblasts may react with keratin and have a diploid DNA content (Jones et al., 1993; Sonobe et al., 1999).

Cytology. Cytologic descriptions of inflammatory pseudotumors of the urinary bladder are exceedingly rare. Some of these tumors may ulcerate, and atypical spindle cells of various sizes and shapes, resembling those of sarcoma, may be present in urine samples (Fig. 5–4A–C) (Sonobe et al., 1999). Eosinophilic cytoplasm may be seen in some spindle cells (Nochomovitz and Orenstein, 1985) (Box 5–8).

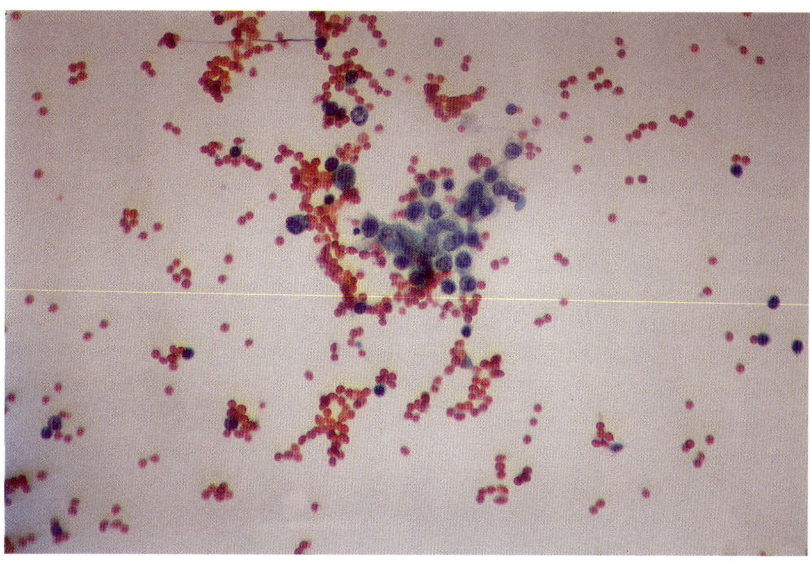

Figure 5–3. *Follicular cystitis.* This bladder washing shows an aggregate of polymorphous lymphoid cells and macrophages characteristic of the germinal center of a lymphoid follicle. (Papanicolaou stain, ×600). Courtesy of Dr. Linda Green, Associate Professor of Pathology, Baylor College of Medicine, Houston, TX.

Differential diagnosis. The differential diagnosis includes spindle cell carcinoma and spindle cell sarcoma. Because myofibroblasts may be positive for keratin, the final diagnosis requires careful clinical and tissue biopsy correlation. Awareness of the possible aberrant expression of cytokeratin and epithelial membrane antigen helps to avoid a misdiagnosis of sarcomatoid carcinoma (Sonobe et al., 1999).

Cystitis Glandularis and Cystitis Cystica

Some consider cystitis glandularis and cystitis cystica to be urothelial variants, and others, believe them to be urothelial metaplasias. These lesions are present in more than 60% of normal bladders. Their frequency increases with age, and they are commonly located in the trigone (Cotran et al., 1999a; Young and Able, 1997). Similar lesions occur in the ureter and renal pelvis.

Histologically, in cystitis glandularis there are glands located in the lamina propria and lined with columnar or cuboidal cells surrounded by urothelial cells. Focal mucin production and goblet cells may be present in the lining epithelium. The presence of an extensive mucin-producing lesion with extensive goblet cell hyperplasia, known as *cystitis glandularis of the intestinal type,* carries a high risk for the development of bladder adenocarcinoma (Delnay et al., 1999). In contrast to cystitis glandularis, which has mostly microscopic lesions, cystitis cystica shows cystoscopically visible cysts. The cysts are lined with urothelial cells and contain a proteinaceous fluid with a few inflammatory cells (Young and Able, 1997).

Clinically, intestinal metaplasia may resemble cancer, and histologic distinction from in situ and invasive well-differentiated bladder adenocarcinoma may be problematic (Jacobs et al., 1997; Young and Bostwick, 1996). The sharing of antigenic re-

Box 5–8.

Inflammatory pseudotumor is a benign stromal spindle cell proliferation.

In the presence of ulceration, smears show atypical spindle cells of various sizes and shapes, some with eosinophilic cytoplasm.

Variable numbers of plasma cells, lymphocytes, and red blood cells are present.

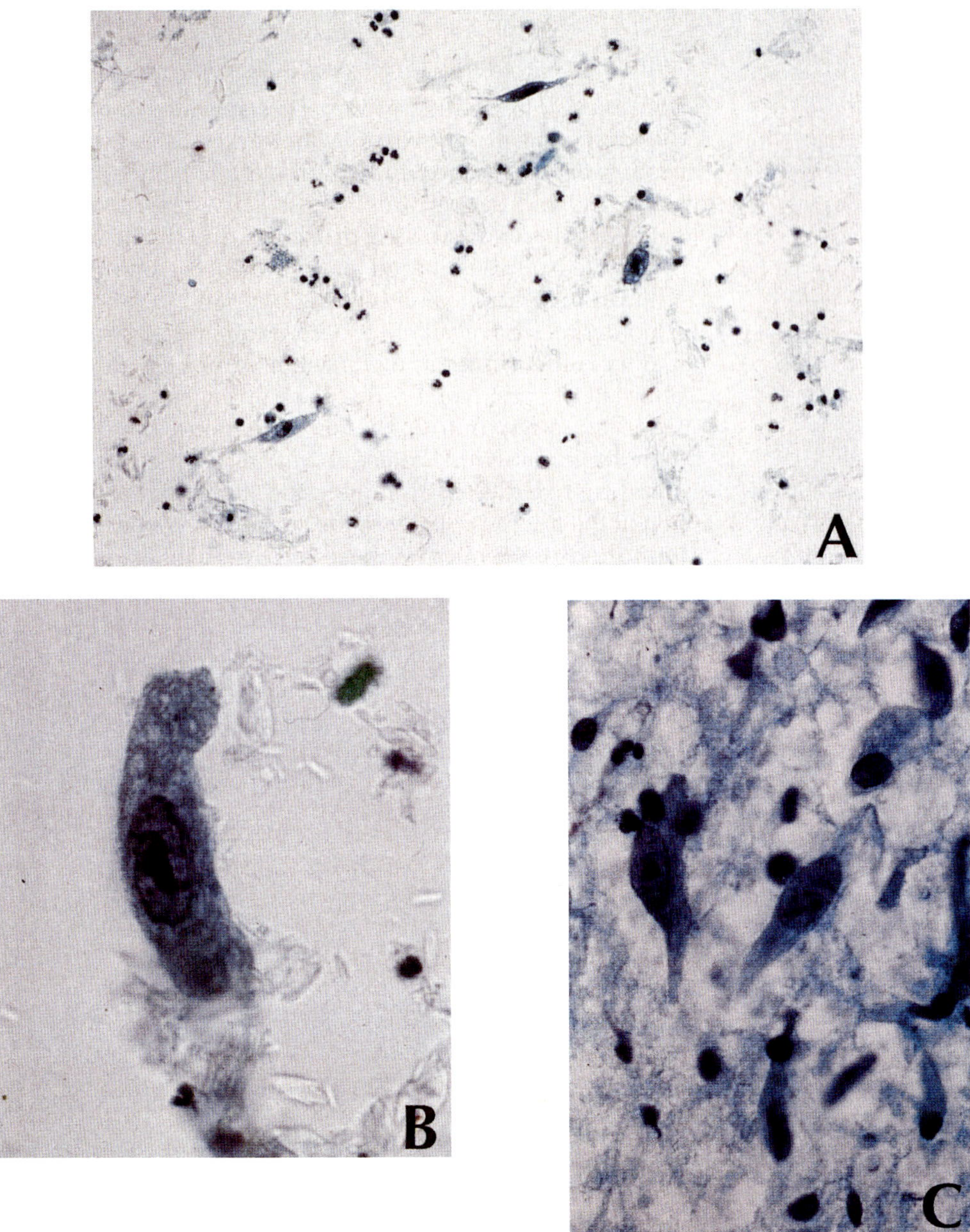

Figure 5–4. *Urine cytology of inflammatory pseudotumor of the bladder.* (*A*) Spindle cells in a background of inflammatory cells are seen. (*B,C*) Large spindle cells with finely vacuolated cytoplasm, a large hyperchromatic nucleus, and a prominent irregular nucleolus mimicking malignancy are identified at a higher magnification. (Papanicolaou stain, (*A*) ×40; (*B*) ×1000; (*C*) ×400). Reproduced with permission from *Acta Cytol* 1999; 43:259.

activity of these lesions with prostate specific antigen detected by immunoperoxidase stain compounds the distinction from prostatic adenocarcinoma (Nowels et al., 1988).

Cytology. Findings include low cellularity, with cohesive groups of columnar cells with bland nuclear morphology. Goblet cells may be admixed. The smear background is clean (Fig. 5–5A–C). Nuclear characteristics

Box 5–9.

Smears show scant cellularity and clusters of columnar or cuboidal epithelium admixed with goblet cells. Cells have bland nuclear characteristics.

Normal or reactive urothelial cells are also present.

Necrosis is absent.

are important for the correct identification of these lesions (Box 5–9).

Differential diagnosis. Primary low-grade bladder adenocarcinoma of the intestinal type may show some similarities to cystitis glandularis and cystitis cystica. A necrotic background, particularly nuclear atypia or anaplasia, is helpful in making the differentiation.

Benign adenomatous polyps of the prostatic urethra may show cytologic characteristics similar to those of cystitis glandularis and cystitis cystica in urine cytology specimens. Clusters of columnar cells with uniform oval nuclei with powdery chromatin are present. Goblet cells and extracellular mucin indicative of intestinal metaplasia may be also identified (Schnadig et al., 2000). Cystoscopic–histologic correlation is necessary for a final diagnosis.

Squamous Metaplasia

Nonkeratinizing squamous epithelium is described in Chapter 2. Keratinizing squamous metaplasia or leukoplakia occurs more often in the bladder than in the renal pelvis and ureter and is associated with inflammation, chronic infection, or squamous carcinoma. Squamous carcinoma is present in 10% to 20% of the cases at diagnosis and has been seen to develop years after the diagnosis of bladder leukoplakia (Reece and Koontz, 1975; Roehrborn et al., 1988). Urethral involvement coexisting with bladder and upper urinary tract involvement has been described (Benson et al., 1984).

Cytology. Squamous metaplastic cells in urine need to be interpreted in light of the clinical setting and the presence of nuclear abnormalities. Nonkeratinized squamous cells without atypia are commonly seen in women of childbearing age and in patients receiving hormonal therapy for prostatic adenocarcinoma. The presence of squamous cells in these patients does not have a preneoplastic connotation. However, squamous metaplastic cells with varying degrees of atypia and keratinization have been found in a significant number of patients with chronic indwelling urinary catheters and schistosomiasis. A high incidence of squamous cell carcinoma of the bladder has been found in these patients (Delnay et al., 1999; Godwin and Hanash, 1984). Serial urine cytology is a useful tool in the surveillance for bladder malignancy in these patients (Stonehill et al., 1997).

In summary, the presence of keratinized squamous metaplastic cells with nuclear atypia should be investigated to rule out a preneoplastic or neoplastic process by means of a tissue biopsy. In contrast, nonkeratinized squamous cells are commonly found in urine samples, and their presence does not carry a preneoplastic connotation (Fig. 5–6A,B; Box 5–10).

Nephrogenic Adenoma

Nephrogenic adenoma is a neoplastic lesion that represents a urothelial response to chronic inflammation, intravesical BCG therapy, irritation, or previous trauma. It is most common in adult males, but may also be seen in children, particularly girls (Heidenreich et al., 1999; Oyama et al., 1998). The bladder is the most common site, but this adenoma has also been described in the urethra, ureters, and renal pelvis (Young, 1997). Renal transplant patients are at high risk of developing nephrogenic adenoma of the bladder (Banyai-Falger et al., 1998).

Cystoscopically, this adenoma may mimic either papillary or flat urothelial carcinoma. Grossly, the lesion is small and papillary or polypoid, usually measuring less than 1 cm, although larger lesions have been described. The diagnosis is rarely suspected on clinical grounds. Patients may have hematuria or urinary frequency, although a significant number are asymptomatic and the lesion is diagnosed incidentally (Peeker et al., 1997).

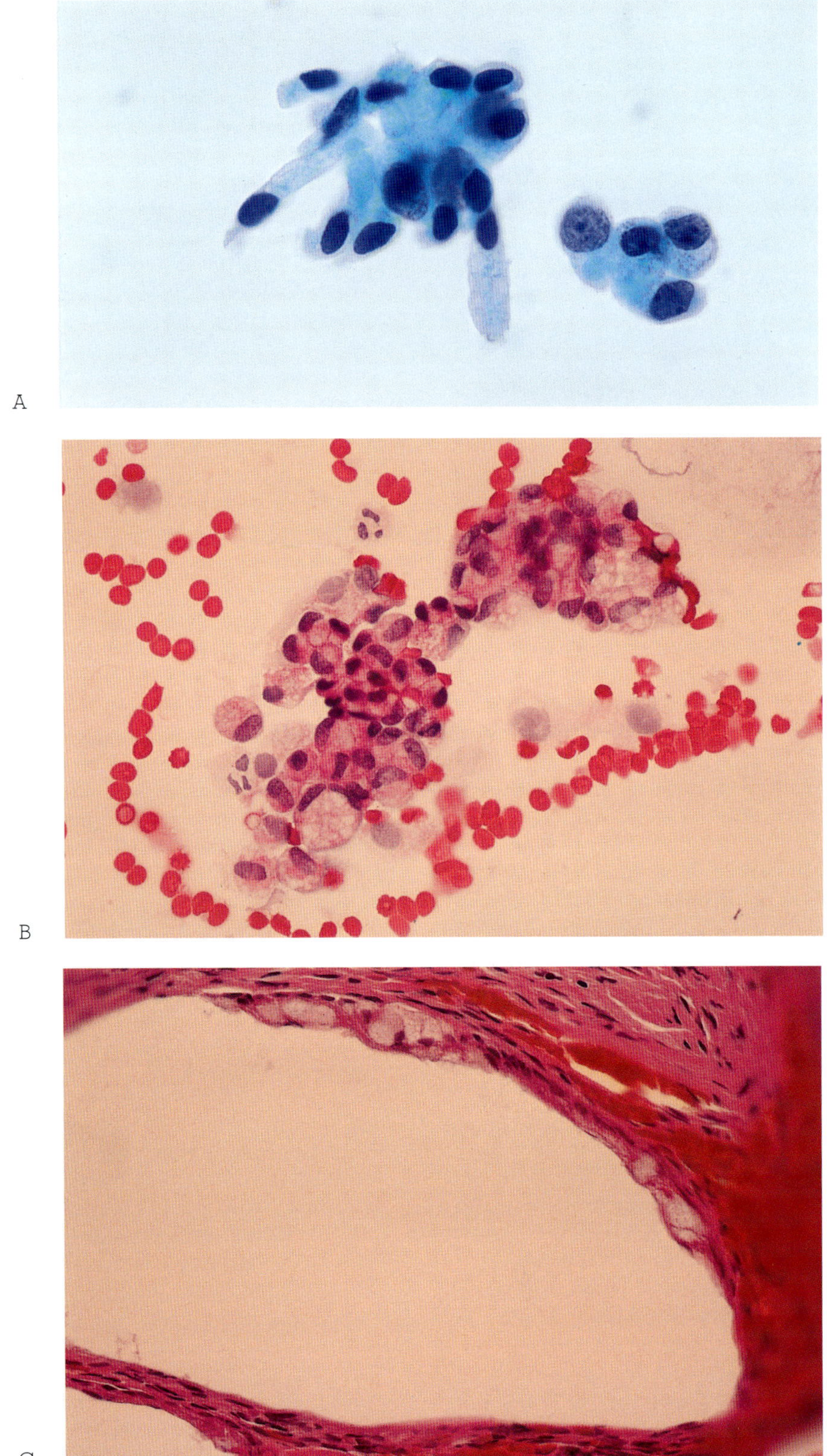

Figure 5–5. *Cystitis cystica and cystitis glandularis.* (*A*) This bladder washing specimen shows a small aggregate of columnar cells with basally located uniform nuclei and vacuolated cytoplasm. (*B*) A cluster of cuboidal intestinal-type cells with finely vacuolated cytoplasm admixed with globlet cells is also noted. (*C*) The bladder biopsy shows a cystically dilated gland lined with goblet cells. (*A*, Papanicolaou stain, ×600; B,C, Hematoxylin & eosin stain (*B*) ×600, (*C*) ×200).

Box 5–10.

Bladder leukoplakia is associated with chronic inflammation, schistosomiasis, and squamous cell carcinoma of the bladder.

The presence of atypical squamous metaplastic cells with nuclear atypia in urine samples carries a pre-neoplastic connotation.

The precancerous potential of this benign lesion is debatable.

Diverse architectural patterns are found in varying combinations, tubular (hence its name because of the resemblance to renal tubules), cystic, and papillary patterns being common, with decreasing frequency. Polypoid, micropapillary, and pseudoinfiltrative growth patterns may be present (Oliva and Young, 1995). The last pattern may be con-

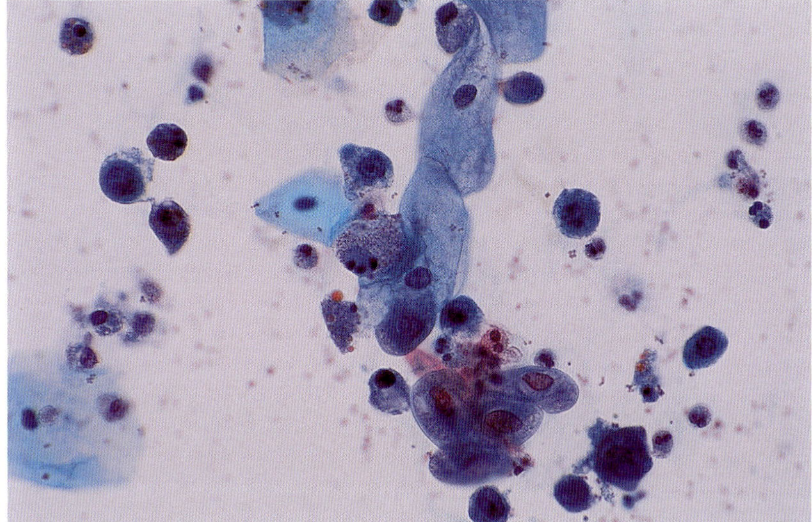

A

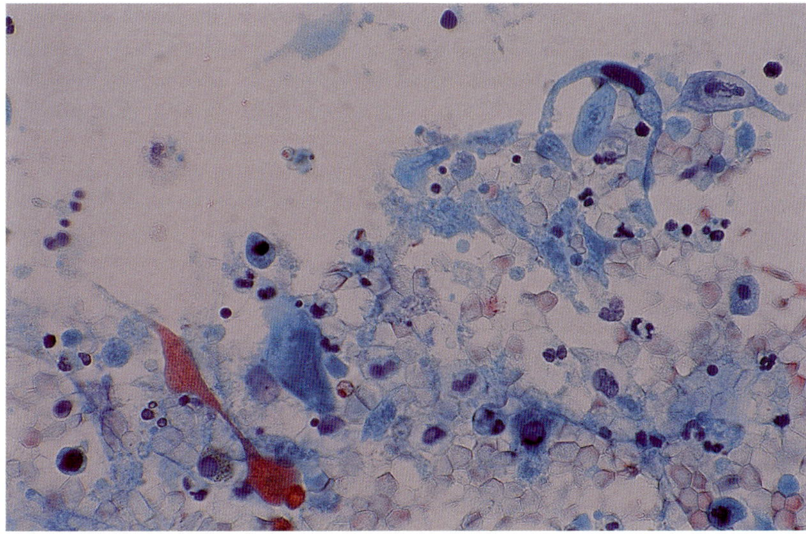

B

Figure 5–6. *Squamous metaplasia.* (*A*) Urothelial cells showing nonkeratinizing squamous metaplasia are seen in a background of cell degeneration and inflammation in a bladder washing of a patient with chronic cystitis. (*B*) Keratinizing squamous metaplasia with cytologic atypia and tissue damage is seen in the bladder washing of a patient with a large bladder stone. These changes regressed after surgical removal of the stone. (Papanicolaou stain, ×600)

Box 5–11.

Nephrogenic adenoma is a metaplastic lesion associated with trauma or inflammation.

Cystoscopically, it resembles urothelial carcinoma.

Cuboidal to columnar hobnail cells arranged in tubules or papillae are characteristic features on tissue sections.

The smear shows clear cells with variable atypia and acute inflammation. Nuclear anaplasia, mitosis, and necrosis are absent.

fused with a malignant process. Although the histologic arrangement is important, the cytologic characteristics are crucial for a correct diagnosis. Cuboidal to low-columnar hobnail cells with rare mitotic figures and a fine chromatin pattern are characteristic of nephrogenic adenoma (Young, 1997).

Cells with clear cytoplasm and a signet-ring appearance may be seen in nephrogenic adenoma and must be interpreted carefully before a diagnosis of clear cell or signet-ring cell adenocarcinoma is made. Histopathologic features that favor clear cell adenocarcinoma over nephrogenic adenoma include a predominance of clear cells, severe cytologic atypia, a high mitotic rate, necrosis, and a high proliferation rate as measured with MIB-1 and p53 (Gilcrease et al., 1998). DNA analysis has shown a diploid phenotype with low proliferative activity, but aneuploidy has been reported (Gaylis et al., 1993; Kontothanassis et al., 1996).

All patients respond well to transurethral resection of the tumor with fulguration of the base. Periodic cystoscopy with urine cytology is used in follow-up of these patients (Box 5–11).

Table 5–2. Cytologic Features of Nephrogenic Adenoma: Review of the Literature

*Samples**	*Cytology*	*Reference*
10 patients (6 VU, 5 BW, 7 CU)	*Pattern*: Loosely or tightly cohesive papillary clusters of up to 15 cells and single cells. Small glandular structures. *Cells*: Columnar to polygonal, with basophilic or amphophilic cytoplasm. Single or multiple small vacuoles indenting the nucleus, some with PMNs. Vacuolar content: clear, slightly basophilic, or, rarely, eosinophilic. Cell-within-cell formations. Signet-ring-cell appearance. *Nucleus*: Round/oval, fine chromatin, regular membrane, small nucleolus. Slightly high NCR, irregular membrane, and notching in a few cases. Coarse chromatin and prominent nucleoli in one case. *Background*: Inflammatory, predominantly neutrophils. *Initial diagnosis*: Negative ($n = 11$), atypical ($n = 5$), suspicious ($n = 1$) *Differential diagnosis*: Low-grade papillary urothelial carcinoma.	Rutgers and Young (1988)
4 patients (8 VU)	*Pattern*: Papillary clusters of variable size, mostly 6–10 vacuolated cells. *Cells*: Columnar with single or multiple cytoplasmic vacuoles, some with PMNs. Signet-ring cells. *Nucleus*: Round, regular membrane, fine chromatin. Irregular membrane, clumped and irregular chromatin, and prominent, round, regular, and eosinophilic nucleolus in some cases. *Background*: Variable inflammation. Occasional RBCs. Degenerating urothelial cells. *Initial diagnosis*: Atypia ($n = 2$), suspicious ($n = 2$). *Differential diagnosis*: Clear cell adenocarcinoma.	Stilmant et al. (1986)
1 patient (1 VU)	*Pattern*: Papillary clusters with peripheral cell palisading. *Cells*: Abundant basophilic cytoplasm. *Nucleus*: Single, round/oval, uniform membrane, finely granular chromatin. No nucleoli. *Background*: Clean, few urothelial cells, occasional neutrophils. *Initial diagnosis*: Not mentioned. *Differential diagnosis*: Low-grade papillary urothelial carcinoma.	Troster et al. (1986)

*All filter preparations.

Abbreviations: VU = voided urine; BW = bladder wash; CU = catheterized urine; NCR = nuclear-cytoplasmic ratio; RBCs = red blood cells; PMNs = polymorphonuclear cells.

Cytology. A summary of cytology findings in 27 urine samples from 15 patients with nephrogenic adenoma from the literature is presented in Table 5–2. Briefly, cuboidal cells showing fine cytoplasmic vacuolization are found in aggregates of varying cohesiveness. Vacuoles do not indent the nuclei. Columnar cells and occasional glandular structures may be seen. The nucleus is round to oval and centrally placed, with a fine chromatin pattern (Fig. 5–7A,B). A small percentage of cases have moderate nuclear atypia. The background shows acute inflammation (Rutgers and Young, 1988; Stilmant et al., 1986; Troster et al., 1986). The most common initial cytologic diagnosis was negative or atypical. Adenocarcinoma of the clear or signet-ring cell type and low-grade urothelial carcinoma were suspected in three cases. No prospective diagnosis of nephrogenic adenoma was rendered.

Differential diagnosis. Reported cases have been mistaken for low-grade urothelial carcinoma and clear cell adenocarcinoma of the bladder. In addition, vacuolated cells of

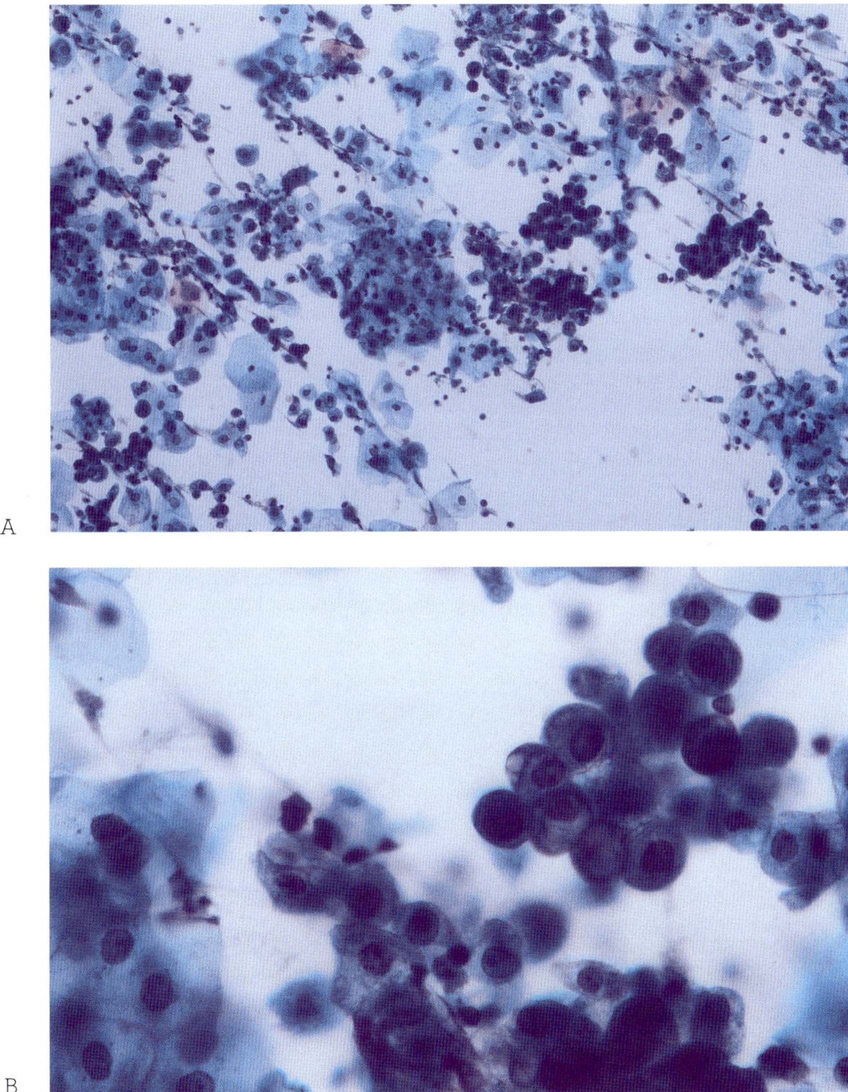

Figure 5–7A,B. *Nephrogenic adenoma of the bladder.* This bladder washing shows an inflammatory pattern with clusters of round cells exhibiting vacuolated cytoplasm, round and uniform nuclei, and small nucleoli. (Papanicolaou stain (*A*) ×200, (*B*) ×600)

Table 5–3. Differential Diagnosis of Nephrogenic Adenoma

Entity	*Differential Features*
Clear cell adenocarcinoma	High mitotic rate, necrosis, nuclear anaplasia. Clinical features: women in the fifth decade or older.
Prostate adenocarcinoma	Acini. Large cuboidal cells with thin, vacuolated cytoplasm and high NCR. Necrosis. Clinical features: men, sixth decade or older. High PSA.
Renal cell carcinoma	Bloody smears with necrosis. Large cells, vacuolated cytoplasm, nuclear hyperchromasia, large nucleolus. Clinical feature: renal mass.
Papillary urothelial carcinoma	Papillary clusters without vacuoles in low-grade lesions. Papillary clusters with nuclear anaplasia and necrosis in high-grade lesons.
Chemotherapy changes	Clinical history. Large cells. Nuclear atypia.
Instrumentation effect	Papillary clusters. Light gray cytoplasm. Cytoplasmic vacuoles without nuclear indentation.

Abbreviations: NCR = nuclear-cytoplasmic ratio; PSA = prostate-specific antigen.

nephrogenic adenoma must be distinguished from those of degeneration, instrumentation effect, chemotherapy and radiotherapy, and other malignancies (e.g., renal, prostatic). A clinical history and evaluation of nuclear anaplasia are helpful. The main distinguishing features are summarized in Table 5.3.

Endometriosis

The urinary bladder is the most common site of endometriosis affecting the urinary tract. Endometriosis is the most frequent proliferation of mullerian epithelium found in the urinary bladder and may be present either in isolation or in conjunction with endocervical (endocervicosis) and/or tubal (endosalpingiosis) epithelium (Young, 1997). *Mullerianosis* is the term applied to the presence of two or more of these patterns in the same lesion.

The most reliable diagnostic signs of endometriosis are the catamenial onset or worsening of bladder symptoms (frequency, urgency, dysuria, or tenesmus) and coexistence with dysmenorrhea. Hematuria is uncommon, and many patients are asymptomatic. This is not surprising because endometriosis rarely infiltrates and ulcerates the mucosal layer of the bladder.

The typical endometriotic bluish nodules are visible by cystoscopy in only a minority of patients (Vercellini et al., 1996). The preoperative diagnosis requires a high index of suspicion based on the clinical setting, supported by radiographic studies that help define the extent of bladder involvement (Fedele et al., 1997). The pathologic diagnosis is based on the finding of endometrial glands, as seen in endometriosis involving other organs (Box 5–12).

Box 5–12.

Endometriosis, endosalpingiosis (tubal epithelium), and endocervicosis (endocervical epithelium) are proliferations of mullerian epithelium that may be present in the bladder.

Patients with bladder endometriosis may complain of bladder symptoms coexistent with dysmenorrhea, but hematuria is rare.

The smear shows single cells and clusters of small cells with variable cellular molding, hyperchromatic nuclei, and a high nuclear-cytoplasmic ratio. Red blood cells and lysed blood are also present.

Cytology. Reports of the urine cytology of bladder endometriosis are rare. Schneider et al. (1980) described one case diagnosed as such by urine cytology in a patient with a known history of endometriosis. Bohlmeyer and Shroyer (1996) described another case in a 32-year-old woman with hematuria diagnosed by cytology as "suspicious for papillary urothelial carcinoma."

In the cases described, smears show three-dimensional papillary clusters of variable cellularity and varying degrees of cellular molding. Cells are cuboidal, are smaller than basal-intermediate urothelial cells, and a show scant cytoplasm with indistinct cytoplasmic boundaries, a high nuclear-cytoplasmic ratio, hyperchromatic nuclei, and inconspicuous nucleoli. Anisonucleosis is not a feature, and nuclear contours, although irregular, are smooth (Bohlmeyer and Shroyer, 1996; Schneider et al., 1980). Red blood cells and varying degrees of cellular degeneration are seen in the background.

Differential diagnosis. The differential diagnosis includes small cell carcinoma, either primary or metastatic, and other small blue cell tumors such as hematolymphoid malignancies, metastatic neuroblastoma, and sarcoma. The association of hematuria in a woman of childbearing age with irritative voiding symptoms at the time of menstruation, associated with the finding of small blue cells with regular nuclear contours in the urine, should alert the cytopathologist to entertain the possibility of endometriosis. However, it should be stressed that without proper clinical information, it is difficult to make a cytologic diagnosis of endometriosis.

Endocervicosis and Endosalpingiosis

It is the peritoneum where commonly benign secondary mullerian proliferations, such as, endometriosis, endosalpingiosis, and endocervicosis occur (Lauchlan, 1994). These lesions have been found in the urinary bladder, and similar designations have been adopted (Clement and Young, 1992). Histologically, endocervicosis, the mucinous analogue of endometriosis, shows benign-appearing or mildly atypical endocervical-type glands, some of which are cystically dilated. Endosalpingiosis is defined as an ectopic tubal epithelium. The glands are lined predominantly by tubal-type epithelium (including ciliated, intercalated, and peg cells). Mullerianosis is defined as the presence of two or more of the epithelia described above (Young, 1997). No psammoma bodies have been described in these lesions.

Women have irritative voiding symptoms, hematuria, and suprapubic pain, alone or in combination. Grossly, these lesions are primarily intramural or submucosal, but they may protrude into the bladder lumen and replace the mucosal epithelium focally (Rodriguez and Alfert, 1997; Young and Clement, 1996).

Awareness of the presence of these lesions in the bladder, and attention to their typical histologic features, should facilitate their distinction from adenocarcinoma. Mullerianosis can be distinguished from adenocarcinoma and various nonneoplastic glandular lesions of the urinary bladder on the basis of a variety of architectural and cytologic features (Young and Clement, 1996).

Cytology. To my knowledge, no cases of endocervicosis or endosalpingiosis have been described in urine cytology. The scarcity of the lesions and the intramural localization, with rare and focal epithelial involvement, account for its rarity in urine samples. However, endosalpingiosis, as described in peritoneal washings, may theoretically be detected in bladder washings.

In fluid cytology, endosalpingiosis usually is characterized by dense papillary structures composed of bland cuboidal epithelial cells surrounding psammoma bodies (Sneige and Fanning, 1992). Distinct ciliated cell borders have also been described (Carlson et al., 1986). In contrast to the finding of peritoneal or pelvic endosalpingiosis, psammoma bodies seem to be rare in endosalpingiosis of the bladder. The identification of papillary clusters with ciliated cells may suggest the diagnosis of endosalpingiosis in urine samples.

URETHRA

Malacoplakia has been described in the urethra. The histologic and cytologic findings are identical to those described in the urinary bladder (Karaiossifidi and Kouri, 1992).

Nephrogenic adenoma of the urethra is uncommon, and the histologic characteristics are identical to those described in the bladder. A cytologic report of this inflammatory lesion as seen in bladder washings emphasizes the potential misdiagnosis as urothelial carcinoma (Henke et al., 2000). Cytologic findings are similar to those described for the urinary bladder (Fig. 5–8A,B) (Rutgers and Young, 1988; Stilmant et al., 1986; Troster et al., 1986).

Urethral caruncle is an inflammatory polypoid lesion of the distal urethra in women (Bolduan and Farah, 1981). The prominence of atypical round stromal cells appears to be a feature of some urethral caruncles and should not be misinterpreted as a neoplastic process (Young et al., 1996). Infections, extragenital malignancies, and reactive processes masquerading as urethral caruncles have been reported (Atalay et al., 1998; Balogh and O'Hara, 1986; Hammadeh et al., 1996; Indudhara et al., 1992; Lopez et al., 1993). Therefore, careful follow-up and/or biopsy are usually indicated in these cases.

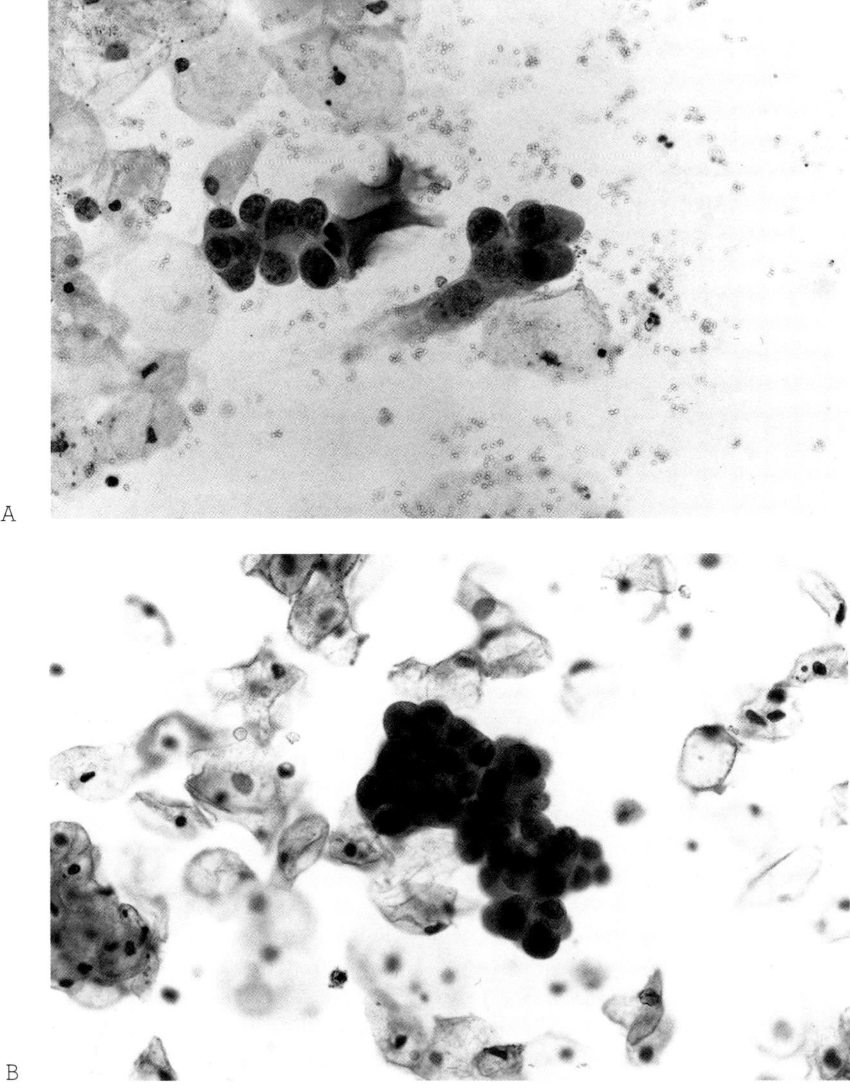

Figure 5–8A,B. *Nephrogenic adenoma of the urethra.* Pseudopapillary aggregates are seen in this bladder washing. Mild hyperchomasia, an increased nuclear-cytoplasmic ratio, and dense cytoplasm are identified, resembling low-grade papillary urothelial carcinoma. (Papanicolaou stain, ×600). Courtesy of Dr. Andrew C. Henke, The University of Iowa Hospitals and Clinics, Department of Pathology, Iowa City, IA.

UROTHELIAL CHANGES SECONDARY TO THERAPY

Table 5–4 presents a classification of therapeutic agents capable of producing cellular changes in the normal urothelium. Some of these agents are used to treat urothelial carcinoma. Cellular changes elicited in benign urothelial cells can mimic to perfection those of high-grade urothelial carcinoma, causing significant diagnostic difficulties in the interpretation of urine cytology specimens.

The standard approach to the management of patients diagnosed with flat carcinoma in situ (Tis), high-grade papillary urothelial neoplasms confined to the mucosa (Ta), and papillary neoplasms that infiltrate the lamina propria (T_1) is intravesical immunotherapy with BCG. Intravesical chemotherapy, for example, with mitomycin C, thiotepa, and doxorubicin, may be used to prolong the interval to recurrence or to prevent recurrences of Tis, Ta, and T_1 bladder cancer. Surveillance is indicated for low-grade Ta (Droller, 1998; Smith et al., 1999). Intravesical therapy is advantageous because the instilled agent covers the entire mucosal surface of the bladder and includes neoplastic and preneoplastic growths. It should be noted that normal urothelium is exposed to the therapeutic agent and exhibits cellular changes as well.

The aim of the clinician in the evaluation of these patients is to monitor the progression or recurrence of the primary tumor and to exclude an associated urothelial carcinoma in situ. The aim of the pathologist is to make an accurate diagnosis of these high-grade tumors and to differentiate them from changes present in nonneoplastic cells secondary to the effect of therapy.

Therapy-associated cytologic abnormalities remain the most common cause of false-positive reports (Lopez-Beltran, 1999; Maier et al., 1995). This is due in part to the pathologist's lack of awareness of the patient's clinical history and of the cystoscopic findings at the time of urine cytology examination. Thus, careful evaluation of cellular changes and correlation with the clinical history are of utmost importance for an accurate interpretation of the urothelial changes (Box 5–13).

Table 5–4. Common Therapeutic Agents Producing Urothelial Changes

Agents
Intravesical immunotherapy
Bacillus Calmette-Guerin (BCG)
Interferon-α
Intravesical chemotherapy
Thiotepa
Mitomycin C
Doxorubicin
Systemic chemotherapy
Cyclophosphamide*
Busulfan*
External radiotherapy

*Used in the treatment of nonurologic malignancies.

Box 5–13.

Therapy-associated urothelial cell abnormalities are the most common cause of false-positive cancer diagnoses in urine cytology.

Awareness of the clinical history at the time of cytology interpretation is crucial for diagnostic accuracy.

Intravesical Immunotherapy with Bacillus Calmette-Guerin

After transurethral resection, adjuvant intravesical immunotherapy with BCG remains the most effective treatment and prophylaxis for noninvasive bladder cancer, particularly

Box 5–14.

Intravesical BCG is the most effective treatment for patients with noninvasive urothelial carcinoma, and urine cytology remains the most important tool for the follow-up of these tumors.

The precise antitumor mechanism of BCG is unknown.

Histologic changes include degeneration of the urothelium and noncaseating granulomatous inflammation of the bladder wall.

The smear shows reactive urothelial cells and variable numbers of chronic inflammatory cells. Granulomas and single histiocytes may be present in urine specimens.

carcinoma in situ (Dalbagni and Herr, 2000; Droller, 1998). Postoperative cytology is a significant prognostic indicator after transurethral resection and intravesical BCG treatment (Badalament et al., 1987; Baltaci et al., 1996; Schwalb et al., 1994).

Bacillus Calmette-Guerin, a vaccine containing freeze-dried, live attenuated tuberculosis organisms, is supplied in ampules containing approximately 5 to 8×10^8 colony-forming units. Several strains have been used, including Montreal, Tice, and Connaught. This modality of treatment presumably stimulates an immune response against the cancer by prompting the immune system to destroy the tumor cells (Soloway, 1987). Immunomodulatory substances including interleukins, tumor necrosis factor, and interferon-γ are believed to play a role, but the precise antitumor mechanism remains unknown (Morales and Nickel, 1992).

When BCG is instilled in the bladder, it produces an intense inflammatory response. Cystitis with dysuria and urinary frequency occurs in 90% of patients and hematuria in 30%. Fever, chills, lethargy, and malaise may be seen. Occasionally, major extravesical complications and death have been reported, particularly when BCG is administered in the presence of urothelial trauma or hematuria (Lamm, 1992).

Pathologic bladder changes associated with BCG administration include urothelial denudation, eosinophilic cystitis, bladder wall noncaseating granulomatous inflammation, and urothelial-cell degeneration (Lopez-Beltran, 1999). The presence of granulomas in tissue biopsy specimens ranges from 14% to 78% of samples (Bassi et al., 1992; Betz et al., 1993; Bhan et al., 1998).

Cytology. Cytologic changes after BCG therapy are nonspecific (Betz et al., 1993). At the start of topical therapy with BCG, an increase in the number of inflammatory cells is seen. Urothelial cells may show enlarged, hyperchromatic nuclei with prominent nucleoli, anisokaryosis, and a distorted nuclear-cytoplasmic ratio. These changes may last for variable periods of time (Mack and Frick, 1994).

Granulomatous inflammation is noted in some but not all urine cytology specimens. Granulomas are present in 1.3% to 76% of specimens (Betz et al., 1993; Bhan et al., 1998). Urine obtained at different time intervals after intravesical BCG therapy and by different sampling techniques may account for the difference in the yield of granulomas. In my experience, in urine specimens obtained by instrumentation, granulomas are rarely seen.

Granulomas can be recognized by the loosely cohesive nature of the epithelioid histiocytes, finely vacuolated or thin cytoplasm, round/oval or reniform nuclei, indistinct cytoplasmic borders, and uniform powdery chromatin with thin nuclear contours.

Degenerated urothelial cells, debris, free histiocytes, and multinucleated giant cells may be present. Free histiocytes have been reported in 64% of patients and multinucleated histiocytic giant cells in 56% (Fig. 5–9A–D) (Betz et al., 1993). After termination of BCG therapy, it takes at least four months for cytologic normalization to occur (Adolphs et al., 1984) (Box 5–14).

Differential diagnosis. The cytologic characteristics of bladder BCG changes are indistinguishable from those of tuberculous cystitis; the distinction is based on the clinical history. Misinterpretation of granulomatous inflammation as malignant is rare (Betz et al., 1993). Features distinguishing epithelial changes resulting from BCG treatment and high-grade superficial cancer include a preserved nuclear-cytoplasmic ratio, smooth nuclear contours, and the lack of nuclear pleomorphism, nucleoli, or cytomegaly (Lage et al., 1986). However, atypical urothelial cells, particularly at the start of therapy, have been reported in rare instances (Mack and Frick, 1994).

When cytology results are uncertain, the recurrence of a carcinoma in situ under maintenance therapy with BCG must be proved by biopsy (Mack and Frick, 1994).

Intravesical Immunotherapy with Interferon-α

Interferon-α, an immune system stimulating agent, is active in superficial bladder cancer, but its role in the treatment of this disease remains to be established (Nseyo and Lamm, 1996). Interferon-α mediates an immune response through antiviral, antiproliferative, and immunoregulatory activities (Nseyo and Lamm, 1996). Clinical experience suggests that recombinant interferon-α may be useful as second-line therapy after failure of BCG or chemotherapy, and it may have synergistic effects when combined with chemotherapy or BCG (Belldegrun et al., 1998).

Interferon-α elicits an edematous reaction of the lamina propria and mild perivascular inflammation with neutrophils and eosinophils (Lopez-Beltran, 1999). Changes in the urothelium are minimal (Polat et al., 1996).

Intravesical Chemotherapy

Intravesical chemotherapy with mitomycin C, thiotepa, and doxorubicin is commonly used for the treatment of noninvasive blad-

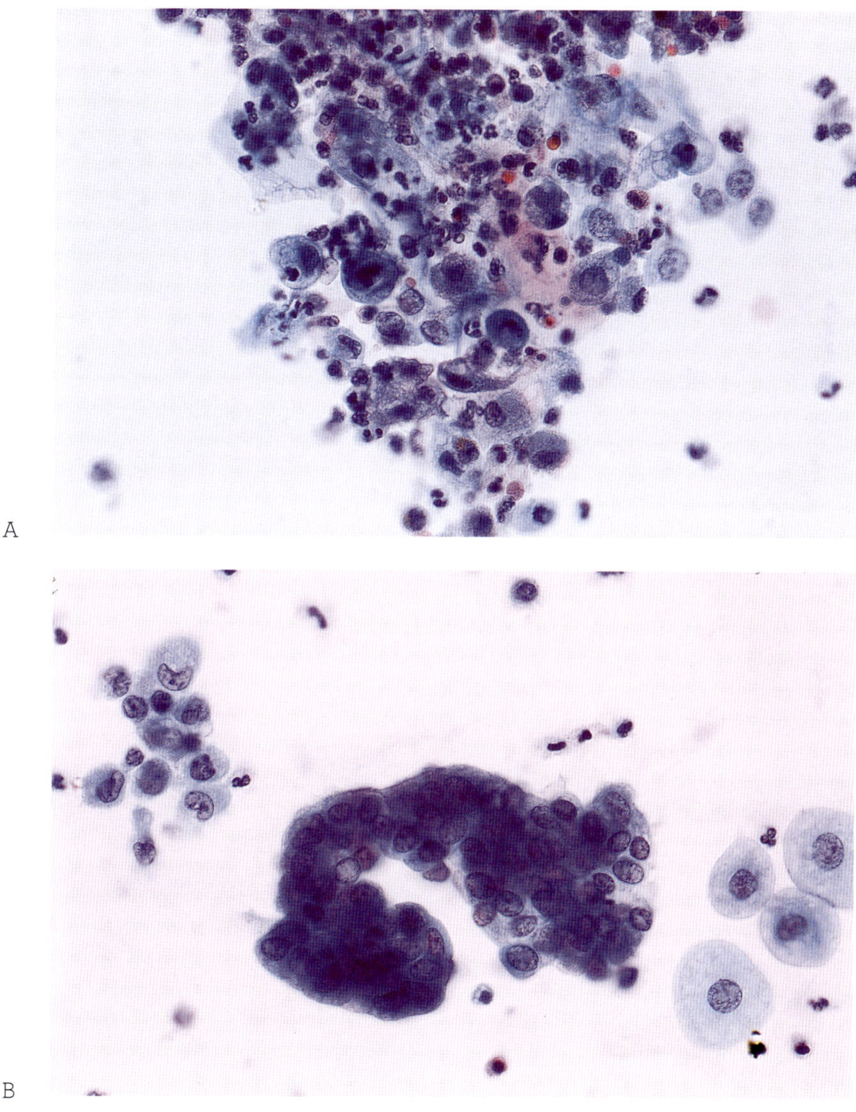

Figure 5–9. *Cytologic changes associated with intravesical BCG therapy.* Bladder washings show a cluster of degenerated urothelial cells admixed with acute inflammation (*A*). Umbrella cells and one small aggregate of histiocytes surround a cluster of reactive urothelial cells (*B*). Granulomas composed of epithelioid histiocytes with reniform nuclei and vacuolated cytoplasm are also present (*C,D*), as well as a single multinucleated Langhan's-type giant cell (*D*). (Papanicolaou stain, ×600)

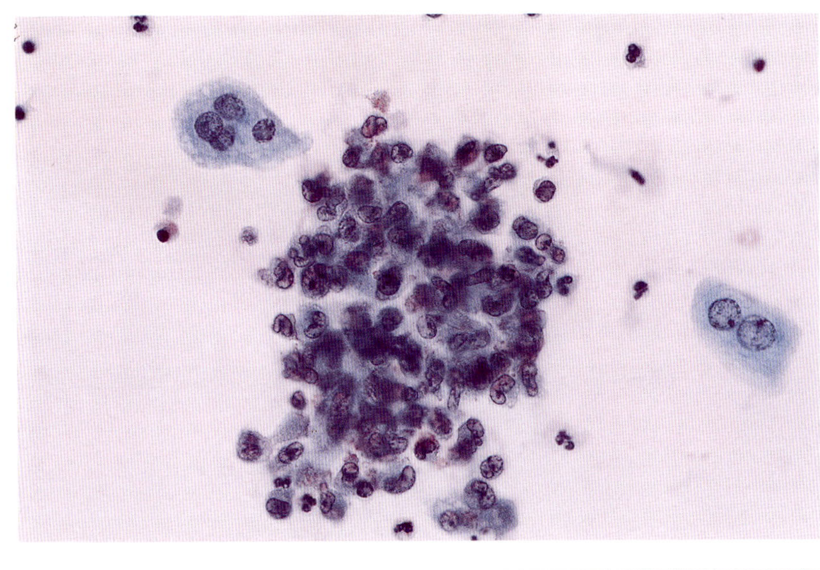
C

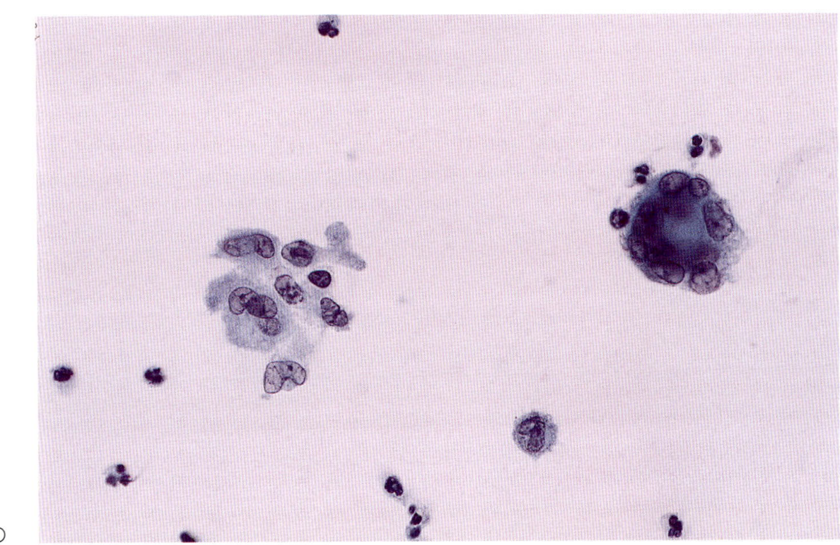
D

Figure 5–9. (*Continued*)

der urothelial carcinoma. These substances may be used to prolong the interval to recurrence or to prevent recurrences (Droller, 1998).

Thiotepa is an alkylating agent that is chemically related to nitrogen mustard, which causes inhibition of nucleic acid synthesis (Richie, 1992). The principal side effect of thiotepa is myelosuppression, which is related to bladder absorption due to its low molecular weight (Soloway, 1987). Chemical cystitis and bladder contracture can occur in rare cases. The risk of acute leukemia may be increased after long-term use (Thrasher and Crawford, 1992).

Mitomycin C is an antitumor antibiotic that results in the production of an alkylating agent. Although its mechanism of action has not been totally clarified, some evidence suggests that it cross-links DNA, resulting in inhibition of DNA synthesis (Dalbagni and Herr, 2000; Richie, 1992). Myelosuppression is rare, and systemic side effects are uncommon because of the limited absorption related to its high molecular weight. Occasionally, a palmar, genital, or diffuse rash

Box 5–15.

Mitomycin C, thiotepa, and doxorubicin are used intravesically for the treatment of noninvasive urothelial carcinoma.

Necrosis, urothelial denudation, and inflammation are commonly seen on tissue sections.

Cellular changes occur in the nonneoplastic urothelium and are almost identical to those of high-grade urothelial carcinoma.

The smear shows variable cellular and nuclear enlargement, smudged chromatin, and prominent nucleoli. Hyperchromasia, coarse chromatin, and a high nuclear-cytoplasmic ratio are less prominent.

Cellular changes associated with mitomycin C are more intense than those induced by thiotepa, and are more remarkable late in the course and after the end of therapy.

occurs, possibly secondary to a hypersensitivity reaction (Thrasher and Crawford, 1992). The most likely side effect of intravesical use is cystitis.

In contrast to cyclophosphamide, mitomycin C and thiotepa are not inhibitors of DNA replication and do not produce the marked cellular atypia seen after cyclophosphamide therapy. The two drugs produce identical histologic and cytologic changes.

Necrosis, denudation, and acute inflammation are the first responses to the administration of topical mitomycin C or thiotepa. Histologically, urothelial cell changes are more evident in the superficial cell layer than in the basal-intermediate cell layer. It is not unusual to find a single-cell layer of markedly atypical urothelial cells similar to the denuded pattern of carcinoma in situ. The presence of an edematous lamina propria and atypical stromal cells provides helpful histologic tips for differentiation from the effects of chemotherapy, particularly those associated with mitomycin C (Lopez-Beltran, 1999). Similar changes have been described in other organs (Salomao et al., 1999). These changes tend to subside after discontinuation of the drug (Box 5–15).

Cytology. Urothelial cell changes induced by the intravesical administration of mitomycin C or thiotepa are not specific and occur predominantly in the nonneoplastic urothelium. Studies in rats have shown that the cytologic changes associated with mitomycin C occur more frequently and are more intense that those induced by thiotepa (Friedman et al., 1991).

Urothelial cell changes associated with thiotepa administration consist of slight to moderate nuclear enlargement, with slight hyperchromasia and smudged chromatin lacking a sharply detailed pattern. The round or oval nuclei have smooth or irregular contours and contain multiple small nucleoli. A degenerated and vacuolated cytoplasm is seen, and the nuclear-cytoplasmic ratio is low. Large, multinucleated cells may be present (Droller and Erozan, 1985).

Urothelial cell changes associated with intravesical administration of mitomycin C include foamy cytoplasm, nuclear and cytoplasmic enlargement, multiple prominent nucleoli, anisokaryosis, and occasional giant nuclei. Urine specimens are highly cellular and contain acute inflammatory cells (Fig. 5–10A,B). Such cytologic changes are particularly remarkable late in the course and after the end of therapy, and they may confuse the cytologic interpretation. Notorious cytologic changes are observed starting four months after the beginning of one-year intravesical therapy with mitomycin C and persist for at least 18 months after its completion (Koshikawa et al., 1989). Hyperchromatic nuclei, a coarse chromatin pattern, and a high nuclear-cytoplasmic ratio are the most important cytologic criteria of malignancy in the presence of cancer relapse. These features are uncommon in relapse-free, mitomycin C–treated patients (Koshikawa et al., 1989).

It has been suggested that the morphologic changes of malignancy, whether low- or high-grade, are not altered by these forms of therapy and that the diagnosis of malignancy can be accepted with assurance (Cant et al., 1986; Droller and Erozan, 1985; Koshikawa et al., 1989; Lopez-Beltran, 1999; Murphy, 1990). However, analysis of the cytologic diagnosis in larger series indicates that the effects of intravesical administration of mitomycin C and thiotepa induce a variety of cytologic effects (toxic and/or meta-

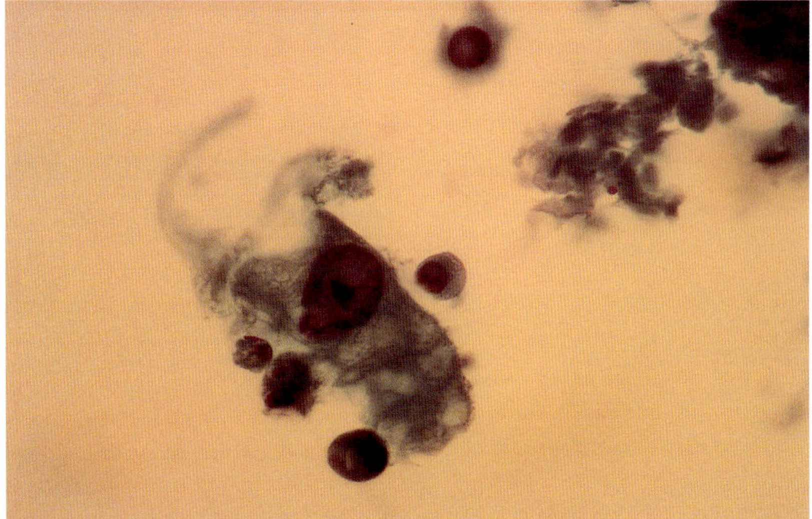

A

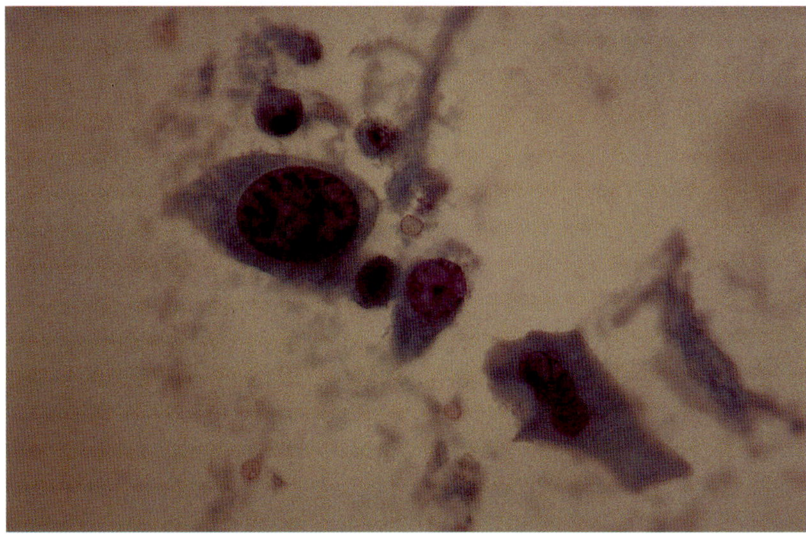

B

Figure 5–10. *Cytologic changes associated with intravesical mitomycin C.* (*A*) This bladder washing shows large cells with a preserved nuclear-cytoplasmic ratio, vacuolated cytoplasm, a large nucleus with degenerated chromatin, and a large, irregular nucleolus. (*B*) In contrast, these large cells have a round, uniform nucleus, coarse chromatin, a high nuclear-cytoplasmic ratio, and vacuolated cytoplasm mimicking malignancy. (Papanicolaou stain, ×1000)

bolic) that may lead to cytologic misinterpretations on follow-up (Borgmann et al., 1993; Koshikawa et al., 1989). In these cases, the clinical history is the most relevant piece of information. The nuclear DNA is usually euploid (Borgmann et al., 1993; Koshikawa et al., 1989).

Differential diagnosis. These forms of therapy do not alter the cytomorphology of low- or high-grade carcinoma. Thus, distinction of the effect of therapy from carcinoma may not be a significant problem. However, changes may be detected early via urinary cytology by an experienced cytopathologist, but only in the light of clinical information. Cytologic changes are not specific and do not differ from those produced by chronic inflammation or instrumentation (Lopez-Beltran, 1999).

Systemic Chemotherapy

Chemotherapeutic agents (e.g., alkylating agents such as cyclophosphamide or busulfan) used in the treatment of various nonurologic tumors and nonneoplastic diseases, as well as in bone marrow transplant patients, may injure the urothelium and produce a severe regenerative atypia.

Long-term oral cyclophosphamide therapy for immunosuppression or treatment for hematologic malignancies is associated with substantial urotoxicity, including the development of urothelial carcinoma of the urinary bladder, an uncommon but not rare event (Fernandes et al., 1996; Kaldor et al., 1995). The estimated incidence of bladder cancer after the first exposure to cyclophosphamide is 5% at 10 years and 16% at 15 years (Talar-Williams et al., 1996). It has been suggested that nonglomerular hematuria, a frequent manifestation of cyclophosphamide-induced cystitis, identifies a subgroup of patients at high risk for the development of bladder cancer (Talar-Williams et al., 1996).

Urothelial carcinoma of the bladder is the most common type associated with cyclophosphamide therapy, although malignancies of other types and in the renal pelvis and ureters have been reported (Fernandes et al., 1996; McDougal et al., 1981; Pedersen-Bjergaard et al., 1995; Schiff et al., 1982; Siddiqui et al., 1996). There is no consensus as to which major cyclophosphamide metabolite (e.g., acrolein or phosphoramide mustard) leads to bladder carcinogenesis (Khan et al., 1998).

Box 5–16.

Alkylating agents produce severe regenerative urothelial cell atypia.

Long-term cyclophosphamide therapy is associated with the development of urothelial carcinoma.

The smear shows cytologic changes similar to those of high-grade urothelial carcinoma. Red blood cells, neutrophils, and cellular debris are also present.

Cytology. Busulfan and cyclophosphamide are toxic to the urinary bladder and can cause hemorrhagic cystitis. The changes produced in the urothelium mimic those seen in high-grade urothelial carcinoma. The cytologic changes observed most commonly are cellular and nuclear enlargement, anisocytosis, and hyperchromasia with an irregular chromatin distribution (Fig. 5–11A,B). The nuclei are round, eccentrically placed, and have regular borders. Nucleoli, when visible, are single or double and occasionally large and distorted, with irregular edges. High cellularity, scattered neutrophils, numerous red blood cells, and cellular debris are commonly found. Cytoplasmic microvacuolization and nuclear pyknosis and karyorrhexis may be present in some cases, but mitoses are usually absent (Lopez-Beltran, 1999; Stella et al., 1990).

A comparative morphometric study has been carried out showing differences in the cell area, nuclear area, nuclear perimeter, and nuclear-cytoplasmic ratio, especially between low-grade urothelial carcinoma and urothelial toxicity. The differences from high-grade urothelial carcinoma were subtle (Stella et al., 1992) (Box 5–16).

Differential diagnosis. Changes are similar to those described in high-grade bladder cancer, due to radiation therapy, and due to the effect of polyomavirus (Koss et al., 1965; Stella et al., 1990). In contrast to the effect of chemotherapy, high-grade urothelial carcinoma usually exhibits less cellular but more nuclear pleomorphism, prominent nucleoli in most cells, and mitoses. Human polyomavirus infection produces similar cell and nuclear alterations, but in contrast to systemic chemotherapy, polyomavirus exhibits nuclei that are smooth or homogeneous, with a thin, clear peripheral periinclusion rim. A history of cyclophosphamide or busulfan exposure is useful so that a misdiagnosis can be avoided.

Regional Radiotherapy

Radiotherapy used for primary invasive urothelial carcinoma or extravesical pelvic disease produces various bladder abnormalities. Although epithelial changes are most prominent in early disease and stromal changes in late disease, there is a con-

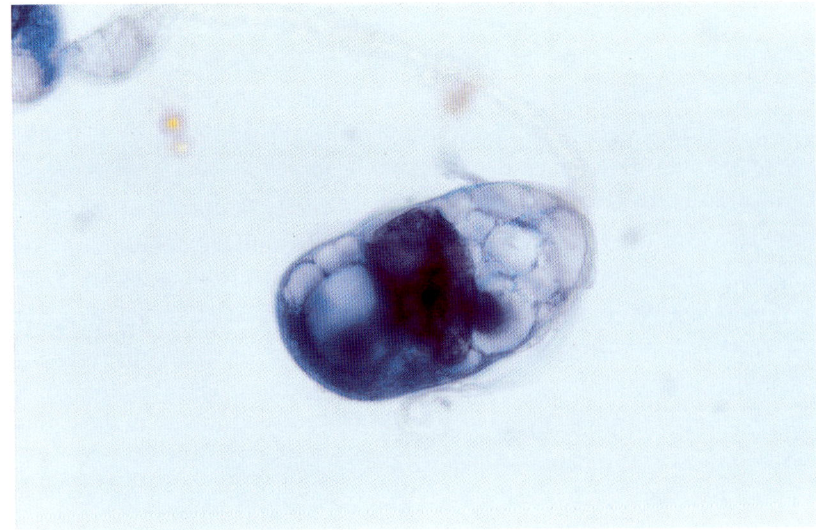

A

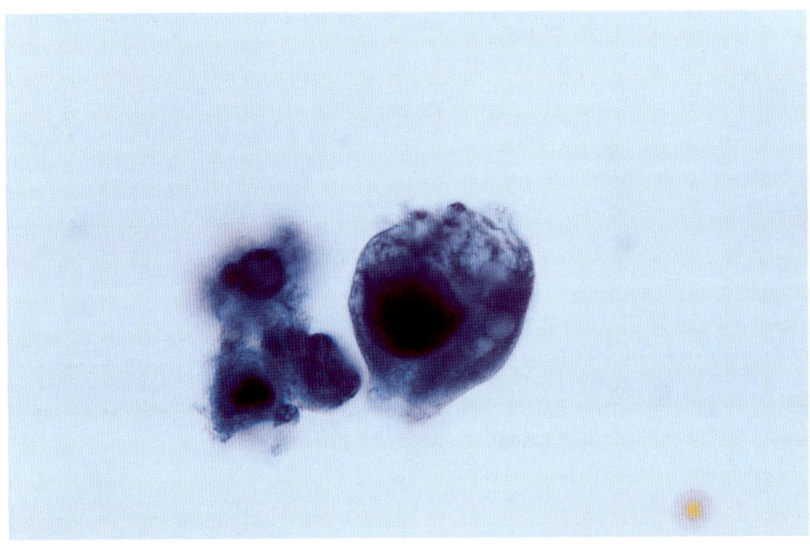

B

Figure 5–11. *Cytologic changes associated with systemic chemotherapy.* (*A,B*) Bladder washing obtained from a 35-year-old man, after bone marrow transplantation for multiple myeloma who developed hemorrhagic cystitis. Urothelial cells show marked enlargement, pronounced vacuolated cytoplasm, and hyperchromatic nuclei. (Papanicolaou stain, ×1000)

tinuing spectrum of epithelial damage that persists many years after the initial irradiation (Suresh et al., 1993). Changes usually develop within four to six weeks after therapy. Radiotherapy may initiate vascular damage, with consequent ischemia of the neoplastic and nonneoplastic epithelium. Radiation therapy acts on neoplastic cells by creating free radicals, binding DNA, and causing gene mutation (Lopez-Beltran, 1999).

Histologically, in acute radiation exposure (less than six months), the urothelium may be extensively denuded and show marked cytologic atypia. The lamina propria exhibits edema and plump, sometimes multinucleated stromal and endothelial cells, capillary congestion, and often thrombosis. Changes due to chronic irradiation (more than two years) include a thickened bladder wall with luminal stricture, extensive fibrosis, and blood vessel wall hyalinization. In general,

Box 5–17.

Urothelial cell changes develop four to six weeks after radiation therapy and persist for many years.

In acute radiation exposure, smears show marked cytologic atypia and tissue damage.

The smear shows large urothelial cells with a large nucleus, vacuolated cytoplasm, and an unchanged nuclear-cytoplasmic ratio.

the histologic urothelial and stromal abnormalities are sufficient to lead one to suspect the diagnosis (Lopez-Beltran, 1999).

Cytology. Cytologic manifestations depend on the severity of the injury and the duration of therapy. In general, the smears contain cellular debris, degenerated or necrotic urothelial cells, and blood and exhibit acute inflammation (Lopez-Beltran, 1999). Urothelial cells are large, with an overall unchanged nuclear-cytoplasmic ratio, vacuolated cytoplasm, and polychromasia. The nuclei are hyperchromatic and pleomorphic, showing degenerated and smudged chromatin, but are not indented by the cytoplasmic vacuoles (Lopez-Beltran, 1999) (Box 5–17).

Two basic conditions for avoiding a false-positive cytologic diagnosis are knowledge of the cytomorphologic changes associated with radiation therapy and a clear statement of previous radiotherapy on the request slip (Esposti et al., 1970).

Photodynamic Therapy

Intravesical photodynamic therapy is being used in the treatment of recurrent superficial bladder cancer (McClellan, 2000). It is based on photochemical sensitization of the urothelium, with subsequent cell death. Significant side effects include irritative voiding symptoms, microscopic hematuria, dermal sensitivity, and loss of bladder volume due to fibrosis of the bladder wall (McClellan, 2000).

In experimental animals, the treatment produces extensive tumor necrosis, with sloughing of tumor cells into the urinary bladder. At the ultrastructural level, tumor cells exhibit extensive cell damage, including mitochondrial destruction, swelling of endoplasmic reticulum, polyribosome disaggregation, and plasma membrane blebbing (Klaunig et al., 1985).

Cytologic findings in urine specimens from patients treated with this modality have not been described.

UROLITHIASIS

Urinary crystals may not be detected in all patients with urolithiasis, and some crystals may be identified in the absence of calculi. The characteristics of crystals present in urine are described in Chapter 2.

Calculi may form at any level of the urinary tree from the urethra to the renal pelvis, but most frequently are present in the urinary bladder and renal pelvis. They may be clinically silent or manifested by abdominal or lumbar pain, with frequent irradiation to the flank and hematuria that may be gross or microscopic and, on occasion, related to the episode of pain. Men are affected more often than women at any age, but predominantly in the third to fifth decades of life (Box 5–18).

The most common types of stones are those containing calcium (calcium oxalate), which comprise three-fourths of cases. Triple phosphate (magnesium ammonium phosphate or struvite) and uric acid com-

Box 5–18.

The development of urolithiasis is a multifaceted process; the precise mechanisms are unknown.

Calculi develop most frequently in the bladder and renal pelvis, and the most common types of stones are those of calcium, triple phosphate, and uric acid.

Urinary crystals may not be detected in all patients with urolithiasis, and some crystals may be identified in the urine in the absence of calculi.

prise most of the remaining types of stones (Pack, 1993).

Predisposing factors for calculus formation may be unknown or include conditions associated with high levels of serum or urinary calcium, oxalate, uric acid, or some amino acids. Briefly, calcium oxalate stones are associated with hypercalcemia of hyperparathyroidism and with neoplasia, intestinal surgery, chronic renal failure, enteric diseases, heredity, diet, toxic agents, or unknown causes. Triple phosphate calculi are usually large "staghorn" renal pelvis stones, more frequent in women than in men and usually associated with urea-splitting (ammonia-forming) bacteria (e.g., *Proteus* spp.) and XGP. Uric acid stones are most often seen in association with gout, dehydration, and malignant tumors with rapid cell turnover and an acidic urinary pH. In contrast to triple phosphate and calcium-containing stones, which are radiopaque, uric acid stones are radiolucent (Cotran et al., 1999b).

The development of urolithiasis is a multifaceted process, starting with urine supersaturation and ending with the formation of mature renal calculi. The precise mechanisms of stone formation are not well established. It is believed that loss of the delicate balance between supersaturation of crystals in the urine (Lieske and Toback, 1996), nucleation, or crystal aggregation in the renal tubules (Khan et al., 1996; Kohjimoto et al., 1996), integrity of the tubular epithelium (Bigelow et al., Mandel, 1997; Mandel, 1994), and inhibitors of crystal growth and aggregation (Ebisuno et al., 1999; Wesson et al., 1998) play important roles in the pathogenesis of urolithiasis. Recent findings suggest an association between the induction of nephrolithiasis and the expression of inhibitors of crystal aggregation genes (Iida et al., 1999).

Rarely, calculi may form as a result of urinary diversions or urinary augmentation procedures and are associated with foreign bodies, infections, or others unknown factors (Drach, 2000). Foreign bodies involved in the formation of bladder calculi include Foley balloon fragments, silk structures, hairpins, chewing gum, and hairs. The various kinds of foreign-body stones, anatomic locations, and categorization according to the manner of introduction into the urinary tract has been described in a classic article by Dalton et al. (1975).

Cytology. The abrasive effect of the stone produces exfoliation of single urothelial cells and cell clusters, with cellular abnormalities that range from undetectable to marked (Fig. 5–12A–D; Box 5–19).

Cellularity in urolithiasis varies, depending on the location of the calculi. A stone confined to a small physical space, such as, the ureteropelvic region, produces a constant rubbing effect on the urothelium, resulting in high cellularity. In contrast, a bladder stone usually sheds fewer cells, and often there is no detectable cytologic abnormality.

Most cases show clusters of umbrella cells with smooth, round borders, a normal to slightly increased nuclear-cytoplasmic ratio, abundant vacuolated cytoplasm, a round eccentric nucleus, uniform nuclear contours, and powdery chromatin with round, conspicuous nucleoli. These clusters have been found reliable for predicting calculi in two-thirds of routine urine specimens (Highman and Wilson, 1982). Basal and intermediate-cell clusters are usually round, with smooth borders, but an occasional cluster may display frayed edges and cells with irregular nuclear contours (Kannan and Gupta, 1999).

Box 5–19.

Smears show cell clusters and single cells with variable cellular abnormalities.

Red blood cells, acute inflammation, crystals, and evidence of tissue damage may be present.

Cellularity depends on the location and the abrasive effect of the calculi.

Most cases show cohesive umbrella cells with smooth, round borders, vacuolated cytoplasm, a low nuclear-cytoplasmic ratio, regular nuclear borders, and powdery chromatin.

Basal and intermediate cells may show irregular nuclear contours, larger nuclei, and hyperchromasia.

The differential diagnosis includes low-grade papillary urothelial tumors and instrumentation effect.

These features are indistinguishable from those of low-grade urothelial carcinoma.

Evidence of tissue damage with lysed blood and extensive cellular degeneration, secondary to the constant stone-rubbing effect against the urothelium, may be present. The constant long-term abrasive effect of staghorn stones in the renal pelvis may be associated with a high yield of squamous metaplastic cells and numerous multinucleated cells in urine samples (Beyer-Boon et al., 1978). The finding of multinucleated cells may not be common in urolithiasis except in XGP, where the association with

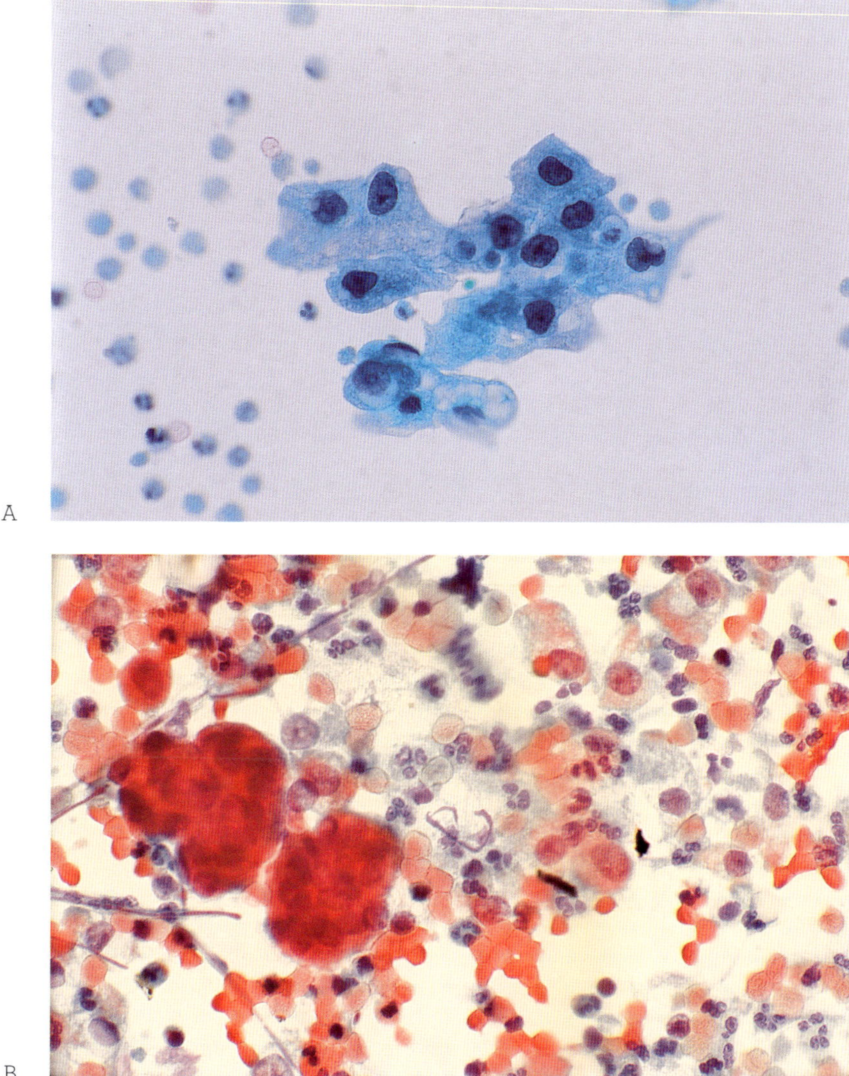

Figure 5–12. *Cytologic changes associated with urolithiasis.* These bladder washings show various cytologic changes present in urolithiasis. (*A*) Small sheet of benign reactive superficial urothelial cells showing degenerated, hyperchromatic nuclei and vacuolated cytoplasm. (*B*) Pseudopapillary clusters and single markedly reactive urothelial cells are seen together with red blood cells. Prominent round nucleoli and vacuolated cytoplasm are present. (*C*) This fragment of urothelial cells is indistinguishable from that of a low-grade papillary urothelial carcinoma. (*D*) Marked keratinizing squamous metaplasia and tissue damage are seen in this case of lithiasis of the urinary bladder. These cytologic changes resolved after surgical removal of the stone. (Papanicolaou stain, ×600)

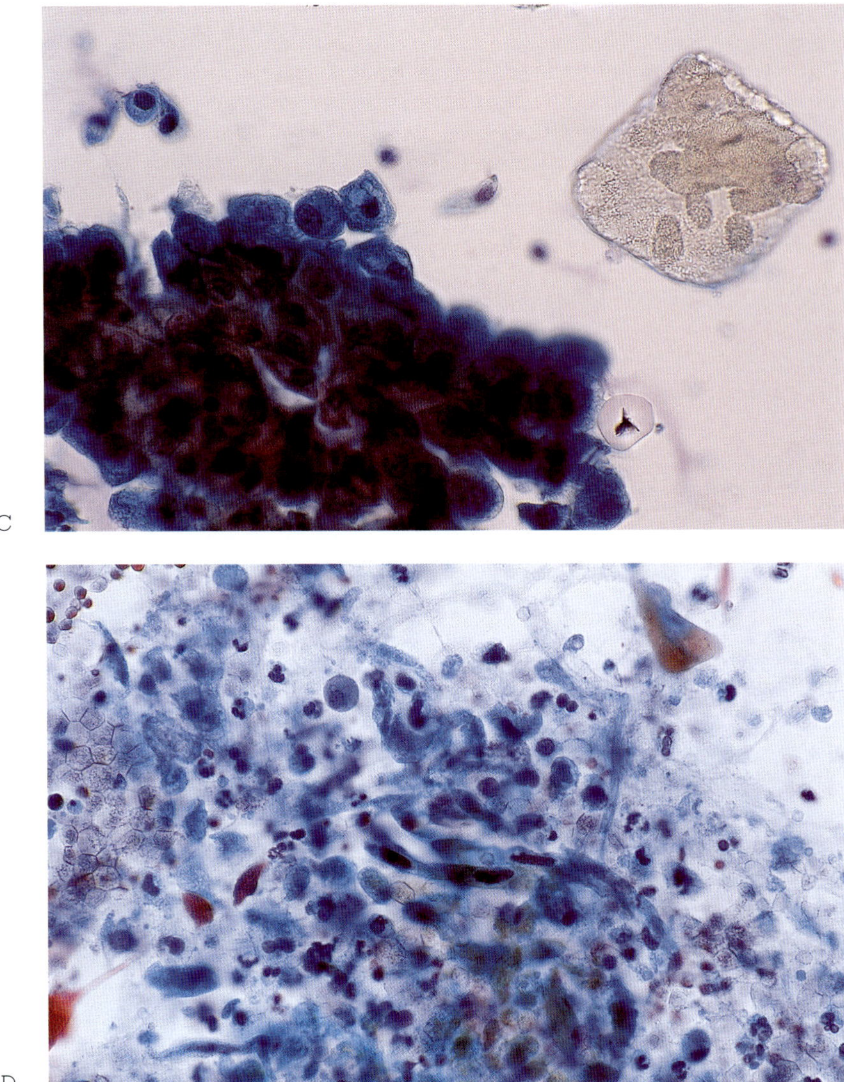

Figure 5–12. *(Continued)*

staghorn calculi is of considerable diagnostic importance. Acute inflammation is seen in cases associated with cystitis or pyelonephritis.

Urothelial cell changes similar to those of low-grade urothelial carcinoma can be identified in the urine sediment of patients with urolithiasis associated with indinavir sulfate, a protease inhibitor of human immunodeficiency virus. These changes include nuclear pleomorphism and hyperchromasia and are reversible after the patient is calculi-free (Hamm et al., 2000).

Detectable cytologic changes may not be present in all urine specimens from patients with urolithiasis (Kannan and Gupta, 1999). A compilation of 216 cases of urine cytology in patients with urolithiasis shows that 12% had features comparable to or suspicious for malignancy, particularly low-grade urothelial carcinoma (Beyer-Boon et al., 1978). These cytologic abnormalities are the source of false-positive diagnoses, estimated to occur in 0.9% of urologic patients (Nakatsu et al., 1991). Occasionally, when lithiasis and carcinoma coexist, the presence of calculi influences the clinician to set up a high threshold for diagnosing malignancy, resulting in a false-negative diagnosis of approximately 0.01% (Nakatsu et al., 1991). When high-grade urothelial carcinoma coexists with urolithiasis, the changes of malignancy must prevail when strict malignancy criteria are applied, although severe cellular changes, comparable to those of carcinoma in situ, may be seen in the epithelium adjacent to a calculus (Beyer-Boon et al., 1978). Not one but a constellation of cytologic features permits the differentiation between urolithiasis and urothelial carcinoma.

All cytologic and histologic changes of the urothelium adjacent to calculi, including hyperplasia and severe epithelial atypia, revert to normal once the calculi are removed. A suggested relationship between a long-term clinical history of urolithiasis and the development of urothelial malignancy of the upper urinary tract has not been proved (Beyer-Boon et al., 1978; de Ruiter et al., 1975). Controversy remains regarding the association between lithiasis and urothelial carcinoma in situ of the bladder.

Differential diagnosis. The cytology of urolithiasis is not specific. Nevertheless, the presence of intact red blood cells, crystals, and clusters of urothelial cells with features of an instrumentation effect in voided urine specimens strongly suggests urolithiasis. In contrast, in the absence of crystals and in urine obtained by instrumentation, the distinction among low-grade urothelial carcinoma, urolithiasis, and an instrumentation effect is practically impossible. Cell clustering in the instrumentation effect may contain cells with more rounded nuclear contours and an external rim of finely vacuolated cytoplasm.

In some cases, close examination of nuclear atypia and the number of single cells is more rewarding than the evaluation of urothelial cell clusters. Smears from patients with urolithiasis may show significantly fewer single and clustered urothelial cells with abnormal nuclear morphology than those from patients with low-grade urothelial carcinoma (Highman and Wilson, 1982). Single- cell dishesion with subtle nuclear abnormalities may also be evident in some cases of low-grade urothelial carcinoma.

Markedly reactive urothelial cells in a background of lysed blood and acute inflammation mimic invasive carcinoma. However, the nuclear anaplasia seen in high-grade urothelial carcinoma is absent in urolithiasis. Therefore, the discrepancy between the intensity of tissue damage and the degree of cellular changes provides a strong argument against the diagnosis of invasive carcinoma.

REFERENCES

Adolphs, H. D., Schwabe, H. W., Helpap, B., Volz, C. (1984). Cytomorphological and histological studies on the urothelium during and after chemoimmune prophylaxis. *Urol Res* 12: 129–33.

al-Sulaiman, M. H., al-Khader, A. A., Mousa, D. H., al-Swailem, R. Y., Dhar, J., Haleem, A. (1993). Renal parenchymal malacoplakia and megalocytic interstitial nephritis: clinical and histological features. Report of two cases and review of the literature. *Am J Nephrol* 13: 483–8.

Armora Mani, J., Munoz Segui, J., Perez Cespedes, M., Aguilo Lucia, F., Torrecilla Ortiz, C., Serrallach Mila, N. (1992). [Leukoplakia of the upper urinary tract]. *Arch Esp Urol* 45: 65–7.

Atalay, A. C., Karaman, M. I., Basak, T., Utkan, G., Ergenekon, E. (1998). Non-Hodgkin's lymphoma of the female urethra presenting as a caruncle. *Int Urol Nephrol* 30: 609–10.

August, C., Holzhausen, H. J., Schroder, S. (1994). Renal parenchymal malakoplakia: ultrastructural findings in different stages of morphogenesis. *Ultrastruct Pathol* 18: 483–91.

Badalament, R. A., Gay, H., Cibas, E. S., Herr, H. W., Whitmore, W. F., Jr., Fair, W. R., Melamed, M. R. (1987). Monitoring intravesical bacillus Calmette-Guerin treatment of superficial bladder carcinoma by postoperative urinary cytology. *J Urol* 138: 763–5.

Balogh, K., O'Hara, C. J. (1986). Myeloid metaplasia masquerading as a urethral caruncle. *J Urol* 135: 789–90.

Baltaci, S., Suzer, O., Ozer, G., Beduk, Y., Gogus, O. (1996). The efficacy of urinary cytology in the detection of recurrent bladder tumours. *Int Urol Nephrol* 28: 649–53.

Banyai-Falger, S., Maier, U., Susani, M., Wiener, H., Watschinger, B., Horl, W. H., Banyai, M. (1998). High incidence of nephrogenic adenoma of the bladder after renal transplantation. *Transplantation* 65: 511–14.

Bassi, P., Milani, C., Meneghini, A., Garbeglio, A., Aragona, F., Zattoni, F., Dalla Palma, P., Rebuffi, A., Pagano, F. (1992). Clinical value of pathologic changes after intravesical BCG therapy of superficial bladder cancer. *Urology* 40: 175–9.

Belldegrun, A. S., Franklin, J. R., O'Donnell, M. A., Gomella, L. G., Klein, E., Neri, R., Nseyo, U. O., Ratliff, T. L., Williams, R. D. (1998). Superficial bladder cancer: the role of interferon-alpha [published erratum appears in J Urol 1998 Oct;160(4):1444]. *J Urol* 159: 1793–1801.

Benson, R. C., Jr., Swanson, S. K., Farrow, G. M. (1984). Relationship of leukoplakia to urothelial malignancy. *J Urol* 131: 507–11.

Betz, S. A., See, W. A., Cohen, M. B. (1993). Granulomatous inflammation in bladder wash specimens after intravesical bacillus Calmette-Guerin therapy for transitional cell carcinoma of the bladder. *Am J Clin Pathol* 99: 244–8.

Beyer-Boon, M. E., Cuypers, L. H., de Voogt, H. J., Brussee, J. A. (1978). Cytological changes due to urinary calculi: a consideration of the relationship between calculi and the development of urothelial carcinoma. *Br J Urol* 50: 81–9.

Bhan, R., Pisharodi, L. R., Gudlaugsson, E., Bedrossian, C. (1998). Cytological, histological, and clinical correlations in intravesical Bacillus Calmette-Guerin immunotherapy. *Ann Diagn Pathol* 2: 55–60.

Bigelow, M. W., Wiessner, J. H., Kleinman, J. G., Mandel, N. S. (1997). The dependence on membrane fluidity of calcium oxalate crystal attachment to IMCD membranes. *Calcif Tissue Int* 60: 375–9.

Bohlmeyer, T. J., Shroyer, K. R. (1996). Endometriosis of the bladder: cytologic findings and differentiation from transitional cell carcinoma [letter]. *Acta Cytol* 40: 383–4.

Bolduan, J. P., Farah, R. N. (1981). Primary urethral neoplasms: review of 30 cases. *J Urol* 125: 198–200.

Bonsib, S. M. (1997). Acute tubulointerstitial disease. In *Urologic Surgical Pathology* (D. G. Bostwick, J. N. Eble, eds.), pp. 64–9. Mosby, St. Louis.

Borgmann, V., al-Abadi, H., Friedrichs, R., Nagel, R. (1993). Effect of different local and systemic therapy upon urinary bladder cytology. *Urol Int* 50: 21–6.

Bossen, E. H., Johnston, W. W. (1977). Exfoliative cytopathologic studies in organ transplantation. V. The diagnosis of rejection in the immediate postoperative period. *Acta Cytol* 21: 502–7.

Bullock, P. S., Thoni, D. E., Murphy, W. M. (1987). The significance of colonic mucosa (intestinal metaplasia) involving the urinary tract. *Cancer* 59: 2086–90.

Byrd, R. B., Viner, N. A., Omell, G. H., Trunk, G. (1976). Leukoplakia associated with renal tuberculosis in the chemotherapeutic era. *Br J Urol* 48: 377–81.

Cant, J. D., Murphy, W. M., Soloway, M. S. (1986). Prognostic significance of urine cytology on initial follow-up after intravesical mitomycin C for superficial bladder cancer. *Cancer* 57: 2119–22.

Carlson, G. J., Samuelson, J. J., Dehner, L. P. (1986). Cytologic diagnosis of florid peritoneal endosalpingiosis. A case report. *Acta Cytol* 30: 494–6.

Carson, C. C., Malek, R. S., Remine, W. H. (1978). Urologic aspects of vesicoenteric fistulas. *J Urol* 119: 744–6.

Casella, F. J., Allon, M. (1993). The kidney in sarcoidosis. *J Am Soc Nephrol* 3: 1555–62.

Clement, P. B., Young, R. H. (1992). Endocervicosis of the urinary bladder. A report of six cases of a benign mullerian lesion that may mimic adenocarcinoma. *Am J Surg Pathol* 16: 533–42.

Corey, H. E., Alfonso, F., Hamele-Bena, D., Greenstein, S. M., Schechner, R., Tellis, V., Geva, P., Koss, L. G. (1997). Urine cytology and the diagnosis of renal allograft rejection. I. Studies using conventional staining. *Acta Cytol* 41: 1732–41.

Corwin, H. L., Bray, R. A., Haber, M. H. (1989). The detection and interpretation of urinary eosinophils. *Arch Pathol Lab Med* 113: 1256–8.

Corwin, H. L., Haber, M. H. (1987). The clinical significance of eosinophiluria. *Am J Clin Pathol* 88: 520–2.

Cotran, R. S., Kumar, V., Collins, T. (1999a). Urinary bladder. In *Robbins Pathologic Basis of Disease* (R. S. Cotran, V. Kumar, T. Collins, eds.), 6th ed., pp. 1000–8. W. B. Saunders, Philadelphia.

Cotran, R. S., Kumar, V., Collins, T. (1999b). Urolithiasis. In *Robbins Pathologic Basis of Disease* (R. S. Cotran, V. Kumar, T. Collins, eds.), 6th ed., pp. 989–90. W. B. Saunders, Philadelphia.

Curran, F. T. (1987). Malakoplakia of the bladder. *Br J Urol* 59: 559–63.

Dalbagni, G., Herr, H. W. (2000). Current use and questions concerning intravesical bladder cancer group for superficial bladder cancer. *Urol Clin North Am* 27: 137–46.

Dalton, D. L., Hughes, J., Glenn, J. F. (1975). Foreign bodies and urinary stones. *Urology* 6: 1–5.

de Ruiter, F., Beyer-Boon, M. E., de Voogt, H. J. (1975). The effect of calculi on transitional epithelium. A clinical and cytological study. *Urol Res* 3: 67–72.

Delnay, K. M., Stonehill, W. H., Goldman, H., Jukkola, A. F., Dmochowski, R. R. (1999). Bladder histological changes associated with chronic indwelling urinary catheter. *J Urol* 161: 1106–8; discussion 1108–9.

Dodd, L. G., Tello, J. (1998). Cytologic examination of urine from patients with interstitial cystitis. *Acta Cytol* 42: 923–7.

Drach, G. W. (2000). Secondary and miscellaneous urolithiasis. Medications, urinary diversions, and foreign bodies. *Urol Clin North Am* 27: 269–73.

Droller, M. J. (1998). Bladder cancer: state-of-the-art care [see comments]. *CA Cancer J Clin* 48: 269–84.

Droller, M. J., Erozan, Y. S. (1985). Thiotepa effects on urinary cytology in the interpretation of transitional cell cancer. *J Urol* 134: 671–4.

Ebisuno, S., Umehara, M., Kohjimoto, Y., Ohkawa, T. (1999). The effects of human urine on the adhesion of calcium oxalate crystal to Madin-Darby canine kidney cells. *BJU Int* 84: 118–22.

Eggensperger, D., Schweitzer, S., Ferriol, E., O'Dowd, G., Light, J. A. (1988). The utility of cytodiagnostic urinalysis for monitoring renal allograft injury. A clinicopathological analysis of 87 patients and over 1,000 urine specimens. *Am J Nephrol* 8: 27–34.

Esposti, P. L., Moberger, G., Zajicek, J. (1970). The cytologic diagnosis of transitional cell tumors of the urinary bladder and its histologic basis. A study of 567 cases of urinary-tract disorder including 170 untreated and 182 irradiated bladder tumors. *Acta Cytol* 14: 145–55.

Fairchild, W. V., Spence, C. R., Solomon, H. D., Gangai, M. P. (1979). The incidence of bladder cancer after cyclophosphamide therapy. *J Urol* 122: 163–4.

Fedele, L., Bianchi, S., Raffaelli, R., Portuese, A. (1997). Pre-operative assessment of bladder endometriosis. *Hum Reprod* 12: 2519–22.

Fernandes, E. T., Manivel, J. C., Reddy, P. K., Ercole, C. J. (1996). Cyclophosphamide associated bladder cancer—a highly aggressive disease: analysis of 12 cases. *J Urol* 156: 1931–3.

Friedman, D., Mooppan, U. M., Rosen, Y., Kim, H. (1991). The effect of intravesical instillations of thiotepa, mitomycin C, and adriamycin on normal urothelium: an experimental study in rats. *J Urol* 145: 1060–3.

Garcia-Ureta, E., Llorente Apat, P., Martinez-Penuela, J. M., Sarmiento Gomez, C. (1978). [Cytology of urine in eosinophilic cystitis. Apropos of 19 cases]. *Arch Esp Urol* 31: 337–50.

Gaylis, F. D., Keer, H. N., Bauer, K. D., Kozlowski, J. M., Grayhack, J. T. (1993). DNA profile of nephrogenic adenoma assessed by flow cytometry. *Urology* 41: 160–1.

Gilcrease, M. Z., Delgado, R., Vuitch, F., Albores-Saavedra, J. (1998). Clear cell adenocarcinoma and nephrogenic adenoma of the urethra and urinary bladder: a histopathologic and immunohistochemical comparison. *Hum Pathol* 29: 1451–6.

Godwin, J. T., Hanash, K. (1984). Pathology of bilharzial bladder cancer. *Progr Clin Biol Res* 162A: 95–143.

Goldstein, M. L., Whitman, T., Renshaw, A. A. (1998). Significance of cell groups in voided urine. *Acta Cytol* 42: 290–4.

Hamm, M., Wawroschek, F., Rathert, P. (2000). Urinary cytology changes in protease inhibitor induced urolithiasis. *J Urol* 163: 1249–50.

Hammadeh, M. Y., Thomas, K., Philp, T. (1996). Urethral caruncle: an unusual presentation of ovarian tumour. *Gynecol Obstet Invest* 42: 279–80.

Hanno, P. M. (1994). Diagnosis of interstitial cystitis. In *The Urologic Clinics of North America* (P. M. Hanno, ed.), Vol. 21, pp. 63–6. W. B. Saunders, Philadelphia.

Hansson, S., Hanson, E., Hjalmas, K., Hultengren, M., Jodal, U., Olling, S., Svanborg-Eden, C. (1990). Follicular cystitis in girls with untreated asymptomatic or covert bacteriuria. *J Urol* 143: 330–2.

Heidenreich, A., Zirbes, T. K., Wolter, S., Engelmann, U. H. (1999). Nephrogenic adenoma: a rare bladder tumor in children. *Eur Urol* 36: 348–53.

Henke, A. C., Callaghan, E. J., Timmerman, T. G., Hughes, J. H. (2000). Nephrogenic adenoma mimicking carcinoma: a potential pitfall in cytodiagnosis. *Diagn Cytopathol* 22: 49–51.

Highman, W., Wilson, E. (1982). Urine cytology in patients with calculi. *J Clin Pathol* 35: 350–6.

Ho, K. L., Rassekh, Z. S., Nam, S. H. (1979). Bilateral renal malakoplakia. *Urology* 13: 321–3.

Iida, S., Peck, A. B., Johnson-Tardieu, J., Moriyama, M., Glenton, P. A., Byer, K. J., Khan, S. R. (1999). Temporal changes in mRNA expression for bikunin in the kidneys of rats during calcium oxalate nephrolithiasis. *J Am Soc Nephrol* 10: 986–96.

Indudhara, R., Vaidyanathan, S., Radotra, B. D. (1992). Urethral tuberculosis. *Urol Int* 48: 436–8.

Jacobs, L. B., Brooks, J. D., Epstein, J. I. (1997). Differentiation of colonic metaplasia from adenocarcinoma of urinary bladder. *Hum Pathol* 28: 1152–7.

Johansson, S. L., Fall, M. (1994). Pathology of interstitial cystitis. In *The Urologic Clinics of North America* (P. M. Hanno, ed.), Vol. 21, pp. 55–62. W. B. Saunders, Philadelphia.

Jones, E. C., Clement, P. B., Young, R. H. (1993). Inflammatory pseudotumor of the urinary bladder. A clinicopathological, immunohistochemical, ultrastructural, and flow cytometric study of 13 cases [see comments]. *Am J Surg Pathol* 17: 264–74.

Kaldor, J. M., Day, N. E., Kittelmann, B., Pettersson, F., Langmark, F., Pedersen, D., Prior, P., Neal, F., Karjalainen, S., Bell, J., et al. (1995). Bladder tumours following chemotherapy and radiotherapy for ovarian cancer: a case-control study. *Int J Cancer* 63: 1–6.

Kannan, V., Gupta, D. (1999). Calculus artifact. A challenge in urinary cytology. *Acta Cytol* 43: 794–800.

Karaiossifidi, H., Kouri, E. (1992). Malacoplakia of the urethra: a case of unique localization with followup. *J Urol* 148: 1903–4.

Khan, M. A., Travis, L. B., Lynch, C. F., Soini, Y., Hruszkewycz, A. M., Delgado, R. M., Holowaty, E. J., van Leeuwen, F. E., Glimelius, B., Stovall, M., Boice, J. D., Jr., Tarone, R. E., Bennett, W. P. (1998). p53 mutations in cyclophosphamide-associated bladder cancer. *Cancer Epidemiol Biomarkers Prev* 7: 397–403.

Khan, S. R., Atmani, F., Glenton, P., Hou, Z., Talham, D. R., Khurshid, M. (1996). Lipids and membranes in the organic matrix of urinary calcific crystals and stones. *Calcif Tissue Int* 59: 357–65.

Klaunig, J. E., Selman, S. H., Shulok, J. R., Schafer, P. J., Britton, S. L., Goldblatt, P. J. (1985). Morphologic studies of bladder tumors treated with hematoporphyrin derivative photochemotherapy. *Am J Pathol* 119: 236–43.

Kohjimoto, Y., Ebisuno, S., Tamura, M., Ohkawa, T. (1996). Adhesion and endocytosis of calcium oxalate crystals on renal tubular cells. *Scanning Microsc* 10: 459–68.

Kontothanassis, D., Christophis, J., Tamvakis, N., Becopoulos, T. (1996). Bladder washing flow cytometry assessment of nephrogenic adenoma of the bladder. *Int Urol Nephrol* 28: 495–7.

Koshikawa, T., Leyh, H., Schenck, U. (1989). Difficulties in evaluating urinary specimens after local mitomycin therapy of bladder cancer. *Diagn Cytopathol* 5: 117–21.

Koss, L. G., Melamed, M. R., Mayer, K. (1965). The effect of busulfan on human epithelia. *Am J Clin Pathol* 44: 385–97.

Koziol, J. A. (1994). Epidemiology of interstitial cystitis. In *The Urologic Clinics of North America* (P. M. Hanno, ed.), Vol. 21, pp. 7–20. W. B. Saunders, Philadelphia.

Lage, J. M., Bauer, W. C., Kelley, D. R., Ratliff, T. L., Catalona, W. J. (1986). Histological parameters and pitfalls in the interpretation of bladder biopsies in bacillus Calmette-Guerin treatment of superficial bladder cancer. *J Urol* 135: 916–19.

Lamm, D. L. (1992). Complications of bacillus Calmette-Guerin immunotherapy. *Urol Clin North Am* 19: 565–72.

Lauchlan, S. C. (1994). The secondary mullerian system revisited. *Int J Gynecol Pathol* 13: 73–9.

Lieske, J. C., Toback, F. G. (1996). Interaction of urinary crystals with renal epithelial cells in the pathogenesis of nephrolithiasis. *Semin Nephrol* 16: 458–73.

Long, J. P., Jr., Althausen, A. F. (1989). Malacoplakia: a 25-year experience with a review of the literature. *J Urol* 141: 1328–31.

Lopez, J. I., Angulo, J. C., Ibanez, T. (1993). Primary malignant melanoma mimicking urethral caruncle. Case report. *Scand J Urol Nephrol* 27: 125–6.

Lopez-Beltran, A. (1999). Bladder treatment. Immunotherapy and chemotherapy. *Urol Clin North Am* 26: 535–54.

Mack, D., Frick, J. (1994). Diagnostic problems of urine cytology on initial follow-up after intravesical immunotherapy with Calmette-Guerin bacillus for superficial bladder cancer. *Urol Int* 52: 204–7.

Maier, U., Simak, R., Neuhold, N. (1995). The clinical value of urinary cytology: 12 years of experience with 615 patients. *J Clin Pathol* 48: 314–17.

Mandal, A. K. (1988). Analysis of urinary sediment by transmission electron microscopy. An innovative approach to diagnosis and prognosis in renal disease. *Clin Lab Med* 8: 463–81.

Mandel, N. (1994). Crystal-membrane interaction in kidney stone disease. *J Am Soc Nephrol* 5: S37–S45.

McClellan, M. W. (2000). The role of photodynamic therapy in the treatment of recurrent superficial bladder cancer. *Urol Clin North Am* 27: 163–70.

McDougal, W. S., Cramer, S. F., Miller, R. (1981). Invasive carcinoma of the renal pelvis following cyclophosphamide therapy for nonmalignant disease. *Cancer* 48: 691–5.

Morales, A., Nickel, J. C. (1992). Immunotherapy for superficial bladder cancer. A developmental and clinical overview. *Urol Clin North Am* 19: 549–56.

Murphy, W. M. (1990). Current status of urinary cytology in the evaluation of bladder neoplasms. *Hum Pathol* 21: 886–96.

Nakatsu, H., Masai, M., Okano, T., Isaka, S., Shimazaki, J., Matsuzaki, O., Horiuchi, F. (1991). [Urinary cytology in patients with urolithiasis]. *Nippon Hinyokika Gakkai Zasshi—Jpn J Urol* 82: 1281–5.

Nochomovitz, L. E., Orenstein, J. M. (1985). Inflammatory pseudotumor of the urinary bladder—possible relationship to nodular fasciitis. Two case reports, cytologic observations, and ultrastructural observations. *Am J Surg Pathol* 9: 366–73.

Nolan, C. R., Anger, M. S., Kelleher, S. P. (1986). Eosinophiluria—a new method of detection and definition of the clinical spectrum. *N Engl J Med* 315: 1516–19.

Nolan, C. R., Kelleher, S. P. (1988). Eosinophiluria. *Clin Lab Med* 8: 555–65.

Nowels, K., Kent, E., Rinsho, K., Oyasu, R. (1988). Prostate specific antigen and acid phosphatase-reactive cells in cystitis cystica and glandularis. *Arch Pathol Lab Med* 112: 734–7.

Nseyo, U. O., Lamm, D. L. (1996). Therapy of superficial bladder cancer. *Semin Oncol* 23: 598–604.

Okafo, B. A., Jones, H. W., Dow, D., Kiruluta, H. G. (1985). Eosinophilic cystitis: pleomorphic manifestations. *Can J Surg* 28: 17–18.

Oliva, E., Young, R. H. (1995). Nephrogenic adenoma of the urinary tract: a review of the microscopic appearance of 80 cases with emphasis on unusual features. *Mod Pathol* 8: 722–30.

Oyama, N., Tanase, K., Akino, H., Mori, H., Kanamaru, H., Okada, K. (1998). Nephrogenic adenoma in a patient with transitional cell carcinoma of the bladder receiving intravesical bacillus Calmette-Guerin. *Int J Urol* 5: 185–7.

Pack, C. T. (1993). Urolithiasis. In *Diseases of the Kidney* (R. W. Schrier, C. W. Gottschalk, eds.), 5th ed., pp. 729–43. Little, Brown, Boston.

Pedersen-Bjergaard, J., Jonsson, V., Pedersen, M., Hou-Jensen, K. (1995). Leiomyosarcoma of the urinary bladder after cyclophosphamide [letter]. *J Clin Oncol* 13: 532–3.

Peeker, R., Aldenborg, F., Fall, M. (1997). Nephrogenic adenoma—a study with special reference to clinical presentation. *Br J Urol* 80: 539–42.

Polat, O., Gundogdu, C., Demirel, A., Bayraktar, Y., Gul, O., Ozbilge, M. (1996). The effects of thiotepa, mitomycin C, BCG and interferon instillation on urothelium. *Int Urol Nephrol* 28: 43–8.

Reece, R. W., Koontz, W. W., Jr. (1975). Leukoplakia of the urinary tract: a review. *J Urol* 114: 165–171.

Richie, J. P. (1992). Intravesical chemotherapy. Treatment selection, techniques, and results. *Urol Clin North Am* 19: 521–7.

Rodriguez, R., Alfert, H. (1997). Endocervicosis of the bladder: a rare mucinous analogue of endometriosis. *J Urol* 157: 1355.

Roehrborn, C. G., Teigland, C. M., Spence, H. M. (1988). Progression of leukoplakia of the bladder to squamous cell carcinoma 19 years after complete urinary diversion. *J Urol* 140: 603–4.

Rutgers, J. L., Young, R. H. (1988). Nephrogenic adenoma of the urinary bladder: a comparison of its cytologic and histopathologic features in ten cases. *Diagn Cytopathol* 4: 210–16.

Salomao, D. R., Mathers, W. D., Sutphin, J. E., Cuevas, K., Folberg, R. (1999). Cytologic changes in the conjunctiva mimicking malignancy after topical mitomycin C chemotherapy. *Ophthalmology* 106: 1756–60; discussion 1761.

Sandison, A., Jones, P. A. (1993). Urine culture in the diagnosis of colovesical fistula. *BMJ* 307: 1588.

Sarma, K. P. (1970). On the nature of cystitis follicularis. *J Urol* 104: 709–14.

Schiff, H. I., Finkel, M., Schapira, H. E. (1982). Transitional cell carcinoma of the ureter associated with cyclophosphamide therapy for benign disease: a case report. *J Urol* 128: 1023–4.

Schillinger, F., Delclaux, B., Milcent, T., Massia, D., Montagnac, R., Tisserand, P., Meekel, P., Schillinger, D., Mehaut, S. (1999). [Renal sarcoidosis with a pseudotumoral pyelic localization]. *Presse Med* 28: 683–5.

Schnadig, V. J., Adesokan, A., Neal, D., Jr., Gatalica, Z. (2000). Urinary cytologic findings in patients with benign and malignant adenomatous polyps of the prostatic urethra. *Arch Pathol Lab Med* 124: 1047–52.

Schneider, V., Smith, M. J., Frable, W. J. (1980). Urinary cytology in endometriosis of the bladder. *Acta Cytol* 24: 30–3.

Schumann, G. B., Burleson, R. L., Henry, J. B., Jones, D. B. (1977). Urinary cytodiagnosis of acute renal allograft rejection using the cytocentrifuge. *Am J Clin Pathol* 67: 134–40.

Schwalb, M. D., Herr, H. W., Sogani, P. C., Russo, P., Sheinfeld, J., Fair, W. R. (1994). Positive urinary cytology following a complete response to intravesical bacillus Calmette-Guerin therapy: pattern of recurrence. *J Urol* 152: 382–7.

Segasothy, M., Lau, T. M., Birch, D. F., Fairley, K. F., Kincaid-Smith, P. (1988). Immunocytologic dissection of the urine sediment using monoclonal antibodies. *Am J Clin Pathol* 90: 691–6.

Siddiqui, A., Melamed, M. R., Abbi, R., Ahmed, T. (1996). Mucinous (colloid) carcinoma of urinary bladder following long-term cyclophosphamide therapy for Waldenstrom's macroglobulinemia. *Am J Surg Pathol* 20: 500–4.

Smith, J. A., Jr., Labasky, R. F., Cockett, A. T., Fracchia, J. A., Montie, J. E., Rowland, R. G. (1999). Bladder cancer clinical guidelines panel summary report on the management of nonmuscle invasive bladder cancer (stages Ta, T1 and TIS). The American Urological Association. *J Urol* 162: 1697–701.

Sneige, N., Fanning, C. V. (1992). Peritoneal washing cytology in women: diagnostic pitfalls and clues for correct diagnosis. *Diagn Cytopathol* 8: 632–40; discussion 640–2.

Soloway, M. S. (1987). Evaluation and management of patients with superficial bladder cancer. *Urol Clin North Am* 14: 771–80.

Sonobe, H., Okada, Y., Sudo, S., Iwata, J., Ohtsuki, Y. (1999). Inflammatory pseudotumor of the urinary bladder with aberrant expression of cytokeratin. Report of a case with cytologic, immunocytochemical and cytogenetic findings. *Acta Cytol* 43: 257–62.

Stella, F., Battistelli, S., Marcheggiani, F., De Santis, M., Donnorso, R. P., Baronciani, D., Durazzi, S. M., Troccoli, R., Lucarelli, G. (1992). Urothelial toxicity following conditioning therapy in bone marrow transplantation and bladder cancer: morphologic and morphometric comparison by exfoliative urinary cytology. *Diagn Cytopathol* 8: 216–21.

Stella, F., Battistelli, S., Marcheggiani, F., De Santis, M., Giardini, C., Baronciani, D., Manenti, F., Mattioli, S., Troccoli, R. (1990). Urothelial cell changes due to busulfan and cyclophosphamide treatment in bone marrow transplantation. *Acta Cytol* 34: 885–90.

Stillwell, T. J., Benson, R. C., Jr. (1988). Cyclophosphamide-induced hemorrhagic cystitis. A review of 100 patients. *Cancer* 61: 451–7.

Stilmant, M., Murphy, J. L., Merriam, J. C. (1986). Cytology of nephrogenic adenoma of the urinary bladder. A report of four cases. *Acta Cytol* 30: 35–40.

Stonehill, W. H., Goldman, H. B., Dmochowski, R. R. (1997). The use of urine cytology for diagnosing bladder cancer in spinal cord injured patients [see comments]. *J Urol* 157: 2112–4.

Suresh, U. R., Smith, V. J., Lupton, E. W., Haboubi, N. Y. (1993). Radiation disease of the urinary tract: histological features of 18 cases. *J Clin Pathol* 46: 228–31.

Synhaivsky, A., Malek, R. S. (1985). Isolated pneumaturia. *Am J Med* 78: 617–20.

Talar-Williams, C., Hijazi, Y. M., Walther, M. M., Linehan, W. M., Hallahan, C. W., Lubensky, I., Kerr, G. S., Hoffman, G. S., Fauci, A. S., Sneller, M. C. (1996). Cyclophosphamide-induced cystitis and bladder cancer in patients with Wegener granulomatosis [see comments]. *Ann Intern Med* 124: 477–84.

Thrasher, J. B., Crawford, E. D. (1992). Complications of intravesical chemotherapy. *Urol Clin North Am* 19: 529–39.

Troster, M., Wyatt, J. K., Alen-Halagah, J. (1986). Nephrogenic adenoma of the urinary bladder. Histologic and cytologic observations in a case. *Acta Cytol* 30: 41–4.

Tsung, S. H. (1982). Urinary sediment cytology: potential diagnostic tool for malakoplakia. *Urology* 20: 546–7.

Vercellini, P., Meschia, M., De Giorgi, O., Panazza, S., Cortesi, I., Crosignani, P. G. (1996). Bladder detrusor endometriosis: clinical and pathogenetic implications. *J Urol* 155: 84–6.

Walser, A. C., Klotz, T., Schoenenberger, A., Ammann, J. (1999). Fifty years of faecaluria and pneumaturia. *BJU Int* 83: 517.

Ward, A. M. (1971). Glandular neoplasia within the urinary tract. The aetiology of adenocarcinoma of the urothelium with a review of the literature. I. Introduction: the origin of glandular epithelium in the renal pelvis, ureter and bladder. *Virchows Arch A Pathol Pathol Anat* 352: 296–311.

Wesson, J. A., Worcester, E. M., Wiessner, J. H., Mandel, N. S., Kleinman, J. G. (1998). Control of calcium oxalate crystal structure and cell adherence by urinary macromolecules. *Kidney Int* 53: 952–7.

Winkelmann, M., Grabensee, B., Pfitzer, P. (1985). Differential diagnosis of acute allograft rejection and CMV-infection in renal transplantation by urinary cytology. *Pathol Res Pract* 180: 161–8.

Young, R. H. (1997). Pseudoneoplastic lesions of the urinary bladder and urethra: a selective review with emphasis on recent information. *Semin Diagn Pathol* 14: 133–46.

Young, R. H., Able, J. N. (1997). Non-neoplastic disorders of the urinary bladder. In *Urologic Surgical Pathology* (D. G. Bostwick, J. N. Able, eds.), pp. 167–212. Mosby, St. Louis.

Young, R. H., Bostwick, D. G. (1996). Florid cystitis glandularis of intestinal type with mucin extravasation: a mimic of adenocarcinoma. *Am J Surg Pathol* 20: 1462–8.

Young, R. H., Clement, P. B. (1996). Mullerianosis of the urinary bladder. *Mod Pathol* 9: 731–7.

Young, R. H., Oliva, E., Garcia, J. A., Bhan, A. K., Clement, P. B. (1996). Urethral caruncle with atypical stromal cells simulating lymphoma or sarcoma—a distinctive pseudoneoplastic lesion of females. A report of six cases. *Am J Surg Pathol* 20: 1190–5.

Zipkin, J. W., Spreen, S. A., Evans, A. T. (1985). Hyperkeratotic squamous epithelium of pelvis and ureter in solitary kidney. *Urology* 25: 625–6.

6

PRIMARY TUMORS OF THE URINARY TRACT

EMBRYOLOGY AND HISTOLOGY

THE urinary bladder forms during the first trimester of gestation. The trigone develops from the mesonephric ducts and the dome, posterior wall, and portions of the lateral wall from the mesenchyme surrounding the urogenital sinus. The rest of the bladder forms at the time of closure of the abdominal wall. The urachus, a cord-like structure between the umbilicus and the bladder dome, is not involved in the formation of the bladder; however, remnants persist in the anterior abdominal wall and in the dome of the bladder, and these are lined by various types of epithelium, most often urothelial.

The bladder wall is composed of mucosa, submucosa, muscularis propria, and adventitia, with peritoneum on the superior surface only. The exact number of cell layers of the pseudostratified urothelium is not known, but ranges between three and seven and includes one layer of "umbrella" cells in the luminal surface in contact with the urine and the underlying basal-intermediate urothelial cells in contact with the basement membrane. The basement membrane separates the urothelium from the lamina propria that contains the discontinuous muscularis mucosae, which is well developed in less than 5% of human urinary bladders. This concept is useful in separating lamina propria-invasive from muscle-invasive bladder cancers, particularly in bladder biopsy specimens. The muscularis propria is composed of large bundles of muscle fibers (Young and Able, 1997).

The ureters have a basic structure similar to that of the bladder. However, the urothelial lining is thinner and is composed of four to five cell layers. The urothelium of the renal pelvis is even thinner and is composed of two or three cell layers.

The male urethra is lined with urothelium similar to that of the bladder only in the proximal (prostatic) portion, with columnar and squamous epithelia in the middle and distal portions. The mucosa of the shorter female urethra is lined with columnar and squamous epithelia.

CLASSIFICATION OF URINARY TRACT NEOPLASMS

The World Health Organization (WHO) classification of neoplasms of the urinary bladder is summarized in Table 6–1. Of these, 95% are primary urothelial, squamous, and glandular neoplasms, and 5% are primary mesenchymal neoplasms and miscellaneous malignancies. Similar neoplasms develop at other levels of the urinary tract (Murphy et al., 1994).

Urothelial neoplasms comprise approximately 75% to 90% of urinary tract neoplasms. Among the remainder, squamous cell carcinomas comprise 2% to 15%, mixed carcinomas 4% to 6%, adenocarcinomas less than 2%, and undifferentiated carcinomas less than 0.5% (Murphy et al., 1994). With the exception of the rare and controversial urothelial papilloma, most urothelial neoplasms are carcinomas of varying grades (Cheng et al., 1999c).

Because the urothelium covers the renal pelvis, ureters, and part of the urethra, the

Table 6–1. Abbreviated World Health Organization Classification of Bladder Neoplasms

Epithelial tumors
Papillomas: transitional (urothelial), inverted, squamous types.
Transitional (urothelial) cell carcinoma: papillary, flat.
Transitional (urothelial) cell carcinoma with squamous or glandular elements, unusual variants.
Squamous cell carcinoma: verrucous, sarcomatoid.
Adenocarcinoma: intestinal, mucinous, signet-ring cell, clear types.
Undifferentiated carcinoma: small cell, lymphoepithelioma-like.
Mesenchymal tumors
Benign: leiomyoma, neurofibroma.
Malignant: leiomyosarcoma, rhabdomyosarcoma.
Miscellaneous tumors
Lymphoma, melanoma, carcinosarcoma, pheochromocytoma.
Metastatic tumors and secondary extensions
Renal, prostatic, colonic, genital tract, breast, lung, gastric carcinomas.
Epithelial abnormalities
Polypoid cystitis, von Brunn's nests, cystitis cystica, squamous and glandular metaplasia, nephrogenic adenoma.
Tumor-like lesions
Follicular cystitis, malakoplakia, amyloidosis, fibroepithelial polyp, endometriosis, hamartoma, cysts.

histologic and cytologic description of urothelial tumors arising in the urinary bladder applies to tumors at other locations as well. The cytology of secondary malignancies, reactive urothelial changes, and tumor-like conditions is covered in other chapters of this book.

HISTOLOGIC CLASSIFICATION OF UROTHELIAL NEOPLASMS

In 1973, the WHO proposed a three-tiered system (1, 2, and 3) to grade papillary urothelial carcinoma of the urinary bladder. Grading was based on the degree of architectural abnormalities of the urothelium and cellular anaplasia (Mostofi et al., 1973). This system is widely accepted and is used by pathologists and urologists around the world.

In 1998, pathologists representing the WHO and the International Society of Urologic Pathologists (ISUP) presented a WHO/ISUP consensus report with standardized terminology to classify and grade papillary and flat urothelial neoplasms (Epstein et al., 1998). Some of the most relevant recommendations are as follows:

1. The terms *urothelial* or *urothelium* and *transitional* can be used synonymously. However, the former is preferred. As a result, the diagnosis "urothelial carcinoma" is preferable to "transitional cell carcinoma."
2. The term *papillary urothelial neoplasm of low malignant potential* (PUNLMP) should replace the term *grade 1 papillary urothelial carcinoma.* The recommendation is based on the preference of some pathologists not to label a low-grade, noninvasive neoplasm as carcinoma.
3. Papillary urothelial carcinoma grades 2 and 3 are designated low-grade and high-grade papillary urothelial carcinoma, respectively. The PUNLMP, low-grade, and high-grade papillary urothelial carcinomas have progressively increasing architectural and cytologic abnormalities.
4. Objective histologic and cytologic criteria for distinguishing between severe dysplasia and carcinoma in situ and mild to moderate dysplasia are unreliable for the diagnosis of flat urothelial neoplasms. Therefore, the first two should be combined into a single category of carcinoma in situ. Furthermore, subclassification of dysplasia into mild and moderate should be avoided, and the term *urothelial dysplasia* should be used to describe changes short of carcinoma in situ but beyond reactive or inflammatory changes.
5. The misleading term *superficial papillary carcinoma* should be avoided be-

cause it combines three biologically different lesions: carcinoma in situ, mucosal noninvasive (Ta), and lamina propria invasive (T1) papillary urothelial carcinomas.

6. Urine cytology should be recommended as an integral part of any program monitoring patients with urothelial neoplasms of the urinary tract.

The WHO/ISUP goal is that pathologists and urologists adopt this terminology to permit a valid comparison of results from around the world. However, critiques and controversies over the proposed classification of papillary neoplasms have developed (Cheng and Bostwick, 2000; Cheng et al., 1999d). Furthermore, the most recent 1999 WHO classification adds more confusion because of the inclusion of PUNLMP to the previous three-tiered 1973 WHO classification, resulting in a four-tiered system consisting of PUNLMP and grade 1, 2, and 3 papillary urothelial carcinomas (Mostofi et al., 1999).

Equally controversial is the histologic classification of flat urothelial neoplasms of the urinary bladder. Approximately ten classifications have been proposed in the last 25 years (Cheng et al., 2000a). Unanimous consensus has been reached to provide criteria for the diagnosis of carcinoma in situ. However, the classification of lesions short of that diagnosis is highly discrepant. Thus, investigators have proposed terms such as *atypical hyperplasia* (Koss, 1975), three grades of dysplasia (Grignon, 1997; Nagy et al., 1982), three grades of carcinoma in situ (Mostofi and Sesterhenn, 1984; Reuter and Melamed, 1989), and three grades of intraurothelial neoplasia (Friedell et al., 1986; Koss, 1996) to classify flat urothelial neoplasms. Additional classifications that include atypia and one or two grades of dysplasia have also been proposed recently (Cheng et al., 2000a; Epstein et al., 1998).

A summary of the most recent histologic classifications for papillary and flat urothelial neoplasms is given in Table 6–2 (Cheng et al., 2000a, 2000b; Epstein et al., 1998; Mostofi et al., 1973, 1999; Murphy et al., 1994).

THE CONCEPT OF UROTHELIAL NEOPLASMS

The following can be concluded as a result of studies on urothelial neoplasms summarized in recent reports (Bhan et al., 1998; Bostwick et al., 1999; Cheng and Bostwick, 2000; Lee and Droller, 2000; Murphy, 1999; Renshaw, 2000a; Reuter, 1999):

- The *field defect* and *clonality* implant theories have been suggested to explain the clinical presentation and progression of urothelial carcinoma. The first theory suggests a tumoral susceptibility of the entire urothelium when it is exposed to carcinogenic factor(s). The clonality theory suggests that cells derived from a stem cell implant at different sites in the bladder to originate a urothelial tu-

Table 6–2. Histologic Grading and Classification of Urothelial Neoplasms

WHO (1973)	*AFIP (1994)*	*WHO/ISUP (1998)*	*WHO (1999)*	*Cheng et al. (2000)*
Papillary				
Papilloma	Papilloma Papilloma	Papilloma	Papilloma PUNLMP	Papilloma
Grade 1	Most low grade	PUNLMP	Grade 1	Grade 1
Grade 2	Most high grade	Low grade	Grade 2	Grade 2
Grade 3	High grade	High grade	Grade 3	Grade 3
Flat				
	Reactive	Reactive atypia Atypia of unknown significance		Reactive atypia
	Dysplasia	Dysplasia		Dysplasia
	CIS	CIS/severe dysplasia		CIS/severe dysplasia

Abbreviations: WHO = World Health Organization; ISUP = International Society of Urologic Pathology; AFIP = Armed Forces Institute of Pathology; PUNLMP = papillary urothelial neoplasm of low malignant potential; CIS = carcinoma in situ.

mor. Evidence favors the field defect theory.

- There are two types of urothelial neoplasms, papillary and flat, which have distinct clinical implications. Papillary neoplasms are usually of low grade and are less ominous, having a 75% incidence of recurrence and less than a 5% progression to a higher-grade papillary neoplasm. In contrast, flat carcinoma in situ is always of high grade and has a 75% risk of recurrence as well, but the risk of progression is up to 50%. Carcinoma in situ has a more aggressive behavior than papillary urothelial carcinoma and frequently accompanies invasive urothelial carcinoma. Genetic changes account for the variable biologic activity of these two types of neoplasms.
- With the exception of carcinoma in situ, there is a strong correlation between the aneuploid DNA content of tumor cells with a high histologic grade and a high clinical stage. One hundred percent of grade 1 and grade 3 urothelial carcinomas are diploid and aneuploid, respectively, with intermediate values for grade 2.
- The grade and stage of tumors at the time of initial diagnosis predict the risk of progression of the disease or the development of synchronous or metachronous urothelial neoplasms of the urinary tracts of these patients. This risk is higher than that of the general population.
- Pathologic cancer staging is the most important predictor of patient survival. The Marshall modification of Jewett-Strong and the American Joint Committee on Cancer (AJCC) are the most common cancer staging systems used in clinical practice. Briefly, bladder carcinoma is staged as follows:

  Ta: Noninvasive papillary urothelial carcinoma.

  Tcis: "Flat" carcinoma in situ.

  T1: Invasion of subepithelial connective tissue (lamina propria).

  T2: Invasion of muscularis propria: T2a, inner half; T2b, outer half.

  T3: Invasion of perivesical tissue.

  T4: Invasion of adjacent organs, that is, the prostate, vagina, uterus, pelvic or abdominal wall.
- The term *carcinoma in situ* should be reserved for flat lesions and should not be used to designate mucosal grade 2 or 3 papillary urothelial carcinomas (Ta).
- Treatment of urothelial carcinoma by stages is outlined as follows:

  Ta: Transurethral resection with fulguration of tumors followed by adjuvant intravesical chemotherapy. Immunotherapy, photoradiation, or laser therapy is considered for patients at high risk for recurrence and progression.

  Tcis: Local resection followed by adjuvant intravesical therapy.

  T1: Transurethral resection followed by adjuvant intravesical therapy. Cystectomy may be considered for grade 3 tumors because of the high risk of progression.

  T2 or higher: Total cystectomy or cystoprostatectomy.
- Cytology of voided or catheterized urine specimens and washes is valuable and should be included in follow-up and monitoring of the therapeutic response of patients with bladder cancer. Likewise, urinary cytology should be considered as a screening tool in the evaluation of patients suspected of having or at risk of developing for urothelial carcinoma. It is accepted that the diagnostic accuracy of urine cytology for low-grade papillary neoplasms is low and variable. However, in high-grade tumors, including invasive carcinoma and carcinoma in situ, urine cytology is highly accurate.

EPIDEMIOLOGY, ETIOLOGY, AND PATHOGENESIS OF BLADDER CANCER

Bladder cancer is the fourth most common neoplasm of men in the Western Hemisphere, with the highest incidence in North America and the lowest in China, middle Africa, and Melanesia (Parkin et al., 1999). In the United States, bladder cancer is the fourth most common cancer in men, is three times more frequent in men than in women, and is more common in whites than in African Americans (Greenlee et al., 2000). The median age at diagnosis is 65 years, although bladder cancer can develop in young individuals and children (Grignon, 1997).

Cigarette smoking is the most important and preventable cause of bladder cancer. An average of 45% of these cancers occurring in men are attributed to tobacco smoking, and continuing smokers have a worse disease-associated outcomes than do patients who stop smoking (Fleshner et al., 1999; Fortuny et al., 1999; Landis et al., 1999; Parkin et al., 1999). Tobacco smoke contains compounds that act in vitro as both initiators and promoters of bladder tumors. Other carcinogenic factors pathogenetically associated with bladder, ureter, and renal pelvis tumors include occupational exposures (to aniline dyes and aromatic amines), chronic inflammation, renal or ureteral stones, excessive use of phenacetin, parasitic infestations, urolithiasis, cyclophosphamide therapy, and pelvic irradiation (Chow et al., 1997; Cohen and Johansson, 1992; Johansson and Cohen, 1997; Lee and Droller, 2000; Pedersen-Bjergaard et al., 1988; Stewart et al., 1999). Nonurothelial bladder cancer is also associated with smoking and with specific occupations such as field-crop and vegetable-farm work. The risk increases with higher tobacco consumption, particularly for squamous cell carcinoma than for other types of nonurothelial cancer (Fortuny et al., 1999).

External carcinogens or spontaneous genetic events trigger intraepithelial neoplastic transformation. Research suggests that hypervascularity has an inductive oncogenic effect on the urothelium and precedes hyperplasia and neoplasia (Sarma, 1981). Urothelial hyperplasia appears to be a precursor of urothelial neoplasms (Koss, 1979; Taylor et al., 1996). Additional cellular proliferation with concomitant stromal, fibroblastic, and marked vascular proliferation accounts for the histologic recognition of noninvasive papillary urothelial neoplasms and flat carcinoma in situ.

Ta papillary urothelial neoplasms are neoplasms of low to intermediate grade and have a better prognosis than does Tcis, which is composed of high-grade cells and is more likely to progress to muscle invasion than does Ta. Similarly, T1 invasive tumors are of higher grade and are much more prone to progression than Ta neoplasms. Ta and Tcis, although confined to the urothelial surface, have different behavioral patterns linked to specific genetic mutations that suggest different pathways of tumorigenesis (Lee and Droller, 2000).

Genetic changes associated with the development of low-grade papillary urothelial carcinoma, stage Ta, are linked to a defect on chromosome 9. Changes in chromosomes 7, 11, and 17 (p53) have been associated with high grade and invasiveness in urothelial carcinoma. Increased expression of angiogenesis growth factors plays a role in the genesis of low-grade tumors and the progressive behavior of muscle-invasive tumors (Lee and Droller, 2000; Sandberg and Berger, 1994).

PATHOLOGIC DIAGNOSIS OF UROTHELIAL NEOPLASMS

The assessment of architectural and cellular abnormalities in tissue sections is the gold standard for the diagnosis of urothelial neoplasms. These two parameters are complementary, although architectural abnormalities predominate over the cellular characteristics for both high- and low-grade urothelial neoplasms. The histologic diagnosis of high-grade urothelial neoplasms is based on pronounced architectural and cellular disarray, and the diagnosis of low-grade urothelial neoplasms considers mainly architectural abnormalities with nondetectable or mild cellular changes.

When these urothelial abnormalities are evaluated by urine cytology, the architectural parameter is absent or minimally represented, and cell morphology becomes the main diagnostic clue. However, cytologic examination of exfoliated urothelial cells may offer interpretive difficulties due to the cell morphology and the manner in which the urine sample is obtained. Because of their high cellularity and excellent cell preservation, catheterized urine specimens are generally preferable to voided specimens. However, occasionally operator-dependent and inherent patient differences influence the yield of cellularity, resulting in hypocellular specimens (Kusuda et al., 1993). Bladder wash or barbotage as well as cystoscopically obtained urine from the ureter have a high diagnostic yield in cases of bladder and renal-pelvis or ureteral neoplasms, respectively (Kannan, 1990).

In spite of such limitations, urine cytology is an excellent method for detecting urinary tract abnormalities (Badalament et al., 1987; Morrison et al., 1984; Schwalb et al., 1994).

CYTOLOGIC CLASSIFICATION OF UROTHELIAL NEOPLASMS

The cytologic classification of bladder neoplasms must be simple, reproducible, of high histologic correlation, and clinically significant. A tentative classification of urine cytology diagnoses of neoplasms of the urinary tract is given in Table 6–3. Briefly, a urine sample submitted for cytologic examination should be categorized as "negative for malignancy," "atypical urothelial cells," "suspicious for malignancy," or "positive for malignant cells." Histologic diagnoses of urothelial neoplasms according to various classifications and their corresponding cytologic diagnoses are listed in Table 6–4.

The diagnosis "negative for malignancy" should be followed by terms such as "normal urothelial cells" or "reactive urothelial cells." A description of cellular details should be included when necessary. Specific diagnoses included in this category are described in previous chapters.

The diagnosis "positive for malignant cells" should address the specific type of malignancy, such as urothelial carcinoma in most instances, or adenocarcinoma, squamous carcinoma, small cell carcinoma, and so on. The specific diagnosis of "low-grade papillary urothelial carcinoma," "high-grade urothelial carcinoma," "carcinoma in situ," or other malignancies should follow the main urothelial carcinoma diagnosis when possible. Features inherent in the cytomorphology of high-grade carcinomas, including papillary and in situ urothelial and other types, may preclude their accurate classification. The diagnosis "suspicious for malignancy" should be used for those cases showing malignant cells quantitatively insufficient for a diagnosis of malignancy.

The diagnosis "atypical urothelial cells" should be used only for those lesions with cytologic changes short of carcinoma. Such lesions need to be investigated exhaustively. Because this endeavor inevitably increases the health care cost, the use of an atypical diagnosis should be issued with judgment and caution. Clinical correlation and familiarity with the cytomorphology of normal and reactive urothelial cells, as well as confidence in the interpretation of a malignant process, will result in fewer atypical diagnoses.

Table 6–3. Report Format for the Cytologic Diagnosis of Urinary Tract Neoplasms

Categorization	*Diagnosis**	*Comment*
• "Negative for malignancy"	• Normal urothelial cells • Reactive urothelial cells	• Description of additional cellular and noncellular elements i.e., inflammation, blood, crystals, casts.
• "Atypical urothelial cells"†	• Descriptive diagnosis	• Description of additional cellular and noncellular elements i.e., inflammation, blood, crystals, casts.
• "Positive for malignant cells"	• Low-grade papillary urothelial carcinoma • High-grade papillary urothelial carcinoma • Carcinoma in situ • Invasive urothelial carcinoma • Adenocarcinoma • Squamous carcinoma • Small cell carcinoma • Other malignancies	• Description of additional cellular and noncellular elements i.e., inflammation, blood, necrosis.

*A description of cellular abnormalities should be included in particular for the "atypical" and "positive" cases.

†Includes the following diagnoses: papillary urothelial neoplasm of low malignant potential (grade 1, WHO 1973), some low-grade papillary urothelial carcinomas (grade 2, WHO 1973), atypical and dysplastic flat urothelial lesions.

Table 6–4. Corresponding Histologic and Cytologic Diagnoses of Urothelial Neoplasms

Histology (Various Terminologies)	*Cytology*
Papillary lesions	
Papilloma	Normal
PUNLMP*	Normal Urothelial cell atypia
Grade 1 papillary carcinoma†	Normal Urothelial cell atypia
Grade 2 papillary carcinoma†	Low-grade urothelial carcinoma Urothelial cell atypia
Low-grade urothelial carcinoma*	Low-grade urothelial carcinoma Urothelial cell atypia
Grade 3 papillary carcinoma†	High-grade urothelial carcinoma
High-grade urothelial carcinoma*	High-grade urothelial carcinoma
Flat lesions	
Urothelial hyperplasia	Normal
Reactive atypia*	Normal, reactive
Atypia of unknown significance*	Atypia (mild)
Dysplasia	Atypia (marked)
Carcinoma in situ	Carcinoma in situ

Abbreviation: PUNLMP = papillary urothelial neoplasm of low malignant potential.

*WHO/ISUP (1998).

†WHO (1973, 1999).

A focused description of cytologic findings is always appropriate and welcomed by the clinical team, and should be included in all malignant, atypical, and reactive diagnoses.

PAPILLARY UROTHELIAL NEOPLASMS

Urothelial Papilloma

Cytology: "Negative for Malignancy. Normal Urothelial Cells"

Urothelial papilloma is at the normal-appearing end of the phenotypic spectrum of urothelial neoplasms. It consists of a delicate fibrovascular core decorated with architecturally and cytologically normal urothelium (Mostofi et al., 1999). Urothelial papilloma is rare, representing less than 3% of papillary urothelial neoplasms.

Patients with urothelial papilloma have a low risk of recurrence and little or no risk of developing urothelial carcinoma (Cheng et al., 1999c). Some investigators believe that this entity should be included under the category of grade 1 papillary urothelial carcinoma (PUNLMP; WHO/ISUP, 1998) (Jordan et al., 1987). However, patients with PUNLMP are at increased risk of local recurrence, progression, and death from bladder carcinoma (Cheng et al., 1999d).

Cytology. The diagnosis of urothelial papilloma is strictly histologic (Box 6–1). The lack of architectural and cytologic abnormalities is compatible with the cytologic diagnosis of "Negative for malignancy. Benign urothelial cells." However, urine cytology specimens from patients with papilloma

Box 6–1.

Urothelial papilloma is rare and is composed of both architecturally and cytologicaly normal urothelium.

Smears show increased numbers of benign-appearing urothelial cells.

The diagnosis is strictly histologic.

show an increased number of exfoliated cells likely to be derived from the surface urothelium of the papilloma, rare cells with minimal atypia, and a small number of red blood cells (Wolinska et al., 1985). Similar findings associated with cellular degeneration have been described in inverted papillomas, but a cytologic diagnosis of urothelial neoplasm was not possible (Highman, 1978).

The Cytologic Diagnosis of Atypia

The diagnosis of atypia (Box 6–2) has strong clinical implications. As much as we dislike this term, there is a place for it in urine cytology, as there is in examining other organs that are subject to cell exfoliation. The term *atypia* is highly subjective and should be used wisely. In cytology, atypia implies cellular changes beyond those of reactive processes, but that fall short of those seen in malignancy.

The cytologic diagnosis of atypia should trigger the investigation of a significant urothelial lesion. Ideally, all cases should correspond to one of the following histologic diagnoses: almost all PUNLMP (grade 1 papillary urothelial carcinoma), some cases of low-grade papillary urothelial carcinoma (grade 2 papillary urothelial carcinoma), and dysplastic flat lesions. In all instances, a good clinical and cystoscopic correlation will make the cytologic interpretation significant.

The urologist is aware of the low diagnostic accuracy of urine cytology for the diagnosis of PUNLMP and some low-grade papillary urothelial carcinomas. These lesions are usually seen cystocopically, and tissue diagnosis confirms the cytologic findings. However, the cytologic diagnosis of atypia as the result of an underlying dysplastic flat lesion is more significant because, cystoscopically, these lesions may go undetected or appear as erythematous/inflamed lesions similar in appearance to carcinoma in situ, and they have an increased risk of progressing to carcinoma in situ or invasive carcinoma (Cheng et al., 2000a). This fact reinforces the strong line of communication that should exist between the pathologist and the clinician.

The pathologist's experience in the identification of subtle but significant cytologic features results in the diagnosis of truly atypical cases. However, often the term *atypia* is liberally used to describe cytologic features of variable significance (Fig. 6–1A–D). These features include vacuolated cytoplasm and prominent round nucleoli (they indicate reactivity and should be differentiated from high-grade urothelial carcinoma or adenocarcinoma), cell clusters in instrumented urine specimens (they carry an increased risk of low-grade papillary urothelial neoplasm), and cells with a round nucleus and a high nuclear-cytoplasmic ratio (they may indicate a low-grade urothelial neoplasm), cells with irregular nuclear contours and a high nuclear-cytoplasmic ratio (they may indicate a high-grade urothelial carcinoma) (Renshaw, 2000b).

The diagnosis of atypia must be based on the evaluation of well-preserved cells (Fig. 6–2A–E). A meticulous microscopic examination of the smear is recommended to identify rare nondegenerated cells with dark, coarse chromatin, which is diagnostic of high-grade urothelial carcinoma but is commonly misinterpreted as degenerated atypical cells (Fig. 6–3). When the evidence is quantitatively insufficient, the diagnosis "suspicious for malignancy" (usually high-grade carcinoma) should be rendered (Fig. 6–4).

Box 6–2.

A cytologic diagnosis of "atypia" implies cellular changes beyond reactivity but short of malignancy.

It is a diagnosis that encompasses most PUNLMPs, some low-grade papillary urothelial carcinomas, and dysplatic flat lesions.

The underlying papillary neoplasm can be detected cystoscopically; however, flat lesions may go undetected.

Papillary Urothelial Neoplasm of Low Malignant Potential (Grade 1 Papillary Urothelial Carcinoma)

Cytology: "Negative for Malignancy" or "Urothelial Cell Atypia"

Histologically, once we exclude papilloma, the remainder of papillary urothelial neo-

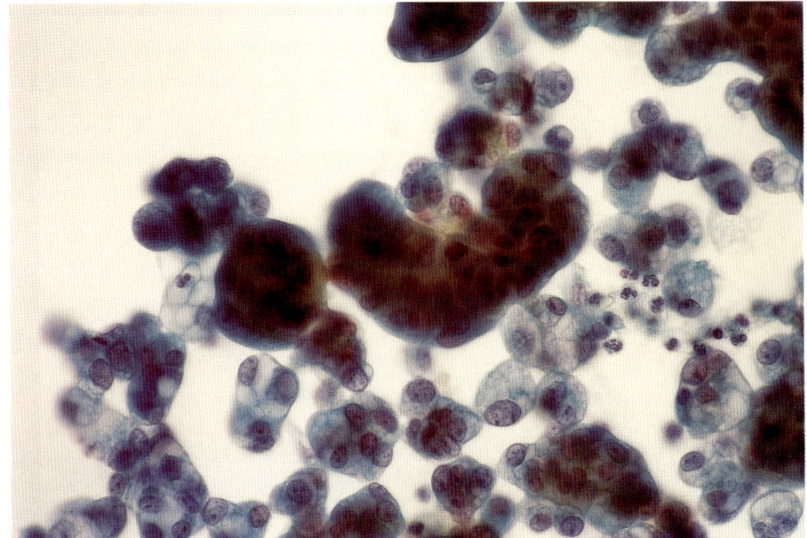

A

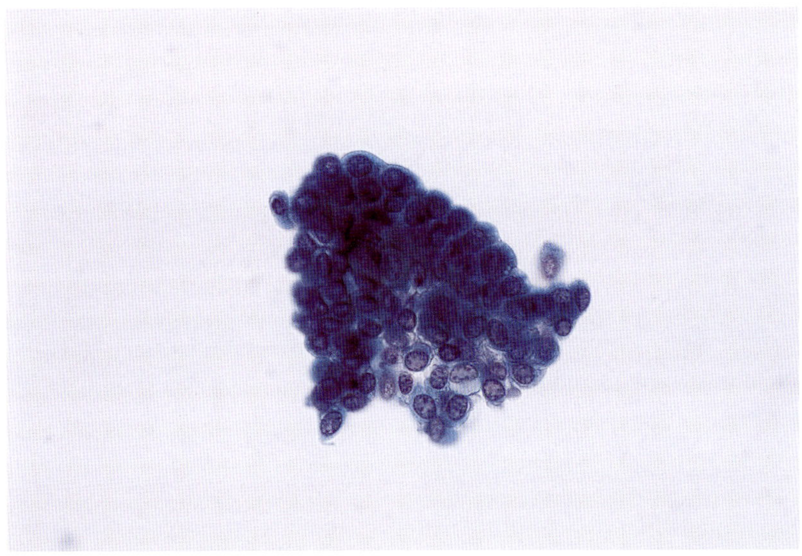

B

Figure 6–1. *Atypical urothelial cells.* (*A*) Numerous cell clusters and single cells with a round nucleus, vacuolated cytoplasm, and prominent nucleoli, seen as the result of cystitis and instrumentation, may be misinterpreted as atypia. (*B*) Cohesive cell clusters with a frayed edge, found in instrumented urine, may be a significant finding and should be correlated with the cystoscopic findings to rule out a papillary urothelial neoplasm. (*C*) A cluster of cells with a high nuclear-cytoplasmic ratio and a round nucleus may indicate a low-grade papillary urothelial neoplasm. (*D*) The presence of rare cells with a high nuclear-cytoplasmic ratio, irregular nuclear contours, and a prominent nucleolus indicates a high-grade urothelial carcinoma and should not be interpreted as atypia. All specimens are bladder washings. (Papanicolaou stain, ×600)

plasms may be difficult to grade consistently because there is no clear dividing line between grade 1 and grade 2 or grade 2 and grade 3 of the 1973 WHO classification (Carbin et al., 1991; Cheng and Bostwick, 2000; Olsen et al., 1993; Ooms et al., 1983; Robertson et al., 1990). As a result, it is difficult to determine the biologic potential of low-grade papillary urothelial neoplasms because of a lack of standardization of the pathologic classification and definition (Cheng et al., 1999d).

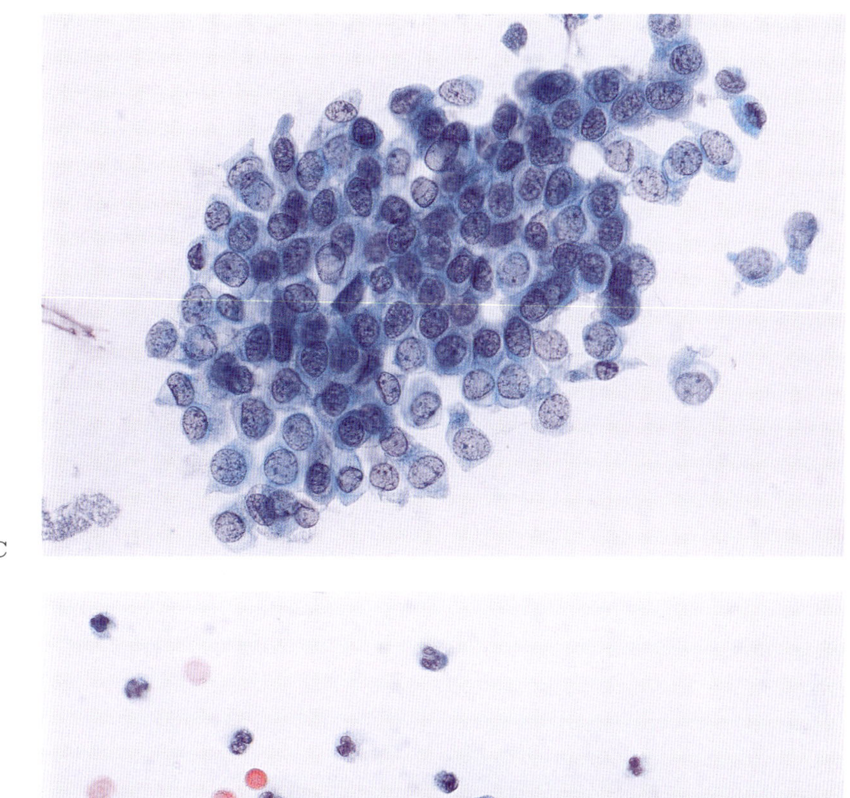

C

D

Figure 6–1. (*Continued*)

Histologically, PUNLMP or grade 1 papillary urothelial carcinoma of the 1973 WHO classification has more than seven cell layers (Fig. 6–5) and is defined as a "papillary urothelial lesion with orderly arranged cells within papillae with minimal architectural abnormalities and minimal nuclear atypia" (Epstein et al., 1998). Most patients have hematuria, and some have irritative and obstructive symptoms. PUNLMP appears to have a predilection for the ureteric orifice, and the most common cystoscopic appearance is that of a papillary tumor or, less commonly, an erythematous or normal urothelium.

The short-term prognosis for patients with PUNLMP is excellent. However, these patients are at significant risk of local recurrence (30%), progression (3%), and death from bladder cancer (3%–4%) after ten years (Bostwick et al., 1999; Cheng et al., 1999d; Greene et al., 1973). Based on these findings, the diagnosis of PUNLMP has been questioned and the elimination of the term has been suggested (Cheng et al., 1999d).

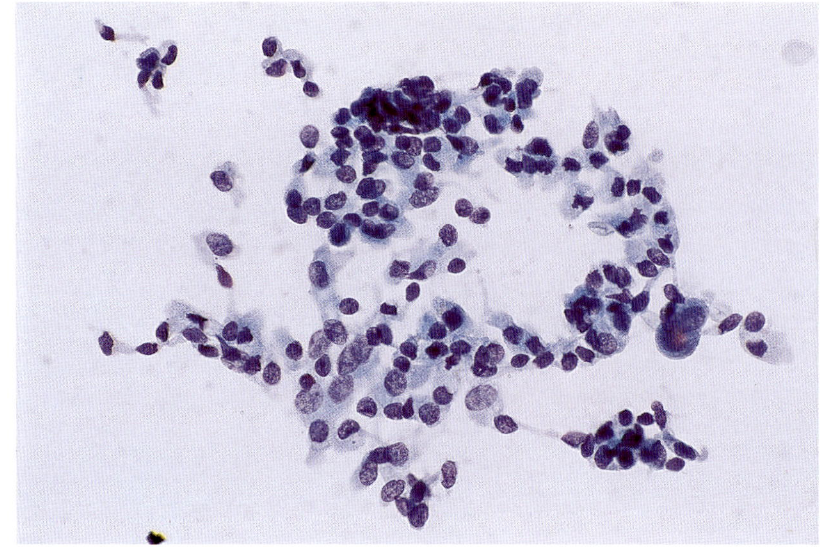

A

B

Figure 6–2. *Urothelial cell degeneration and atypia.* (*A*) Degenerated cells exhibiting nuclear hyperchromasia with an indiscernible chromatin pattern and a low nuclear-cytoplasmic ratio should not be considered atypical. (*B*) Likewise, degenerated elongated cells with an oval/elongated nucleus, granular chromatin, and a preserved nuclear-cytoplasmic ratio often are associated with repair, inflammation, or infection and should not be interpreted as atypical. (*C*) A finding of urothelial cells with a globular body and cytoplasmic elongation is significant because these cells are frequently seen in low-grade urothelial neoplasms. Biopsy diagnosis in this case showed a low-grade papillary urothelial carcinoma of the ureter. (*D*) These keratinized, highly atypical squamous cells were the only evidence of a high-grade urothelial carcinoma diagnosed in a subsequent urine sample. (*E*) A well-preserved cluster of atypial urothelial cells with large nuclei, slight anisonucleosis, and a low-nuclear cytoplasmic ratio is seen in this papillary neoplasm of low malignant potential; two umbrella cells are also present. All specimens are bladder washings. (Papanicolaou stain, ×600)

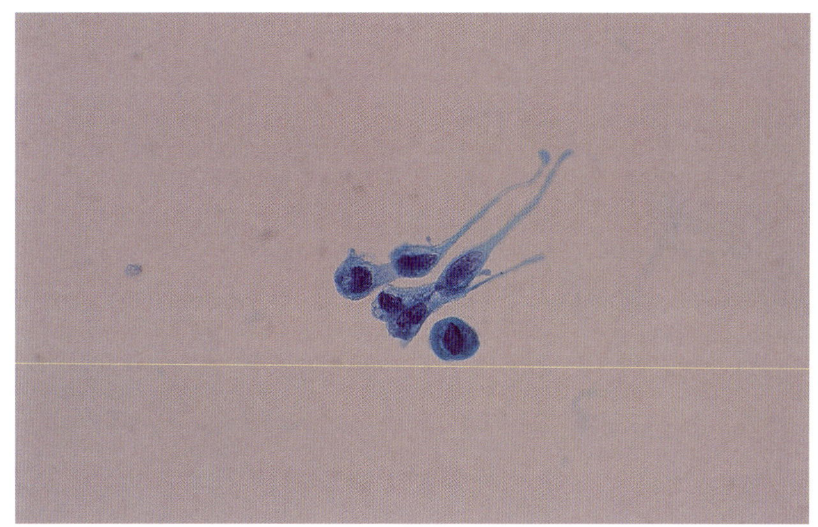

C

D

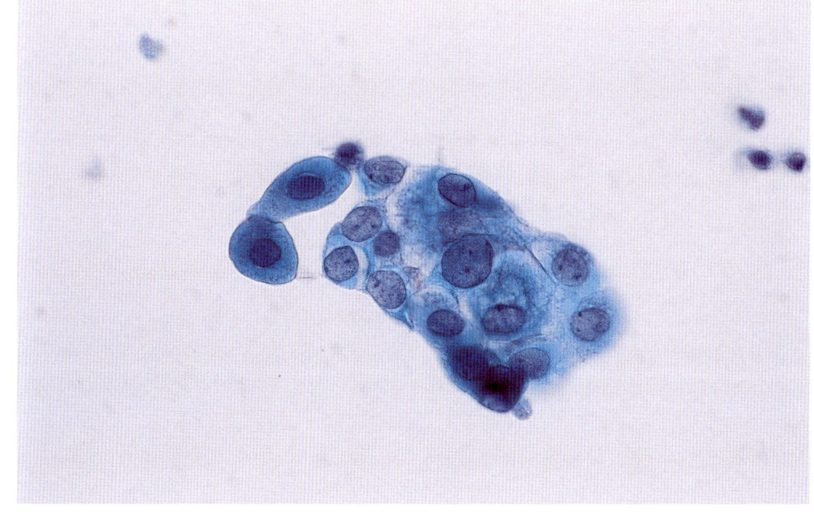

E

Figure 6–2. (*Continued*)

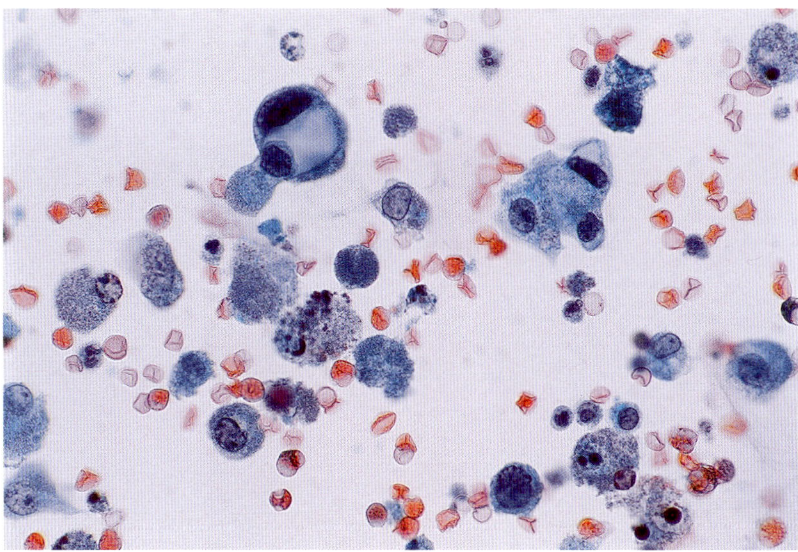

Figure 6–3. *Marked cellular degeneration.* An exhaustive search for viable malignant cells avoids the diagnosis of atypia in cases of invasive carcinoma that usually exhibit necrosis and cell degeneration with inflammation and blood. Bladder washing. (Papanicolaou stain, ×600)

The diagnosis of PUNLMP is cystoscopic and histologic. If the lesion remains undetected, the slow progression and the high recurrence rate with repeated bouts of hematuria or urinary symptoms will prompt cystoscopy and biopsy. If PUNLMP progresses to a higher grade, urine cytology is an effective and inexpensive method for detecting these changes (Morrison et al., 1984) (Box 6–3).

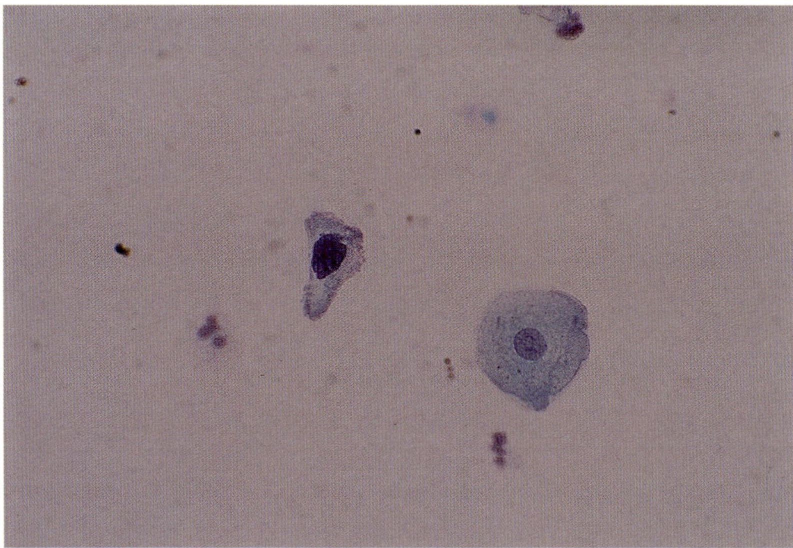

Figure 6–4. *Urothelial cells suspicious for carcinoma.* This voided urine specimen showed a single cell with a large nucleus and irregular nuclear contours, hyperchromasia, and irregular chromatin clumping. Subsequent workup showed a high-grade urothelial carcinoma of the ureter. (Papanicolaou stain, ×600)

Box 6–3.

PUNLMP has more than seven layers of urothelial cells, in an orderly arrangement and with minimal architectural and cytologic atypia.

Smears show urothelial cells, often with homogeneous cytoplasm, slight nuclear enlargement, and a slightly high nuclear-cytoplasmic ratio.

The cytologic features are similar to those seen in the instrumentation effect and urolithiasis.

The diagnosis remains elusive by urine cytology and should be made histologically.

Cytology. These neoplasms exhibit bland cytomorphologic features, and their DNA content is usually diploid when assessed by image analysis or flow cytometry. Lesions classified as PUNLMP are very difficult to diagnose in urine samples because of the striking similarity of the cells to normal and reactive urothelial cells (Fig. 6–6A–D). Most cytologic features are so subtle that a high degree of interpretive subjectivity is inevitable, resulting in numerous studies of low reproducibility. Because of the low sensitivity of urine cytology for diagnosing this neoplasm, much effort has been placed on the evaluation of cytologic criteria and on adjunctive cytologic methods for improving its diagnostic accuracy.

The cytologic features of PUNLMP (Table 6–5) include slight to moderate cellular and nuclear enlargement, eccentric nuclei, homogeneous, slightly basophilic cytoplasm, a slightly high nuclear-cytoplasmic ratio, and lack of nuclear pleomorphism. Nuclear features include smooth nuclear contours, slight hyperchromasia with slightly denser granular chromatin in occasional cells, and an occasional nucleolus (Fig. 6–7A,B). These features are not present in all cells, and not all PUNLMPs share identical cell characteristics (Esposti and Zajicek, 1972; Kern, 1975; Murphy et al., 1984; Rife et al., 1979).

A homogeneous cytoplasm, irregular nuclear contours, and a high nuclear-cytoplasmic ratio as primary criteria, and eccentrically placed nuclei and nuclear hyperchromasia as secondary criteria, are useful

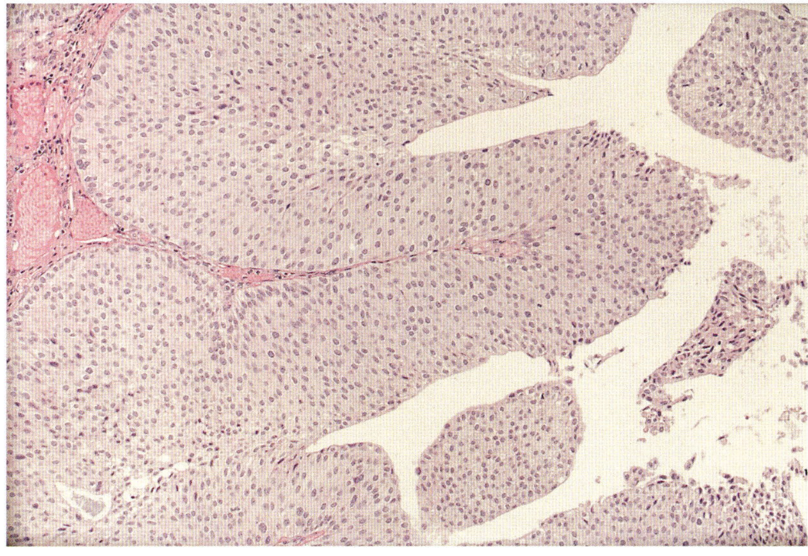

Figure 6–5. *Papillary urothelial neoplasm with low malignant potential (grade 1 papillary urothelial carcinoma).* Tissue biopsy showed a papillary tumor with fibrovascular cores decorated by numerous layers of neoplastic cells with bland cytologic features. (Hematoxylin and eosin stain, ×100)

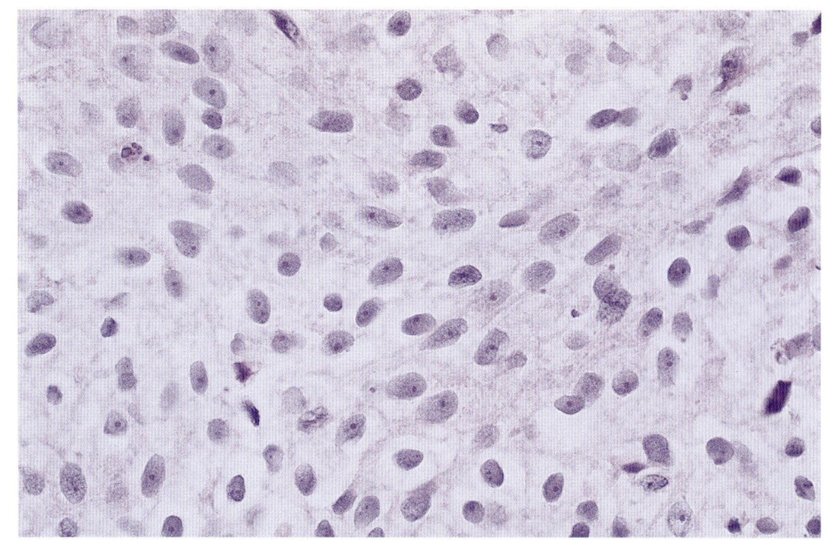

A

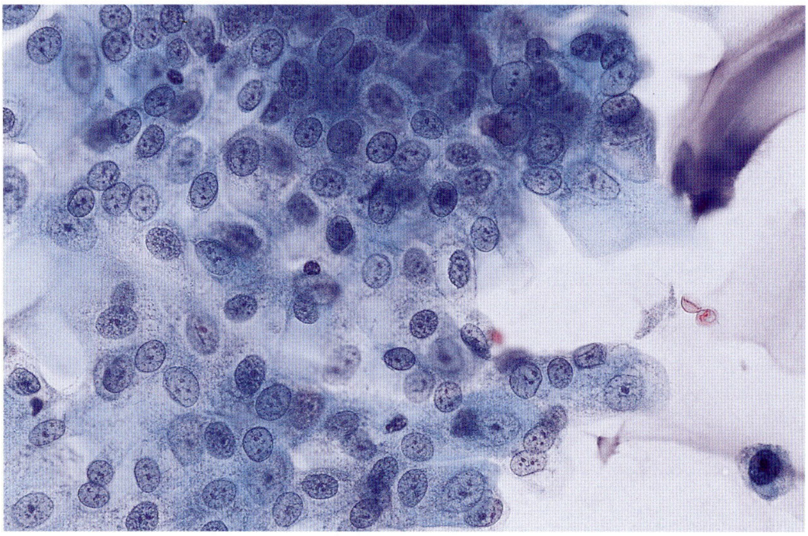

B

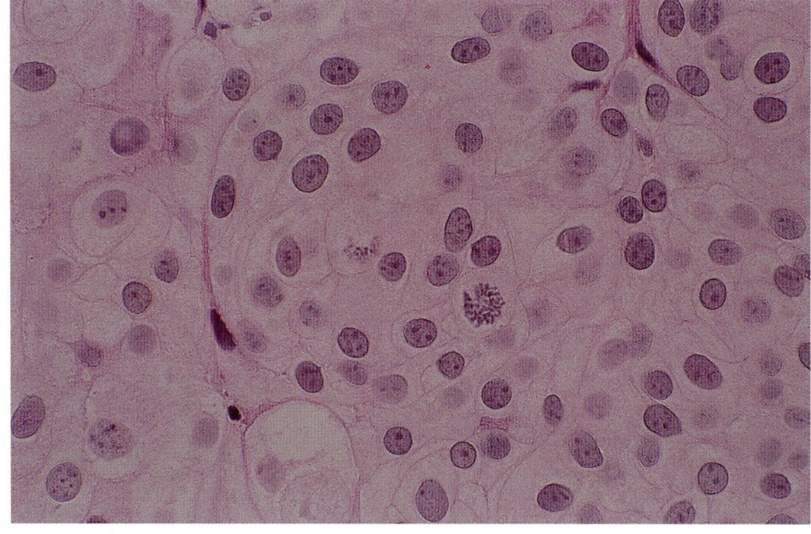

C

Figure 6–6. (*Continued on next page*)

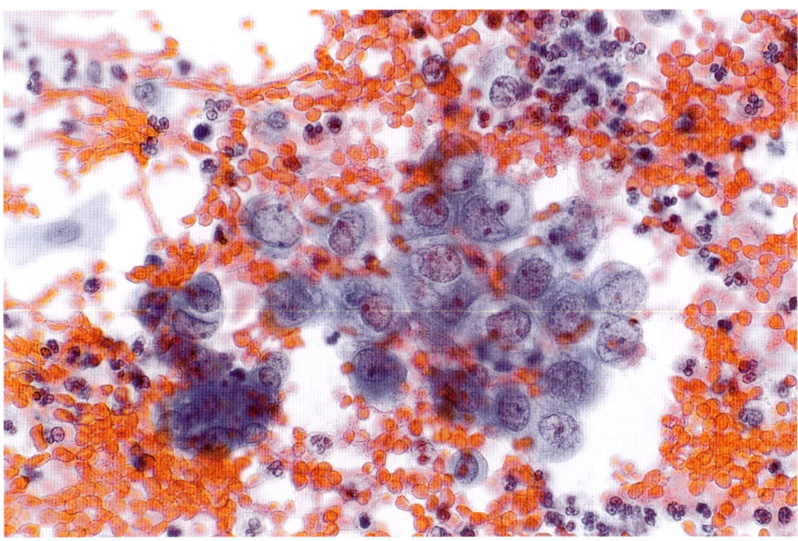

Figure 6–6. *Cytohistologic correlation of PUNLMPs.* These two cases illustrate the bland cytologic features that characterize these neoplasms. (*B*) The nuclear-cytoplasmic ratio is not high, the cytoplasm is thin and semitransparent, and the nuclei are round/oval with powdery chromatin, uniform "pencil-drawn" nuclear contours, and small nucleoli. (*D*) Inflammation and intact red blood cells may be present and make it more difficult to distinguish these neoplastic cells from reactive cellular changes. (*A,C*) Tissue biopsy specimens show bland cytologic features with a rare mitotic figure and ample cytoplasm. Bladder washings. (A,C, Hematoxylin stain, ×600; B,D, Papanicolaou stain, ×600)

Table 6–5. Comparative Cytology of Papillary Urothelial Neoplasms

	Reactive	*Atypia**	*LGPUC†*
Pattern			
Cellularity	Variable	Variable	Variable/high
Single cells	Rare	Rare	Variable
Cell arrangement	Papillary	Papillary	Papillary and loose
Cell size	Increased & uniform	Increased & uniform	Increased & variable
Cytoplasm	**Vacuolated**	**Usually homogeneous**	**Homogeneous**
NCR	**Preserved**	**Slightly increased**	**Increased**
Nuclear			
Position	Central/eccentric	Eccentric	Eccentric
Enlargement	Slight	Slight/moderate	Moderate
Size variation	Absent/slight	Slight	Moderate
Shape	Oval	Oval/round	Round & variable
Shape variation	Absent (uniform)	Slight	Moderate
Contours	**Smooth**	**Smooth**	**Few notches**
Pleomorphism	Absent	Absent	Mild to moderate
Hyperchromasia	Absent	Absent/mild	Mild/moderate
Chromatin	Fine, uniform	Fine, uniform	Granular, variable
Nucleolus	Large, round	Inconspicuous	Variable

Abbreviation: LGPUC = low grade papillary urothelial carcinoma; NCR = nuclear-cytoplasmic ratio.

*Includes papillary urothelial neoplasms of low malignant potential (WHO/ISUP, 1998) or grade 1 papillary urothelial carcinomas (WHO, 1973).

†Includes low-grade papillary urothelial carcinomas (WHO/ISUP, 1998) and grade 2 papillary urothelial carcinoma (WHO, 1973).

The most significant features appear in boldface type.

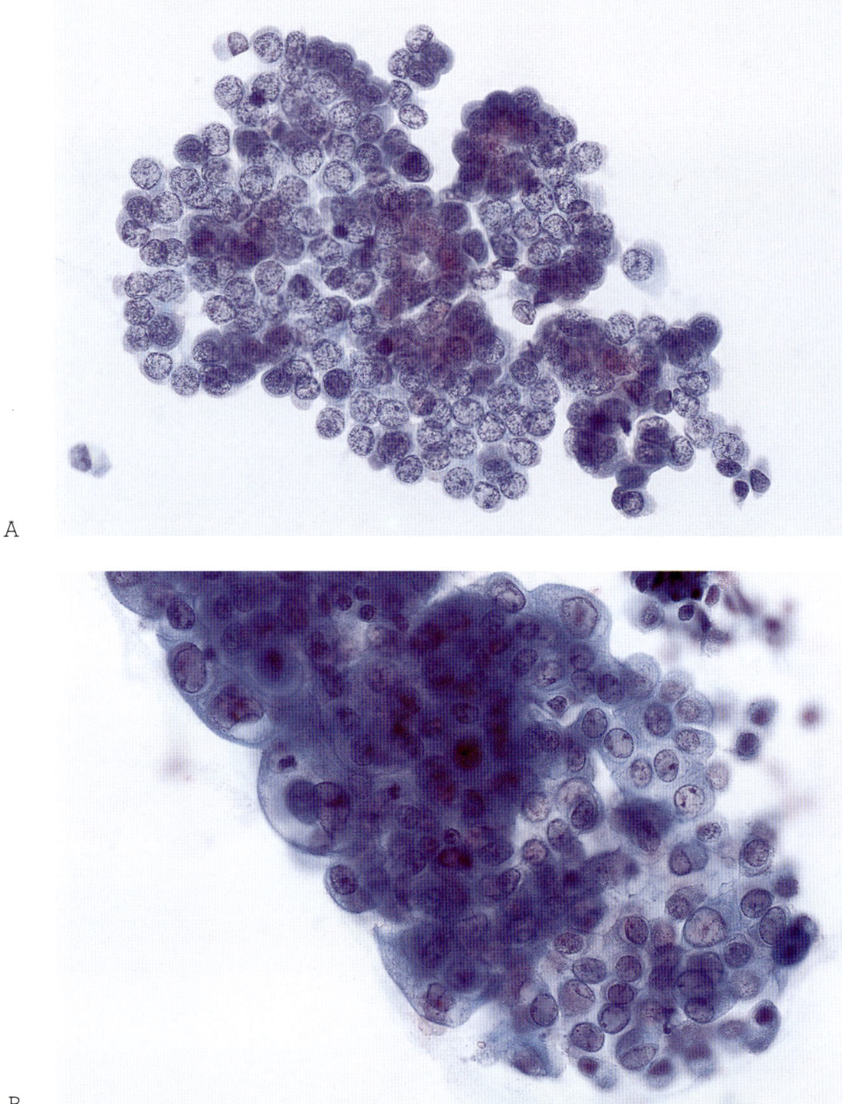

Figure 6–7. *Clusters of PUNLMP.* (*A*) This cohesive cell cluster is composed of a uniform population of neoplastic cells with a high nuclear-cytoplasmic ratio. (*B*) Admixture with umbrella cells is not uncommonly seen in PUNLMP and low-grade carcinomas. Mild anisonucleosis and a low nuclear-cytoplasmic ratio are present in the neoplastic cells. Bladder washings. (Papanicolaou stain, ×600)

for the diagnosis of low-grade papillary urothelial neoplams in a study that combines grades 1 and 2 papillary urothelial carcinomas. These criteria have been applied to the diagnosis of PUNLMP only, and the results have not been encouraging (Fig. 6–8A–D) (Hughes et al., 2000; Raab et al., 1994; Renshaw et al., 1996). These data emphasize the fact that the diagnosis of PUNLMP based on cytologic criteria only remains elusive (Koss, 1996).

The value of papillary clusters as a diagnostic criterion for PUNLMP is controversial. In instrumented urine, cell groups of any type or number are no more common in patients with PUNLMP than in those with nonneoplastic conditions such as medullary sponge kidney, hemorrhagic cystitis, cystitis not otherwise specified, or urolithiasis (Goldstein et al., 1998). In voided urine, papillary clusters are more common in nonneoplastic conditions, and only occasionally voided urine shows papillary clusters of PUNLMP. Because the cytologic atypia present in the papillary clusters in some cases of urolithiasis further complicates the differential diagnosis, the finding of a fibrovascular core is the only diagnostic feature for

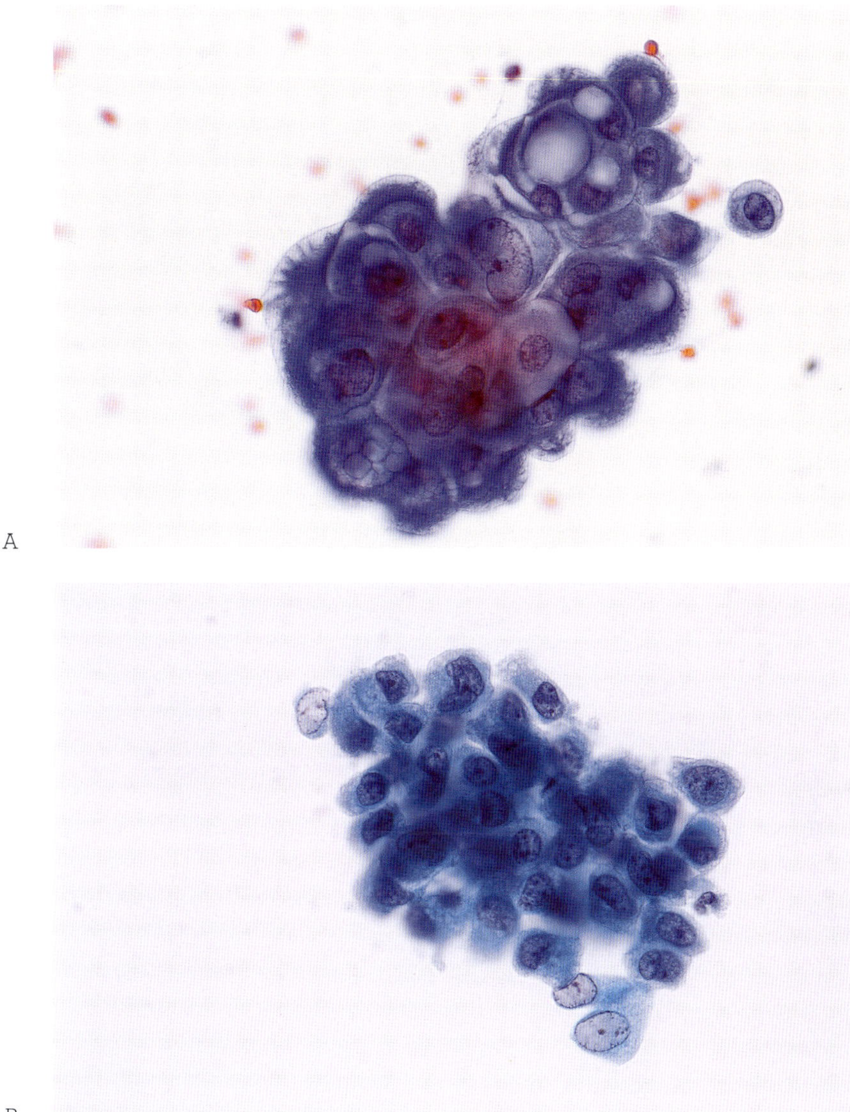

Figure 6–8. *Diverse cytomorphologic features of PUNLMP.* (*A*) Only rare isolated neoplastic cells are present in PUNLMP. Cellular features can best be evaluated at the edges of the clusters. A low nuclear-cytoplasmic ratio, vacuolated cytoplasm, and slight anisonucleosis are seen in this case. (*B*) A low nuclear-cytoplasmic ratio, homogeneous cytoplasm, irregular nuclear contours, and small chromocenters are seen in this voided urine specimen. (*C*) This cluster, which exhibits cells with homogeneous cytoplasm and a low nuclear-cytoplasmic ratio, features a few cells with irregular nuclear contours, a high nuclear-cytoplasmic ratio, and cell phagocytosis. (*D*) Hyperchromatic nuclei with a regular chromatin distribution and an occasional larger nucleus are present. Note the absence of inflammation in all cases. All specimens except B are bladder washings. (Papanicolaou stain, ×600)

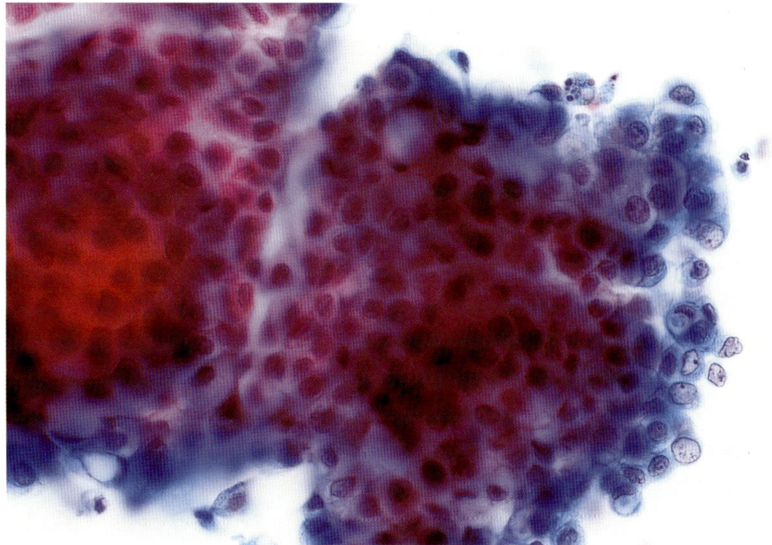
C

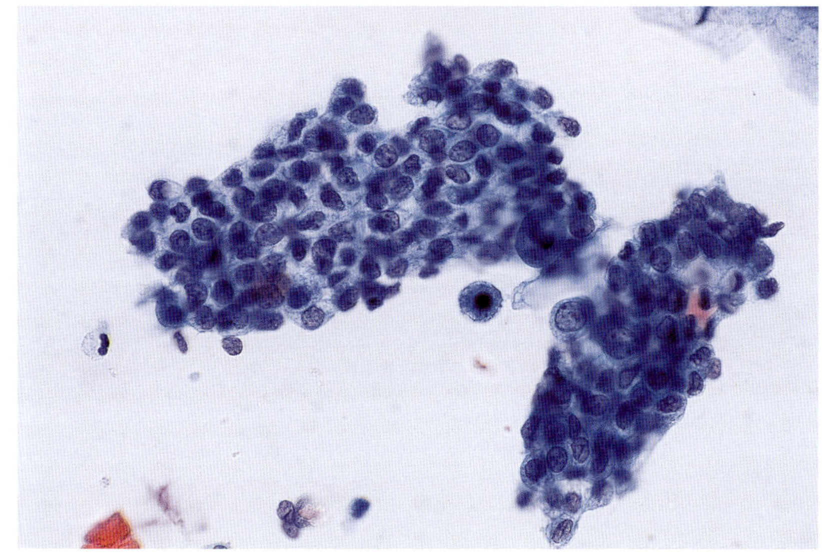
D

Figure 6–8. (*Continued*)

papillary urothelial carcinoma (Fig. 6–9) (de Ruiter et al., 1975; Highman and Wilson, 1982). Unfortunately, this feature is reliably assessed commonly in a cell block prepared from urine samples or tissue sections (Harris et al., 1971). With the exception of the irregular clusters seen in urothelial carcinomas of higher grade, the identification of cell groups of any type in voided urine has little diagnostic utility (Goldstein et al., 1998).

In summary, cytomorphology in most cases of PUNLMP overlaps with that of reactive conditions, and only a rare PUNLMP can be identified cytologically as a malignancy. Thus, most cases will be given the cytologic diagnosis of "Negative for malignancy. Reactive urothelial cells" or "Atypical urothelial cells," followed by a descriptive diagnosis.

If anaplastic urothelial cells are detected in urine cytology and the biopsy specimen shows only PUNLMP, then it is reasonable to assume that the patient harbors a lesion of higher grade and that fur-

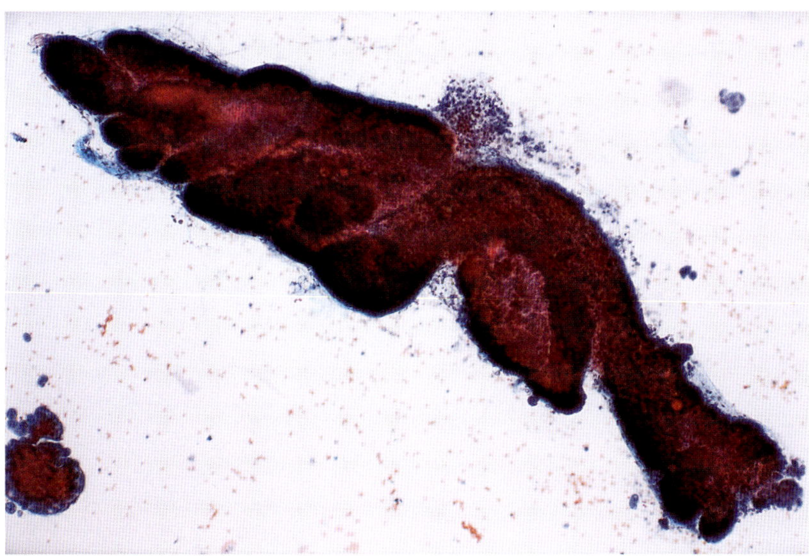

Figure 6–9. *Papillary cluster of PUNLMP.* Papillary clusters are rare findings in urine specimens. The fibrovascular core can be evaluated best in cell block sections. Bladder washing. (Papanicolaou stain, ×100)

ther clinical investigation is warranted (Fig. 6–10).

Differential diagnosis. A few cell clusters may be a diagnostic feature of particular importance only in voided urine specimens, as long as urolithiasis, recent trauma or instrumentation, and inflammatory conditions of the urinary tract are excluded. Single round or oval cells with round or oval nuclei and irregularly shaped cells are pres-

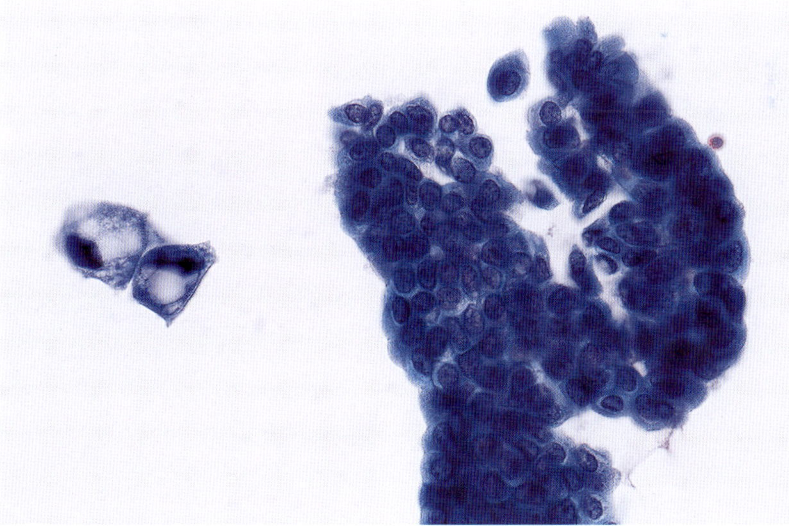

Figure 6–10. *High-grade urothelial carcinoma and PUNLMP.* Cytologic identification of high-grade carcinoma cells along with a low-grade papillary neoplasm indicates tumor dedifferentiation and illustrates the advantages of urine cytology evaluation in the follow-up of these patients. Bladder washing. (Papanicolaou stain, ×600)

ent in both benign and malignant cases (Kalnins et al., 1970).

A calculous artifact can produce papillary clusters similar if not identical to those of PUNLMP and low-grade papillary urothelial carcinoma. Although the less cohesive nature of clusters, overlapping nuclei, ragged edges, and peripheral palisading favor a neoplastic process, the two cannot be separated conclusively on the basis of urine cytology alone (Highman and Wilson, 1982). Urolithiasis may be associated with significant but reversible nuclear atypia, and to avoid diagnostic pitfalls, further investigation is suggested after removal of the calculus (Kannan and Gupta, 1999).

The cohesive cell clusters observed in instrumented urine and bladder wash specimens resulting from traumatic exfoliation of normal urothelial cells are similar to the cell clusters of PUNLMP and low-grade papillary urothelial carcinoma (Koss, 1996). The specimens are cellular and show cohesive, ball-shaped papillary clusters with smooth, even borders lined with a densely staining cytoplasmic collar that may be helpful in distinguishing instrumented from neoplastic clusters (Kannan and Bose, 1993). The significance of this feature remains to be tested in a larger series of cases (Renshaw et al., 1996).

Reactive urothelial cell changes seen in inflammatory conditions affecting the urinary tract may produce cell exfoliation with features similar to those seen in PUNLMP and low-grade urothelial carcinoma. Therefore, in the presence of acute inflammation, urothelial cell changes must be regarded with caution, and the diagnosis of atypia or malignancy should not be made unless frank nuclear anaplasia is present. Cells with vacuolated cytoplasm, smooth nuclear borders, isonucleosis, and a prominent round nucleolus in a background of inflammation, crystals, and red blood cells indicate a reactive process.

Low-Grade Papillary Urothelial Carcinoma (Grade 2 Papillary Urothelial Carcinoma)

Cytology: "Positive for Malignancy. Low-Grade Papillary Urothelial Carcinoma"

At the time of initial diagnosis, 50% to 75% of noninvasive papillary urothelial carcinomas are PUNLMP (grade 1 papillary urothelial carcinoma) or low-grade papillary urothelial carcinoma (grade 2 papillary urothelial carcinoma). These are also the

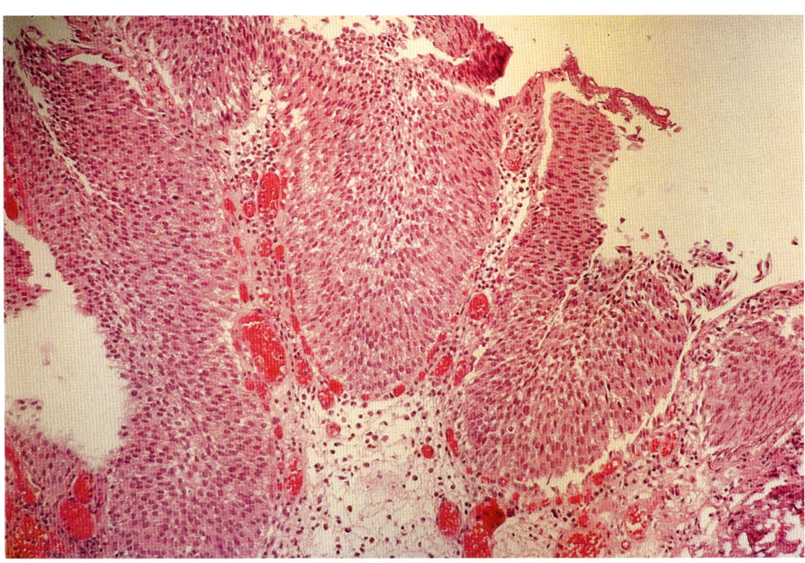

Figure 6–11. *Low-grade (grade 2) papillary urothelial carcinoma.* Tumor biopsy specimen shows papillary fronds decorated by neoplastic cells with hyperchromatic nuclei. (Hematoxylin and eosin stain, ×100)

most common papillary neoplasms diagnosed in patients between the ages of 15 and 20 years.

Because histologic features such as superficial cell layer preservation, mitotic activity, and cellular crowding prevail over cytologic features of irregular nuclear contours and the chromatin pattern, an exact cytologic–histologic correlation may not be possible (Fig. 6.11) (Murphy et al., 1984). However, in contrast to most PUNLMPs, most of these urothelial tumors have cellular features that permit their ready identification in urine cytology specimens (Fig. 6–12A,B). Problems in the

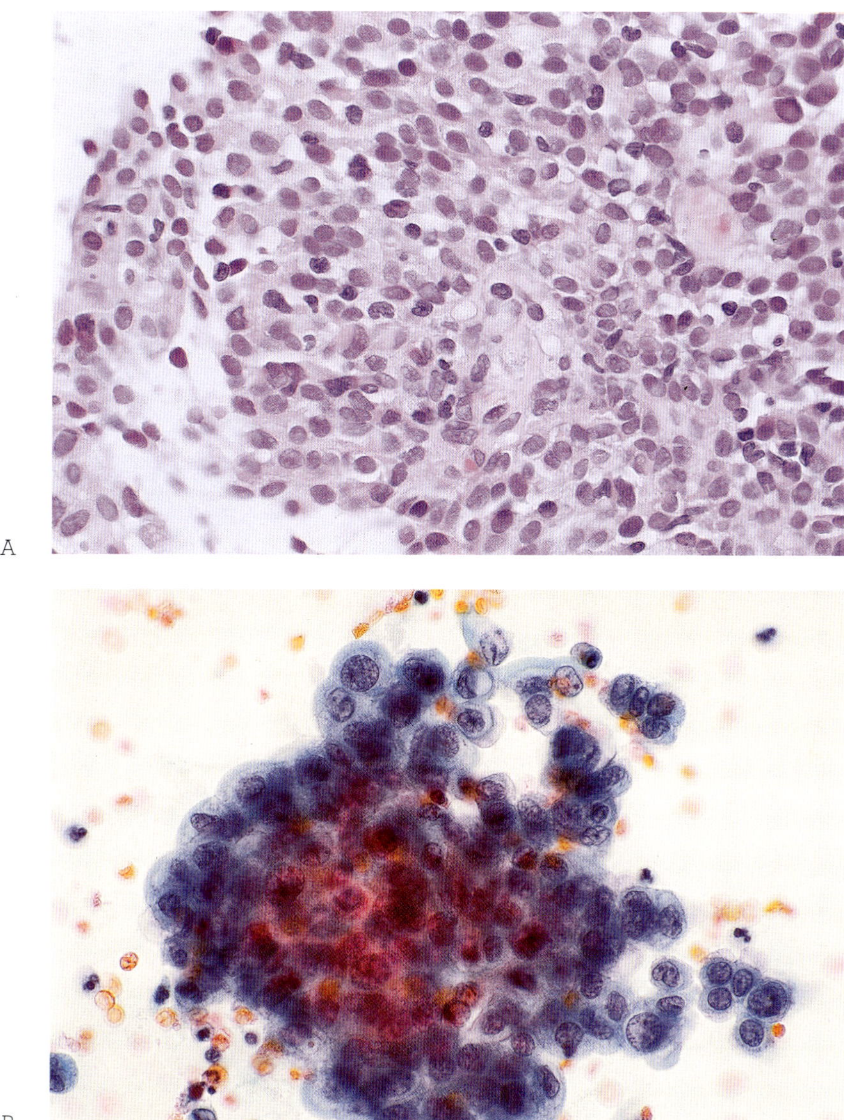

Figure 6–12. *Cytohistologic features of low-grade papillary urothelial carcinoma.* (*A*) Tissue section shows cellular enlargement, moderate pleomorphism, and nuclear overlapping with scattered, hyperchromatic nuclei. (*B*) Despite the fine vacuolization of the cytoplasm, neoplastic cells show a moderately increased nuclear-cytoplasmic ratio, mild to moderate hyperchromasia and pleomorphism, and slight nuclear membrane irregularity. (A, Hematoxylin and eosin, ×400; B, Papanicolaou stain, ×600)

identification of these neoplasms in urine cytology include the presence of small, inconspicuous clusters of malignant cells that may be missed on low-power examination (Bastacky et al., 1999).

Cytology. The majority of patients harboring low-grade papillary urothelial carcinoma yield abnormal cells in urine specimens (Epstein et al., 1998). The cytologic interpretation in such cases varies, and diagnoses of "atypia" in a few cases, "low-grade papillary urothelial carcinoma" in most cases, and "high-grade carcinoma" on occasion are rendered (Fig. 6–13A,B). This diversity of cytologic diagnoses reflects the heterogeneity of this tumor group (Di-Bonito et al., 1992; Kern, 1975).

The cytologic features of low-grade papillary urothelial carcinoma are summarized in Table 6–5. Smears show predominantly large clusters with a few small clusters and single cells. The cells are enlarged, with enlarged nuclei and an increment in the nuclear-cytoplasmic ratio. There is a slight variation in cellular and nuclear size and shape, with an overall mild to moderate pleomorphism. The cytoplasm is homogeneous, and the nuclei show slightly irregular contours with a few notches, mild to moderate hyperchromasia, granular chromatin, and frequently a small nucleolus (Fig. 6–14A–D) (Esposti and Zajicek, 1972; Kern, 1975; Murphy et al., 1984; Rife et al., 1979).

The presence of cercariform cells, that is, urothelial cells with a nucleated globular body and a variable slender cytoplasmic process with a nontapered and flattened end, has been described as a useful marker of urothelial differentiation that is present in primary low-grade papillary urothelial carcinomas and metastatic urothelial carcinomas (Fig. 6–15) (Moriyama, 1989; Powers and Elbadawi, 1995).

Unfortunately, the lack of consensus in the diagnostic terminology for reporting the histologic diagnosis of low-grade papillary urothelial neoplasms (PUNLMP and grade 2 papillary urothelial carcinoma) makes the cytologic diagnosis more confusing. In a cytologic study of low-grade papillary urothelial neoplasms, Raab and coworkers identified a homogeneous cytoplasm, an irregular nuclear membrane, and an increased nuclear-cytoplasmic ratio (defined as greater than 1 to 3) as the three criteria most useful in the separation of low-grade urothelial neoplasms from nonneoplastic processes in bladder wash specimens. They concluded that, when any two of these criteria are present, the sensitivity of the test is 85% and the specificity is 96% (Raab et al., 1994, 1996). Nuclear hyperchromasia and eccentrically placed nuclei were helpful when used in conjunction with the former three criteria. These criteria have been found useful in the evaluation of voided and instrumented urine specimens processed by conventional methods. Studies must be conducted to evaluate these criteria in the diagnosis of low-grade urothelial carcinoma in urine specimens processed using monolayer technologies (Hughes et al., 2000).

Correlation of the cytologic interpretation with clinical and cystoscopic findings is helpful in diagnosing equivocal cases. As in cases of PUNLMP, the finding of high-grade malignant cells is significant and merits further investigation to prove the coexistence of a high-grade or invasive carcinoma (Box 6–4).

Differential diagnosis. Occasionally, urothelial injury associated with instrumentation, urolithiasis, and polypoid cystitis results in exfoliation of atypical urothelial fragments with a high nuclear-cytoplasmic ratio, leading to a false interpretation of low-grade papillary urothelial carcinoma. The presence of smooth nuclear contours and a vacuolated cytoplasm favors a nonneoplastic condition.

Box 6–4.

Most low-grade papillary urothelial carcinomas yield abnormal cells in urine specimens.

Smears show urothelial cells with homogeneous cytoplasm, a moderate increment in the nuclear-cytoplasmic ratio, slightly irregular nuclear contours, and moderate hyperchromasia.

The finding of high-grade urothelial carcinoma cells in cases with tissue biopsy of a low-grade neoplasm is significant and indicates a coexistent high-grade lesion.

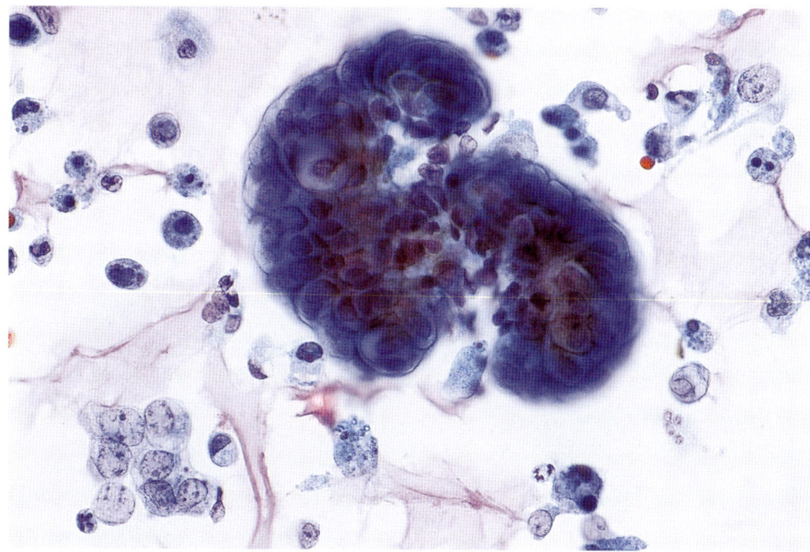

A

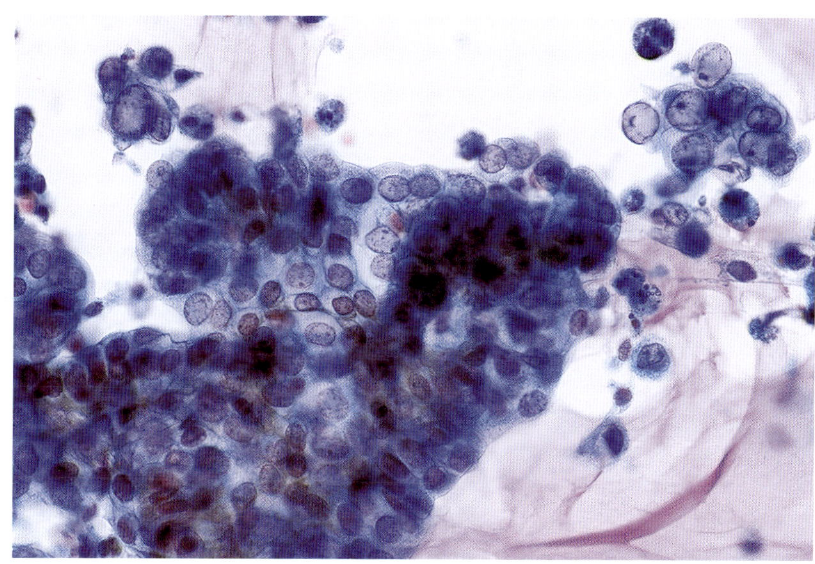

B

Figure 6–13. *Low-grade papillary urothelial carcinoma.* (*A*) Neoplastic cells are seen (left lower corner) along with a cluster of benign urothelial cells with features of the instrumentation effect (center). (*B*) A large cluster of PUNLMP cells are present together with low-grade papillary urothelial carcinoma cells (right upper corner). Note the less cohesive nature, larger size, and higher nuclear-cytoplasmic ratio of the neoplastic cells compared with the benign (A) and PUNLMP (B) cells. Bladder washings. (Papanicolaou stain, ×600)

High-Grade Papillary Urothelial Carcinoma (Grade 3 Papillary Urothelial Carcinoma)

Cytology: "Positive for Malignancy. High-Grade Urothelial Carcinoma"

The WHO/ISUP diagnosis "Positive for malignancy. High-grade papillary urothelial carcinoma" designates grade 3 papillary urothelial carcinoma of the 1973 WHO classification.

High-grade papillary urothelial carcinomas exhibit pronounced histologic and cytologic abnormalities characterized by loss of cellular polarity, variation in cell size, pleomorphic and often hyperchromatic nuclei, a high nuclear-cytoplasmic ratio, and

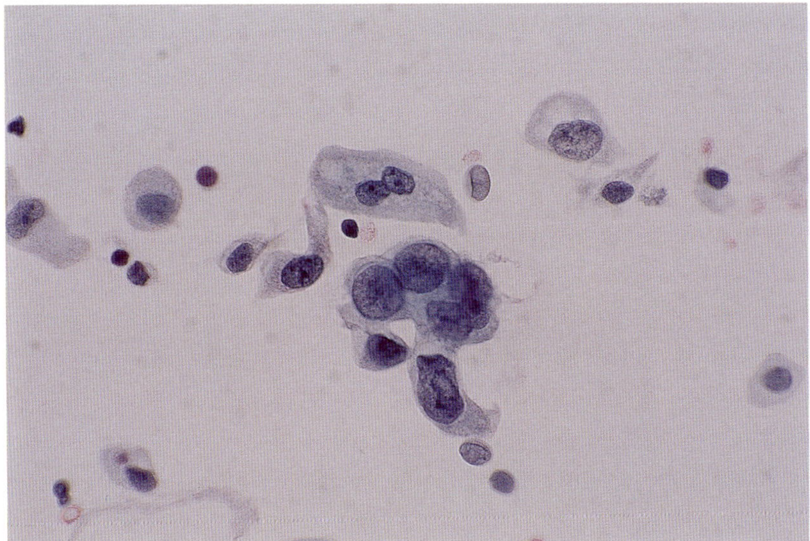

A

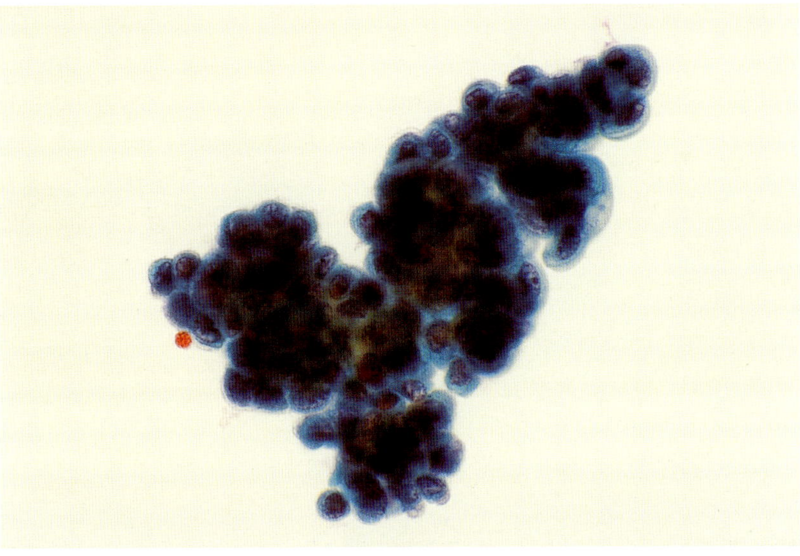

B

Figure 6–14. *Low-grade papillary urothelial carcinoma.* (*A*) Clusters of neoplastic cells showing large nuclei, an increased nuclear-cytoplasmic ratio, homogeneous cytoplasm, and irregularity of the nuclear contours are present. (*B–D*) Dense clusters of neoplastic cells with an increased nuclear-cytoplasmic ratio, dense cytoplasm, and moderate hyperchromasia (*B*), a few cells with nuclear notches and moderate pleomorphism (*C*), and occasional cell phagocytosis (*D*) are also seen. All specimens are bladder washings. (Papanicolaou stain, ×600)

high mitotic activity (Fig. 6–16). Foci of glandular, small cell, oncocytic, and particularly squamous differentiation may be identified in these tumors but have no apparent clinical significance (Grignon, 1997). Squamous differentiation and mucin-containing cells can be present in up to 20% and 60% of high-grade urothelial carcinomas, respectively (Donhuijsen et al., 1992; Sakamoto et al., 1992).

Cellular pleomorphism and an aneuploid DNA content are characteristic of these high-grade neoplasms. Thus, urine cytology has high accuracy in making the diagnosis and is the preferred method for screening and surveillance of patients with these neo-

C

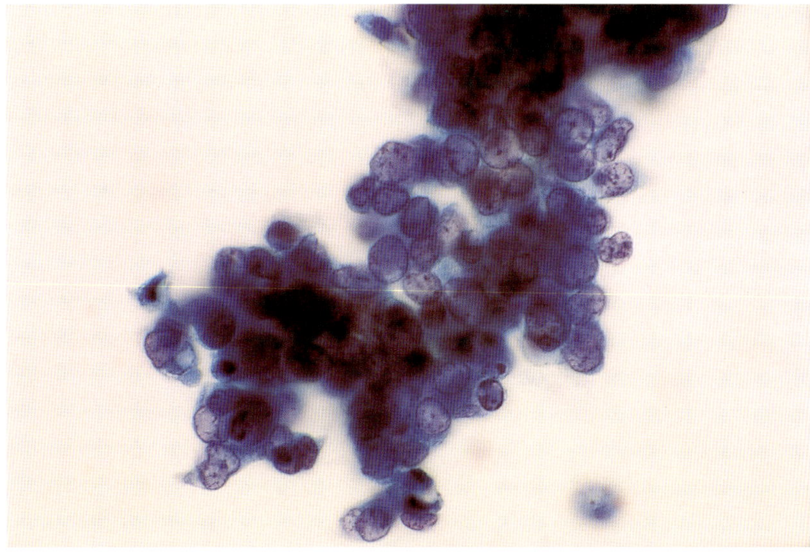

D

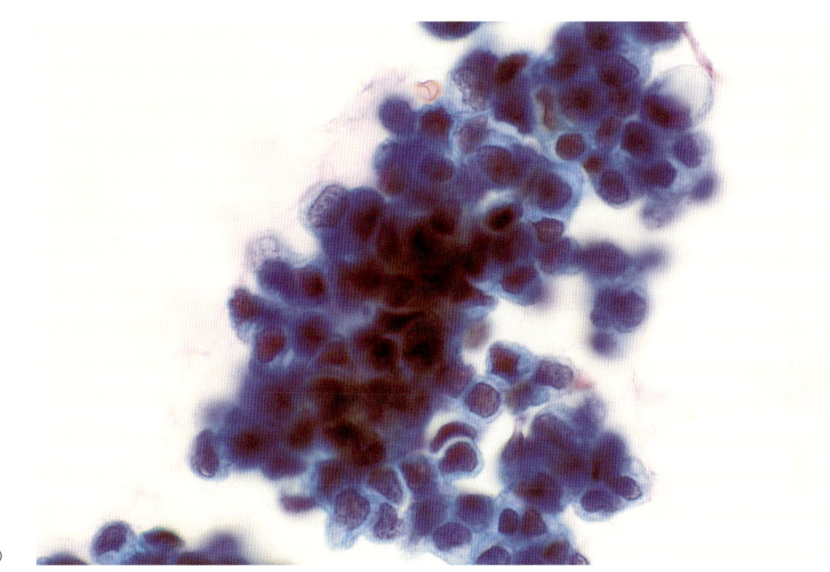

Figure 6–14. *(Continued)*

plasms. Shedding of tumor cells may not be constant, and evidence of malignancy may not be present in all voided urine specimens. In this regard, instrumented urine specimens are more suitable and yield higher cellularity than does voided urine for the interpretation of high-grade urothelial tumors (Epstein et al., 1998).

Cytology. The cytologic features of high-grade papillary urothelial carcinoma are summarized in Table 6–6. Smears usually exhibit high cellularity and a dishesive cellular pattern of high-grade tumor cells, both singly and in clusters (Fig. 6–17A–C).

Malignant cells are larger than normal and show pleomorphic nuclei with prominent nucleoli. The cytoplasm ranges from homogeneous and scant to vacuolated and abundant, with a resulting variable nuclear-cytoplasmic ratio. Nuclear contours are irregular, and the nucleus is usually hyperchromatic with clumped, irregular chromatin, although it may appear vesicular. The nucleoli are large, with irregular contours, and may be multiple. In comparison

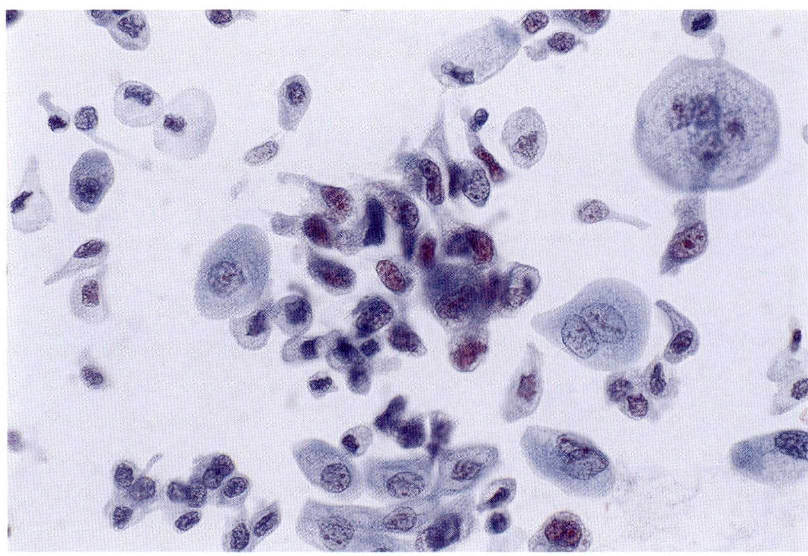

Figure 6–15. *Cercariform cells.* Cells with cytoplasmic elongations are more evident in PUNLMPs and low-grade papillary urothelial carcinoma than in high-grade carcinomas. Large reactive superficial umbrella cells with vacuolated cytoplasm are also present. Bladder washing specimen. (Papanicolaou stain, ×600)

to tissue biopsy, specimens, urine cytology smears show fewer mitotic figures. The smear background is usually bloody, but necrosis is absent. However, coexistent urolithiasis, a permanent bladder indwelling catheter, or any other cause of mucosal tissue damage may hinder the evaluation of the smear background, making the distinction from invasive carcinoma difficult, if not impossible (Esposti and Zajicek, 1972; Kern, 1975; Murphy et al., 1984; Rife et al., 1979).

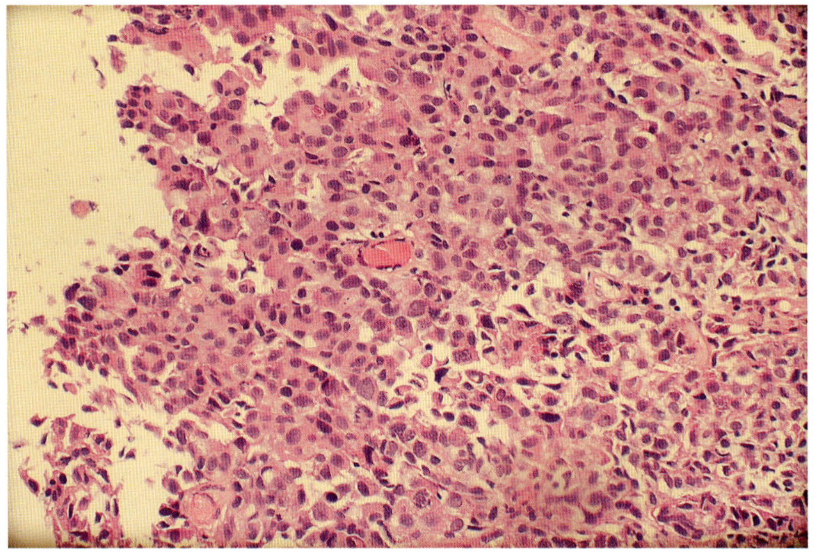

Figure 6–16. *High-grade papillary urothelial carcinoma.* Bladder biopsy specimen shows residual papillary architecture and marked cellular pleomorphism. (Hematoxylin and eosin stain, ×100)

Table 6–6. Comparative Cytology of High-Grade Papillary Urothelial Carcinoma and Invasive Urothelial Carcinoma

	*HGPUC**	*Invasive UC*
Pattern		
Cellularity	Usually high	Usually high
Single cells	Numerous	Numerous
Cell arrangement	**Tight and loose clusters**	**Loose clusters**
Cell size	Large, pleomorphic	Large, pleomorphic
Cytoplasm	Homogeneous/vacuolated	Homogeneous/vacuolated
NCR	Variable	Variable
Nuclear		
Position	Eccentric	Eccentric
Enlargement	Marked	Marked
Size variation	Marked	Marked
Pleomorphism	Moderate/marked	Moderate/marked
Contours	Irregular, notches	Irregular, notches
Shape	Marked variation	Marked variation
Hyperchomasia	Marked	Marked
Chromatin	Irregular, coarse	Irregular, coarse
Nucleolus	Variable/large, irregular	Variable/large, irregular
Abnormal mitosis	May be present	May be present
Necrosis	**Absent**	**Present**

Abbreviation: HGPUC = high-grade papillary urothelial carcinoma; UC = urothelial carcinoma; NCR = nuclear-cytoplasmic ratio.

*Includes grade 3 papillary urothelial carcinoma (WHO, 1973).

The most significant features appear in boldface type.

Cytophagocytosis or "cannibalism," a characteristic cytologic feature associated with malignancy, appears to be an indicator of both high nuclear grade and invasiveness of urothelial carcinoma (Fig. 6–18A,B) (Kojima et al., 1998).

Box 6–5.

Because shedding of high-grade tumor cells may not be constant, instrumented urine specimens are more reliable than voided urine.

Smears show pleomorphic cells with anaplastic nuclei and prominent nucleoli.

"Cannibalism" appears to be an indicator of high grade and invasiveness.

The smear background may be bloody, but necrosis is absent.

Squamous and glandular features are not uncommon and should not be interpreted as indicative of adenocarcinoma or squamous cell carcinoma.

Glandular differentiation manifested by the presence of mucin-producing cells is not uncommon in high-grade urothelial carcinoma. Likewise, malignant urothelial cells may show squamous features that should not be interpreted as indicative of squamous carcinoma.

In cases of scant cellularity and in the rare events in which marked cell degeneration prevails, a diagnosis of "suspicious for malignancy" may be warranted (Box 6–5).

Differential diagnosis. The diagnosis of high-grade urothelial carcinoma can be made accurately by means of urine cytology examination. Malignant cells present in instrumented urine specimens are usually mixed with clusters of traumatically exfoliated reactive urothelial cells. Occasionally, the distinction may be difficult, particularly in the presence of a vacuolated cytoplasm and prominent nucleoli. However, close examination of the anaplastic nuclear and nucleolar characteristics makes the distinction from the instrumentation effect possible.

The distinction of high-grade papillary urothelial carcinoma from flat carcinoma in situ may be possible because of the presence

A

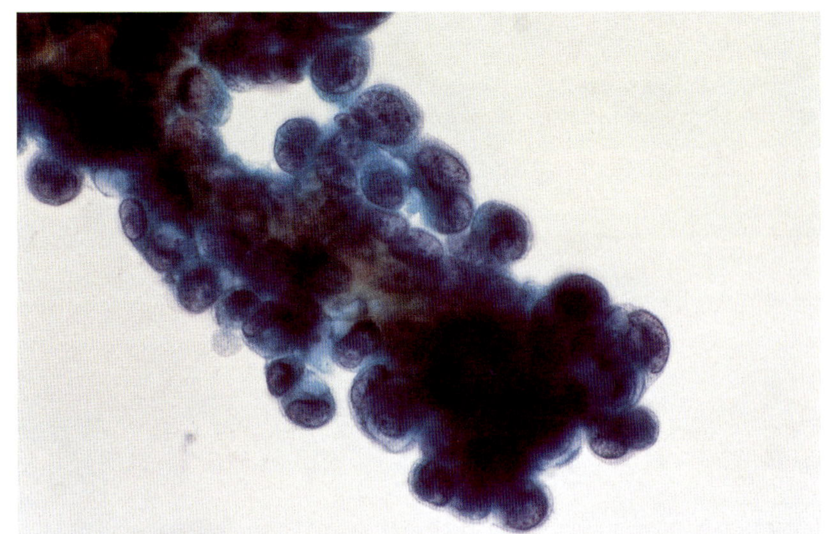

B

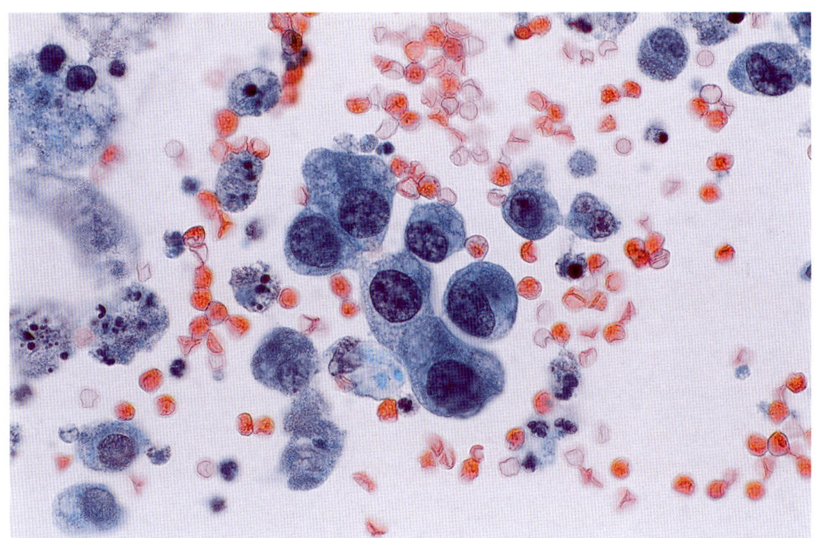

C

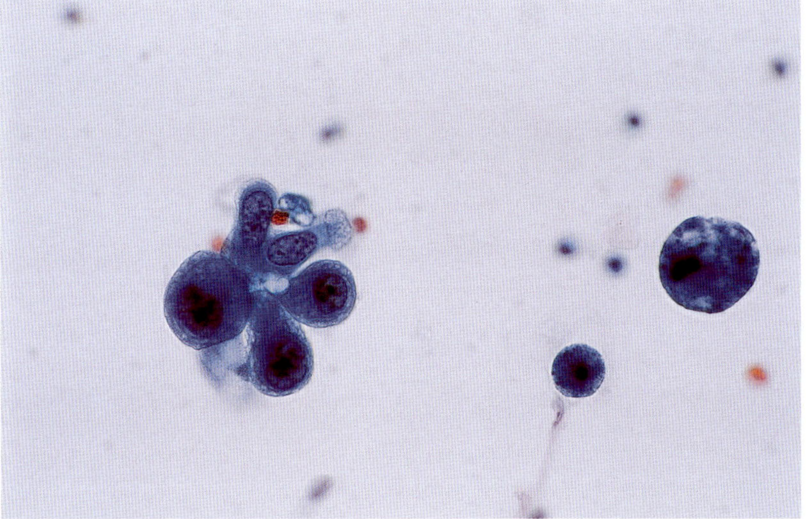

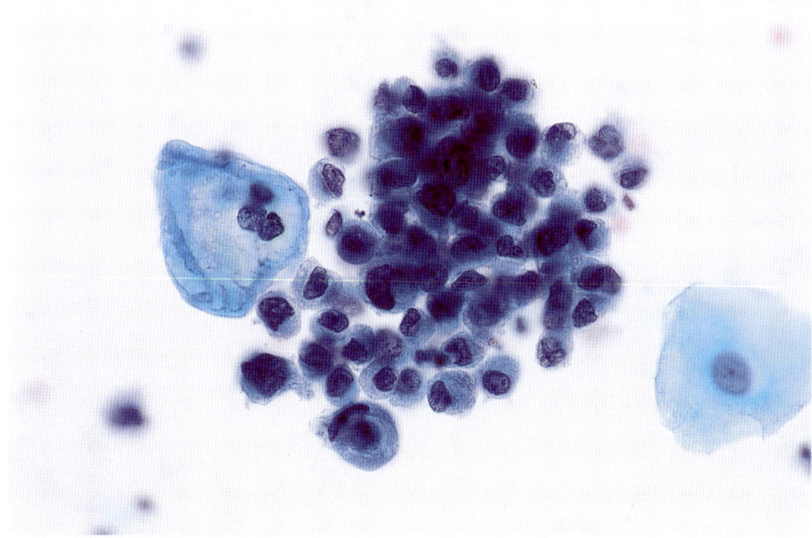

A

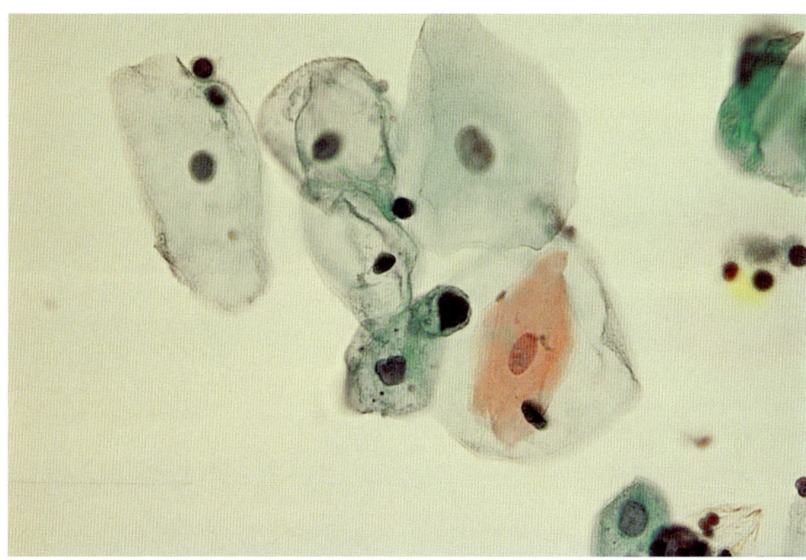

B

Figure 6–18. *High-grade papillary urothelial carcinoma in voided urine.* Voided urine specimens are reliable for diagnosing high-grade carcinoma. The cellular degeneration present in these two cases does not prevent the pathologist from making such a diagnosis. (*A*) Pleomorphism, hyperchromatic nuclei, coarse chromatin, and cell phagocytosis are seen. (*B*) An angulated and irregular nucleus is present in the single malignant cell. (Papanicolaou stain, ×600)

Figure 6–17. *Cytologic features of high-grade papillary urothelial carcinoma.* (*A*) Cluster of neoplastic cells showing large hyperchromatic nuclei, an irregular chromatin pattern, and a high nuclear-cytoplasmic ratio. (*B*) Small cluster of malignant cells exhibits a low-nuclear cytoplasmic ratio, vacuolated cytoplasm, hyperchromatic nuclei, and coarse chromatin. Individual necrotic cells and blood are also seen. (*C*) Concentric arrangement of cells with hyperchromatic nuclei, irregular nuclear contours, coarse chromatin, and prominent nucleoli resembles the acinar configuration of glandular neoplasms. All specimens are bladder washings. (Papanicolaou stain, ×600)

of more uniform tumor cells in carcinoma in situ, in contrast to the more pleomorphic population of larger cells in high-grade papillary urothelial carcinoma. Necrosis is absent in both instances; however, inflammation without other evidence of tissue damage may be present in both carcinomas.

FLAT UROTHELIAL NEOPLASMS

Reactive Atypia and Atypia of Unknown Significance

Cytology: "Reactive Urothelial Cells"

The histologic features of lesions with the two recently named diagnoses "reactive atypia" and "atypia of unknown significance" may be difficult to separate from each other or from those of urothelial dysplasia (Oyasu, 2000). Varying degrees of nuclear crowding, nuclear irregularity, loss of nuclear polarity, overlapping nuclei, coarse chromatin, and lack of cell maturation, described as characteristic of these lesions, may be interpreted differently by different pathologists (Epstein et al., 1998). Thus, the histologic diagnosis provided by pathologists may range from reactive changes to dysplasia.

The significance of the diagnoses of "reactive atypia" and "atypia of unknown significance" in patient management is questionable (Bostwick et al., 1999; Cheng et al., 2000a). A recent study suggests that patients with these entities do not have adverse clinical outcomes (Cheng et al., 2000a). In contrast, patients with urothelial dysplasia of the bladder have an increased risk of developing carcinoma in situ and urothelial carcinoma (Cheng et al., 1999a).

Therefore, in keeping with the fact that excessive subclassification of lesions may not have clinical significance and complicates the report format, the term *reactive atypia* should be avoided in histologic and cytologic diagnoses. Instead, the term *reactive urothelial cell changes*, which indicates no neoplastic significance, should be used. Further studies of patients with the diagnosis of "atypia of unknown significance" are necessary for a full assessment of the clinical implications of this entity.

Box 6–6.

The tissue diagnosis and significance of "reactive atypia" and "atypia of unknown significance" in patient management are questionable.

In urine cytology reports, the term *reactive atypia* should be avoided; instead, *reactive urothelial cell changes* should be used.

The cytologic characteristics of "atypia of unknown significance" are subjective and interpreted differently, i.e., as "reactive urothelial cell changes" or as "urothelial cell atypia" by different pathologists.

Cytology. The cytologic features of reactive atypia are those of reactive cellular changes described in previous chapters. Cells are slightly enlarged and show a slightly enlarged nucleus, bland chromatin, a prominent nucleolus, and a vacuolated cytoplasm. Often there is a history of instrumentation, lithiasis, intravesical therapy, or chronic inflammation, and the cellular changes are proportional to the magnitude and duration of the underlying nonneoplastic disorder.

Attempts to make a cytologic diagnosis of "atypia of unknown significance" should be discouraged until the significance of this term is fully assessed. Most of the cases will fall in the "reactive" category and few in the "atypical" category (Box 6–6).

Differential Diagnosis. The main differential-diagnostic considerations are discussed in Chapter 5 in the sections on urolithiasis, intravesical therapy, and instrumentation effect.

Urothelial Dysplasia

Cytology: "Atypical Urothelial Cells"

Urothelial dysplasia is the putative precursor of in situ and invasive urothelial carcinoma (Amin and Young, 1997; Cheng et al., 1999a). After exposure to chemical carcinogens, experimental models have shown that urothelial changes occur prior to the development of carcinoma in situ (Ohtani et al., 1986). Furthermore, urothelial dysplasia shares certain morphologic and genetic features with carcinoma in situ (Epstein et al., 1998).

Some clinical studies show that 15% of patients with urothelial dysplasia develop invasive carcinoma, in contrast to none of the patients with the histologic diagnosis of "atypia." It is generally agreed that *atypia* connotes a reactive process (Cheng et al., 1999a). In addition, dysplasia in the urothelial mucosa adjacent to papillary urothelial carcinoma is associated with an increased risk of cancer recurrence and progression (Bostwick et al., 1999). However, some authors believe that the histologic diagnosis of "dysplasia" should be avoided because this term lacks clear clinical significance and specific connotation of disease in the urothelium (Koss, 1996; Reuter and Melamed, 1989). Despite these controversies, recent histologic classifications support the use of the term *dysplasia* (Cheng et al., 2000a; Epstein et al., 1998).

The histologic diagnosis of dysplasia is notoriously subjective, and practicing pathologists may have difficulty distinguishing "dysplasia" from "atypia of unknown significance" or carcinoma in situ (Oyasu, 2000). Histologically, dysplastic lesions are defined as those that exhibit significant nonreactive architectural and cytologic changes but lack the full constellation of changes of carcinoma in situ (Cheng et al., 1999a). Cytologic changes are usually confined to the basal and intermediate layers, and the histological distinction from "reactive atypia" is said to rest on the nuclear and architectural abnormalities (Bostwick et al., 1999).

Because the histologic diagnosis of dysplasia is problematic, I believe that we do not have a sufficient foundation to support such a diagnosis in urine cytologic examinations. The term *urothelial dysplasia* in urine cytology is highly subjective and not reproducible, and its use should be discouraged. Instead, the term *urothelial cell atypia* should be used to describe nonreactive cellular features that fall short of urothelial carcinoma in situ. Thus, most dysplastic lesions may be classified as urothelial cell atypia, although some inevitably will be called carcinoma in situ. In trying to reproduce histologic classifications, some investigators have tried to subdivide this category into mild, moderate, or severe atypia or various grades of intraurothelial neoplasm (Bostwick et al., 1999; Friedell et al., 1986; Koss, 1996).

In summary, it is impossible to make the diagnosis of dysplasia by urine cytologic examination. The cytologic diagnosis of "atypia" has to be correlated with cystoscopic and biopsy results to exclude a significant flat lesion, that is, "dysplasia" or carcinoma in situ. In practice, all patients with Ta, T1, and Tcis are treated with intravesical therapy, and patients with dysplasia may not be treated any differently, particularly those with a history of previous carcinoma in situ.

Cytology. The main cytologic features of lesions considered histologically as low-grade dysplasia include a loose cellular arrangement, clear cytoplasm, slight nuclear enlargement, irregular nuclear contours with few notches, mild hyperchromasia, an irregular chromatin distribution without clumping, and mild pleomorphism (Amin and Young, 1997; Cheng et al., 1999a).

These changes occur in the absence of inflammation, or are not in proportion to the degree of inflammation or other underlying nonneoplastic processes, and are below the level of carcinoma in situ. As recommended by the WHO/ISUP consensus meeting, if inflammation is present, these patients need to be reevaluated once the inflammation subsides.

The cytologic features of atypia are summarized in Table 6–7 and illustrated in Figure 6–19A–D. I must emphasize that the cy-

Box 6–7.

The histologic diagnosis of "urothelial dysplasia" is subjective.

In urine cytology examination, the term *dysplasia* should be discouraged and the term *urothelial cell atypia* should be used instead.

Smears show a loose cell arrangement, a slightly increased nuclear-cytoplasmic ratio, mild hyperchromasia, irregular nuclear contours, and mild pleomorphism.

Table 6–7. Comparative Cytology of Flat Urothelial Lesions

	*Reactive**	*Atypia†*	*CIS*
Pattern			
Cellularity	Variable	Scant	Usually high
Single cells	Rare	Scattered	Numerous
Cell arrangement	**Papillary & loose**	**Loose**	**Loose**
Cell Size	Increased & uniform	Increased & uniform	Increased & pleomorphic
Cytoplasm	**Vacuolated**	**Clear/homogeneous**	**Homogeneous**
NCR	**Preserved**	**Slightly increased**	**Markedly increased**
Nuclear			
Position	Central/eccentric	Eccentric	Eccentric
Enlargement	Slight	Slight	Moderate, marked
Size variation	Absent	Slight	Moderate/marked
Shape	Oval	Oval/round	Variable
Shape variation	Absent (uniform)	Slight	Moderate/marked
Contours	**Regular/smooth**	**Irregular/notches**	**Pleomorphic**
Pleomorphism‡	Absent	Mild	Moderate, severe
Hyperchromasia	Absent	Mild	Moderate, marked
Chromatin	**Fine, regular**	**Granular, irregular**	**Coarse, irregular**
Nucleolus	Large, round	Absent/small	Variable/large

Abbreviations: CIS = carcinoma in situ; NCR = nuclear-cytoplasmic ratio.

*Includes "reactive atypia" (WHO/ISUP, 1998).

†Includes "atypia of uncertain significance" and low-grade dysplasia (WHO/ISUP, 1998).

‡Refers to uniform nuclear enlargement in all cells.

The most significant features appear in boldface type.

tologic diagnosis of atypia implies a significant lesion that merits further investigation (Box 6–7).

Differential diagnosis. At times, it may be difficult to differentiate atypical changes from reactive urothelial cell changes associated with instrumentation, inflammation, lithiasis, infections, or intravesical or systemic chemotherapy. Reactive urothelial cells are enlarged, with a vacuolated cytoplasm, a central nucleus with regular contours, a low nuclear-cytoplasmic ratio, and a prominent nucleolus. Thus, careful evaluation of nuclear and nucleolar characteristics and the smear background helps to make the distinction.

The cytologic differentiation from PUNLMP may be difficult. The greater degree of cell dishesion in addition to the presence of small but loose cell clusters, favors a urothelial dysplasia (Murphy et al., 1984). However, it may be difficult to reach this level of detail without compromising the diagnostic accuracy. Thus, the report should be signed out as "atypical urothelial cells," followed by a description of the cytologic findings and enumeration of diagnostic considerations.

Urothelial Carcinoma in Situ

Cytology: "Carcinoma in Situ"

Carcinoma in situ, a flat urothelial neoplasm, is a documented precursor of invasive cancer and encompasses lesions designated as severe dysplasia. Most cases of carcinoma in situ occur in patients with high-grade papillary or invasive urothelial carcinoma, and less than 10% are primary (Amin and Young, 1997). Patients with carcinoma in situ of the bladder are at significant risk of cancer progression and death from bladder carcinoma (Bostwick et al., 1999; Cheng et al., 1999b).

The incidence of carcinoma in situ increases with the grade and stage of the associated urothelial neoplasm. Extensive carcinoma in situ of the bladder with involvement of the ureters and prostatic urethra is present in 60% of invasive urothelial carcinomas (Bostwick et al., 1999; Mahadevia et al., 1986). Because of the frequent involvement of the trigone and the floor and neck of the bladder, patients with carcinoma in situ complain of dysuria, frequency, pain, hematuria, and nocturia and have sterile pyuria (Bostwick et al., 1999). Carcinoma in situ is usually multifocal and exhibits a red, granular, velvety, erythematous-appearing mucosa on cystoscopic examination. However, in some instances the lesion is not visualized (Bostwick et al., 1999).

Marked histologic and cytologic abnormalities are characteristic of carcinoma in situ. Considerable nuclear anaplasia is invariably present in a single-layer, normal-thickness, or hyperplastic urothelium. A full-thickness abnormality is usually present but

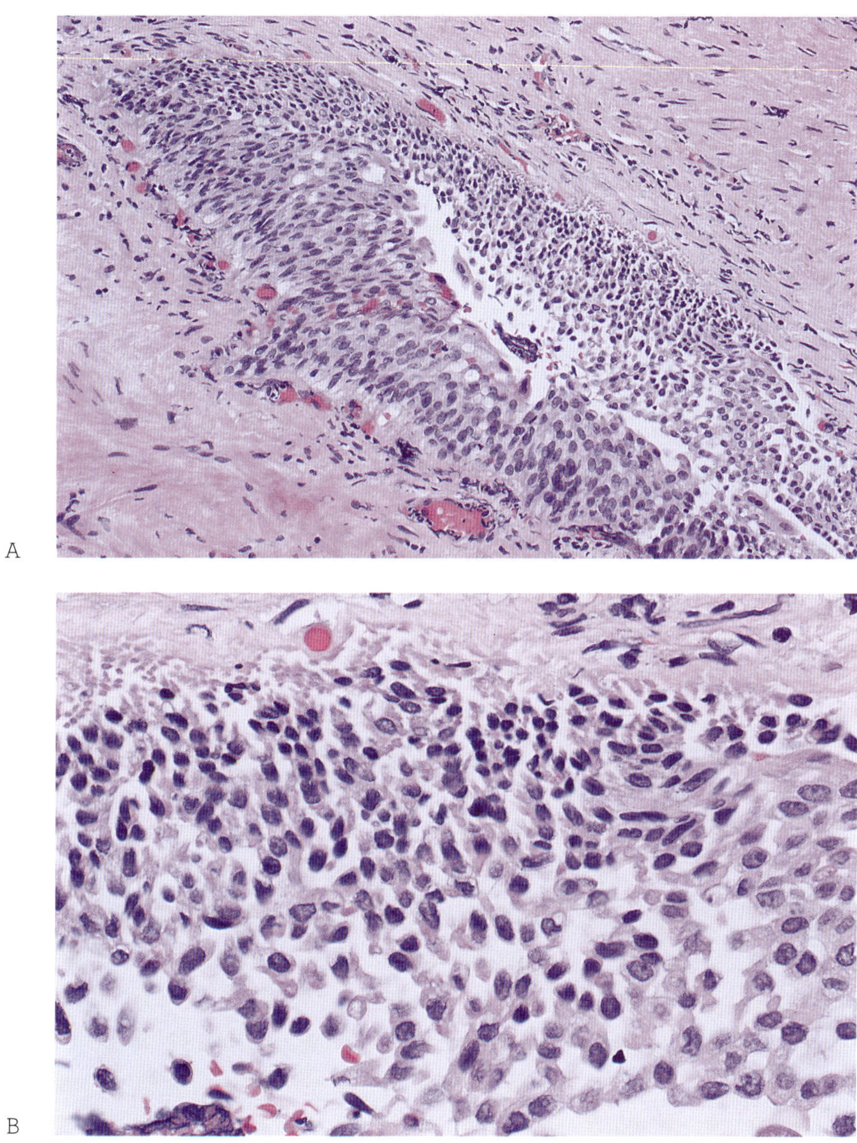

Figure 6–19. *Urothelial dysplasia.* (*A*) Bladder biopsy specimen shows a flat urothelial lesion. (*B*) Architectural and cytologic urothelial cell changes fall short of those present in carcinoma in situ. (*C*) Corresponding bladder washing cytology shows loosely cohesive clusters of large cells with mild anisonucleosis, slightly irregular nuclear contours, a preserved nuclear-cytoplasmic ratio, and slight hyperchromasia, admixed with smaller, hyperchromatic nuclei. (*D*) Cohesive cluster of atypical urothelial cells with enlarged nuclei, overlapping, loss of polarity, and mild anisonucleosis are seen; these cells can be identified in the bladder biopsy specimen (B). (Hematoxylin and eosin stain, ×200 (A), ×600 (B); C,D, Papanicolaou stain, ×600)

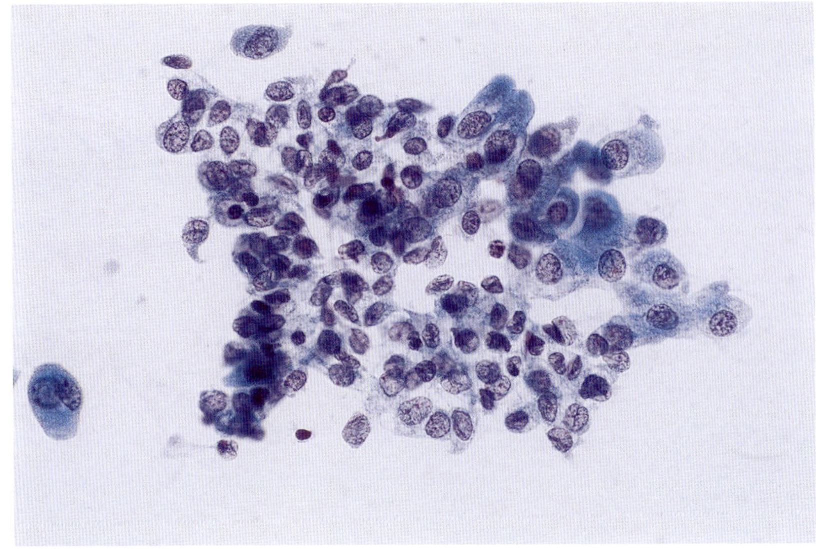

C

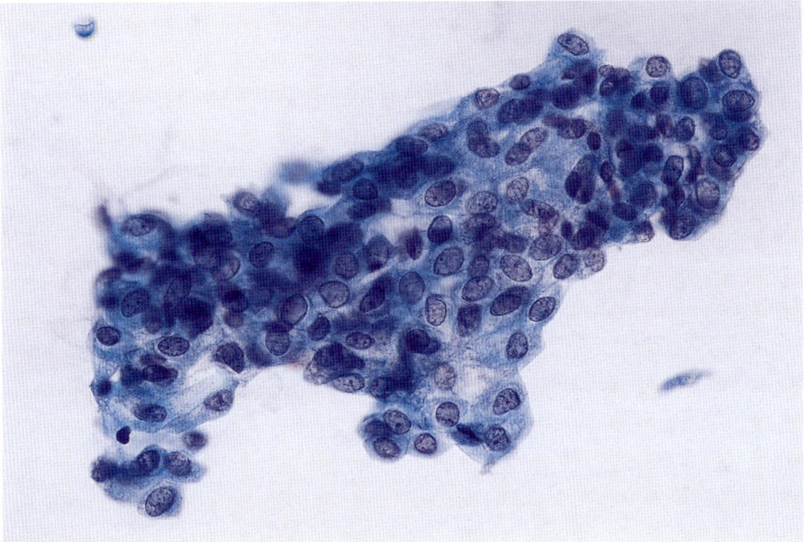

D

Figure 6–19. (*Continued*)

is not required for the diagnosis. Architectural changes include loss of polarity, disorganization, and loss of cohesiveness. However, and in contrast to papillary urothelial neoplasms, progressive qualitative and quantitative cytologic abnormalities predominate over the histologic pattern (Fig. 6–20) (Amin and Young, 1997).

Characteristically, carcinoma in situ shows a high frequency of aneuploid cells, exhibits carcinoembryonic antigen positivity, and lacks blood group isoantigens (Grignon, 1997).

Cytology. Urine cytology is useful in the diagnosis of carcinoma in situ. Smears are of variable cellularity and show single cells and small cell clusters. Urothelial cells are usually uniform and exhibit unequivocal features of malignancy. A nucleolus is variably present and is irregular, angulated, and sometimes multiple. The cytoplasm varies from vacuolated to dense. Red blood cells and occasional inflammatory cells may be identified, but necrosis is absent. The main cytologic features of

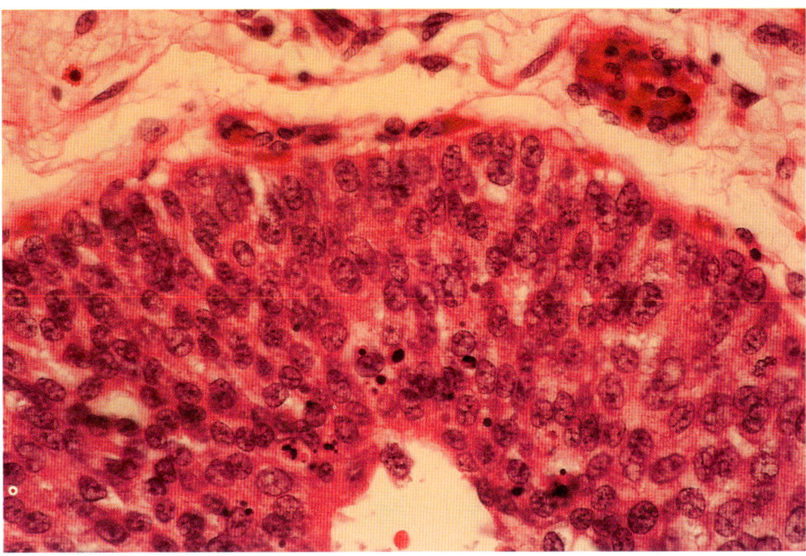

Figure 6–20. *Carcinoma in situ.* Bladder biopsy specimen shows a flat lesion that replaces the urothelium and is composed of small to medium-sized cells with marked anaplasia. (Hematoxylin and eosin stain, ×400)

carcinoma in situ are summarized in Table 6–7.

Small-cell urothelial carcinoma in situ shows smaller and fairly uniform cells with a high nuclear-cytoplasmic ratio and hyperchromatic nuclei. The nuclear contours are irregular, with coarse chromatin and inconspicuous or absent nucleoli. Mitotic figures, although present in tissue sections, are uncommon in cytologic preparations (Fig. 6–21A–C).

Occasionally, carcinoma in situ may exhibit large, pleomorphic cells with abundant cytoplasm, a variable nuclear-cytoplasmic ratio, and large vesicular nuclei with irregular nuclear contours, coarse chromatin, and large nucleoli (Fig. 6–22A,B; Box 6–8).

Box 6–8.

Cystoscopically, carcinoma in situ shows a red, granular, velvety-appearing mucosa.

Smears of small-cell carcinoma in situ show a uniform population of highly dysplastic cells with a homogeneous cytoplasm, a high nuclear-cytoplasmic ratio, coarse chromatin, and inconspicuous nucleoli.

Smears of large-cell carcinoma in situ show large, pleomorphic cells with a variable nuclear-cytoplamic ratio, vesicular nuclei, coarse chromatin, and large nucleoli.

The background may have red blood cells, but necrosis is absent.

Differential diagnosis. The presence of less pleomorphic, smaller cells with fewer, less prominent nucleoli and a nonnecrotic background favors urothelial carcinoma in situ over invasive urothelial carcinoma. However, this distinction cannot be made with certainty because up to 34% and 10% of cystectomy specimens for carcinoma in situ show foci of microinvasion and muscle-invasive cancer, respectively (Grignon, 1997).

Because carcinoma in situ is treated with intravesical agents and these patients are followed up with urine cytology, the effects of these drugs on urothelial cells need to be considered before the diagnosis of residual or recurrent carcinoma in situ is made. As in the interpretation of bladder biopsy specimens, the separation of these

A

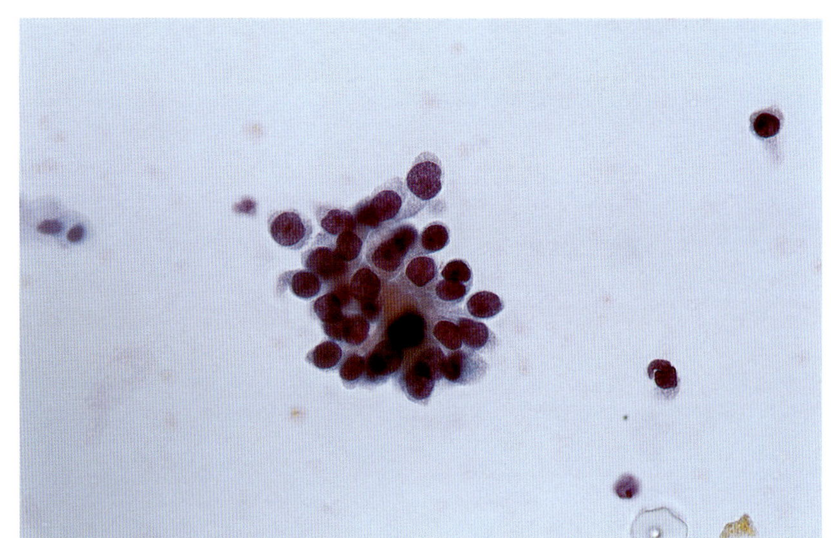

B

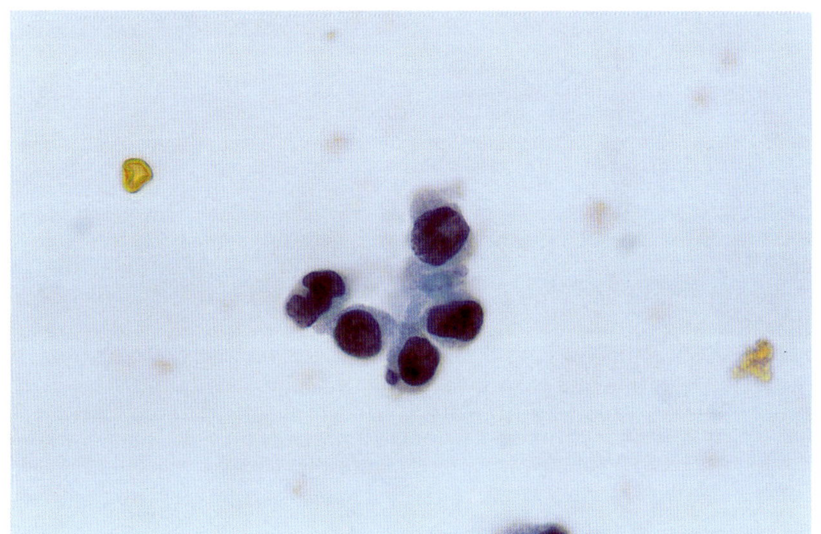

C

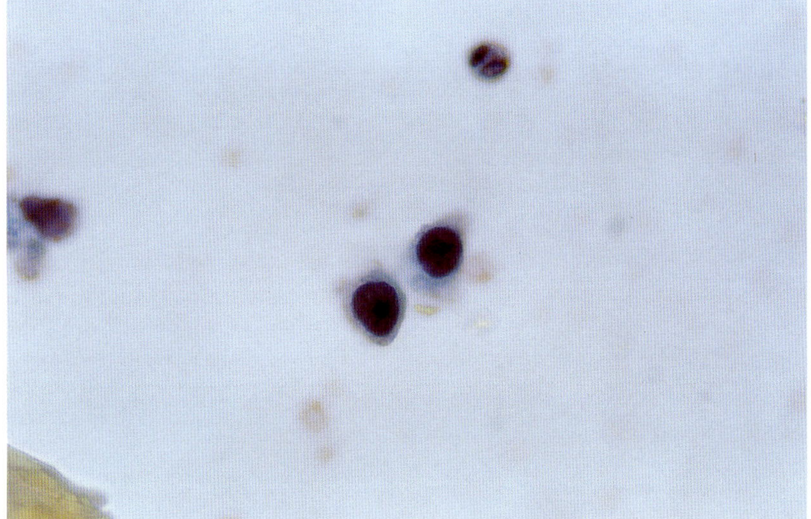

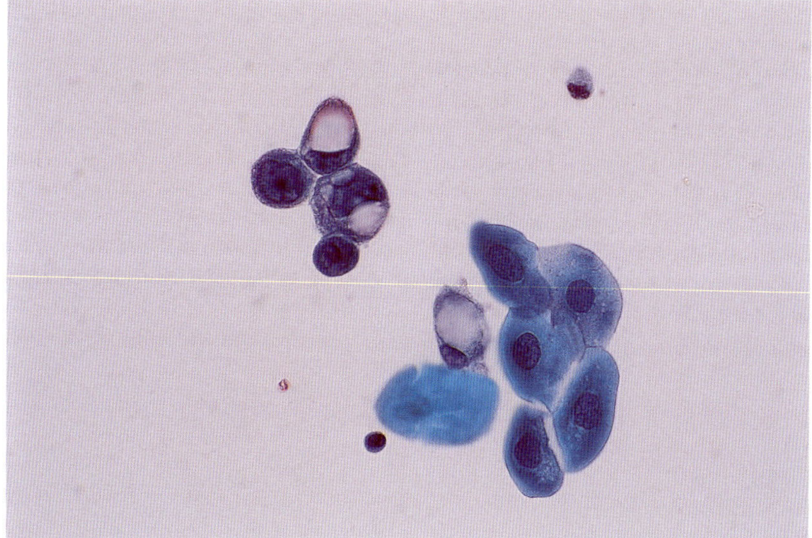

A

B

Figure 6–22. *Carcinoma in situ of large cell type.* Lack of cohesion and small clusters and single cells showing vacuolated cytoplasm (*A*), a large nucleus, irregular nuclear contours, coarse chromatin, and moderate pleomorphism (*B*) are present in the bladder washings of these two patients. Note the lack of necrosis and enlargement in comparison with umbrella cells (A) and red blood cells (B). (Papanicolaou stain, ×600)

Figure 6–21. *Carcinoma in situ of small cell type.* (*A,B*) Loosely cohesive clusters of malignant small cells with marked hyperchromasia, coarse chromatin, and irregular nuclear borders are evident in the cytology of a voided urine specimen obtained from the patient whose biopsy specimen is shown in Figure 6–20. Compare the cell size with that of the red blood cell present in (B). (*C*) Hyperchromasia with irregular, coarse chromatin and irregular nuclear contours permits the distinction from the human polyomavirus cytopathic effect. (Papanicolaou stain (A), ×600; (B,C), ×1000)

entities may be difficult. Cyclophosphamide-induced cystitis causes remarkable changes in the urothelial cells that are indistinguishable from those of high-grade urothelial tumors. Clinical correlation is the only means of separating these entities, underscoring the good communication that must exist between the clinician and the pathologist. Reactive cellular changes of the reepithelialized bladder mucosa may be responsible for false-positive diagnoses in patients who are receiving photodynamic therapy (Broghamer et al., 1989).

Another, no less important entity in the differential diagnosis is human polyomavirus infection, which may be reactivated by the effect of immunosuppressive therapy, including intravesical chemotherapy. Identification of a homogeneous, smooth nuclear basophilic inclusion, nuclear "cracking," and peripheral nuclear chromatin beading are key diagnostic features of polyomavirus infection, in contrast to the irregularly shaped nucleus, coarse chromatin, and nucleoli often seen in malignant urothelial cells.

Invasive Urothelial Carcinoma

Cytology: "Positive for Malignancy. High-Grade Urothelial Carcinoma"

Invasive urothelial carcinoma is usually of high grade. The prognosis for patients with invasive urothelial carcinoma is poor and depends mainly on the clinical stage. The five-year survival rate is less than 50% even with therapy.

Histologically, the single or multiple nests of pleomorphic tumor cells involve the lamina propria (stage T1), muscularis propria (T2), or beyond (T3 and T4). It is not unusual to find other cell types, that is, squamous, glandular, small, and various growth patterns in these tumors, but residual papillary architecture is present only in a minority of cases (Amin et al., 1997).

The clinical significance of squamous differentiation remains unknown; however, it appears to be an unfavorable prognostic factor for patients undergoing radical cystectomy and a poor indicator of the response to radiation therapy (Frazier et al., 1993; Martin et al., 1989). Glandular differentia-

A

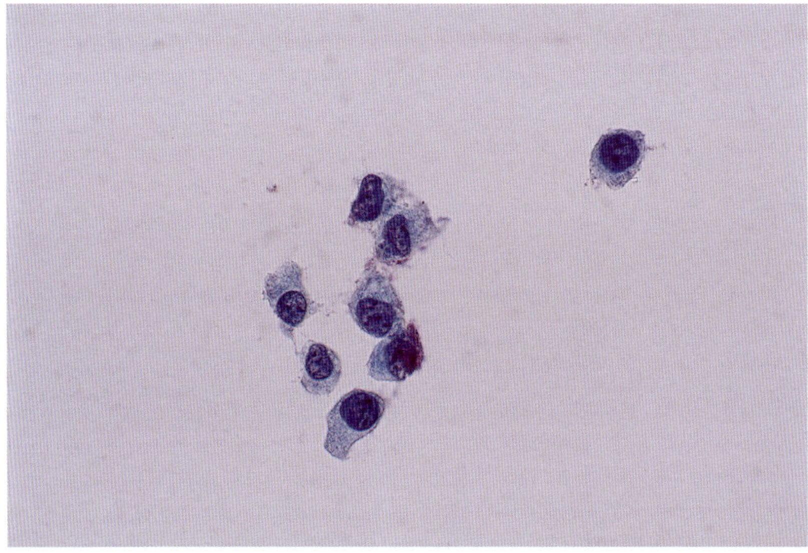

Figure 6–23. *Invasive urothelial carcinoma of the urethra.* Cellular features of invasive carcinoma are identical to those of carcinoma of the bladder and upper urinary tract. Cellularity, cell pleomorphism, necrosis, and inflammation are present in various proportions in these urethral washings. Hyperchromasia and coarse chromatin (*A*) must be present, but necrosis and cellular anaplasia (*B,C*) are important for the diagnosis of invasive carcinoma. (Papanicolaou stain, ×600)

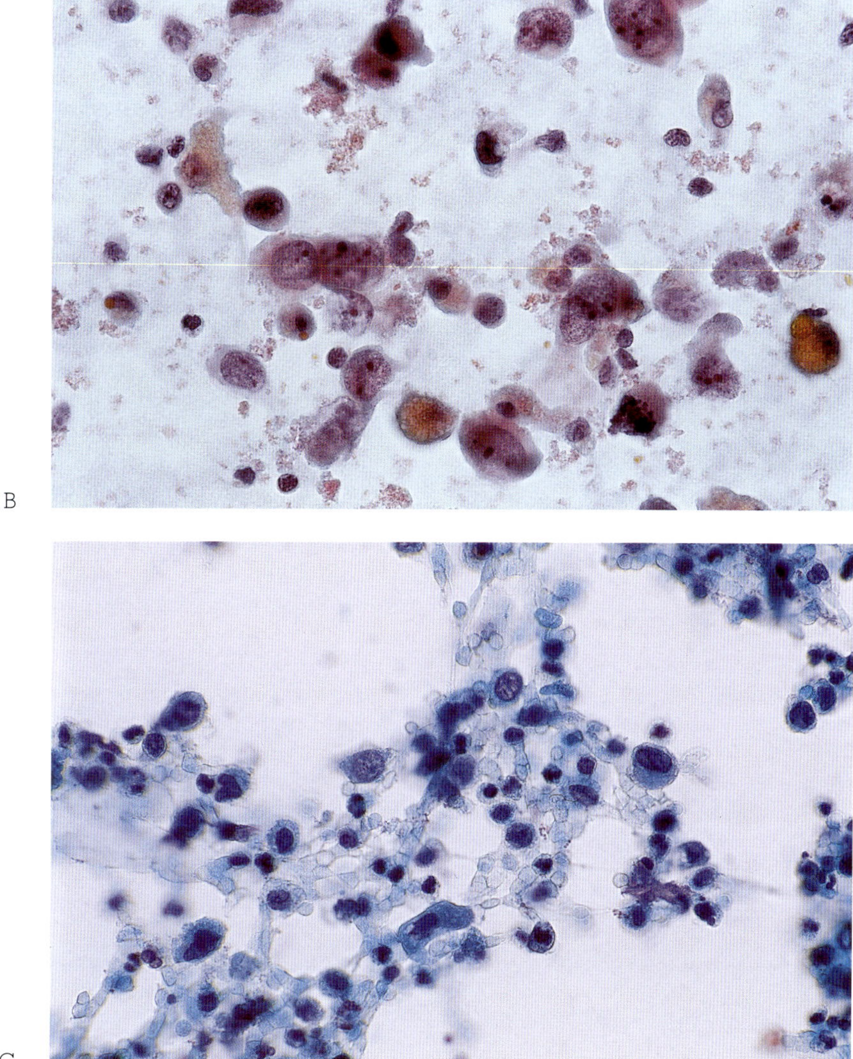

Figure 6–23. (*Continued*)

tion appears to be associated with a poor response to some types of systemic chemotherapy (Logothetis et al., 1989). The coexistence of small cell neuroendocrine carcinoma is associated with a poor prognosis and needs a different therapeutic management. Thus, the main histologic diagnosis in these cases will be small-cell carcinoma followed by the non-small-cell components.

Cytology. The cytologic features of invasive urothelial carcinoma are identical regardless of their primary location, that is, the urinary bladder, urethra, or upper urinary tract (Box 6–9). Those arising in the urethra commonly exhibit marked degeneration and abundant necrosis, and may be of scant cellularity when examined by urethral washing or voided urine (Fig. 6–23A–C). Cytology of urothelial tumors of the bladder is described next, and cytology of those of the upper urinary tract is described separately.

Irregular, three-dimensional cell groups are more common in voided and instrumented urine from patients with invasive urothelial carcinoma than in urine from patients with noninvasive tumors. However, the

Box 6–9.

Smears show cellular dissociation greater than that of noninvasive tumors.

Pleomorphic cells show a variable nuclear-cytoplasmic ratio, anaplastic nuclei, irregular chromatin, and large, irregular nucleoli.

The background shows lysed blood, granular debris, and acute inflammation.

Squamous or glandular differentiation is common in invasive carcinoma.

It may not be possible to distinguish poorly differentiated carcinomas from secondary malignancies, that is, metastases from a distant primary or invasive from an adjacent organ.

degree of cell dissociation is greater in invasive carcinomas (Fig. 6–24A–C). Malignant cells are large and show abundant cytoplasm, enlarged hyperchromatic and pleomorphic nuclei, coarse and irregular chromatin, and a single large, irregular eosinophilic nucleolus or multiple small nucleoli. Abnormal mitotic figures may not be as numerous as in tissue sections. Cell phagocytosis or "cannibalism" is more common than in noninvasive tumors (Fig. 6–25A–C).

Because the cellular makeup is similar to that of high-grade noninvasive carcinoma, the unequivocal diagnosis of invasive urothelial carcinoma should be made only if necrosis, evidenced as lysed blood, granular debris, and degenerated inflammatory cells, is present along with high-grade tumor cells.

Invasive urothelial carcinomas exfoliate more cells of mixed urothelial and squamous types than do low-grade urothelial and noninvasive carcinomas that exfoliate predominantly pure urothelial cells (Fig. 6–26A,B) (Suprun and Bitterman, 1975). However, neoplastic cells of irradiated urothelial carcinoma, in addition to having greater cellular pleomorphism, exhibit more prominent squamous features (Neumann and Limas, 1986). Mucinous differentiation in invasive urothelial carcinoma, seen as intracytoplasmic vacuoles containing mucoid material, is well recognized, and the cells are indistinguishable from those of an invasive adenocarcinoma (Fig. 6–27A–C) (Donhuijsen et al., 1992).

A

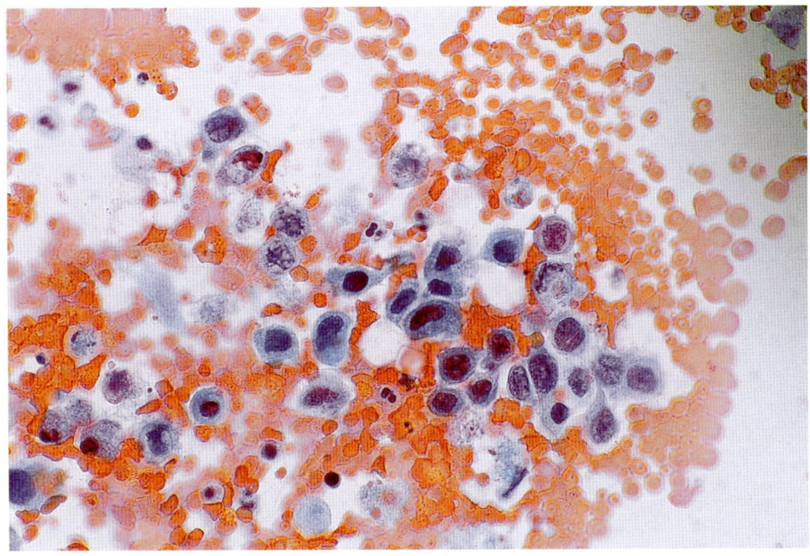

Figure 6–24. *Clusters of invasive urothelial carcinoma.* (*A*) Voided urine smears of a 66-year old man with invasive bladder cancer show anaplastic and individual necrotic cells and a bloody background. (*B*) Pleomorphic tumor cells with vacuolated cytoplasm and hyperchromatic nuclei are noted. (*C*) Clustered and single large anaplastic cells are seen together with blood but without necrosis. (B) and (C) are bladder washings. (Papanicolaou stain, ×600)

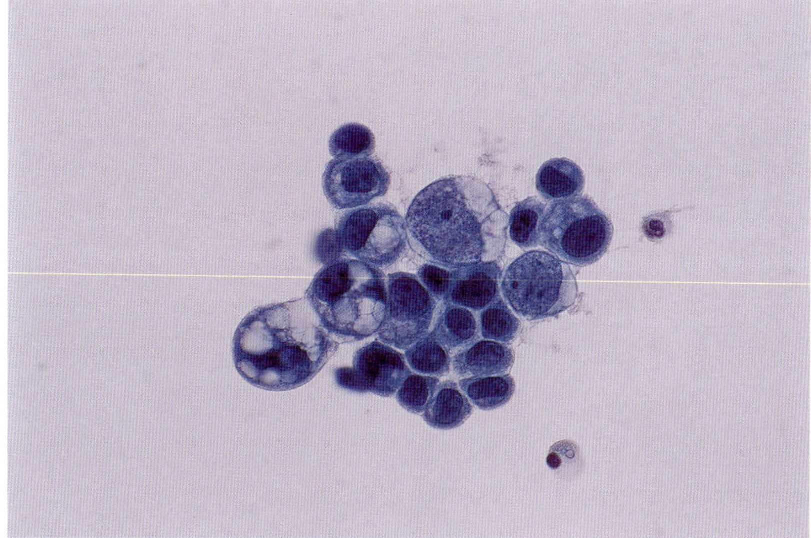

B

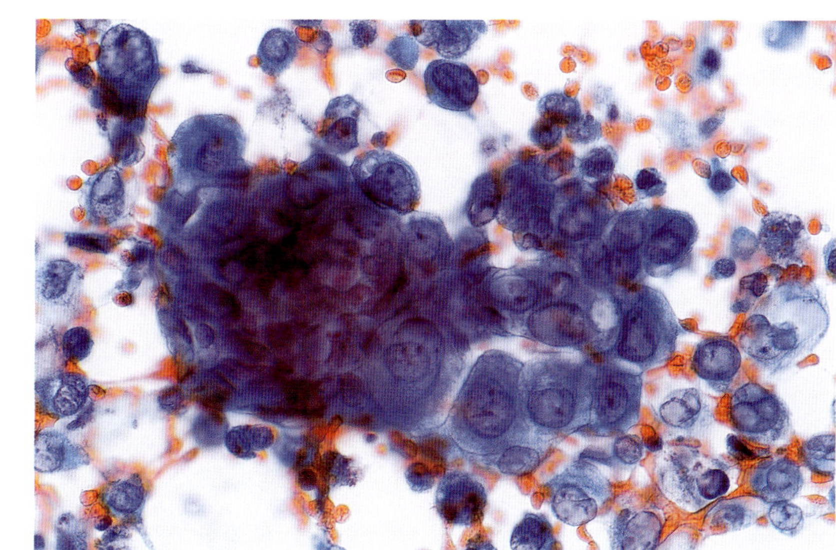

C

Figure 6–24. *(Continued)*

In summary, the cytologic diagnosis of invasive carcinoma in the presence of the triad of cell pleomorphism, necrosis, and inflammation offers little difficulty. However, superficially invasive urothelial carcinomas may exhibit less cellular pleomorphism and minimal necrosis, making the distinction from noninvasive high-grade papillary urothelial carcinoma difficult and its conclusive cytologic diagnosis impossible. Similarly, a less uniform urothelial cell population of carcinoma in situ associated with inflammation, particularly after biopsies, may mimic invasive carcinoma; however, necrosis is absent (Koss, 1996). The cytologic features of invasive urothelial carcinoma are summarized in Table 6–6.

Differential diagnosis. The presence of squamous and glandular differentiation frequently seen in invasive urothelial carcinoma may be prominent and sufficient to consider a diagnosis of squamous carcinoma or adenocarcinoma. However, such diagnoses should be considered cautiously and only in the presence of a uniform popula-

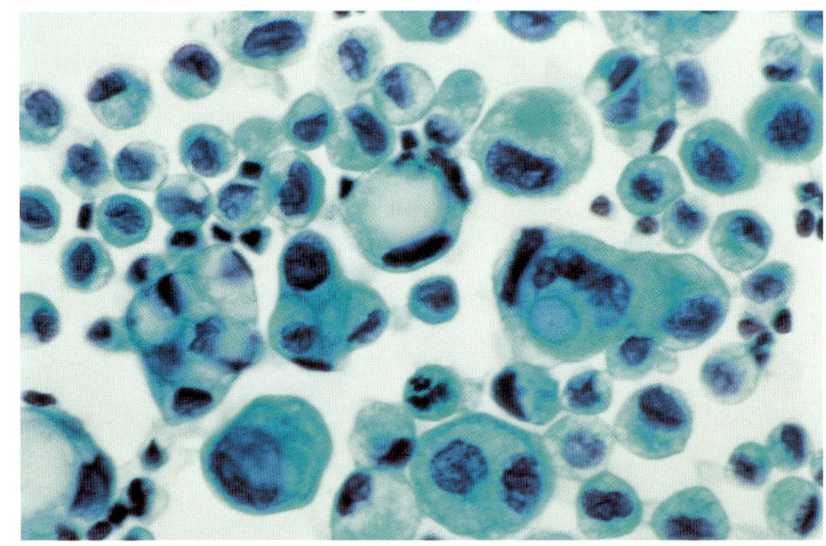
A

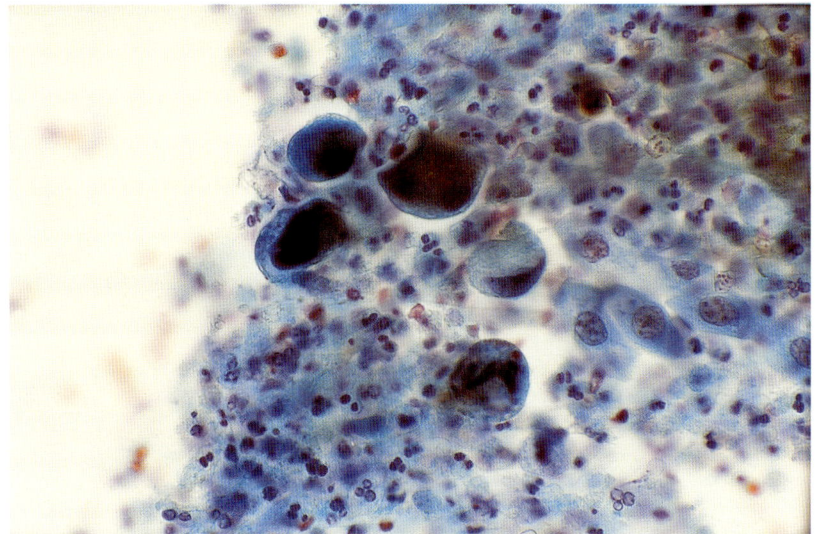
B

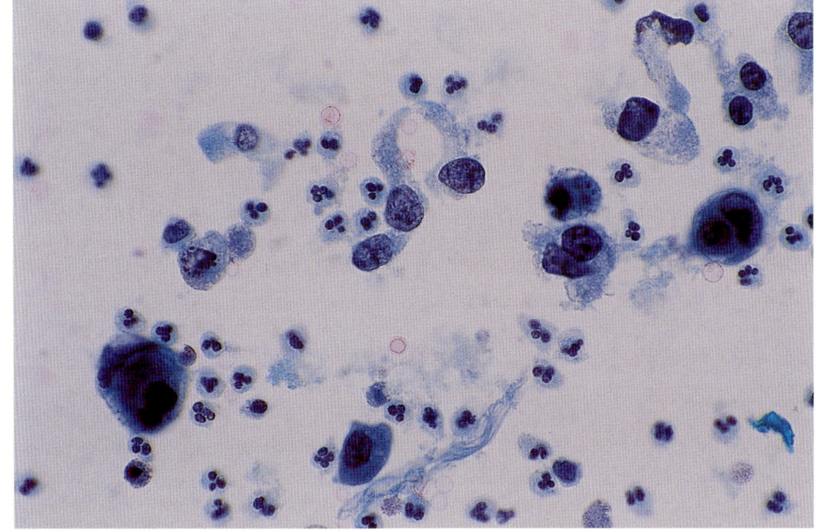
C

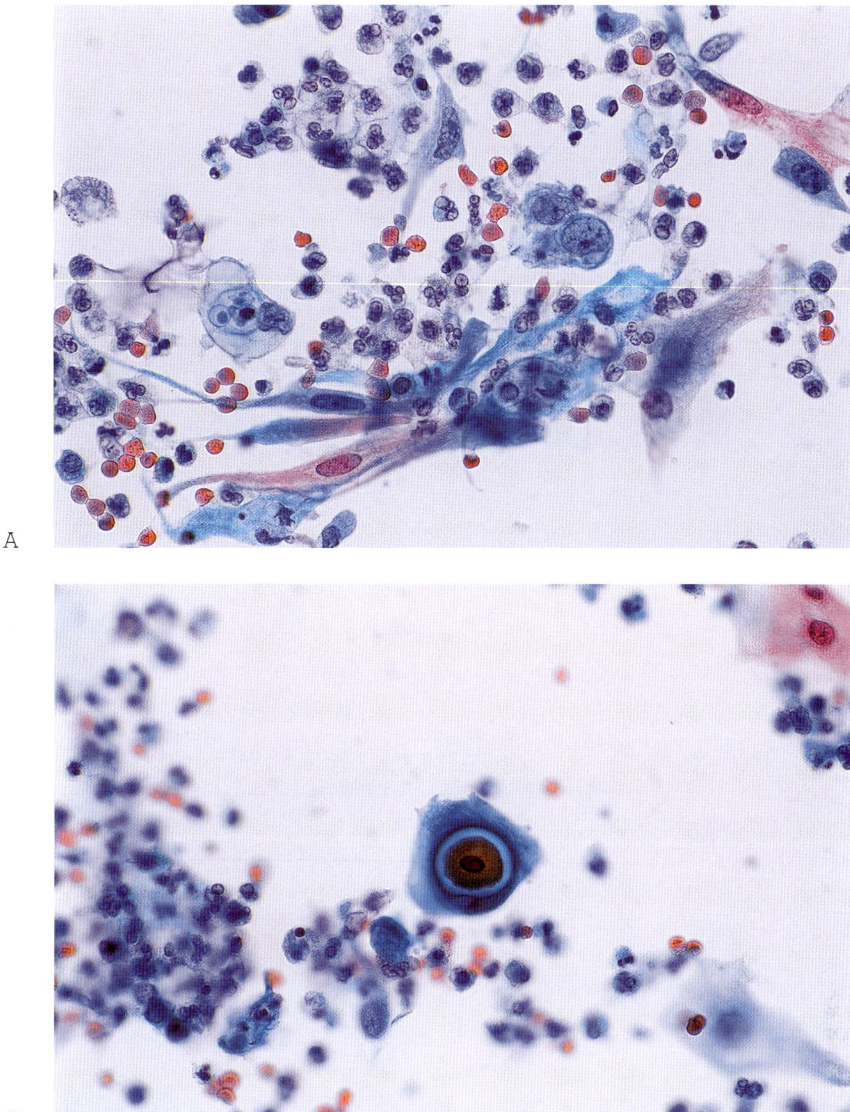

Figure 6–26. *Squamous differentiation in invasive urothelial carcinoma.* Anaplastic spindle tumor cells with focal keratinization are present in the bladder washing of an invasive urothelial bladder carcinoma. Nonsquamous, high-grade urothelial carcinoma cells are also present. Squamous and glandular differentiation is not unusual in high-grade invasive urothelial tumors. (Papanicolaou stain, ×600)

Figure 6–25. *Single cells of invasive urothelial carcinoma.* (*A*) Greater cell dissociation is present in invasive than in noninvasive carcinoma. (*B*) Necrosis, acute inflammation, and anaplastic cancer cells must be present for the cytologic diagnosis of invasion. (*C*) Necrosis, acute inflammation, and phagocytosis of anaplastic cells may be present in the voided urine of patients with invasive bladder cancer. (A) and (B) are bladder washings. (Papanicolaou stain, ×600)

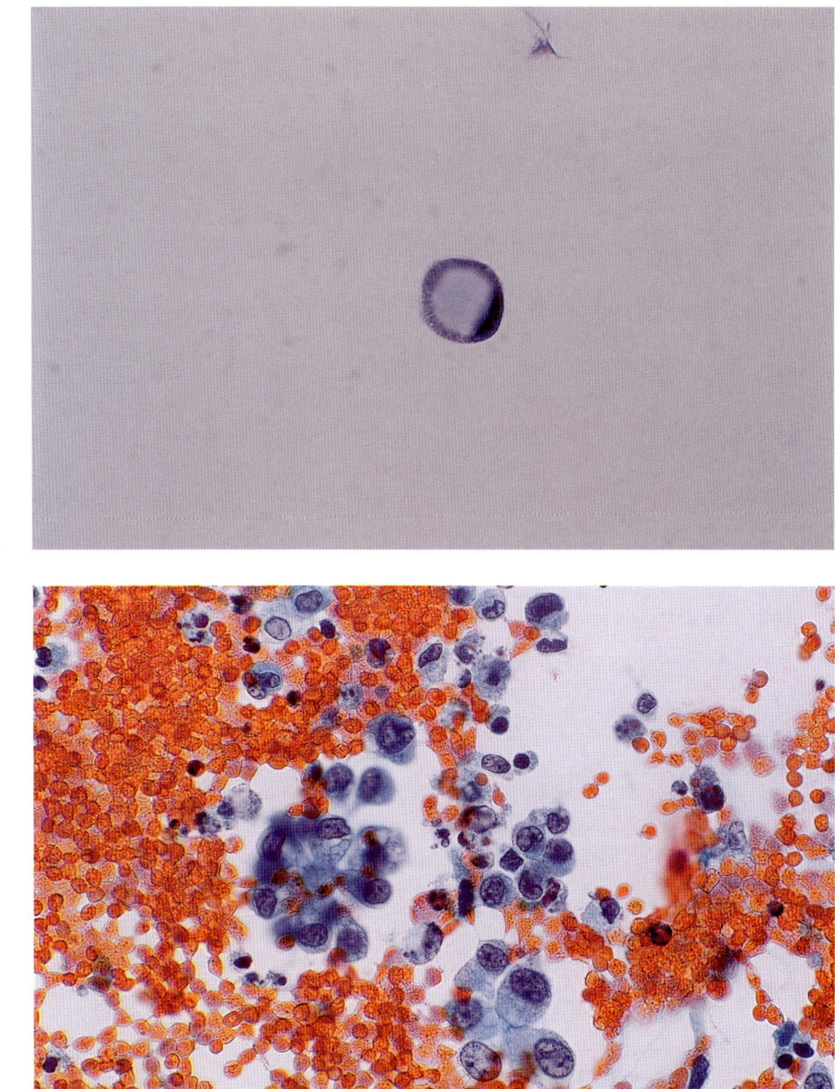

Figure 6–27. *Glandular differentiation in invasive urothelial carcinoma.* (*A*) A cell with a single cytoplasmic vacuole pushing the hyperchromatic nucleus is evidence of glandular differentiation, not uncommonly seen in high-grade carcinoma. (*B*) Gland formation, necrosis, and single cells with anaplastic nuclei are present in this invasive urothelial carcinoma with glandular features. (*C*) A sheet of poorly cohesive round cells with finely vacuolated cytoplasm and hyperchromatic, round nuclei and prominent nucleoli is present in another urothelial carcinoma with glandular features. All specimens are bladder washings. (Papanicolaou stain, ×600)

tion of clearly malignant squamous or glandular cells, respectively. The final diagnosis of adenocarcinoma or squamous cell carcinoma is based on extensive histologic examination of the resected specimen.

Cytoplasmic vacuolization, a low nuclear-cytoplasmic ratio, and prominent nucleoli are often present in high-grade tumor cells and resemble superficially reactive urothelial cell changes (Fig. 6–28A,B). However, mild hyperchromasia with a regular chromatin distribution, round, smooth nucleolar borders, and lack of necrosis characterize a reactive process. In contrast, frank nuclear

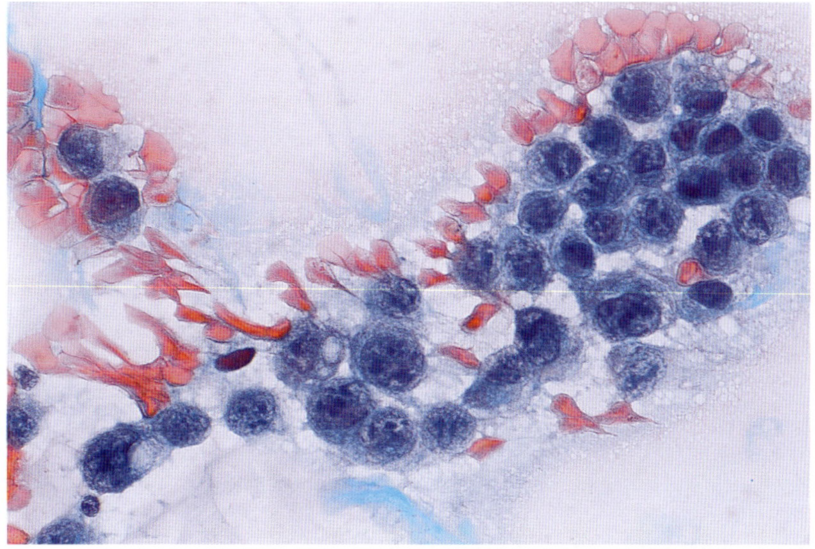

C

Figure 6–27. (*Continued*)

anaplasia and necrosis are present in invasive high-grade malignancies.

It may not be possible to differentiate high-grade invasive urothelial carcinomas from poorly differentiated secondary or primary nonurothelial malignancies on the basis of cytology alone; a final diagnosis requires tissue examination and clinical correlation. Occasionally, well-differentiated carcinomas of prostate, renal, or colon origin show sufficient cytologic features to establish their primary site. In other cases, the cytology smears show intestinal or signet-ring cells that may be from bladder, breast, or gastrointestinal tract origin. The cytologic features of secondary malignancies involving the urinary tract are described in Chapter 7.

Clinical correlation is helpful in the evaluation of poorly differentiated high-grade carcinomas with necrosis, particularly for ruling out involvement of the urinary tract by tumors from an adjacent or distant primary site.

UROTHELIAL CARCINOMA OF THE UPPER URINARY TRACT

Tumors of the ureter and renal pelvis comprise less than 10% of cancers of the urinary tract in adults, with papillary urothelial neoplasms being the most common histologic type (Bonsib and Eble, 1997). Mesenchymal, hematopoietic, and miscellaneous neoplasms are less common.

Renal pelvis and ureteral papillary urothelial neoplasms are identical to those developing in the urinary bladder. However, because of the location, these lesions may be difficult to diagnose pathologically. Because of the technical difficulties inherent in obtaining an adequate endoscopic biopsy specimen by transurethral selective catheterization, the diagnosis is not uncommonly based on brushings and washings of the lesion and on urine cytology. Fine-needle aspiration biopsy is performed occasionally. Another diagnostic exfoliative cytology technique is brush cytology performed with a percutaneous translumbar approach (Bibbo et al., 1974; Dodd et al., 1997; Lang et al., 1978; Leistenschneider and Nagel, 1980; Zaman et al., 1996).

Both brushing and washing techniques are considered complementary in the diagnosis of urothelial neoplasms of the upper urinary tract and are virtually without complications (Kannan, 1990). However, an anecdotal report of rapid spread of urothelial carcinoma of the ureter after retrograde brush biopsy has been presented (Farkas et al., 1983).

Voided urine specimens, although suitable for cytologic evaluation of bladder cancer, are less optimal for evaluation of the up-

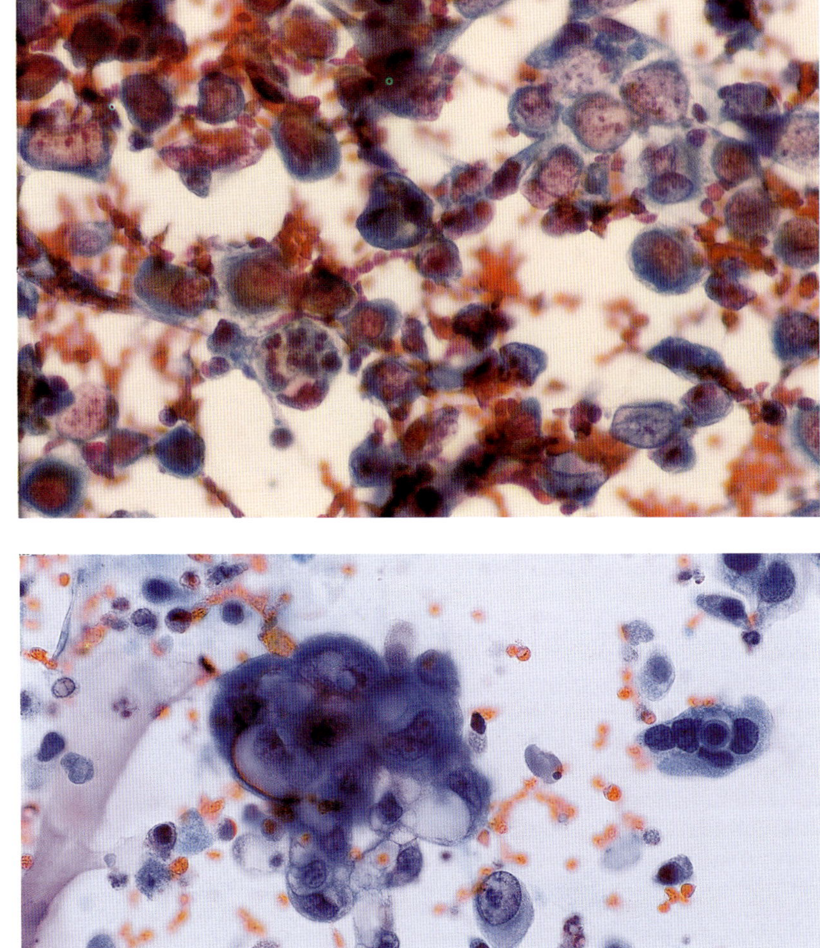

Figure 6–28. *Invasive urothelial carcinoma resembling nonneoplastic conditions.* (*A*) Anaplastic tumor cells with phagocytosis of neutrophils resemble cellular features seen in cystitis. (*B*) Marked cytoplasmic vacuolization and a low nuclear-cytoplasmic ratio resemble those of reactive urothelial cells. Nuclear anaplasia is evident in both cases and permits the distinction between benign and malignant etiology. Papanicolaou stain, ×600)

per urinary tract because of the high false-negative rate due to cellular degeneration (Bian et al., 1995; Eriksson and Johansson, 1976; Kannan, 1990; Seldenrijk et al., 1987).

Cytology. Cytologic examination is of paramount importance in evaluating renal pelvis/kidney tumors to define the therapeutic approach, particularly in cases amenable to conservative surgery. However, voided urine samples exhibit degenerative cell changes, and urine cytology of cystocopically obtained specimens from the upper urinary tract may be difficult to interpret (Fig. 6–29A–C). These specimens are obtained by instrumentation, which may compromise the cytologic interpretation by producing artifacts that mimic a urothelial neoplasm. In

Box 6–10.

If possible, obtaining urine samples from both the left and right upper urinary tracts is most desirable to compare cell morphology.

The cytomorphologic features of urothelial neoplasms or the ureter and renal pelvis are identical to those of the urinary bladder.

Anisonucleosis, nuclear anaplasia, irregular nucleoli, and nuclear overlapping are useful in the differentiation of high-grade neoplasms from benign lesions.

addition, benign cells of the upper urinary tract often have an atypical appearance (Potts et al., 1997) (Box 6–10).

Endoscopic retrograde brushing of the upper urinary tract is a specific and more sensitive sampling method than catheterized urine sampling and offers better cytologic preservation, as long as the sample is smeared and fixed immediately for Papanicolaou stain (Dodd et al., 1997). In addition, the material can be used to perform immediate cytologic interpretation in air-dried smears stained with Diff-Quik, to make smears using thin-layer cytology preparation techniques, and to prepare a cell block (Zaman et al., 1996). Zaman et al., in a review of 17 low to intermediate-grade papillary urothelial carcinomas of the renal pelvis, found 94% diagnostic accuracy of endoscopic renal pelvis brush cytology. Cytologic criteria helpful in the diagnosis included a clean background, tissue fragments and monomorphic single cells with an irregular nuclear membrane, true papillary fragments, and columnar-appearing urothelial cells. True papillary fragments with a fibrovascular core and columnar cells were found in 73% and 53% of cases, respectively (Zaman et al., 1996).

Because malignant cells from low-grade urothelial tumors often display subtle cytologic abnormalities, examination of urine specimens from both the left and right upper urinary tracts is most desirable and useful in the diagnosis of these neoplasms. In general, the criteria mentioned for the diagnosis of PUNLMP (grade 1 papillary urothelial carcinoma) and low-grade (grade 2) papillary urothelial carcinoma of the

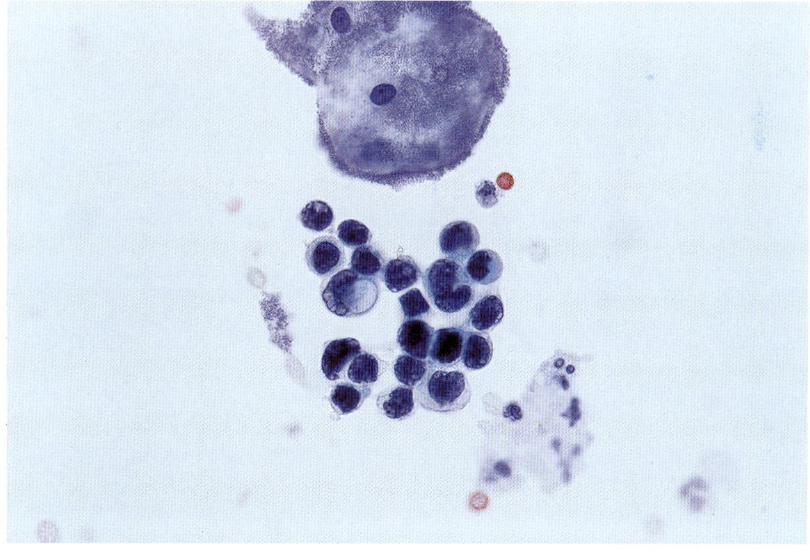

A

Figure 6–29. *Urothelial carcinoma of the renal pelvis.* Voided urine (*A*), a renal pelvis washing (*B*), and a renal pelvis brushing (*C*) show large cells with anaplastic nuclei and a high nuclear-cytoplasmic ratio indicative of high-grade urothelial carcinoma. Note the cellular degeneration present in the voided urine, in contrast to the excellent nuclear detail of the washing and brushing. (Papanicolaou stain, ×600)

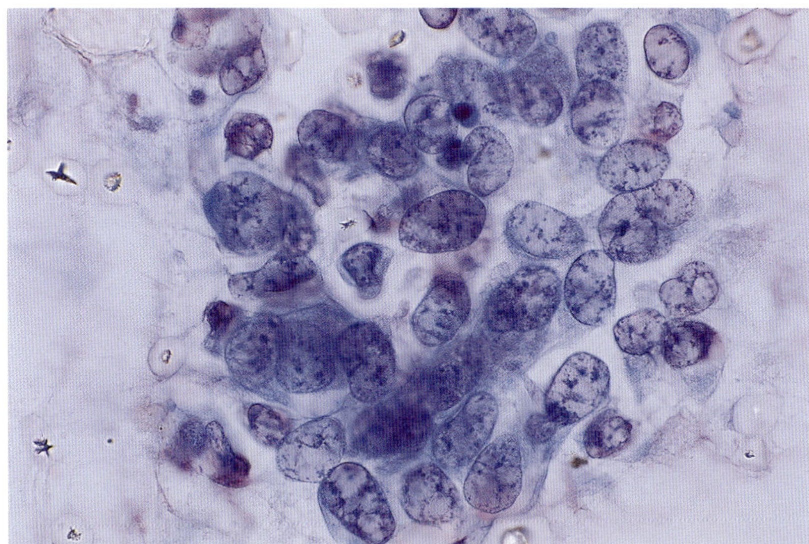

B

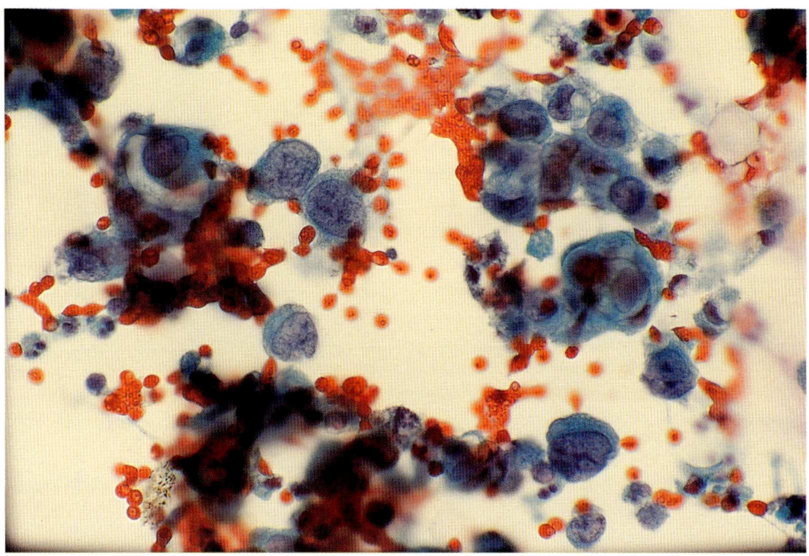

C

Figure 6–29. (*Continued*)

bladder also apply to urothelial neoplasms of the upper urinary tract (Fig. 6–30A–D).

In high-grade urothelial carcinomas, the cytologic findings are also similar to those described above for the urinary bladder (Fig. 6–31A–D). Single cells with nuclear overlap, a high nuclear-cytoplasmic ratio, hyperchromatic nuclei with irregular nuclear contours, coarse chromatin, and prominent irregular nucleoli are useful for the diagnosis. Anaplastic cells, necrotic debris, lysed blood, and degenerated inflammatory cells indicate invasive carcinoma (Dodd et al., 1997; Highman, 1986; Kannan, 1990; Potts et al., 1997; Zaman et al., 1996).

Differential diagnosis. As in the diagnosis of tumors of the urinary bladder, the main distracters are the instrumentation effect, urolithiasis, and atypia and degeneration associated with inflammatory disorders. These disorders cause exfoliation of papillary clusters with cytologic features similar to those of papillary neoplasms, particularly PUNLMP and some low-grade papillary urothelial carcinomas. Papillary clusters with nuclear crowding, overlapping, and peripheral palisading of cells favor a neoplasm over an instrumentation effect (Fig. 6–32) (Highman, 1986).

Nuclear morphology helps in the diagnosis of high-grade papillary urothelial neoplasms. Most grade 2 tumors have irregular nuclear contours, hyperchromasia, dense cytoplasm, and a high nuclear-cytoplasmic ratio. Invasive urothelial carcinomas often show hyperchromasia, prominent nucleoli, and necrosis, features that may resemble those present in degeneration associated with inflammatory conditions, in particular

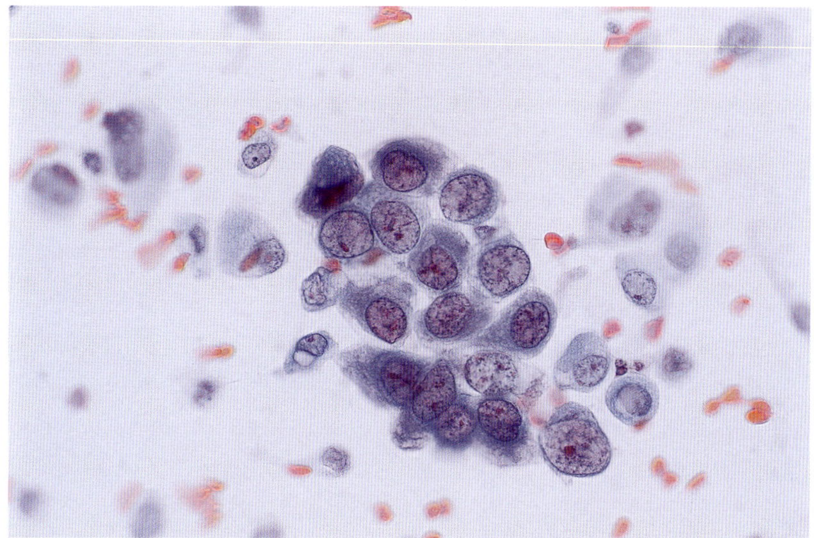

A

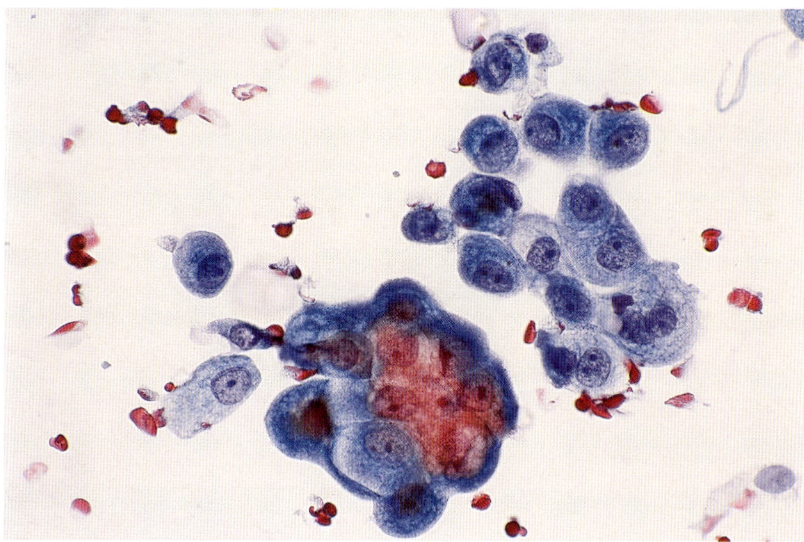

B

Figure 6–30. *Urothelial carcinoma of the ureter.* Comparison of cancer cells with benign cells from the opposite side helps the diagnosis of low-grade neoplasms. Slight anisonucleosis, a mildly increased nuclear-cytoplasmic ratio, and slightly enlarged nuclei indicative of a low-grade papillary tumor (*A*) are contrasted with the reactive urothelial cells from the opposite side (*B*). Cercariform cells and high-grade urothelial carcinoma cells (*C*) are compared with the reactive urothelial cells, characterized by cytoplasmic elongation and a preserved nuclear-cytoplasmic ratio, admixed with urothelial and chronic inflammatory cells of the opposite side (*D*). All specimens are ureteral washings. (Papanicolaou stain, ×600)

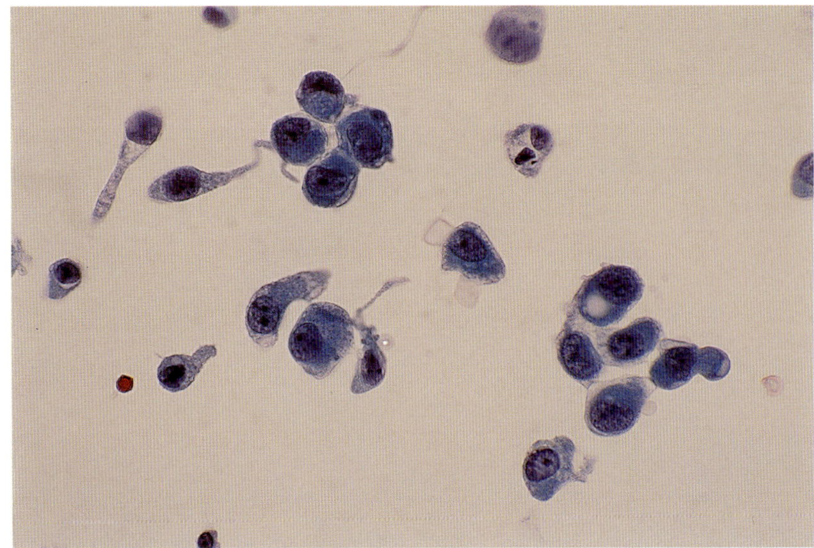

C

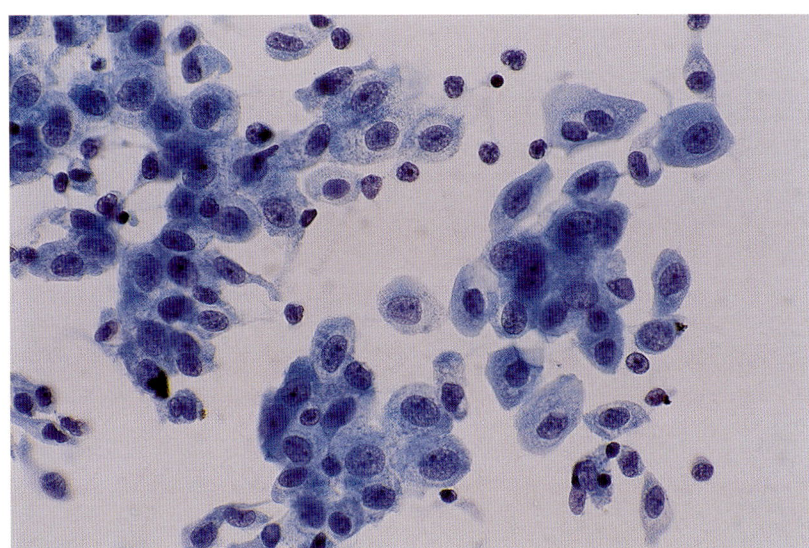

D

Figure 6–30. (*Continued*)

urolithiasis. In addition, vacuolated cytoplasm may be present in both a reactive process and invasive carcinoma (Fig. 6–33). However, degenerated and uniformly dark nuclei with a round, regular nucleolus are seen in reactive nonneoplastic conditions, instead of the coarse, dark chromatin and irregular nucleolar contours characteristic of high-grade invasive urothelial carcinoma. Atypical squamous metaplasia can be seen in patients with the double-J stent permanent catheter placed to alleviate ureteral obstruc-

Figure 6–31. *Urothelial carcinoma of the ureter.* Clusters of malignant cells with varying degrees of anaplasia are present in these cases of urothelial carcinoma of the ureter. (*A*) Cells with vacuolated cytoplasm, eccentric nuclei, slightly irregular nuclear borders, and lack of hyperchromasia are seen in low-grade urothelial carcinoma. Cellular phagocytosis is also present. (*B*) A loosely cohesive cluster of cells with moderate pleomorphism and hyperchromasia is seen. (*C,D*) Marked cellular pleomorphism and necrosis are identified in high-grade urothelial carcinoma. (Papanicolaou stain, ×600)

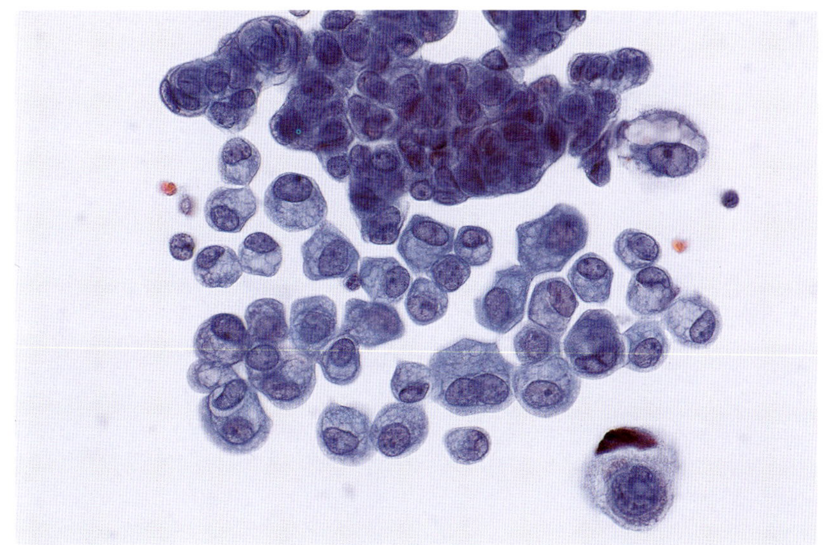

A

B

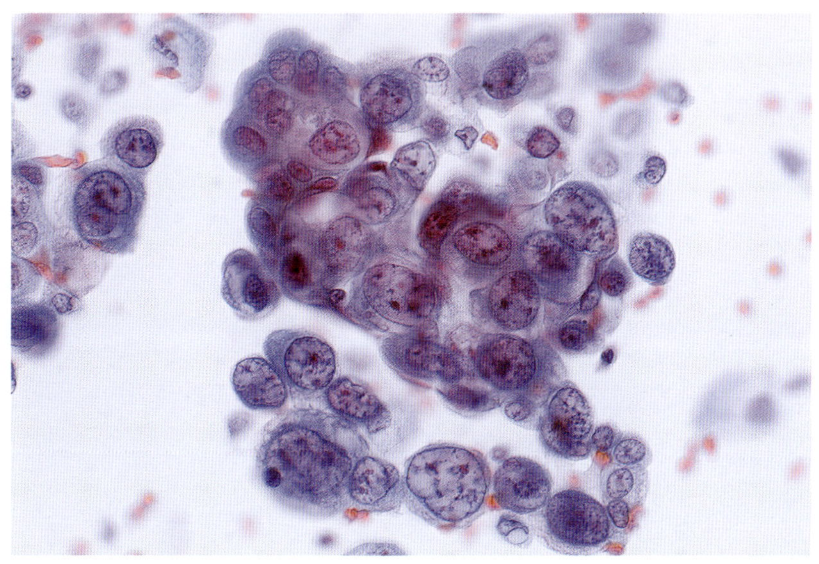

C

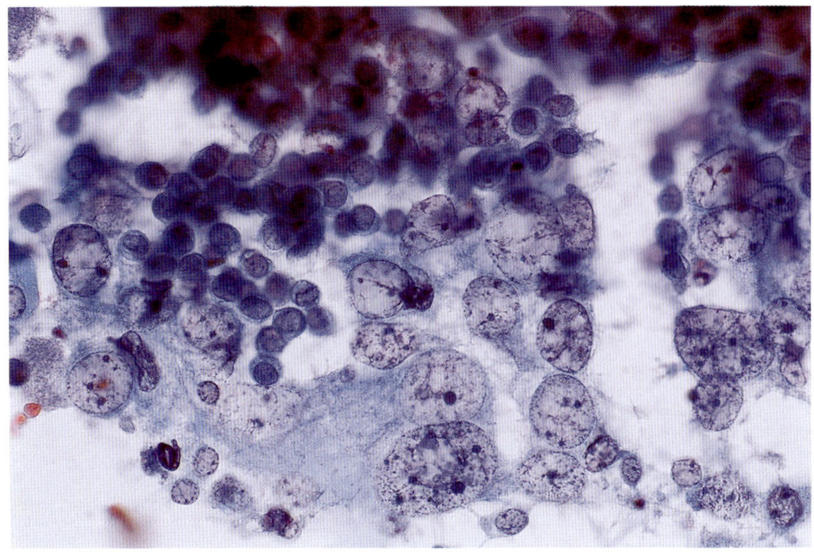

Figure 6–31. *(Continued)*

tion (Fig. 6–34). Regression analysis suggests that anisonucleosis, a high nuclear-cytoplasmic ratio, and nuclear overlapping are useful in the differentiation of high-grade neoplasms from benign lesions (Potts et al., 1997).

VARIANTS OF UROTHELIAL CARCINOMA

These types of malignancies are listed in Box 6–11.

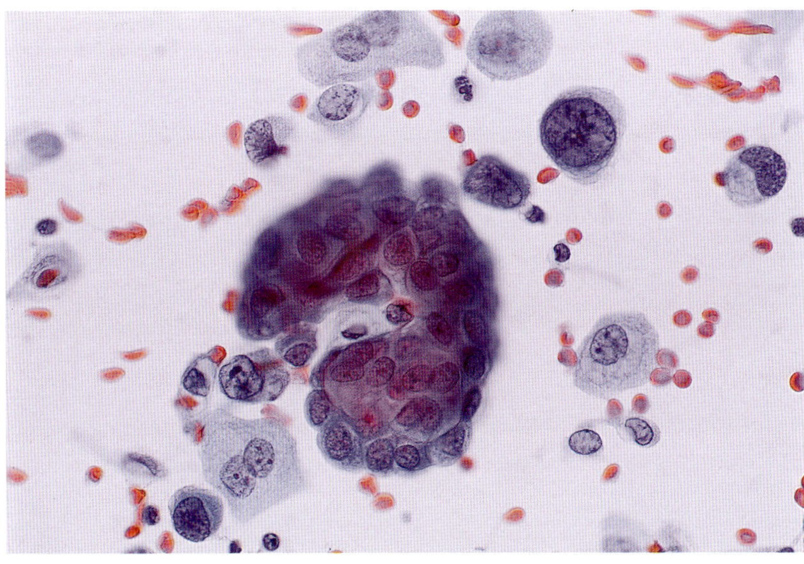

Figure 6–32. *Instrumentation effect and urothelial carcinoma.* A single cluster of cells dislodged by instrumentation is not uncommonly found in washings or brushings of the upper urinary tract and may be difficult to distinguish from cell clusters of low-grade urothelial neoplasms. However, highly anaplastic malignant cells are diagnostic of high-grade carcinoma. Ureteral washing. (Papanicolaou stain, ×600)

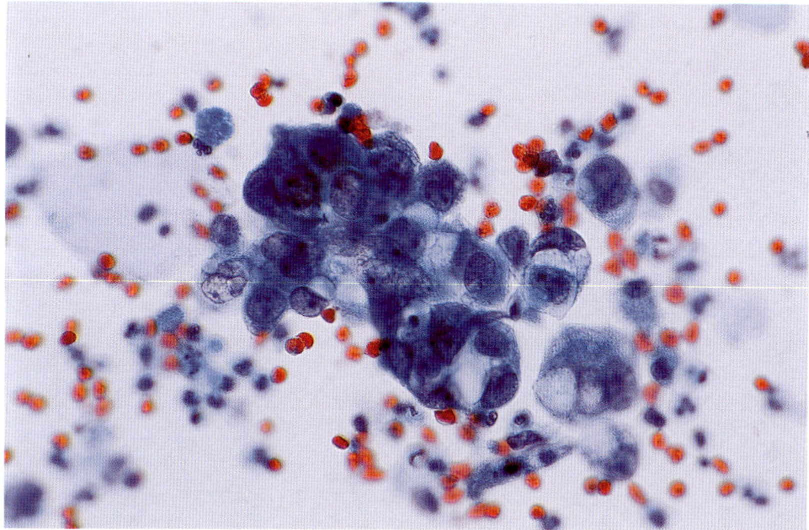

Figure 6–33. *Malignancy mimicking reactive urothelial cells.* Malignant cells with cytoplasmic vacuolization and a low nuclear-cytoplasmic ratio in a background of inflammation may be difficult to distinguish from reactive urothelial cells due to inflammation. Anaplastic nuclei help in the diagnosis of high-grade malignancy. Ureteral washing. (Papanicolaou stain, ×600)

Micropapillary Type

Urothelial carcinoma with a micropapillary component resembling ovarian papillary serous carcinoma is rare. Most tumors have aneuploid DNA content when tested by image analysis, and are associated with carcinoma in situ and high-stage urothelial carcinoma with a tendency to vascular invasion (Amin et al., 1994a).

The micropapillary component is usually

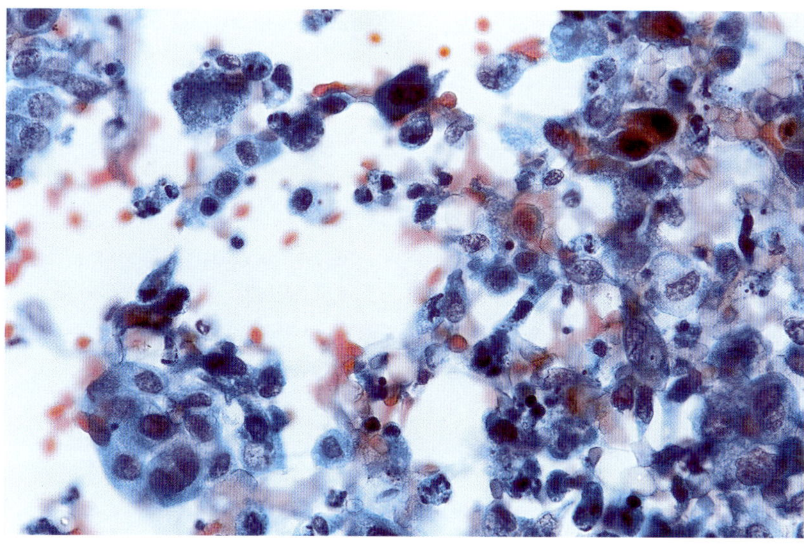

Figure 6–34. *Double-J stent effect.* Marked squamous metaplasia with atypia and a background of tissue damage are present in a urine specimen obtained though a double-J stent placed to alleviate ureteral obstruction secondary to urolithiasis. Follow-up of this patient for three years showed no evidence of urothelial carcinoma. (Papanicolaou stain, ×600)

Box 6–11.

Variants of urothelial carcinoma include

- micropapillary, tubuloglandular, nested, microcystic, and inverted types
- sarcomatoid carcinoma
- urothelial carcinoma with syncytiotrophoblasts
- lymphoma-like urothelial carcinoma.

seen in the mucosal surface of the tumor as slender, delicate filiform processes or small papillary clusters of tumor cells. The cytologic features are those of a high-grade urothelial carcinoma (Amin et al., 1994a).

Nested, Tubuloglandular, Microcystic, and Inverted Types

Nested, tubuloglandular, microcystic, and inverted tumors, deceptively benign variants of urothelial carcinoma, pose more problems histologically than cytologically. The nested type may have histologic patterns varying from tubules to small nests infiltrating the lamina propria, histologically resembling von Brunn's nests. This low-grade carcinoma (WHO grade 2) shows slight cytologic atypia, a low nuclear-cytoplasmic ratio, mildly pleomorphic nuclei, and occasionally prominent nucleoli (Drew et al., 1996). In spite of these features, the nested variant appears to be a persistent and aggressive neoplasm.

The microcystic type resembles cystitis cystica, and the inverted type recapitulates features of inverted papilloma (Eble and Young, 1997). These tumors usually show cytologic features of low-grade (WHO grade 2) or high-grade (WHO grade 3) papillary urothelial carcinoma.

Sarcomatoid Carcinoma

Sarcomatoid carcinoma (Box 6–12) of the bladder has the characteristic gross appearance of a large polypoid or pedunculated solid tumor in patients who are nearly always diagnosed at an advanced pathologic stage and who have gross hematuria.

Histologically, sarcomatoid carcinoma is a mixture of malignant epithelial cells and a prominent and predominant spindle cell or malignant fibrous histiocytoma-like cytomorphology. The epithelial component may be urothelial, squamous, or, less commonly, adenocarcinomatous; in rare instances, it consists only of carcinoma in situ (Eble and Young, 1997; Grignon, 1997). The epithelial origin is supported by positive immunohistochemistry for keratin and epithelial membrane antigen and desmosomes or tonofilaments by ultrastructure, at least focally, in the sarcoma-like component (Grignon, 1997; Perret et al., 1998).

Some authors use the terms *carcinosarcoma* and *sarcomatoid carcinoma* interchangeably to describe such tumors (Grignon, 1997). However, others require the unequivocal presence of leiomyosarcoma, fibrosarcoma, or malignant fibrous histiocytoma for the tumor to be called a *homologous carcinosarcoma* (Perret et al., 1998; Reuter, 1993). The terms *heterologous carcinosarcoma, carcinosarcoma with heterologous differentiation*, and *metaplastic carcinoma* are reserved for those cases showing heterologous (chondrosarcoma, osteosarcoma, rhabdomyosarcoma, liposarcoma) elements in the spindle cell component (Perret et al., 1998). The terms *sarcomatoid carcinoma* and *carcinosarcoma* should be used for those cases in which mesenchymal elements comprise 90% or more of the tumor (Reuter, 1993).

The cellular origin of carcinosarcoma of the bladder is still being disputed. However,

Box 6–12.

Sarcomatoid carcinoma is a mixture of malignant urothelial cells and a spindle cell component.

The smear shows high-grade urothelial carcinoma cells; stromal cells usually do not exfoliate into the urinary stream.

Stromal cells may exfoliate in the presence of surface urothelial ulceration.

microsatellite analysis for loss of heterozygosity in both the carcinomatous and sarcomatous components suggests that both epithelial and nonepithelial components are derived from a common stem cell. Downregulation of E-cadherin, a cell adhesion molecule that is important in cell–cell and cell–stroma interactions, evaluated by immunohistochemistry, may define one of the pathways responsible for conversion of epithelial cells to the sarcomatous phenotype (Halachmi et al., 2000).

Cytology. Urine samples usually show the presence of malignant urothelial cells of high nuclear grade (Colonna et al., 1989). The hyperchromatic pseudosarcomatous or heterologous stromal elements do not exfoliate spontaneously into the urinary stream. Rare stromal cells may be seen in urine specimens in the presence of surface urothelial ulceration. A marked spindling artifact of single cells and loose clusters of cells caused by laser therapy may be a potential pitfall in the diagnosis of carcinosarcoma and mesenchymal neoplasms (Fanning et al., 1993).

The differential diagnosis includes sarcomas, which are far less common than sarcomatoid carcinomas. Immunohistochemistry for cytokeratin, which is usually focally positive in sarcomatoid carcinoma and negative in sarcomas, is helpful in making the diagnosis (Wick et al., 1988).

Urothelial Carcinoma with Syncytiotrophoblasts

The presence of syncytiotrophoblasts in urothelial carcinoma with high levels of circulating chorionic gonadotropin has been associated with a poor prognosis (Gallagher et al., 1984; Young and Eble, 1991). However, human chorionic gonadotropin can be detected in high-grade papillary and urothelial carcinoma in situ by immunohistochemistry in the absence of syncytiotrophoblast giant cells and has no prognostic significance (Jacobsen et al., 1990; Rodenburg et al., 1985).

Giant cells other than syncytiotrophoblasts present in urothelial carcinoma can be associated with BCG therapy as part of the granulomatous response, a foreign body reaction associated with prior tumor resection or biopsy, and urothelial carcinoma with osteoclast-like giant cells. The latter cells are of mesenchymal origin and possibly represent a stromal response to the tumor (Grignon, 1997).

Lymphoma-like Urothelial Carcinoma

Rare carcinomas of the urinary bladder simulate malignant lymphoma or multiple myeloma cytologically (Eble and Young, 1997; Sahin et al., 1991; Zukerberg et al., 1991). Microscopic examination shows a diffuse monomorphic cellular infiltrate involving the wall of the bladder.

Cytology. The cells are usually medium-sized and round to oval, with a moderate amount of cytoplasm. They have eccentric, round nuclei with fine chromatin and distinct nucleoli. In some instances, these cells have a striking plasmacytoid appearance that is difficult to differentiate from that of the plasma cells of multiple myeloma (Grignon, 1997; Sahin et al., 1991).

Differential diagnosis. Immunoreactivity for cytokeratin and carcinoembryonic antigen, and failure to demonstrate lymphoid markers or light chain restriction, support the diagnosis of carcinoma.

SQUAMOUS CELL CARCINOMA

Cytology: "Squamous Cell Carcinoma"

In the United States, squamous cell carcinoma accounts for less than 5% of bladder carcinomas, and it is slightly more common in men than in women, with a mean age at diagnosis of 65 years (Grignon, 1997). Squamous cell carcinoma frequently arises in bladders with schistosomiasis, chronic and severe inflammation, and long-standing lithiasis. It is also frequent in nonfunctioning bladders and in renal transplant patients (Eble and Young, 1997) (Box 6–13).

Schistosoma haematobium is associated with bladder squamous carcinoma in endemic areas, particularly in Egypt. The cercaria, after completing asexual reproduction in water snails, penetrate the human skin and mature in the pelvic and mesenteric venous plexuses,

Box 6–13.

Schistosomiasis, nonfunctioning bladders, and renal transplants are some predisposing factors in the development of squamous cell carcinoma.

Squamous cell carcinoma is frequently invasive and associated with keratinizing squamous metaplasia.

Smears show pleomorphic malignant squamous cells, keratinization, and necrosis.

Final diagnosis is based on exhaustive histologic examination of the resected specimen.

where they reproduce, laying eggs identified in the bladder wall and urine. In endemic areas, 80% of patients with bladder cancer have schistosoma eggs in the bladder, where they elicit fibrosis and calcification of the wall, polyposis and ulceration of the mucosa, and urothelial hyperplasia, squamous metaplasia, dysplasia, and ultimately carcinoma (Khafagy et al., 1972; Smith and Christie, 1986).

In Egypt, squamous cell carcinoma was once the most common cancer in the urinary bladder, constituting about 60% to 70% of all tumors. However, a decline has been recently observed, with a parallel increase in the frequency of urothelial carcinoma. Now squamous and urothelial tumors are nearly equal, with an incidence of 40%–45% each (Smith and Christie, 1986).

Squamous carcinoma is frequently invasive at the time of diagnosis and is associated with keratinizing squamous metaplasia of the surrounding urothelium. The histologic diagnosis of squamous carcinoma requires the presence of a malignant squamous cell population without a urothelial component. The histologic grade is variable, ranging from well to poorly differentiated, with marked nuclear pleomorphism and a few islands of squamous differentiation (Grignon, 1997). Histologic variants include verrucous and sarcomatoid squamous carcinomas (Fig. 6–35A–C).

A

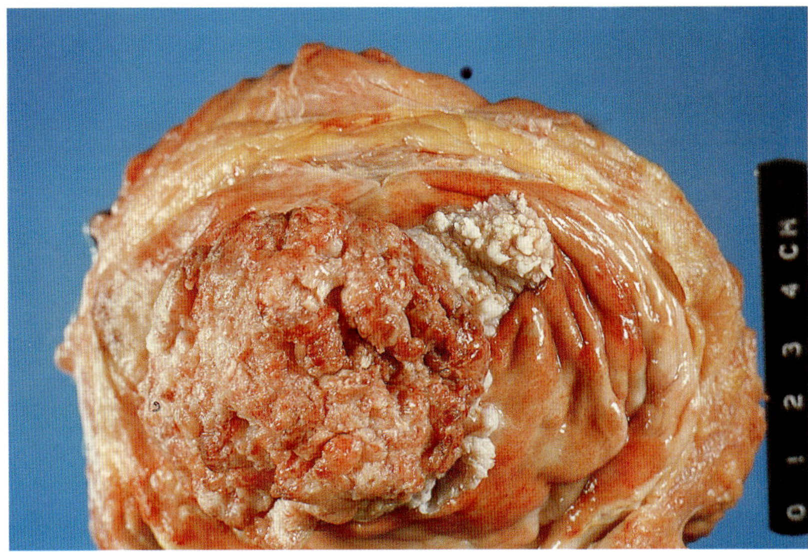

Figure 6–35. *Squamous cell carcinoma and schistosomiasis.* (*A*) The gross appearance of urinary bladder cancer associated with schistosomiasis is shown. The tumor is usually nodular and exophytic, whereas the rest of the mucosa shows leukoplakia and sandy patches (rusty mucosa). (*B*) Verrucous carcinoma of the urinary bladder, a highly differentiated variant of squamous cell carcinoma, has branching hairy projections and well-defined pushing margins. Nodal metastases are rare in such a tumor. (*C*) A well-differentiated squamous cell carcinoma encloses a *Schistosoma* egg. (Hematoxylin and eosin stain, (B) ×100, (C) ×400). Courtesy of Dr. Hassan Nabil Tawfik, Professor and Director of Anatomic and Surgical Pathology, National Cancer Institute, Cairo University, Cairo, Egypt.

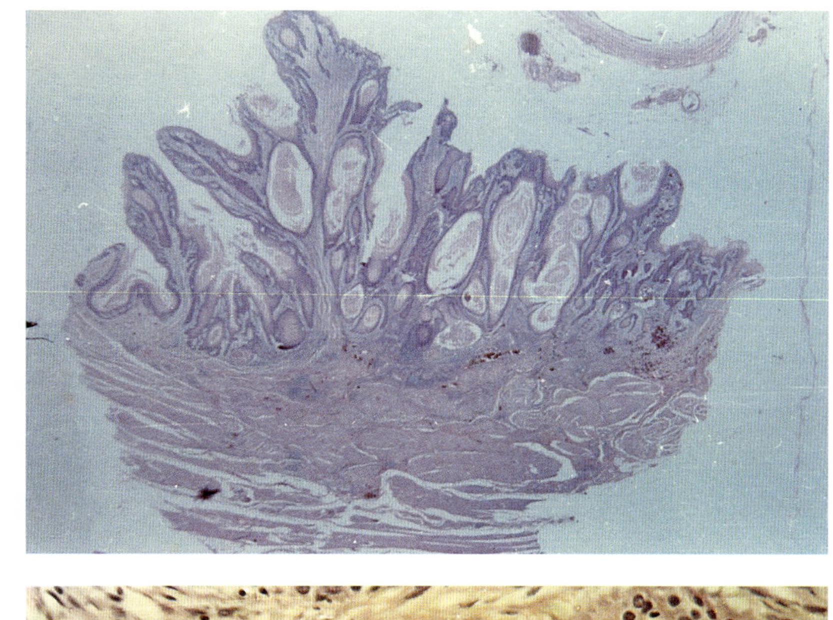
B

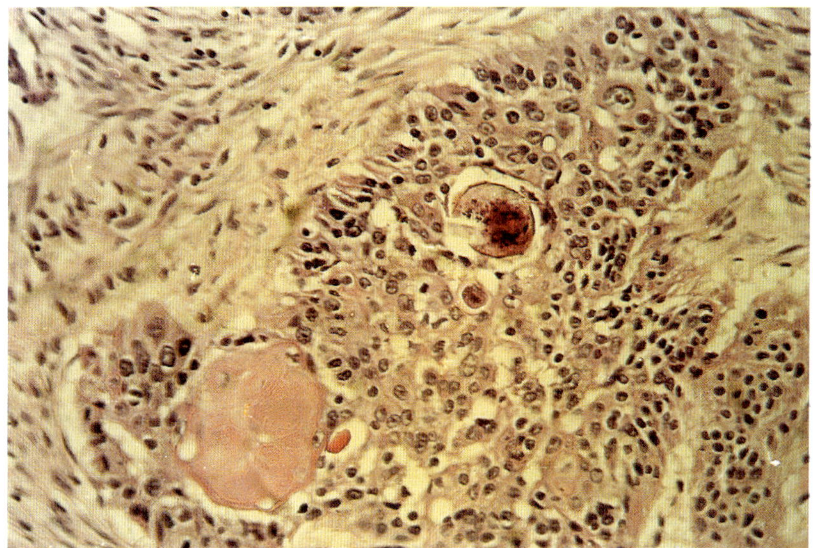
C

Figure 6–35. (*Continued*)

The management of squamous cell carcinoma remains similar to that of urothelial carcinoma. The treatment of choice is radical cystectomy or cystoprostatectomy, with or without preoperative radiotherapy. The prognosis is poor, with a survival rate of 13% at five years. Distant metastases are rare (Serretta et al., 2000).

Cytology. The cytologic appearance varies with tumor differentiation. Cells may have hyperchromatic, small nuclei with bright cytoplasmic keratinization in well-differentiated carcinomas (Fig. 6–36 A–D). In poorly differentiated squamous carcinoma, cells appear highly pleomorphic and impossible to distinguish from those of poorly differentiated urothelial carcinoma. Keratinization and necrosis are commonly present in the smear background.

The rare verrucous carcinoma has a deceptively bland cytomorphology. The histologic diagnosis is reached only after careful evaluation of the deep pushing margins and stroma of the excised tumor.

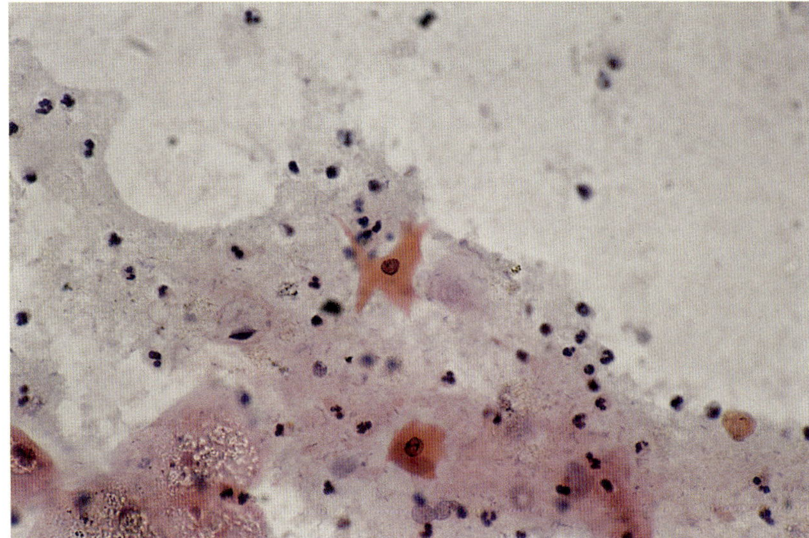
A

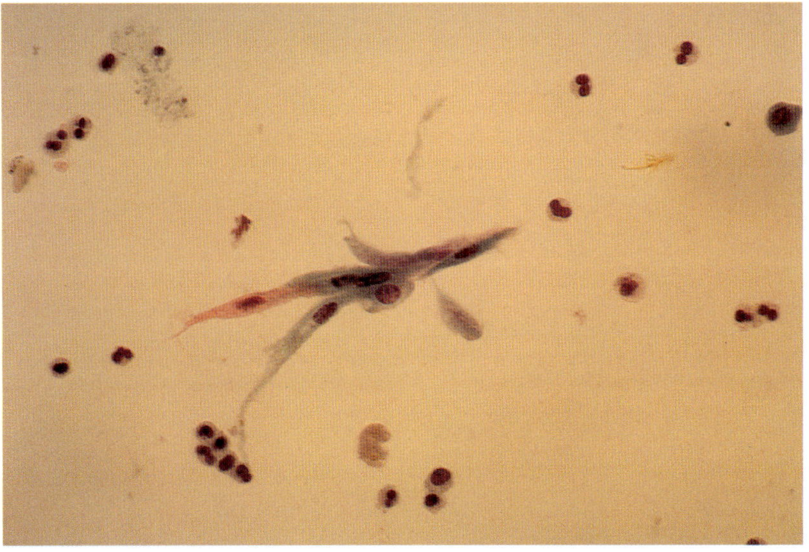
B

Figure 6–36. *Cellular features of squamous cell carcinoma.* (*A*) Cells in well-differentiated squamous cell carcinoma have subtle nuclear abnormalities and brightly keratinized cytoplasm. (*B,C*) Spindle cells with irregular, hyperchromatic nuclei and clusters of anaplastic squamous cells with less keratinized cytoplasm are present in less differentiated tumors. (*D*) In poorly differentiated squamous cell carcinoma, the exfoliated malignant cells lack squamous features and are extremely difficult to distinguish from those of poorly differentiated urothelial carcinoma. Voided urine (A–C) and bladder washing (D). (Papanicolaou stain, ×600). A: Courtesy of Dr. Hassan Nabil Tawfik, Professor and Director of Anatomic and Surgical Pathology, National Cancer Institute, Cairo University, Cairo, Egypt.

As a safe rule in cytology, it is preferable to describe the high-grade nature of the neoplasm and the squamous component, but to refrain from making a diagnosis of squamous carcinoma, which should be made only after careful evaluation of the resected specimen.

Differential diagnosis. The differential diagnosis includes urothelial carcinoma with

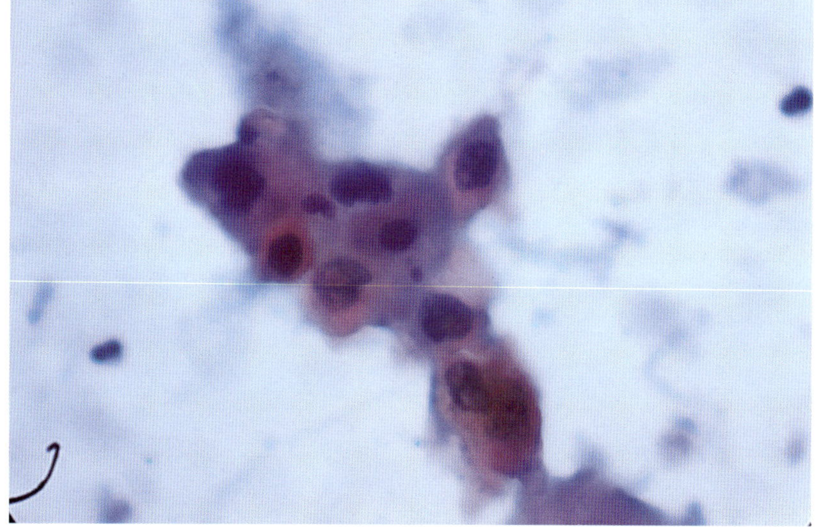

C

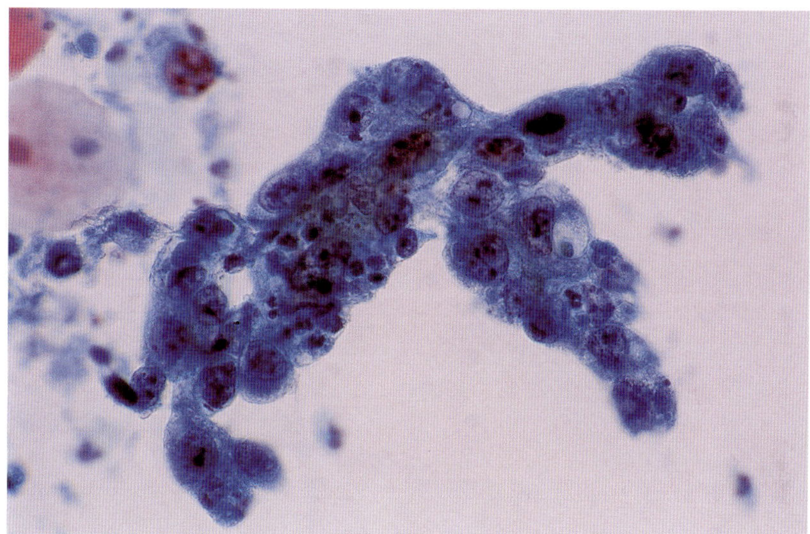

D

Figure 6–36. (*Continued*)

squamous differentiation, metastasis from a distant primary, and invasion from an adjacent organ such as the cervix or vagina. The latter two may be excluded on clinical grounds.

Squamous differentiation occurs in up to 20% of patients with high-grade urothelial carcinoma. Keratinized high-grade carcinoma cells may be difficult, if not impossible, to distinguish from those of squamous cell carcinoma (Fig. 6–37A,B). A search for dishesive high-grade tumor cells, including those of carcinoma in situ, is advisable when the smear shows large numbers of keratinized cells. Occasionally the high-grade carcinoma component may be expressed only as carcinoma in situ.

ADENOCARCINOMA

Cytology: "Carcinoma NOS" or "Adenocarcinoma"

Adenocarcinoma accounts for less than 2% of bladder carcinomas and can arise in the urinary bladder itself or in the urachus. Adenocarcinoma arising in the bladder is

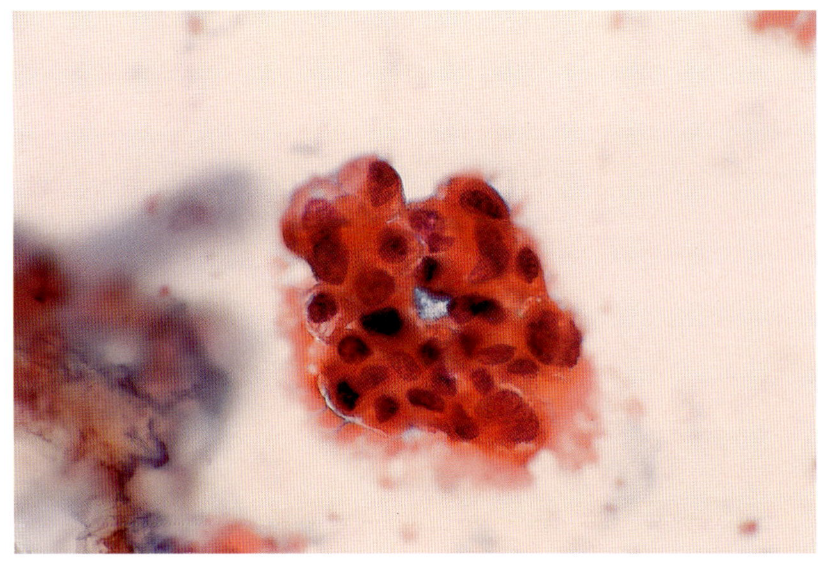

A

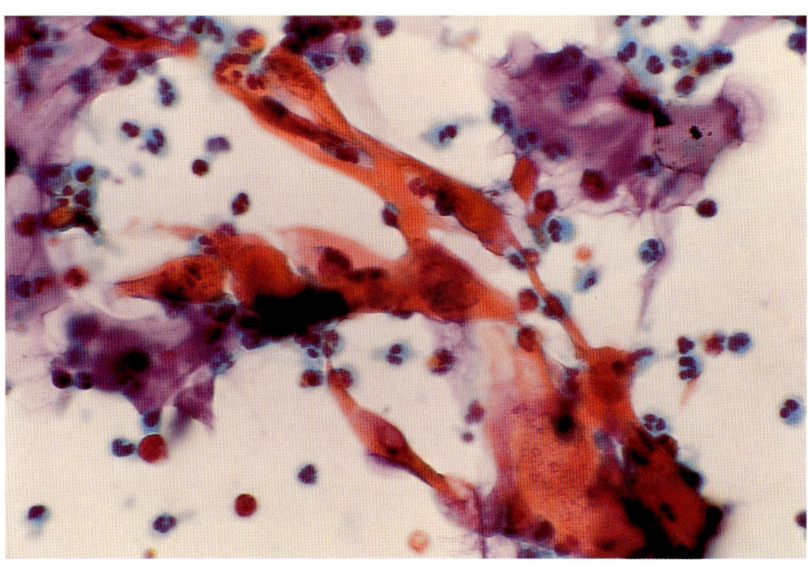

B

Figure 6–37. *Invasive urothelial carcinoma with squamous differentiation.* (*A,B*) Occasionally, high-grade urothelial carcinomas have pronounced squamous differentiation and shed cells with keratinized cytoplasm that are difficult to distinguish from keratinized squamous carcinoma. These two cases exhibited nonkeratinized high-grade urothelial carcinoma cells in the urine samples, not shown in these figures. Bladder washings. (Papanicolaou stain, ×600)

twice as common as urachal adenocarcinoma. Because there are no histologic features that distinguish between the two, the distinction requires correlation with other pathologic and clinical findings. Location in the bladder dome, lack of transition from normal to neoplastic urothelium, and exclusion of a distant primary are the main criteria that support a urachal origin (Eble and Young, 1997; Grignon, 1997; Murphy et al., 1994).

Most bladder adenocarcinomas have no predisposing factors, although some develop in the setting of schistosomiasis, bladder augmentation, or neurogenic bladder. In cases of protracted chronic cystitis, the sequence of chronic irritation–intestinal metaplasia–dysplasia–adenocarcinoma has been pro-

Box 6–14.

Bladder adenocarcinoma has a less favorable prognosis than urothelial carcinoma.

Bladder adenocarcinoma can be of intestinal, mucinous, clear, or signet-ring cell morphology.

Urachal adenocarcinoma is frequently mucinous.

Mucinous and clear-cell types are the most common urethral adenocarcinomas in women.

Renal pelvis adenocarcinomas are similar to intestinal tumors.

posed (Adegboyega and Adesokan, 1999). Clinical symptoms are nonspecific and include hematuria, urgency, frequency, and occasionally mucusuria, that is, the emission of mucinous material at the time of urine evacuation. Mucusuria is seen in up to 25% of patients with urachal adenocarcinoma, and its presence along with a suprapubic mass in a young patient may suggest such a diagnosis (Eble and Young, 1997). Primary adenocarcinoma is generally invasive at the time of diagnosis and is associated with a less favorable prognosis than is urothelial carcinoma (Lamm and Torti, 1996).

Histologically, bladder adenocarcinoma may be of intestinal, mucinous, clear, or signet-ring cell morphology. The intestinal type is the most common and is indistinguishable from colorectal adenocarcinoma; the mucinous (colloid) type has nests of malignant cells embedded in pools of mucin; and the signet-ring cell carcinoma can produce a linitis plastic-like type of carcinoma. Urachal adenocarcinomas are frequently of the mucinous type (Grignon et al., 1991). With the exception of the signet-ring cell type, which has a particularly poor prognosis, histologic subtyping of pure adenocarcinoma has little prognostic significance (Grignon et al., 1991; Murphy et al., 1994). Tumors with more than 10% urothelial carcinoma should be designated as urothelial carcinoma with glandular differentiation.

Columnar/mucinous and clear-cell adenocarcinoma are the most common types of adenocarcinoma of the urethra in women (Meis et al., 1987). Clear-cell adenocarcinoma shows clear cells containing abundant glycogen-rich cytoplasm arranged in tubular, cystic, papillary, and diffuse patterns in various proportions (Eble and Young, 1997; Oliva and Young, 1996). Adenocarcinoma of the male urethra, thought to originate from metaplastic urethral mucosa, is extremely rare, comprising less than 1% of all urologic malignancies (Bostwick et al., 1984).

Adenocarcinomas of the renal pelvis are morphologically similar to intestinal tumors. They are believed to arise in foci of intestinal metaplasia and have the same prognosis as all bladder adenocarcinomas (Spires et al., 1993).

The diagnosis of bladder adenocarcinoma raises the question of whether the lesion is primary, urachal, or secondary from a distant or adjacent organ. Unfortunately, when evaluated by the cytologic, histopathologic, histochemical, immunologic, and ultrastructural techniques currently available, primary adenocarcinomas have no features distinguishing them from those of secondary adenocarcinoma of gastrointestinal origin (Abenoza et al., 1987; Alroy et al., 1981) (Box 6–14).

Cytology. As in urothelial carcinoma in situ, high-grade carcinomas, including ade-

Box 6–15.

Columnar cells and necrosis are present in colonic-type adenocarcinoma.

A large cytoplasmic vacuole displacing a hyperchromatic nucleus is seen in signet-ring cell adenocarcinoma.

Numerous columnar cells with bland nuclear features and abundant extracellular mucin may be found in urachal adenocarcinoma.

Clear-cell adenocarcinoma shows rare clusters of cells with prominent nucleoli and numerous naked nuclei. Tissue diagnosis is necessary.

High-grade adenocarcinoma may not have detectable glandular differentiation.

nocarcinomas, can be detected with high accuracy in urine samples (Gill et al., 1989; Wiener et al., 1993). The diagnosis of malignancy by urine cytology can be made in close to 90% of samples, with a specific diagnosis of adenocarcinoma in two-thirds of them (Fig. 6–38A,B). Thus, approximately one-third of adenocarcinomas may be indistinguishable from poorly differentiated urothelial carcinomas based on urine cytology alone (Bardales et al., 1998). Variables affecting the diagnostic accuracy include the urine collection method, histologic grade, pretreatment and posttreatment status, and the observer's diagnostic experience (Box 6–15).

The cytomorphology of primary urinary bladder adenocarcinoma needs to be correlated with clinical and radiographic findings,

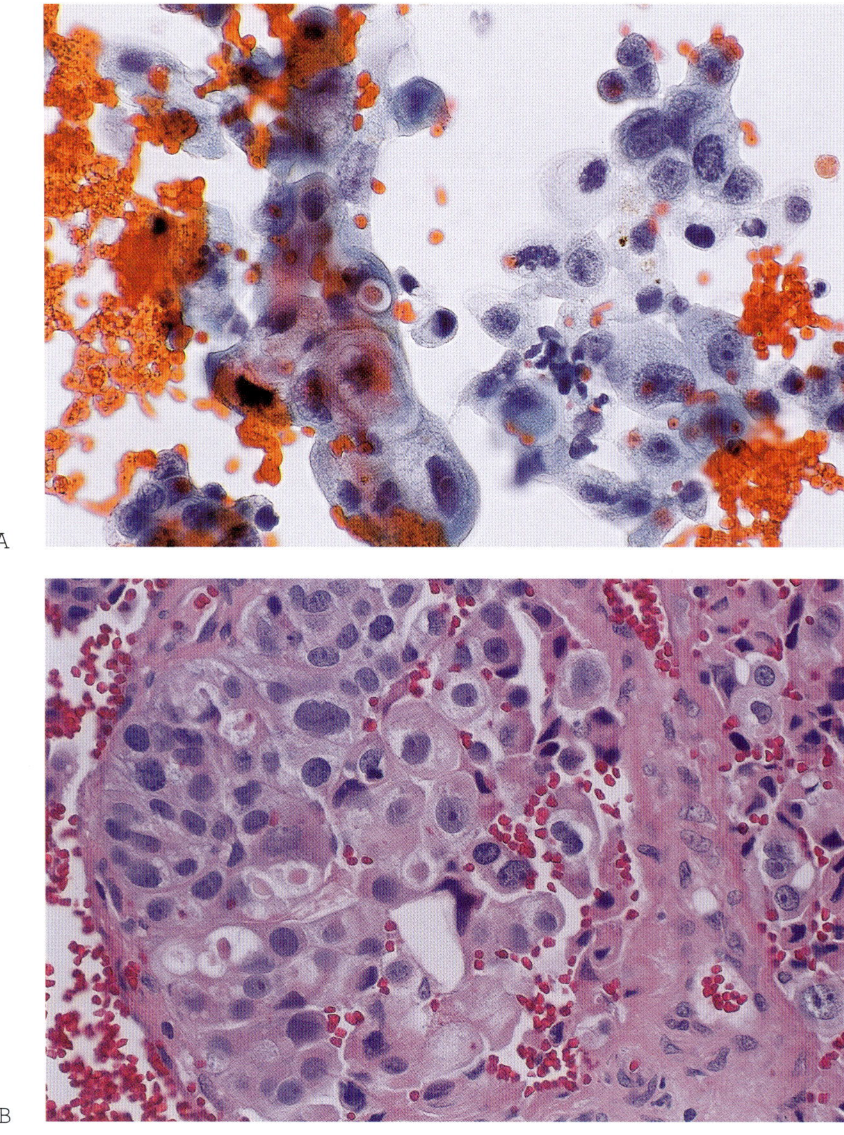

Figure 6–38. *Adenocarcinoma of the bladder.* (*A*) Adenocarcinoma can be accurately diagnosed by urine cytology examination. In bladder washings, cells exhibit abundant, finely vacuolated cytoplasm, a round/oval nucleus with vesicular chromatin, and a prominent nucleolus. (*B*) Tissue section shows similar features. (A, Papanicolaou stain, ×600; B, Hematoxylin and eosin stain, ×400)

and ultimately histologic confirmation is required. The diagnosis of a primary tumor should be considered only when clinical evaluation has ruled out a distant primary, commonly of alimentary tract origin.

The identification of glandular structures in the smear is helpful as long as the single cells show features of adenocarcinoma. These features include thin or finely vacuolated cytoplasm, vesicular chromatin, and prominent nucleoli (Fig. 6–39A–C). However, glandular differentiation and cellular features of adenocarcinoma may be minimal or absent, particularly in primary high-grade adenocarcinoma (Fig. 6–40A,B). Colonic-type adenocarcinoma shows a mucinous background containing rare, poorly differentiated and degenerated pleomorphic malignant cells, a few columnar cells, and a background of necrosis (Fig. 6–41A–C). A few scattered round cells with a large cytoplasmic vacuole pushing a hyperchromatic nucleus are seen in signet-ring cell adenocarcinoma (Fig. 6–42). Numerous columnar cells with "bland" nuclear features and abundant extracellular mucin are found in urachal adenocarcinoma (Fig. 6–43A–D) (Bardales et al., 1998).

Clear-cell adenocarcinoma is diagnosed on the basis of the histologic pattern, not the clarity of the cells. The cytologic diagnosis may be difficult to make because clear cells may be present in other types of adenocarcinoma as well (Eble and Young, 1997). Papillary clear-cell adenocarcinoma of the urethra in urine cytology shows cytologic and ultrastructural features similar to those of clear-cell adenocarcinoma of the female genital tract. Cytologic features include cell clusters with ill-defined cytoplasmic borders and numerous dissociated cells with fragile cytoplasm, resulting in numerous bare nuclei in a background of necrosis. Well-preserved cells are large, with abundant thin or vacuolated cytoplasm, round or oval nuclei, vesicular chromatin, and a prominent nucleolus (Peven and Hidvegi, 1985). These features are sufficient for the generic diagnosis of adenocarcinoma but not for the specific diagnosis of clear-cell adenocarcinoma (Kusuyama et al., 1988).

An occasional report of papillary adenocarcinoma of the renal pelvis describes small cells, arranged singly and in clusters, having mucin-positive cytoplasmic vacuoles, nuclear

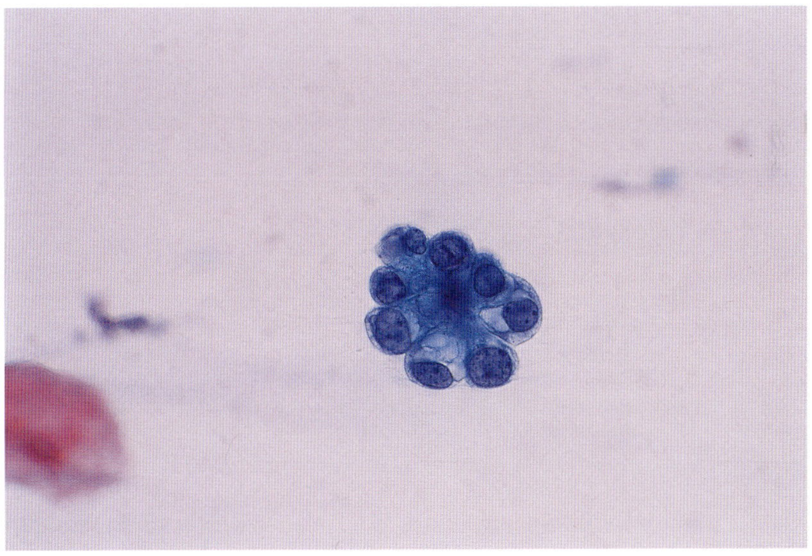

A

Figure 6–39. *Adenocarcinoma of the bladder.* (*A*) In well-differentiated adenocarcinoma, cells exhibit vacuolated cytoplasm and form glandular arrangements with round/oval nuclei and vesicular chromatin. (*B,C*) In less differentiated tumors the glandular configuration is rarely seen, and the nuclei have irregular membranes and coarse chromatin; however, the malignant cells can still be classified as glandular. Cellularity may be scant, particularly in voided urine specimens. (Papanicolaou stain, ×600)

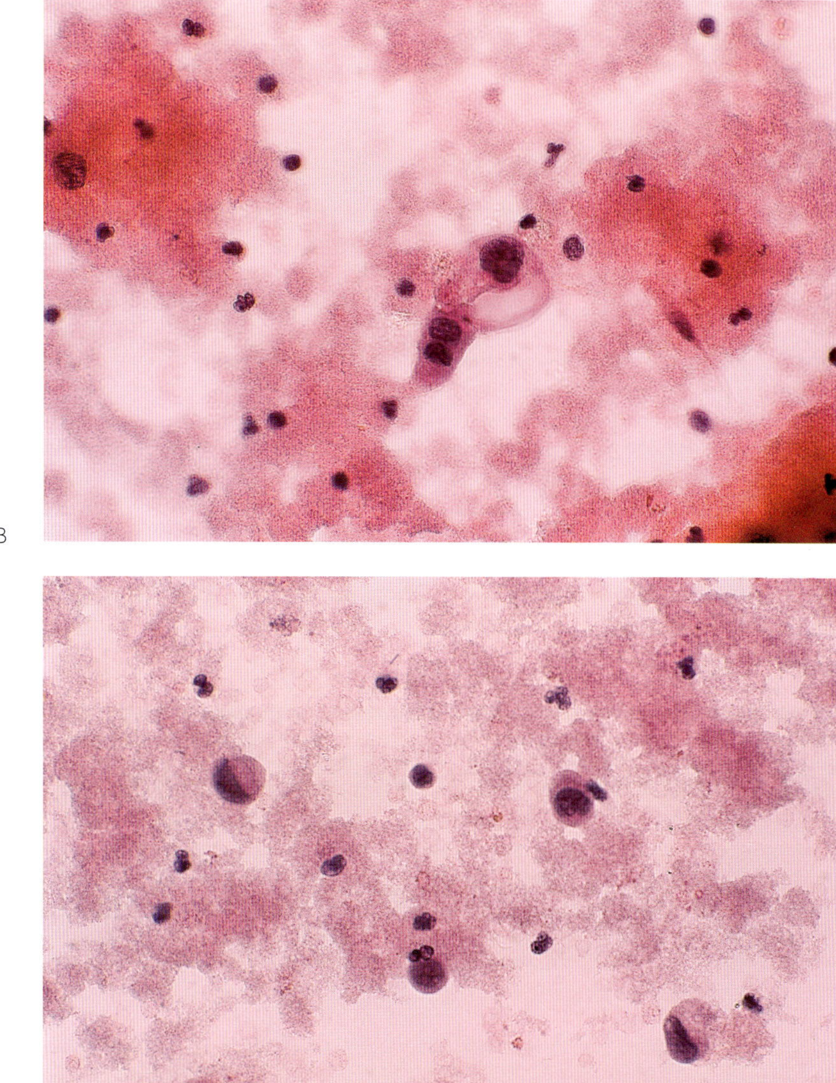

Figure 6–39. (*Continued*)

overlapping, and moderate variations in nuclear size and shape. A population of less pleomorphic columnar cells is also identified (Kobayashi et al., 1985). Intestinal and mucinous renal pelvis adenocarcinomas show a cytomorphogy similar to that of the bladder.

None of these cytologic features is site- or type-specific, and the final tumor diagnosis rests on histologic interpretation and clinical correlation.

Differential diagnosis. The differential diagnosis of primary adenocarcinoma in urine includes benign and malignant tumors and reactive conditions.

Vacuolated cells of nephrogenic adenoma must be distinguished from those of degeneration, instrumentation, and a therapy effect, and from malignancies such as clear-cell carcinoma, prostate adenocarcinoma, renal cell carcinoma, and papillary urothelial carcinoma. Helpful diagnostic

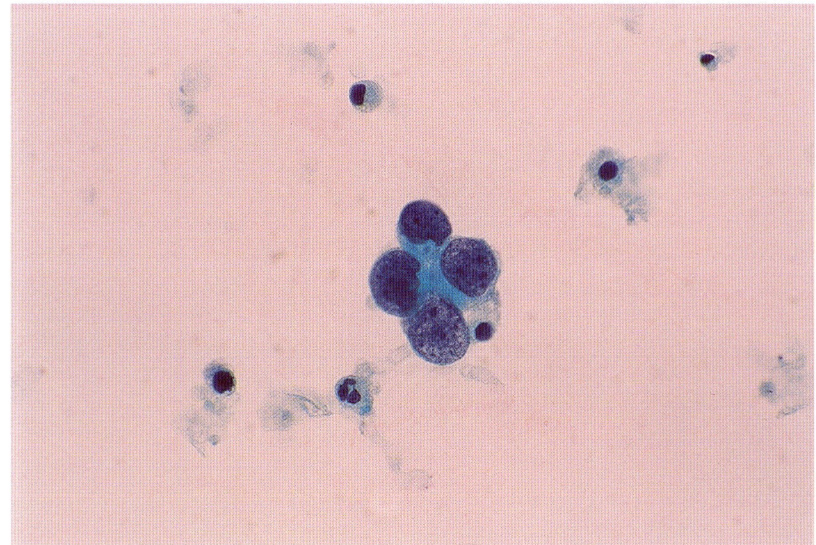

A

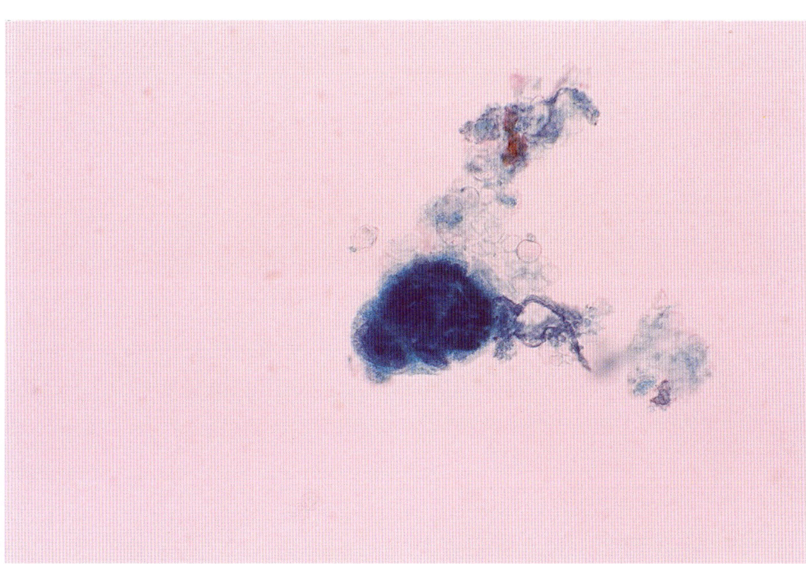

B

Figure 6–40. *Poorly differentiated adenocarcinoma.* In poorly differentiated adenocarcinoma, cells may form small aggregates or may be single. (*A*) The cells have a high nuclear-cytoplasmic ratio, scant and dense or minimally vacuolated cytoplasm, and large pleomorphic nuclei with coarse chromatin. (*B*) The glandular features are less discernible. Both specimens are voided urine. (Papanicolaou stain, ×600)

features are shown in Table 5–4 in Chapter 5.

The distinction of a well-differentiated primary bladder adenocarcinoma from cystitis glandularis may be difficult when cytomorphology alone is considered. Low cellularity and cohesive groups of columnar cells with bland nuclei are reliable indicators of a benign condition. In contrast, necrosis and nuclear anaplasia indicate malignancy. Clusters of columnar cells showing uniform, oval nuclei with powdery chromatin have been described in urine specimens in cases of benign adenomatous polyp of the prostatic urethra. A background of scant mucin may be present in these cases, but necrosis and mild nuclear anaplasia

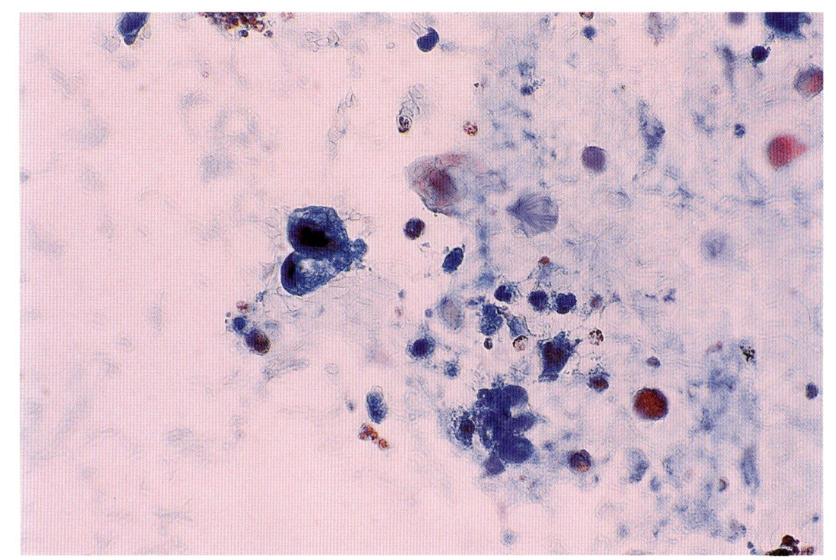
A

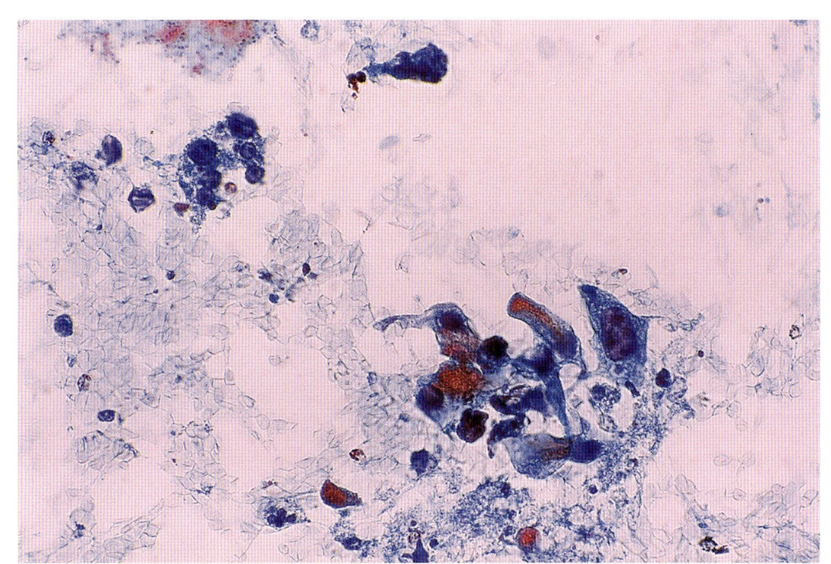
B

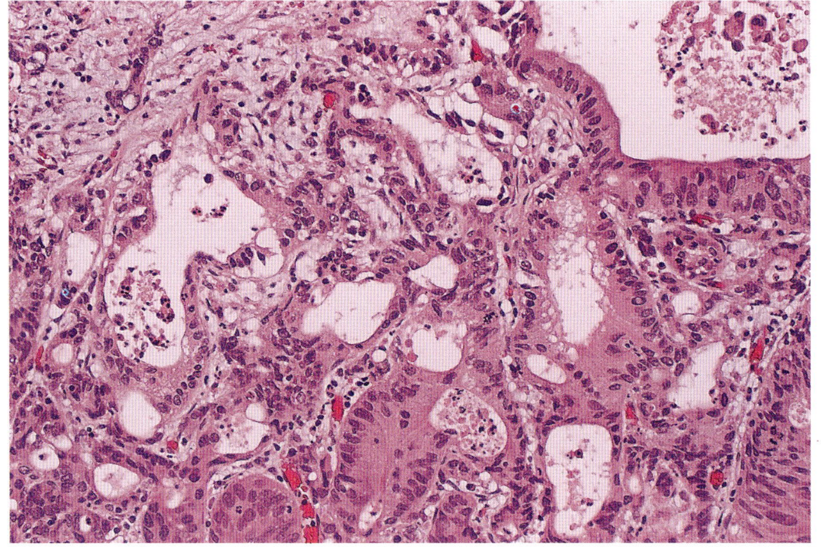
C

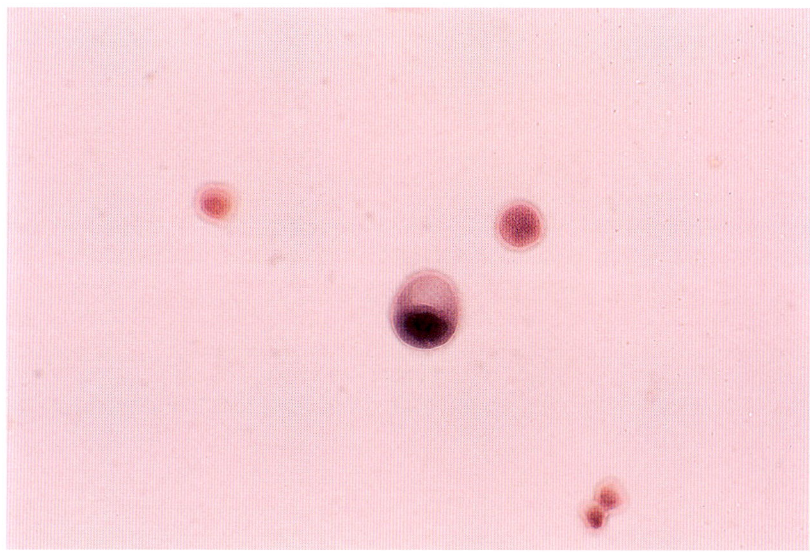

Figure 6–42. *Signet-ring cell type bladder adenocarcinoma.* A single large cytoplasmic vacuole of mucin displacing the hyperchromatic nucleus peripherally is present in these cases. The cytology is identical to that of metastatic signet-ring cell adenocarcinoma of breast or gastric origin. The specimen is voided urine. (Papanicolaou stain, ×600)

are absent and permit the distinction from well-differentiated urachal adenocarcinoma (Schnadig et al., 2000).

A diagnosis of endocervicosis, the mucinous analogue of endometriosis, must be considered in young women who have bladder symptoms and a posterior-wall bladder mass. The cytology of this lesion, characterized histologically by a haphazard proliferation of irregular glands lined by single columnar cells with abundant mucin production, has not been reported.

UNDIFFERENTIATED CARCINOMA

Small-Cell Carcinoma

Cytology: "Small-Cell Carcinoma"

Small-cell carcinoma of the urinary tract is a highly aggressive malignancy, accounting for less than 1% of bladder malignancies, and is exceedingly rare in the renal pelvis. Most cases of small-cell carcinoma of the bladder and renal pelvis occur in combination with urothelial carcinoma (Ali et al., 1997; Essenfeld, 1990; Guillou et al., 1993). An association with adenocarcinoma, and less commonly with squamous carcinoma and sarcoma, has been reported (Mazzucchelli et al., 1992). An origin of this tumor in the multipotential stem cells in the urothelium has been proposed (Eble and Young, 1997).

Small-cell carcinoma is most frequent in men after the sixth decade of life. Patients frequently have hematuria, with or without dysuria, and occasionally paraneoplastic syndromes (Ali et al., 1997; Grignon, 1997).

The histologic pattern is that of small-cell carcinoma of other organs, with cells having neuroendocrine differentiation demonstrable by immunohistochemistry and electron microscopy in most cases. In contrast to adenocarcinoma and squamous carcinoma, small-cell carcinoma does not need to be present in a pure form to be reported. Mixed carcinoma should be reported as small-cell carcinoma followed by a descrip-

Figure 6–41. *Colonic-type bladder adenocarcinoma.* (*A*) Abundant necrosis and degenerated malignant cells are present in these cases. (*B*) Cells are sparse and poorly preserved and exhibit a columnar shape, pleomorphic nuclei, and hyperchromasia with coarse chromatin. (*C*) Histologic and cytologic features are identical to those of adenocarcinoma of colonic origin. The clinical history is crucial to distinguish between the two. Specimens are bladder washings and bladder biopsy. (A,B, Papanicolaou stain, ×600; C, hematoxylin and eosin stain, ×400)

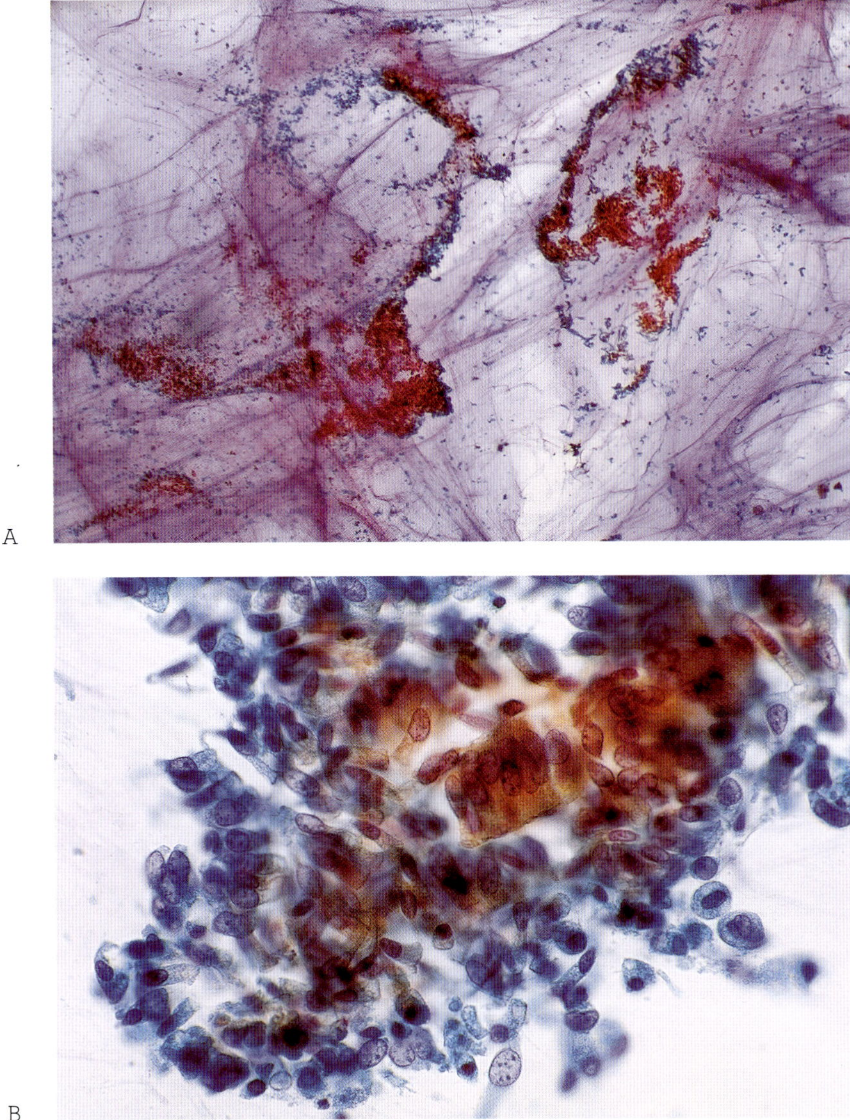

Figure 6–43. *Urachal adenocarcinoma.* A 67-year old man had mucusuria. (*A*) Smears show abundant mucus and both single and clustered columnar cells. (*B,C*) Cells have mild anisonucleosis and bland nuclear features. Minimal necrosis is present. (*D*) This bladder tumor shows a well-differentiated adenocarcinoma. Urine samples are bladder washings. (A–C, Papanicolaou stain, ×100 (A), ×600 (B,C); D, hematoxylin and eosin, stain, ×400)

tion of the other components (Grignon, 1997).

The diagnosis of small-cell carcinoma is important because of its prognostic and therapeutic implications (Weiss, 1993). Based on published reports, small-cell carcinoma of the bladder is characterized by early metastatic spread and a poor prognosis (Lohrisch et al., 1999). Cystectomy followed by adjuvant chemotherapy appears to be the recommended treatment for resectable cases (Grignon et al., 1992). Combined systemic chemotherapy and radiotherapy has been proposed for patients with local pelvic disease without metastases (Lohrisch et al., 1999).

Cytology. Regardless of the specimen preparation technique, that is, smears, cytospin, or Thin-prep (Cytyc Corp., Boxborough, MA), the cytologic pattern is identical

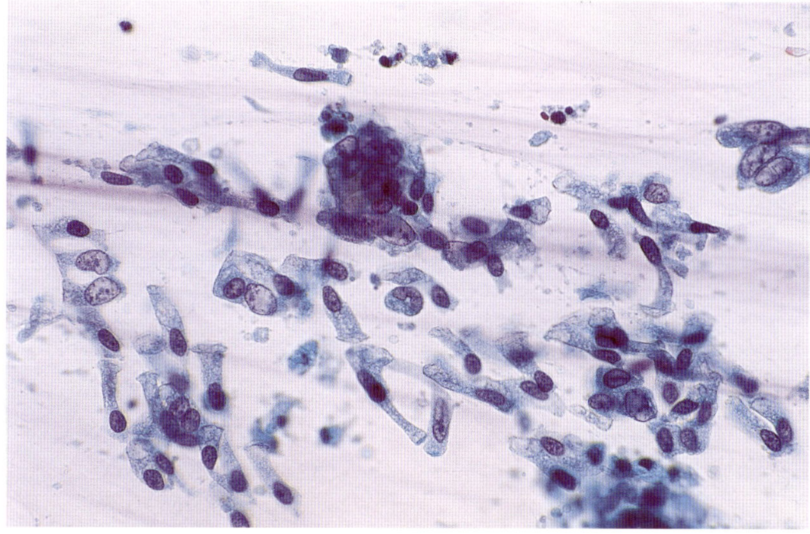

C

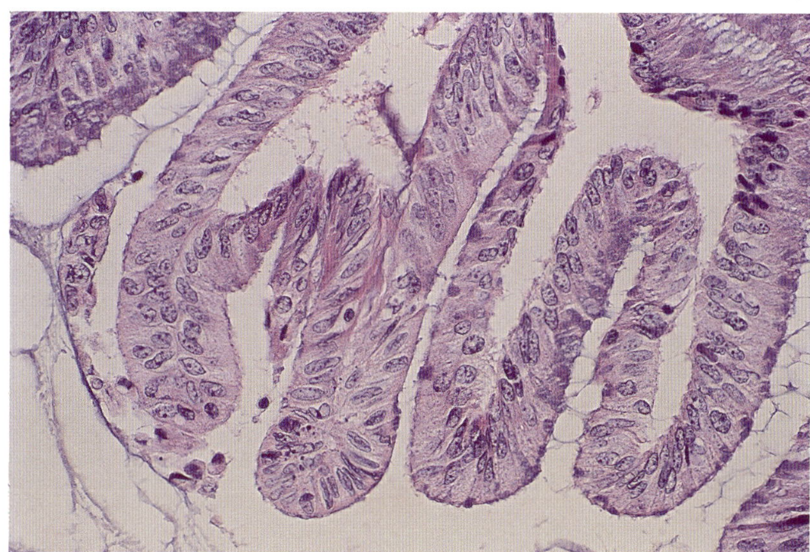

D

Figure 6–43. (*Continued*)

to that of smears of small-cell carcinoma of the lung. In contrast to the intermediate cell type, which shows nuclear pleomorphism, hyperchromasia, and small nucleoli, smears from cases of the oat cell type show minimally pleomorphic tumor cells and less striking hyperchromasia (Ali et al., 1997). In either case, smears are of high cellularity and show small, round to oval cells with scant cytoplasm, hyperchromatic nuclei, and coarse chromatin arranged singly and in clusters of variable cohesiveness. Numerous hyperchromatic naked and karyorrhectic nuclei, nuclear molding, and cytophagocytosis may be found, together with rare abnormal mitotic figures (Fig. 6–44A,B) (Acs et al., 2000; Ali et al., 1997; Borghi et al., 1995; McRae and Garcia, 1997; Rollins and Schumann, 1991; van Hoeven and Artymyshyn, 1996).

The background usually shows considerable acute inflammation, blood, and necrotic debris that may obscure the cellular detail. Blood and necrotic debris are more evident in specimens prepared by smear or cytospin methods than in filter preparations or Thinprep smears.

In cases of mixed carcinomas, urine specimens may show mixed small-cell and non-small-cell carcinoma of urothelial, squa-

Box 6–16.

Most cases of primary small-cell carcinoma occur in combination with urothelial carcinoma.

Smears are highly cellular, showing small cells with scant cytoplasm, hyperchromatic nuclei, and prominent nuclear molding in a background of necrosis and blood.

An inconspicuous nucleolus may be seen in intermediate-type small-cell carcinoma.

The oat-cell type shows cells with less hyperchromasia and mild pleomorphism.

mous, or glandular origin or only the non-small-cell component. In these cases, the non-small-cell component often masks the coexisting small-cell carcinoma component (Acs et al., 2000). Cytologic examination is useful in these cases to guide the examination and tissue sampling of the cystectomy specimen.

The cytologic features of small-cell carcinoma are distinct, and an accurate urinary cytologic diagnosis can be made, with no false-positive or false-negative diagnoses. However, small-cell carcinoma may be misdiagnosed as high-grade transitional cell carcinoma based on cytology (Ali et al., 1997). The diagnosis of primary small-cell carcinoma of the urinary tract is made only after a metastasis from a distant primary is excluded.

Immunohistochemical stains are useful to evaluate small-cell carcinoma. Cells show positive stain for neuron specific antigen, chromogranin, and synaptophysin. Keratin stain is positive in 60% of cases. CD44v6, a cell surface transmembrane glycoprotein, is positive in urothelial carcinoma and absent in small-cell carcinoma (Acs et al., 2000) (Box 6–16).

Differential diagnosis. The identification of urothelial carcinoma in situ indicates a bladder origin for small-cell carcinoma. However, when the smears show small cells only, the differential diagnosis includes metastatic small-cell carcinoma from another source, Merkel cell carcinoma, and other small blue cell neoplasms including hematolymphoid malignancies.

A cytologic distinction from pure small-cell carcinoma of the prostate is not possible, and clinical correlation is required. A negative cystoscopic examination with a normal-appearing bladder mucosa helps to exclude a bladder primary and favors a prostatic origin (Ali et al., 1997). Prostatic small-cell carcinoma is not uncommonly associated with prostatic adenocarcinoma that shows positive immunocytochemistry for prostate specific antigen and prostatic acid phosphatase, thus helping to distinguish between a prostatic and a bladder origin in cases of mixed small- and non-small-cell carcinomas. The small-cell component is negative for these markers (Nguyen-Ho et al., 1994).

Metastasis to the bladder from a small-cell lung carcinoma is rare and usually occurs in association with disseminated disease that is evident on clinical grounds. The cytologic findings are similar to those of primary small-cell carcinoma of the bladder.

Involvement of the urinary tract by lymphomas and leukemias is usually a manifestation of a systemic disease. Such diseases tend to exfoliate as single cells with almost no cell clusters. However, some cases of primary malignant lymphoma of the bladder may be difficult to differentiate from small-cell carcinoma. Immunocytochemistry for cytokeratin, which is positive in 74% of small-cell carcinoma cases, and CD45 (leukocyte common antigen), which is negative in small-cell carcinoma and positive in hematolymphoid malignancies, is helpful in differentiating between the two (van Hoeven and Artymyshyn, 1996).

Occasionally, the cytokeratin stain in small-cell carcinoma of the bladder exhibits a dot-like cytoplasmic positivity similar to that seen in Merkel cell carcinoma, a neuroendocrine carcinoma of the skin that occasionally metastasizes to the urinary bladder (Woo and Kencian, 1995). In addition to the clinical history of disseminated disease that is the most important diagnostic clue, positive immunocytochemistry for cytokeratin 20 strongly predicts a diagnosis of Merkel cell carcinoma (Chan et al., 1997).

Carcinoid tumors have been described in the urinary tract. In these cases, the smear background lacks necrosis and the cells are better preserved, with more abundant cytoplasm, more uniform nuclei, and less hyperchromasia (Rudrick et al., 1995).

A

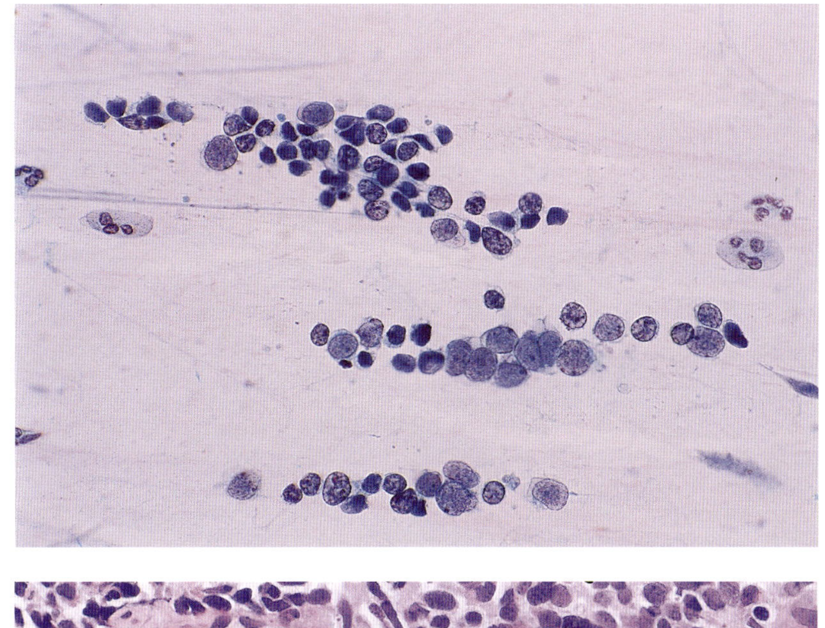

B

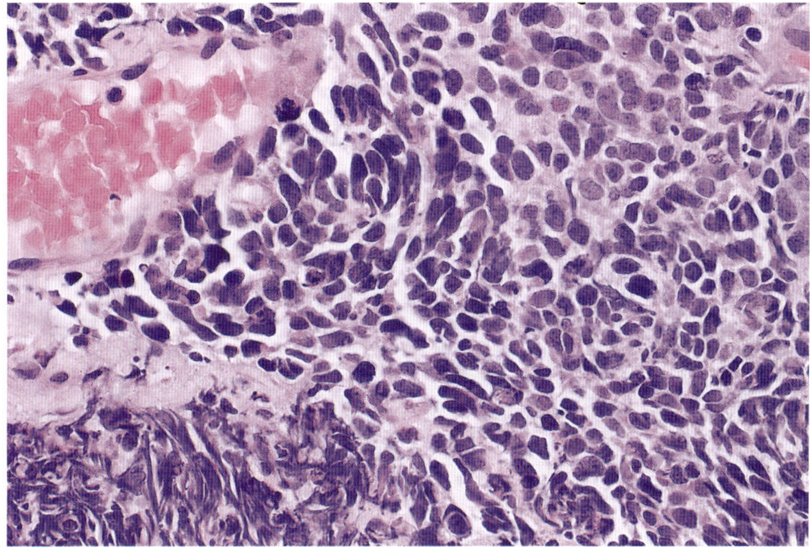

Figure 6–44. *Small cell carcinoma.* (*A*) Bladder washing shows loosely cohesive clusters of small cells with scant cytoplasm and molding. Numerous smaller cells with apoptotic nuclei are also present. Note the cell size in contrast with that of neutrophils. (*B*) Cytologic and histologic features are identical to those of metastatic small-cell carcinoma of the lung. The clinical history helps to distinguish between the two. (A, Papanicolaou stain, ×600; B, hematoxylin and eosin stain, ×400)

Occasionally, high-grade urothelial carcinoma may resemble small-cell carcinoma. However, the larger cells, with a more pleomorphic appearance and often prominent nucleoli, lack of nuclear molding, and less necrotic background are features that distinguish high-grade urothelial carcinoma from small-cell carcinoma. Lack of necrotic background is the main feature that distinguishes small-cell carcinoma from urothelial carcinoma in situ composed of small cells with hyperchromatic nuclei and absent nucleoli.

Lymphoepithelioma-like Carcinoma

Lymphoepithelioma-like carcinoma is morphologically identical to nasopharyngeal carcinoma and is rarely seen in the urinary tract, with only isolated cases involving the urinary bladder, ureter, and renal pelvis (Amin et al., 1994b; Cohen et al., 1999). Histologically, it may be pure or mixed, with urothelial, glandular, or squamous components. The undifferentiated epithelial cells show indistinct cytoplasmic membranes, large, pale pleomorphic nuclei arranged in syncytial sheets, and prominent nucleoli present in a dense lymphoplasmacytic inflammatory infiltrate with histiocytes and occasional segmented leukocytes (Grignon, 1997).

Cytology. The cytology of this tumor in urine sediment has not been reported. However, the cytologic characteristics of lymphoepithelioma-like carcinoma of the cervix in a cervical smear include large, uniform tumor cells with indistinct cell borders, finely granular cytoplasm, round or oval vesicular nuclei with prominent nucleoli, and a background of small lymphocytes (Reich et al., 1999). The inconspicuous epithelial cells become evident on immunohistochemical stain for cytokeratin.

Differential diagnosis. Lymphoepithelioma-like carcinoma should be distinguished from inflammatory lesions including XGP and from malignant conditions such as lymphoma, poorly differentiated urothelial carcinoma, squamous cell carcinoma, and lymphoma, the last being the most important (Cohen et al., 1999; Grignon, 1997).

MESENCHYMAL NEOPLASMS

Benign mesenchymal neoplasms of the urinary bladder are uncommon. They are confined to the bladder wall, with an overlying intact urothelial mucosa. The most common benign neoplasms are leiomyomas, which comprise approximately one-third of cases. Less common benign soft-tissue tumors include hemangiomas, neurofibromas, benign fibrous histiocytomas, ganglioneuromas, and lymphangiomas (Grignon, 1997).

Leiomyosarcoma and rhabdomyosarcoma are the most common types of sarcoma in adults and children, respectively. Other, less common malignant soft-tissue tumors in adults include angiosarcoma, liposarcoma, osteosarcoma, malignant fibrous histiocytoma, and rhabdoid tumor (Grignon, 1997).

In rare instances, malignant soft-tissue neoplasms of the bladder may be detected in urine cytology examination when invasion and ulceration of the urothelial mucosa occur (Box 6–17).

Rhabdomyosarcoma

Rhabdomyosarcoma is the most common bladder tumor in children under the age of 15 years and is rare in adults. Most patients have hematuria and bladder neck obstruction by a polypoid "grape-like" mass (sarcoma botryoides). In children, sarcoma botryoides is characterized by the presence of small, round blue cells with scant cytoplasm. In adults, rhabdoid or strap cells with cytoplasmic striations may be found. The final diagnosis requires immunohistochemistry and ultrastructural studies to prove the rhabdomyoblastic differentiation.

Cytology. Urinary cytology in botryoid rhabdomyosarcoma shows small clusters and single small cells. The cells have scant cytoplasm, hyperchromatic nuclei, and absent nucleoli. Some cells may exhibit a wispy cytoplasm. Adult rhabdomyosarcoma may exhibit cells with ample eosinophilic cytoplasm with cross-striations (Koss, 1996; Krumerman and Katatikarn, 1976; Mincione and Grechi, 1983).

Box 6–17.

Leiomyosarcoma is the most common urinary tract sarcoma in adults.

Rhabdomyosarcoma is the most common bladder tumor in children.

Malignant cells in urine cytology smears may be detectable because of invasion and ulceration of the urothelial mucosa.

Clusters and single small cells with hyperchromatic nuclei and absent nucleoli may be seen in smears from botryoid rhabdomyosarcoma.

Rhabdoid or strap cells with cytoplasmic striations may be seen in adult rhabdomyosarcoma.

Differential diagnosis. Botryoid rhabdomyosarcoma must be differentiated from other small blue cell tumors. Tissue biopsy, immunocytochemistry, ultrastructural studies, and clinical correlation are necessary for a final diagnosis.

LEIOMYOSARCOMA

Leiomyosarcoma is the most common soft-tissue tumor of the bladder in adults. This tumor usually occurs in the sixth decade of life, but occasionally tumors have been described in younger individuals (Grignon, 1997). An association with cyclophosphamide therapy has been suggested (Thrasher et al., 1990).

Grossly, leiomyosarcoma is a polypoid mass that may have mucosal ulceration. The spindle cells are arranged in interwoven fascicles, with variable nuclear pleomorphism and necrosis in high-grade tumors. Immunohistochemistry is helpful in the differential diagnosis with sarcomatoid carcinoma, because leiomyosarcoma is cytokeratin-negative and muscle-specific actin-positive.

Cytology. Leiomyosarcoma is an intramural neoplasm with an intact overlying urothelium. Occasionally, when there is mucosal ulceration, granular and amorphous debris may be seen in urine cytology. However, it is conceivable that malignant cells may be present in urine samples from high-grade leiomyosarcomas with mucosal ulceration. Unless there is a history of known sarcoma, the specific diagnosis is not possible on the basis of urine cytology alone.

OTHER SARCOMAS AND MISCELLANEOUS NEOPLASMS

Histologic diagnoses of malignant fibrous histiocytoma, angiosarcoma, liposarcoma, osteosarcoma, fibrosarcoma, hemangiopericytoma, rhabdoid tumor, and undifferentiated mesenchymal malignancies of the urinary bladder have been described in textbooks (Grignon, 1997). However, the urine cytology of some of these tumors has rarely been reported (Schindler et al., 1999).

HEMATOLYMPHOID NEOPLASMS

Most hematopoietic malignancies of the urinary tract are the result of secondary involvement by a systemic disease (Andrion et al., 1993; Helson and Hajdu, 1972). However, diffuse B-cell malignant lymphoma of the small- and large-cell types and rare cases of primary Hodgkin's disease of the urinary bladder and kidney have been described (Grignon, 1997; Jones, 1989).

MALIGNANT LYMPHOMA

Cytology: "Positive for Malignancy. Hematolymphoid Neoplasm"

Primary malignant lymphoma of the urinary bladder is more frequent in women than in men and usually occurs in the seventh decade of life. The involvement is submucosal, and patients have hematuria (Box 6–18).

Cytology. Positive urine cytology may be the presenting evidence of disseminated lymphoma (Gupta et al., 1989; Mincione, 1982; Mincione and Gasbarre, 1995). Most

Box 6–18.

Evidence of kidney tissue damage by malignant lymphoma is frequently present in urine sediment preparations.

Cytologic features are best seen in Romanovsky-stained preparations.

Smears show variable cellularity, lack of cellular cohesion, necrosis, and well-preserved monomorphic cells as well as damaged malignant cells.

Ki-1-positive lymphoma shows cells with convoluted nuclei similar to those of Hodgkin's disease or a poorly differentiated large-cell epithelial neoplasm.

Inmunocytochemistry helps to classify lymphomas further and to diagnose Hodgkin's disease.

patients with renal lymphoma and positive urine cytology have evidence of renal parenchymal necrosis, renal tubular injury, or pathologic cast formation, compared with only 56% of those with negative cytology (Cheson et al., 1984).

The urine sediment shows numerous dishesive, medium-sized or large mononuclear cells of monomorphic appearance exhibiting large nuclei and coarse chromatin. The tumor cells are fragile and can easily be damaged during smear preparation. Acute inflammatory cells, smudged cells, and tissue debris are commonly observed in the background.

Medium-sized cells about twice the size of red blood cells, with a cleaved nucleus and small nucleoli are seen in malignant lymphoma of the large cleaved cell type. Large cells with a round nucleus and large nucleoli located close to the nuclear membrane are seen in malignant lymphoma of the large noncleaved cell type. Large cells with vacuolated cytoplasm, eccentric nuclei, and prominent central nucleoli are seen in malignant lymphoma of the immunoblastic type.

Cytomorphologic findings in urinary cytology of Ki-1-positive anaplastic large-cell lymphoma and Hodgkin's disease involving the urinary bladder have been described (Bocian et al., 1982; Tanaka et al., 1993). Large cells with multilobated and bizarre nuclei with a "wreath" or "doughnut" shape and a high mitotic rate are characteristic of Ki-1-positive anaplastic large-cell lymphoma, just as Reed-Sternberg cells are of Hodgkin's disease. Presumably, similar cytomorphology is present in the urine when these malignancies develop primarily in the urinary tract. Immunocytochemistry for CD15+, CD30+, and CD45− confirms the diagnosis of Ki-1-positive anaplastic large-cell lymphoma, and immunocytochemistry for CD15+, CD30−, and CD45− confirms the diagnosis of Hodgkin's disease.

Differential diagnosis. Chronic inflammation can occasionally be confused with malignant lymphoma of the small lymphocytic type; however, a more monomorphous type of infiltrate and the presence of coarse chromatin and a small nucleolus indicate a malignant process. Small blue cell tumors including primary or metastatic small-cell carcinoma, Merkel cell carcinoma, and occasionally urothelial carcinoma in situ of the small-cell type may resemble malignant lymphoma. The dishesive cell nature of lymphoma and the aggregated carcinoma cells are strong indicators of the correct diagnosis. In some cases, the use of immunocytochemistry (leukocyte common antigen, cytokeratin) is helpful for a definitive diagnosis (Yam and Janckila, 1985). The clinical history is helpful for the diagnosis of metastatic neuroendocrine tumors because, in most cases, bladder involvement is heralded by systemic metastatic disease. Some variants of primary bladder urothelial carcinoma with marked lymphoid infiltration may pose a problem for the histologic diagnosis (Zukerberg et al., 1991). Positive immunocytochemical stain for cytokeratin is necessary to distinguish poorly differentiated large-cell malignant epithelial neoplasms from Ki-1-positive anaplastic large-cell lymphoma.

Plasma Cell Myeloma

Cytology: "Positive for Malignancy. Plasma Cell Myeloma"

Urinary bladder involvement can be secondary to disseminated disease or can be a primary solitary plasmacytoma. Primary urethral plasmacytomas have been described (Lemos et al., 2000). Characteristically, the overlying mucosa in these tumors is intact. However, an occasional report of positive urine cytology addresses the presence of abnormal plasma cells in the urine (Yang et al., 1982).

The cytomorphology and differential diagnosis of primary and metastatic plasma cell myeloma are similar and are discussed in Chapter 7.

Leukemia

Cytology: "Positive for Malignancy. Hematolymphoid Malignancy"

Leukemias rarely involve the urinary bladder primarily (Chaitin et al., 1984). Although bladder involvement in cases of chronic and acute leukemia ranges from 15% to 25%, it is rarely manifest clinically (Givler, 1971).

The cytology of these malignancies is discussed further in Chapter 7.

MALIGNANT MELANOMA

Cytology: "Positive for Malignancy. Malignant Melanoma"

The urethral meatus and distal urethra are the most common sites of origin in primary malignant melanoma of the genitourinary tract (Katz and Grabstald, 1976). Primary malignant melanoma of the bladder is exceedingly rare. It can occur in women in the sixth decade, usually presenting with hematuria. The cystoscopic gross appearance is that of a fungating or polypoid brown-black mass (Grignon, 1997). The histologic, immunohistochemical, and ultrastructural findings are similar to those in melanoma of the skin.

Malignant melanoma metastatic to the urinary tract from a skin primary is more frequent than primary melanoma of this organ. Approximately 18% of patients who die of metastatic malignant melanoma have bladder metastases (Stein and Kendall, 1984). Therefore, the absence of a history of skin, ophthalmic, or other visceral melanoma, a regional pattern of metastasis consistent with a bladder primary, and the presence of atypical melanocytes in the bladder mucosa adjacent to the tumor mass are strict criteria that must be applied before a diagnosis of primary melanoma is made (Stein and Kendall, 1984).

Box 6–19.

Metastatic malignant melanoma to the bladder is more frequent than primary malignant melanoma.

Strict clinical and histologic criteria must be applied before a diagnosis of primary malignant melanoma of the urinary tract is made.

Smears usually show cells with brown-black pigment in the cytoplasm, a large nucleus, coarse chromatin, and prominent nucleoli.

Irrespective of the type of therapy used, the prognosis for primary malignant melanoma of the bladder is dismal, and patients die of disseminated disease (Oldbring and Mikulowski, 1987).

Cytology. Cytodiagnosis of primary and metastatic malignant melanoma in urine is similar. Cells are large, oval or pleomorphic, and show a single, vesicular, sharply defined, often regularly contoured, and eccentric nucleus with red macronucleoli. Their cytoplasm is cyanophilic, with granular, brown-black melanin pigment. These composite characteristics, which are distinctive for malignant melanoma, eliminate other diagnostic possibilities (Woodard et al., 1978).

The unusual amelanotic malignant melanoma should be considered in the presence of pleomorphic cells with unusual nuclear features, a plasmacytoid appearance, and prominent nucleoli. In spindle cell melanoma, melanin pigment may be inconspicuous and located in the distal end of the elongated cytoplasm (Box 6–19).

Differential diagnosis. The differential diagnosis includes high-grade transitional cell carcinoma, plasmacytoid malignant lymphoma, metastatic malignancies, and occasionally small-cell malignancies including small-cell carcinoma. Clinical correlation, the presence of pigmented cells, and positive immunocytochemistry for S100 and HMB45 are helpful in making the diagnosis of melanoma.

REFERENCES

Abenoza, P., Manivel, C., Fraley, E. E. (1987). Primary adenocarcinoma of urinary bladder. Clinicopathologic study of 16 cases. *Urology* 29: 9–14.

Acs, G., Gupta, P. K., Baloch, Z. W. (2000). Cytomorphology of high-grade neuroendocrine carcinoma of the urinary tract. *Diagn Cytopathol* 23: 92–6.

Adegboyega, P. A., Adesokan, A. (1999). Tubulovillous adenoma of the urinary bladder. *Mod Pathol* 12: 735–8.

Ali, S. Z., Reuter, V. E., Zakowski, M. F. (1997). Small cell neuroendocrine carcinoma of the urinary bladder. A clinicopathologic study with emphasis on cytologic features. *Cancer* 79: 356–61.

Alroy, J., Roganovic, D., Banner, B. F., Jacobs, J. B., Merk, F. B., Ucci, A. A., Kwan, P. W., Coon, J. S. T., Miller, A. W. D. (1981). Primary adenocarcinomas of the human urinary bladder: histochemical, immunological and ultrastructural studies. *Virchows Arch [Pathol Anat]* 393: 165–81.

Amin, M. B., Gomez, J. A., Young, R. H. (1997). Urothelial transitional cell carcinoma with endophytic growth patterns: a discussion of patterns of invasion and problems associated with assessment of invasion in 18 cases. *Am J Surg Pathol* 21: 1057–68.

Amin, M. B., Ro, J. Y., el-Sharkawy, T., Lee, K. M., Troncoso, P., Silva, E. G., Ordonez, N. G., Ayala, A. G. (1994a). Micropapillary variant of transitional cell carcinoma of the urinary bladder. Histologic pattern resembling ovarian papillary serous carcinoma [see comments]. *Am J Surg Pathol* 18: 1224–32.

Amin, M. B., Ro, J. Y., Lee, K. M., Ordonez, N. G., Dinney, C. P., Gulley, M. L., Ayala, A. G. (1994b). Lymphoepithelioma-like carcinoma of the urinary bladder. *Am J Surg Pathol* 18: 466–73.

Amin, M. B., Young, R. H. (1997). Intraepithelial lesions of the urinary bladder with a discussion of the histogenesis of urothelial neoplasia. *Semin Diagn Pathol* 14: 84–97.

Andrion, A., Gaglio, A., Zai, G. (1993). Bladder involvement in disseminated malignant lymphoma diagnosed by voided urine cytology. *Cytopathology* 4: 115–17.

Badalament, R. A., Gay, H., Cibas, E. S., Herr, H. W., Whitmore, W. F., Jr., Fair, W. R., Melamed, M. R. (1987). Monitoring intravesical bacillus Calmette-Guerin treatment of superficial bladder carcinoma by postoperative urinary cytology. *J Urol* 138: 763–5.

Bardales, R. H., Pitman, M. B., Stanley, M. W., Korourian, S., Suhrland, M. J. (1998). Urine cytology of primary and secondary urinary bladder adenocarcinoma. *Cancer* 84: 335–43.

Bastacky, S., Ibrahim, S., Wilczynski, S. P., Murphy, W. M. (1999). The accuracy of urinary cytology in daily practice. *Cancer* 87: 118–28.

Bhan, R., Pisharodi, L. R., Gudlaugsson, E., Bedrossian, C. (1998). Cytological, histological, and clinical correlations in intravesical Bacillus Calmette-Guerin immunotherapy. *Ann Diagn Pathol* 2: 55–60.

Bian, Y., Ehya, H., Bagley, D. H. (1995). Cytologic diagnosis of upper urinary tract neoplasms by ureteroscopic sampling. *Acta Cytol* 39: 733–40.

Bibbo, M., Gill, W. B., Harris, M. J., Thomsen, S., Wied, G. L. (1974). Retrograde brushing as a diagnostic procedure of ureteral, renal pelvic and renal calyceal lesions. A preliminary report. *Acta Cytol* 18: 137–41.

Bocian, J. J., Flam, M. S., Mendoza, C. A. (1982). Hodgkin's disease involving the urinary bladder diagnosed by urinary cytology: a case report. *Cancer* 50: 2482–5.

Bonsib, S. M., Eble, J. N. (1997). Renal pelvis and ureter. In *Urologic Surgical Pathology* (D. G. Bostwick, J. N. Eble, eds.), pp. 150–65. Mosby, St. Louis.

Borghi, L., Bianchini, E., Altavilla, G. (1995). Undifferentiated small-cell carcinoma of the urinary bladder: report of two cases with a primary urinary cytodiagnosis. *Diagn Cytopathol* 13: 61–5.

Bostwick, D. G., Lo, R., Stamey, T. A. (1984). Papillary adenocarcinoma of the male urethra. Case report and review of the literature. *Cancer* 54: 2556–63.

Bostwick, D. G., Ramnani, D., Cheng, L. (1999). Diagnosis and grading of bladder cancer and associated lesions. *Urol Clin North Am* 26: 493–507.

Broghamer, W. L., Jr., Parker, J. E., Harty, J. I., Gilkey, C. M. (1989). Cytohistologic correlation of urothelial lesions secondary to photodynamic therapy. *Acta Cytol* 33: 881–6.

Carbin, B. E., Ekman, P., Gustafson, H., Christensen, N. J., Sandstedt, B., Silfversward, C. (1991). Grading of human urothelial carcinoma based on nuclear atypia and mitotic frequency. I. Histological description. *J Urol* 145: 968–71.

Chaitin, B. A., Manning, J. T., Ordonez, N. G. (1984). Hematologic neoplasms with initial manifestations in lower urinary tract. *Urology* 23: 35–42.

Chan, J. K., Suster, S., Wenig, B. M., Tsang, W. Y., Chan, J. B., Lau, A. L. (1997). Cytokeratin 20 immunoreactivity distinguishes Merkel cell (primary cutaneous neuroendocrine) carcinomas and salivary gland small cell carcinomas from small cell carcinomas of various sites. *Am J Surg Pathol* 21: 226–34.

Cheng, L., Bostwick, D. G. (2000). World Health Organization and International Society of Urological Pathology classification and two-number grading system of bladder tumors: reply [editorial; comment]. *Cancer* 88: 1513–16.

Cheng, L., Cheville, J. C., Neumann, R. M., Bostwick, D. G. (1999a). Natural history of urothelial dysplasia of the bladder. *Am J Surg Pathol* 23: 443–7.

Cheng, L., Cheville, J. C., Neumann, R. M., Bostwick, D. G. (2000a). Flat intraepithelial lesions of the urinary bladder. *Cancer* 88: 625–31.

Cheng, L., Cheville, J. C., Neumann, R. M., Leibovich, B. C., Egan, K. S., Spotts, B. E., Bostwick, D. G. (1999b). Survival of patients with carcinoma in situ of the urinary bladder. *Cancer* 85: 2469–74.

Cheng, L., Darson, M., Cheville, J. C., Neumann, R. M., Zincke, H., Nehra, A., Bostwick, D. G. (1999c). Urothelial papilloma of the bladder. Clinical and biologic implications [see comments]. *Cancer* 86: 2098–101.

Cheng, L., Neumann, R. M., Bostwick, D. G. (1999d). Papillary urothelial neoplasms of low malignant potential. Clinical and biologic implications [see comments]. *Cancer* 86: 2102–8.

Cheng, L., Neumann, R. M., Nehra, A., Spotts, B. E., Weaver, A. L., Bostwick, D. G. (2000b). Cancer heterogeneity and its biologic implications in the grading of urothelial carcinoma [see comments]. *Cancer* 88: 1663–70.

Cheson, B. D., Schumann, J. L., Schumann, G. B. (1984). Urinary cytodiagnostic abnormalities in 50 patients with non-Hodgkin's lymphomas. *Cancer* 54: 1914–19.

Chow, W. H., Lindblad, P., Gridley, G., Nyren, O., McLaughlin, J. K., Linet, M. S., Pennello, G. A., Adami, H. O., Fraumeni, J. F., Jr. (1997). Risk of urinary tract cancers following kidney or ureter stones [see comments]. *J Natl Cancer Inst* 89: 1453–7.

Cohen, R. J., Stanley, J. C., Dawkins, H. J. (1999). Lymphoepithelioma-like carcinoma of the renal pelvis. *Pathology* 31: 434–5.

Cohen, S. M., Johansson, S. L. (1992). Epidemiology and etiology of bladder cancer. *Urol Clin North Am* 19: 421–8.

Colonna, A., Guaitoli, P., Valantig, A., Colonna, M. (1989). [Carcinoma of the bladder with pseudosarcomatous stroma. Presentation of a case]. *Pathologica* 81: 551–8.

de Ruiter, F., Beyer-Boon, M. E., de Voogt, H. J. (1975). The effect of calculi on transitional epithelium. A clinical and cytological study. *Urol Res* 3: 67–72.

DiBonito, L., Musse, M. M., Dudine, S., Falconieri, G. (1992). Cytology of transitional-cell carcinoma of the urinary bladder: diagnostic yield and histologic basis. *Diagn Cytopathol* 8: 124–7.

Dodd, L. G., Johnston, W. W., Robertson, C. N., Layfield, L. J. (1997). Endoscopic brush cytology of the upper urinary tract. Evaluation of its efficacy and potential limitations in diagnosis. *Acta Cytol* 41: 377–84.

Donhuijsen, K., Schmidt, U., Richter, H. J., Leder, L. D. (1992). Mucoid cytoplasmic inclusions in urothelial carcinomas [see comments]. *Hum Pathol* 23: 860–4.

Drew, P. A., Furman, J., Civantos, F., Murphy, W. M. (1996). The nested variant of transitional cell carcinoma: an aggressive neoplasm with innocuous histology. *Mod Pathol* 9: 989–94.

Eble, J. N., Young, R. H. (1997). Carcinoma of the urinary bladder: a review of its diverse morphology. *Semin Diagn Pathol* 14: 98–108.

Epstein, J. I., Amin, M. B., Reuter, V. R., Mostofi, F. K. (1998). The World Health Organization/International Society of Urological Pathology consensus classification of urothelial (transitional cell) neoplasms of the urinary bladder. Bladder Consensus Conference Committee [see comments]. *Am J Surg Pathol* 22: 1435–48.

Eriksson, O., Johansson, S. (1976). Urothelial neoplasms of the upper urinary tract. A correlation between cytologic and histologic findings in 43 patients with urothelial neoplasms of the renal pelvis or ureter. *Acta Cytol* 20: 20–5.

Esposti, P. L., Zajicek, J. (1972). Grading of transitional cell neoplasms of the urinary bladder from smears of bladder washings. A critical review of 326 tumors. *Acta Cytol* 16: 529–37.

Essenfeld, H., Manivel, J. C., Benedetto, P., Albores-Saavedra, J. (1990). Small cell carcinoma of the renal pelvis: a clinicopathological, morphological and immunohistochemical study of 2 cases. *J Urol* 144: 344–7.

Fanning, C. V., Staerkel, G. A., Sneige, N., Thomsen, S., Myhre, M. J., Von Eschenbach, A. C. (1993). Spindling artifact of urothelial cells in post-laser treatment urinary cytology [see comments]. *Diagn Cytopathol* 9: 279–81.

Farkas, A., Moriel, E. Z., Prat, O. (1983). Rapid spread of transitional ureteral tumor: a serious unusual complication of ureteral brush biopsy. *J Urol* 129: 1229–30.

Fleshner, N., Garland, J., Moadel, A., Herr, H., Ostroff, J., Trambert, R., O'Sullivan, M., Russo, P. (1999). Influence of smoking status on the disease-related outcomes of patients with tobacco-associated superficial transitional cell carcinoma of the bladder [see comments]. *Cancer* 86: 2337–45.

Fortuny, J., Kogevinas, M., Chang-Claude, J., Gonzalez, C. A., Hours, M., Jockel, K. H., Bolm-Audorff, U., Lynge, E., Mannetje, A., Porru, S., Ranft, U., Serra, C., Tzonou, A., Wahrendorf, J., Boffetta, P. (1999). Tobacco, occupation and non-transitional-cell carcinoma of the bladder: an international case-control study. *Int J Cancer* 80: 44–6.

Frazier, H. A., Robertson, J. E., Dodge, R. K., Paulson, D. F. (1993). The value of pathologic factors in predicting cancer-specific survival among patients treated with radical cystectomy for transitional cell carcinoma of the bladder and prostate. *Cancer* 71: 3993–4001.

Friedell, G. H., Soloway, M. S., Hilgar, A. G., Farrow, G. M. (1986). Summary of workshop on carcinoma in situ of the bladder. *J Urol* 136: 1047–8.

Gallagher, L., Lind, R., Oyasu, R. (1984). Primary choriocarcinoma of the urinary bladder in association with undifferentiated carcinoma. *Hum Pathol* 15: 793–5.

Gill, H. S., Dhillon, H. K., Woodhouse, C. R. (1989). Adenocarcinoma of the urinary bladder. *Br J Urol* 64: 138–42.

Givler, R. L. (1971). Involvement of the bladder in leukemia and lymphoma. *J Urol* 105: 667–70.

Goldstein, M. L., Whitman, T., Renshaw, A. A. (1998). Significance of cell groups in voided urine. *Acta Cytol* 42: 290–4.

Greene, L. F., Hanash, K. A., Farrow, G. M. (1973). Benign papilloma or papillary carcinoma of the bladder? *J Urol* 110: 205–7.

Greenlee, R. T., Murray, T., Bolden, S., Wingo, P. A. (2000). Cancer statistics, 2000. *CA Cancer J Clin* 50: 7–33.

Grignon, D. J. (1997). Neoplasms of the urinary bladder. In *Urologic Surgical Pathology* (D. G. Bostwick, J. N. Eble, eds.), pp. 215–305. Mosby, St. Louis.

Grignon, D. J., Ro, J. Y., Ayala, A. G., Johnson, D. E., Ordonez, N. G. (1991). Primary adenocarcinoma of the urinary bladder. A clinicopathologic analysis of 72 cases. *Cancer* 67: 2165–72.

Grignon, D. J., Ro, J. Y., Ayala, A. G., Shum, D. T., Ordonez, N. G., Logothetis, C. J., Johnson, D. E., Mackay, B. (1992). Small cell carcinoma of the urinary bladder. A clinicopathologic analysis of 22 cases. *Cancer* 69: 527–36.

Guillou, L., Duvoisin, B., Chobaz, C., Chapuis, G., Costa, J. (1993). Combined small-cell and transitional cell carcinoma of the renal pelvis. A light microscopic, immunohistochemical, and ultrastructural study of a case with literature review [see comments]. *Arch Pathol Lab Med* 117: 239–43.

Gupta, R., Belitsky, P., Stewart, L., Cohen, A. D., Handa, S. P. (1989). Urine cytology in a case of B-cell lymphoma in a transplanted kidney. *Transplant Proc* 21: 3585–7.

Halachmi, S., DeMarzo, A. M., Chow, N. H., Halachmi, N., Smith, A. E., Linn, J. F., Nativ, O., Epstein, J. I., Schoenberg, M. P., Sidransky, D. (2000). Genetic alterations in urinary bladder carcinosarcoma: evidence of a common clonal origin. *Eur Urol* 37: 350–7.

Harris, M. J., Schwinn, C. P., Morrow, J. W., Gray, R. L., Browell, B. M. (1971). Exfoliative cytology of the urinary bladder irrigation specimen. *Acta Cytol* 15: 385–99.

Helson, L., Hajdu, S. I. (1972). The cytology of urine of pediatric cancer patients. *J Urol* 108: 660–2.

Highman, W., Wilson, E. (1982). Urine cytology in patients with calculi. *J Clin Pathol* 35: 350–6.

Highman, W. J. (1978). Urinary cytology in inverted papilloma of the lower urinary tract. *Acta Cytol* 22: 335–8.

Highman, W. J. (1986). Transitional carcinoma of the upper urinary tract: a histological and cytopathological study. *J Clin Pathol* 39: 297–305.

Hughes, J. H., Raab, S. S., Cohen, M. B. (2000). The cytologic diagnosis of low-grade transitional cell carcinoma. *Am J Clin Pathol* 114(Suppl 1): S59–S67.

Jacobsen, A. B., Nesland, J. M., Fossa, S. D., Pettersen, E. O. (1990). Human chorionic gonadotropin, neuron specific enolase and deoxyribonucleic acid flow cytometry in patients with high grade bladder carcinoma. *J Urol* 143: 706–9.

Johansson, S. L., Cohen, S. M. (1997). Epidemiology and etiology of bladder cancer. *Semin Surg Oncol* 13: 291–8.

Jones, M. W. (1989). Primary Hodgkin's disease of the urinary bladder. *Br J Urol* 63: 438.

Jordan, A. M., Weingarten, J., Murphy, W. M. (1987). Transitional cell neoplasms of the urinary bladder. Can biologic potential be predicted from histologic grading? [published erratum appears in Cancer 1988 Apr 1;61(7):1385]. *Cancer* 60: 2766–74.

Kalnins, Z. A., Rhyne, A. L., Morehead, R. P., Carter, B. J. (1970). Comparison of cytologic findings in patients with transitional cell carcinoma and benign urologic diseases. *Acta Cytol* 14: 243–8.

Kannan, V. (1990). Papillary transitional-cell carcinoma of the upper urinary tract: a cytological review. *Diagn Cytopathol* 6: 204–9.

Kannan, V., Bose, S. (1993). Low grade transitional cell carcinoma and instrument artifact. A challenge in urinary cytology. *Acta Cytol* 37: 899–902.

Kannan, V., Gupta, D. (1999). Calculus artifact. A challenge in urinary cytology. *Acta Cytol* 43: 794–800.

Katz, J. I., Grabstald, H. (1976). Primary malignant melanoma of the female urethra. *J Urol* 116: 454–7.

Kern, W. H. (1975). The cytology of transitional cell carcinoma of the urinary bladder. *Acta Cytol* 19: 420–8.

Khafagy, M. M., el-Bolkainy, M. N., Mansour, M. A. (1972). Carcinoma of the bilharzial urinary bladder. A study of the associated mucosal lesions in 86 cases. *Cancer* 30: 150–9.

Kobayashi, S., Ohmori, M., Miki, H., Hirata, K., Shimada, K. (1985). Exfoliative cytology of a primary adenocarcinoma of the renal pelvis. A case report. *Acta Cytol* 29: 1021–5.

Kojima, S., Sekine, H., Fukui, I., Ohshima, H. (1998). Clinical significance of "cannibalism" in urinary cytology of bladder cancer. *Acta Cytol* 42: 1365–9.

Koss, L. G. (1975). *Tumors of the Urinary Bladder. Fascicle 11.* Armed Forces Institute of Pathology, Washington, DC.

Koss, L. G. (1979). Mapping of the urinary bladder: its impact on the concepts of bladder cancer. *Hum Pathol* 10: 533–48.

Koss, L. G. (1996). Tumors of the bladder. In *Diagnostic Cytology of the Urinary Tract with Histopathologic and Clinical Correlations.* (L. G. Koss, ed.), pp. 71–139. J. B. Lippincott, Philadelphia.

Krumerman, M. S., Katatikarn, V. (1976). Rhabdomyosarcoma of the urinary bladder with intraepithelial spread in an adult. *Arch Pathol Lab Med* 100: 395–7.

Kusuda, L., Kimmel, M., Fair, W. R., Melamed, M. (1993). Hypocellular bladder wash specimens and their clinical significance. *J Urol* 150: 1123–5.

Kusuyama, Y., Yoshida, M., Uekado, Y., Ogawa, T., Fujinaga, T., Kuribayashi, K., Saito, K. (1988). Clear cell adenocarcinoma of the female urethra. A case report. *Acta Pathol Jpn* 38: 217–23.

Lamm, D. L., Torti, F. M. (1996). Bladder cancer, 1996. *CA Cancer J Clin* 46: 93–112.

Landis, S. H., Murray, T., Bolden, S., Wingo, P. A. (1999). Cancer statistics, 1999 [see comments]. *CA Cancer J Clin* 49: 8–31, 1.

Lang, E. K., Alexander, R., Barnett, T., Palomar, J., Hamway, S. (1978). Brush biopsy of pyelocalyceal lesions via a percutaneous translumbar approach. *Radiology* 129: 623–7.

Lee, R., Droller, M. J. (2000). The natural history of bladder cancer. Implications for therapy. *Urol Clin North Am* 27: 1–13, vii.

Leistenschneider, W., Nagel, R. (1980). Lavage of cytology of the renal pelvis and ureter with special reference to tumors. *J Urol* 124: 597–600.

Lemos, N., Melo, C. R., Soares, I. C., Lemos, R. R., Lemos, F. R. (2000). Plasmacytoma of the urethra treated by excisional biopsy. *Scand J Urol Nephrol* 34: 75–6.

Logothetis, C. J., Dexeus, F. H., Chong, C., Sella, A., Ayala, A. G., Ro, J. Y., Pilat, S. (1989). Cisplatin, cyclophosphamide and doxorubicin chemotherapy for unresectable urothelial tumors: the M.D. Anderson experience. *J Urol* 141: 33–7.

Lohrisch, C., Murray, N., Pickles, T., Sullivan, L. (1999). Small cell carcinoma of the bladder: long term outcome with integrated chemoradiation. *Cancer* 86: 2346–52.

Mahadevia, P. S., Koss, L. G., Tar, I. J. (1986). Prostatic involvement in bladder cancer. Prostate mapping in 20 cystoprostatectomy specimens. *Cancer* 58: 2096–2102.

Martin, J. E., Jenkins, B. J., Zuk, R. J., Blandy, J. P., Baithun, S. I. (1989). Clinical importance of squamous metaplasia in invasive transitional cell carcinoma of the bladder [see comments]. *J Clin Pathol* 42: 250–3.

Mazzucchelli, L., Kraft, R., Gerber, H., Egger, C., Studer, U. E., Zimmermann, A. (1992). Carcinosarcoma of the urinary bladder: a distinct variant characterized by small cell undifferentiated carcinoma with neuroendocrine features. *Virchows Arch A Pathol Anat Histopathol* 421: 477–83.

McRae, S., Garcia, B. M. (1997). Cytologic diagnosis of a primary pure oat cell carcinoma of the bladder in voided urine. A case report. *Acta Cytol* 41: 1279–83.

Meis, J. M., Ayala, A. G., Johnson, D. E. (1987). Adenocarcinoma of the urethra in women. A clinicopathologic study [published erratum appears in Cancer 1987 Dec 15;60(12):2900]. *Cancer* 60: 1038–52.

Mincione, G. P. (1982). Primary malignant lymphoma of the urinary bladder with a positive cytologic report. *Acta Cytol* 26: 69–72.

Mincione, G. P., Gasbarre, M. (1995). Primary vesical malignant lymphoma of histiocytes detected by urine cytology. *Pathologica* 87: 559–62.

Mincione, G. P., Grechi, G. (1983). Urinary cytology of rhabdomyosarcoma in children. Report of two cases located in the urinary bladder and in the prostate. *Pathologica* 75: 797–801.

Moriyama, M. (1989). [Cellular architecture of papillary and nonpapillary transitional cell carcinoma]. *Nippon Hinyokika Gakkai Zasshi* 80: 1017–24.

Morrison, D. A., Murphy, W. M., Ford, K. S., Soloway, M. S. (1984). Surveillance of stage O, grade I bladder cancer by cytology alone—is it acceptable? *J Urol* 132: 672–4.

Mostofi, F., Sobin, L., Torloni, H. (1973). *Histological Typing of Urinary Bladder Tumors* (Vol. 10). World Health Organization, Geneva.

Mostofi, F. K., Davies, C. J., Sesterhenn, I. A. (1999). *WHO Histologic Typing of Urinary Bladder Tumors.* Springer, Berlin.

Mostofi, F. K., Sesterhenn, I. A. (1984). Pathology of epithelial tumors and carcinoma in situ of bladder. *Prog Clin Biol Res* 162A: 55–74.

Murphy, W. M. (1999). Bladder cancer redefined [editorial; comment]. *Cancer* 86: 1890–2.

Murphy, W. M., Beckwith, J. B., Farrow, G. M. (1994). Tumors of the urinary bladder. In *Tumors of the Kidney, Bladder, and Related Urinary Structures* (J. Rosai, ed.), pp. 193–298. Armed Forces Institute of Pathology, Washington, DC.

Murphy, W. M., Soloway, M. S., Jukkola, A. F., Crabtree, W. N., Ford, K. S. (1984). Urinary cytology and bladder cancer. The cellular features of transitional cell neoplasms. *Cancer* 53: 1555–65.

Nagy, G. K., Frable, W. J., Murphy, W. M. (1982). Classification of premalignant urothelial abnormalities. A Delphi study of the National Bladder Cancer Collaborative Group A. *Pathol Annu* 17: 219–33.

Neumann, M. P., Limas, C. (1986). Transitional cell carcinomas of the urinary bladder. Effects of preoperative irradiation on morphology. *Cancer* 58: 2758–63.

Nguyen-Ho, P., Nguyen, G. K., Villanueva, R. R. (1994). Small cell anaplastic carcinoma of the prostate: report of a case with positive urine cytology. *Diagn Cytopathol* 10: 159–61.

Ohtani, M., Kakizoe, T., Nishio, Y., Sato, S., Sugimura, T., Fukushima, S., Niijima, T. (1986). Sequential changes of mouse bladder epithelium during induction of invasive carcinomas by *N*-butyl-*N*-(4-hydroxybutyl)nitrosamine. *Cancer Res* 46: 2001–4.

Oldbring, J., Mikulowski, P. (1987). Malignant melanoma of the penis and male urethra. Report of nine cases and review of the literature. *Cancer* 59: 581–7.

Oliva, E., Young, R. H. (1996). Clear cell adenocarcinoma of the urethra: a clinicopathologic analysis of 19 cases. *Mod Pathol* 9: 513–20.

Olsen, L. H., Overgaard, S., Frederiksen, P., Ladefoged, C., Ludwigsen, E., Petri, J., Poulsen, J. T. (1993). The reliability of staging and grading of bladder tumours. Impact of misinformation on the pathologist's diagnosis. *Scand J Urol Nephrol* 27: 349–53.

Ooms, E. C., Anderson, W. A., Alons, C. L., Boon, M. E., Veldhuizen, R. W. (1983). Analysis of the performance of pathologists in the grading of bladder tumors. *Hum Pathol* 14: 140–3.

Oyasu, R. (2000). World Health Organization and International Society of Urological Pathology classification and two-number grading system of bladder tumors [editorial; comment] [see comments]. *Cancer* 88: 1509–12.

Parkin, D. M., Pisani, P., Ferlay, J. (1999). Global cancer statistics. *CA Cancer J Clin* 49: 33–64, 1.

Pedersen-Bjergaard, J., Ersboll, J., Hansen, V. L., Sorensen, B. L., Christoffersen, K., Hou-Jensen, K., Nissen, N. I., Knudsen, J. B., Hansen, M. M. (1988). Carcinoma of the urinary bladder after treatment with cyclophosphamide for non-Hodgkin's lymphoma. *N Engl J Med* 318: 1028–32.

Perret, L., Chaubert, P., Hessler, D., Guillou, L. (1998). Primary heterologous carcinosarcoma (metaplastic carcinoma) of the urinary bladder: a clinicopathologic, immunohistochemical, and ultrastructural analysis of eight cases and a review of the literature. *Cancer* 82: 1535–49.

Peven, D. R., Hidvegi, D. F. (1985). Clear-cell adenocarcinoma of the female urethra. *Acta Cytol* 29: 142–6.

Potts, S. A., Thomas, P. A., Cohen, M. B., Raab, S. S. (1997). Diagnostic accuracy and key cytologic features of high-grade transitional cell carcinoma in the upper urinary tract. *Mod Pathol* 10: 657–62.

Powers, C. N., Elbadawi, A. (1995). "Cercariform" cells: a clue to the cytodiagnosis of transitional cell origin of metastatic neoplasms? *Diagn Cytopathol* 13: 15–21.

Raab, S. S., Lenel, J. C., Cohen, M. B. (1994). Low grade transitional cell carcinoma of the bladder. Cytologic diagnosis by key features as identified by logistic regression analysis. *Cancer* 74: 1621–6.

Raab, S. S., Slagel, D. D., Jensen, C. S., Teague, M. W., Savell, V. H., Ozkutlu, D., Lenel, J. C., Cohen, M. B. (1996). Low-grade transitional cell carcinoma of the urinary bladder: application of select cytologic criteria to improve diagnostic accuracy [corrected] [published erratum appears in Mod Pathol 1996 July;9(7):803]. *Mod Pathol* 9: 225–32.

Reich, O., Pickel, H., Purstner, P. (1999). Exfoliative cytology of a lymphoepithelioma-like carcinoma in a cervical smear. A case report. *Acta Cytol* 43: 285–8.

Renshaw, A. A. (2000a). Compassionate conservatism in urinary cytology [editorial]. *Diagn Cytopathol* 22: 137–8.

Renshaw, A. A. (2000b). Subclassifying atypical urinary cytology specimens. *Cancer* 90: 222–9.

Renshaw, A. A., Nappi, D., Weinberg, D. S. (1996). Cytology of grade 1 papillary transitional cell carcinoma. A comparison of cytologic, architectural and morphometric criteria in cystoscopically obtained urine. *Acta Cytol* 40: 676–82.

Reuter, V. E. (1993). Sarcomatoid lesions of the urogenital tract. *Semin Diagn Pathol* 10: 188–201.

Reuter, V. E. (1999). Bladder. Risk and prognostic factors—a pathologist's perspective. *Urol Clin North Am* 26: 481–92.

Reuter, V. E., Melamed, M. R. (1989). The lower urinary tract. In *Diagnostic Surgical Pathology* (S. S. Sternberg, ed.), pp. 1355–92. Raven Press, New York.

Rife, C. C., Farrow, G. M., Utz, D. C. (1979). Urine cytology of transitional cell neoplasms. *Urol Clin North Am* 6: 599–612.

Robertson, A. J., Beck, J. S., Burnett, R. A., Howatson, S. R., Lee, F. D., Lessells, A. M., McLaren, K. M., Moss, S. M., Simpson, J. G., Smith, G. D., et al. (1990). Observer variability in histopathological reporting of transitional cell carcinoma and epithelial dysplasia in bladders. *J Clin Pathol* 43: 17–21.

Rodenburg, C. J., Nieuwenhuyzen Kruseman, A. C., de Maaker, H. A., Fleuren, G. J., van Oosterom, A. T. (1985). Immunohistochemical localization and chromatographic characterization of human chorionic gonadotropin in a bladder carcinoma. *Arch Pathol Lab Med* 109: 1046–8.

Rollins, S., Schumann, G. B. (1991). Primary urinary cytodiagnosis of a bladder small-cell carcinoma. *Diagn Cytopathol* 7: 79–82.

Rudrick, B., Nguyen, G. K., Lakey, W. H. (1995). Carcinoid tumor of the renal pelvis: report of a case with positive urine cytology. *Diagn Cytopathol* 12: 360–3.

Sahin, A. A., Myhre, M., Ro, J. Y., Sneige, N., Dekmezian, R. H., Ayala, A. G. (1991). Plasmacytoid transitional cell carcinoma. Report of a case with initial presentation mimicking multiple myeloma. *Acta Cytol* 35: 277–80.

Sakamoto, N., Tsuneyoshi, M., Enjoji, M. (1992). Urinary bladder carcinoma with a neoplastic squamous component: a mapping study of 31 cases. *Histopathology* 21: 135–41.

Sandberg, A. A., Berger, C. S. (1994). Review of chromosome studies in urological tumors. II. Cytogenetics and molecular genetics of bladder cancer. *J Urol* 151: 545–60.

Sarma, K. P. (1981). Genesis of papillary tumours: histological and microangiographic study. *Br J Urol* 53: 228–36.

Schindler, S., De Frias, D. V., Yu, G. H. (1999). Primary angiosarcoma of the bladder: cytomorphology and differential diagnosis. *Cytopathology* 10: 137–43.

Schnadig, V. J., Adesokan, A., Neal, D., Jr., Gatalica, Z. (2000). Urinary cytologic findings in patients with benign and malignant adenomatous polyps of the prostatic urethra. *Arch Pathol Lab Med* 124: 1047–52.

Schwalb, M. D., Herr, H. W., Sogani, P. C., Russo, P., Sheinfeld, J., Fair, W. R. (1994). Positive urinary cytology following a complete response to intravesical bacillus Calmette-Guerin therapy: pattern of recurrence. *J Urology* 152: 382–7.

Seldenrijk, C. A., Verheggen, W. J., Veldhuizen, R. W., Blok, A. P., Ooms, E. C. (1987). Use of cytomorphometry and cytology in the diagnosis of transitional-cell carcinoma of the upper urinary tract. *Acta Cytol* 31: 137–42.

Serretta, V., Pomara, G., Piazza, F., Gange, E. (2000). Pure squamous cell carcinoma of the bladder in western countries. Report on 19 consecutive cases. *Eur Urol* 37: 85–9.

Smith, J. H., Christie, J. D. (1986). The pathobiology of *Schistosoma haematobium* infection in humans. *Hum Pathol* 17: 333–45.

Spires, S. E., Banks, E. R., Cibull, M. L., Munch, L., Delworth, M., Alexander, N. J. (1993). Adenocarcinoma of renal pelvis [published erratum appears in Arch Pathol Lab Med 1994 Sep;118(9):872]. *Arch Pathol Lab Med* 117: 1156–60.

Stein, B. S., Kendall, A. R. (1984). Malignant melanoma of the genitourinary tract. *J Urol* 132: 859–68.

Stewart, J. H., Hobbs, J. B., McCredie, M. R. (1999). Morphologic evidence that analgesic-induced kidney pathology contributes to the progression of tumors of the renal pelvis. *Cancer* 86: 1576–82.

Suprun, H., Bitterman, W. (1975). A correlative cytohistologic study on the interrelationship between exfoliated urinary bladder carcinoma cell types and the staging and grading of these tumors. *Acta Cytol* 19: 265–73.

Tanaka, T., Yoshimi, N., Sawada, K., Takami, T., Sugie, S., Etori, F., Kachi, H., Mori, H. (1993). Ki-1–positive large cell anaplastic lymphoma diagnosed by urinary cytology. A case report. *Acta Cytol* 37: 520–4.

Taylor, D. C., Bhagavan, B. S., Larsen, M. P., Cox, J. A., Epstein, J. I. (1996). Papillary urothelial hyperplasia. A precursor to papillary neoplasms. *Am J Surg Pathol* 20: 1481–8.

Thrasher, J. B., Miller, G. J., Wettlaufer, J. N. (1990). Bladder leiomyosarcoma following cyclophosphamide therapy for lupus nephritis. *J Urol* 143: 119–21.

van Hoeven, K. H., Artymyshyn, R. L. (1996). Cytology of small cell carcinoma of the urinary bladder. *Diagn Cytopathol* 14: 292–7.

Weiss, M. A. (1993). Small-cell carcinoma of the urinary tract. Important prognostic and therapeutic implications [editorial; comment]. *Arch Pathol Lab Med* 117: 237–8.

Wick, M. R., Brown, B. A., Young, R. H., Mills, S. E. (1988). Spindle-cell proliferations of the urinary tract. An immunohistochemical study. *Am J Surg Pathol* 12: 379–89.

Wiener, H. G., Vooijs, G. P., van't Hof-Grootenboer, B. (1993). Accuracy of urinary cytology in the diagnosis of primary and recurrent bladder cancer. *Acta Cytol* 37: 163–9.

Wolinska, W. H., Melamed, M. R., Klein, F. A. (1985). Cytology of bladder papilloma. *Acta Cytol* 29: 817–22.

Woo, H. H., Kencian, J. D. (1995). Metastatic merkel cell tumour to the bladder. *Int Urol Nephrol* 27: 301–5.

Woodard, B. H., Ideker, R. E., Johnston, W. W. (1978). Cytologic detection of malignant melanoma in urine. A case report. *Acta Cytol* 22: 350–2.

Yam, L. T., Janckila, A. J. (1985). Immunocytochemical diagnosis of lymphoma from urine sediment. *Acta Cytol* 29: 827–32.

Yang, C., Motteram, R., Sandeman, T. F. (1982). Extramedullary plasmacytoma of the bladder: a case report and review of literature. *Cancer* 50: 146–9.

Young, R. H., Eble, J. N. (1997). Non-neoplastic disorders of the urinary bladder. In *Urologic Surgical Pathology* (D. G. Bostwick, J. N. Able, eds.), pp. 167–212. Mosby, St. Louis.

Young, R. H., Eble, J. N. (1991). Unusual forms of carcinoma of the urinary bladder. *Hum Pathol* 22: 948–65.

Zaman, S. S., Sack, M. J., Ramchandani, P., Tomaszewski, J. E., Gupta, P. K. (1996). Cytopathology of retrograde renal pelvis brush specimens: an analysis of 40 cases with emphasis on collecting duct carcinoma and low–intermediate grade transitional cell carcinoma. *Diagn Cytopathol* 15: 312–21.

Zukerberg, L. R., Harris, N. L., Young, R. H. (1991). Carcinomas of the urinary bladder simulating malignant lymphoma. A report of five cases. *Am J Surg Pathol* 15: 569–76.

7

SECONDARY NEOPLASMS OF THE URINARY TRACT

SECONDARY neoplasms involving the urinary tract can be subdivided into two groups: (*1*) direct extension from a primary malignancy located in an adjacent organ and (*2*) lymphatic or hematogenous spread from a primary tumor in a distant organ. The organs and types of malignancies are summarized in Table 7–1.

In adults, contiguous spread of a malignancy in the prostate, colon, and female genital tract is more common than metastasis from a distant primary. The most common distant malignancies that disseminate to the urinary tract include those of the stomach, breast, and lung in order of decreasing frequency (Azimuddin et al., 1999; Malone et al., 1997; Treiger and Marshall, 1991). In children, small round blue-cell neoplasms, in particular neuroblastoma and Wilms' tumor, may be detected in urine specimens (Farr and Hajdu, 1972; Helson and Hajdu, 1972).

The cytomorphology of secondary neoplasms affecting the urinary tract has been covered in case reports, small series, and personal observations, all condensed in general and specialized cytology treatises, but large series are few and are mostly focused on secondary malignancies of the urinary bladder (Bardales et al., 1998; DeMay, 1996; Koss, 1992, 1996; Krishnan and Truong, 2000).

DIRECT EXTENSION FROM A MALIGNANCY IN AN ADJACENT ORGAN

Renal Tumors

Renal cell carcinoma is the predominant histologic type of kidney cancer. In 1993, 86% of reported cases were diagnosed as renal cell carcinomas, urothelial carcinomas comprised 4%, Wilms' tumor 4%, and other histologic types the remainder (Marshall et al., 1997).

Early involvement of the adipose tissue of the renal sinuses and renal sinus veins without involvement of the renal pelvis is present in a significant number of renal cell carcinoma cases, particularly renal cell clear cell carcinoma, and may account for the metastatic disease seen in tumors limited to the kidney (Bonsib et al., 2000). These initial events are followed by involvement of the renal tubules and invasion of the renal pelvis, with violation of the urinary space accounting for the presence of malignant cells in urine cytology specimens (Bunge and Kraushaar, 1950).

However, exfoliation of carcinoma cells into the urinary space may not be constant, explaining why the presence of cancer cells in the urine is intermittent, with negative results occurring among positive ones. In fact, it has been seen that exfoliation of carcinoma cells in the urinary stream does not depend on tumor invasion of the renal pelvis (Piscioli et al., 1985). In contrast, and perhaps because of its lack of cell cohesiveness, more continuous cell exfoliation in the urine is seen in diffuse lymphomatous or myelomatous renal involvement that can be diagnosed and followed more reliably by means of urine cytology (Schumann and Colon, 1980).

Although urine cytology has proved to be of considerable diagnostic value for high-grade urothelial malignancies, its role is considered unsatisfactory in the detection of renal cell carcinoma, with its sensitivity

Table 7–1. Secondary Neoplasms Involving the Urinary Tract

Direct extension from a primary tumor in adjacent organs
Renal tumors
Prostatic carcinomas
Colonic carcinomas
Female genital tract tumors
Testicular tumors
Other retroperitoneal tumors
Metastasis from a distant primary site
Carcinomas
Hematolymphoid malignancies
Melanoma
Germ cell tumors
Neoplasms arising in intestinal loops

believed to be about 50% or less (Meisels, 1963; Piscioli et al., 1983, 1985). Insensitivity of the procedure in the early diagnosis of carcinoma, and the high rate of false-negative diagnoses as result of intermittent tumor cell exfoliation, diminish the practical value of cytologic examination of urinary sediment (Piscioli et al., 1985).

When high-grade tumor cells are present in urine specimens obtained from a patient who has a mass involving both the kidney and the renal pelvis, the diagnostic considerations include high-grade primary renal cell carcinoma, renal pelvis adenocarcinoma, and high-grade papillary urothelial carcinoma. Because the precise diagnosis of the tumor type and the origin of the tumor influence the mode of surgery, the cytologic findings must be correlated with the clinical and radiologic information (Box 7–1).

Box 7–1.

Renal cell clear cell carcinoma exfoliates large cells with clear cytoplasm, a round nucleus, and large nucleoli. Multinucleated cells and necrosis may be present.

Papillary carcinoma shows columnar or cuboidal cells with dense granular cytoplasm, a round/oval nucleus, nuclear grooves, and nucleoli.

Chromophobe carcinoma shows large cells with granular eosinophilic cytoplasm, a clear perinuclear zone, and prominent nucleoli resembling those of oncocytoma.

Renal collecting duct carcinoma exfoliates large cells with hyperchromatic, pleomorphic nuclei and prominent nucleoli.

Cytology. Renal cell clear cell carcinoma exfoliates large cells that have an eccentric nucleus and a large, round, red nucleolus. The nucleus is usually round, with irregularly distributed vesicular chromatin similar to that seen in tissue sections. The cytoplasm is thin, clear, or vacuolated. However, occasional cells with granular and dense cytoplasm may be present, accounting for the cellular heterogeneity seen in these tumors. Pyknotic nuclei and poorly defined cytoplasmic borders are signs of cellular degeneration that may preclude assessment of the nuclear and nucleolar characteristics. Multinucleated cells with anaplastic nuclei may be seen, although occasionally benign-appearing multinucleated cells can be identified (Bunge and Kraushaar, 1950; Meisels, 1963; Piscioli et al., 1983). A necrotic and bloody background is usually present (Fig. 7–1A–C). The value of lipid stain in renal neoplasms as an adjunct to routine exfoliative cytology is controversial (Hajdu et al., 1971; Milsten et al., 1973).

Papillary carcinomas, as defined cytogenetically, are limited to neoplasms with eosinophilic or basophilic cells (chromophil tumors), suggested to originate from the proximal and distal convoluted tubules, respectively. These tumors show columnar to cuboidal cells with dense granular cytoplasm, round/oval nuclei with smooth/irregular contours, nuclear grooves, finely granular chromatin, and variable numbers of nucleoli. Scattered single foamy macrophages may be present (Bardales, 1998).

Chromophobe renal cell carcinoma is a tumor that arises from the intercalated cells of the collecting ducts. The tumor cells bear a marked resemblance to oncocytes, showing a well-defined, abundant, granular, eosinophilic cytoplasm with a clear perinuclear zone, vesicular nuclei, and prominent nucleoli. Positivity with Hale's colloidal iron stain indicates the presence of acidic mucin and distinguishes chromophobe renal cell

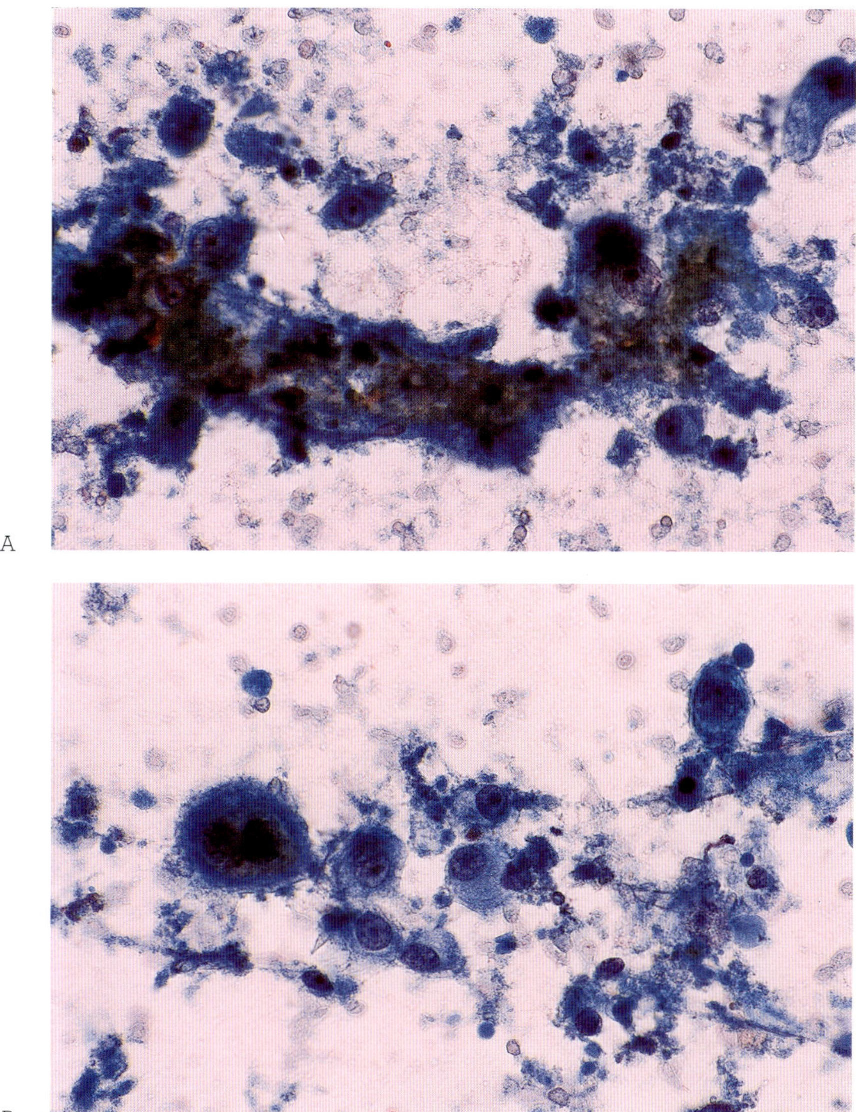

Figure 7–1. *Renal cell carcinoma.* Clear cell carcinoma, the most common type of primary renal tumor, occasionally and intermittently exfoliates cells into the urine. (*A,B*) Urine samples show marked cellular degeneration, blood, and necrosis. Cells are large, with round nuclei and prominent round nucleoli, and some show vacuolated cytoplasm. (*C*) Prominent fine vascularization and clear cells are seen in the corresponding tissue section. (A,B, Papanicolaou stain, ×600; C, hematoxylin and eosin stain, ×400)

carcinoma from oncocytoma (Nguyen and Akin, 1998).

A rare variant of renal cell carcinoma is renal collecting duct carcinoma, a tumor that originates from the collecting ducts of Bellini (Fallick et al., 1997; Mauri et al., 1994; Nguyen and Schumann, 1997). Because of its location in the renal medulla and renal pelvis, this highly aggressive neoplasm can easily exfoliate tumor cells into the urine. Urine cytology examination lacks specific features and includes sheets and clusters of large cells with occasional acinar formation, nonvacuolated or granular cytoplasm, hyperchromatic oval or pleomorphic nuclei, and prominent nucleoli (Fallick et al., 1997; Mauri et al., 1994; Nguyen and Schumann, 1997). Thus, the radiologic evidence of re-

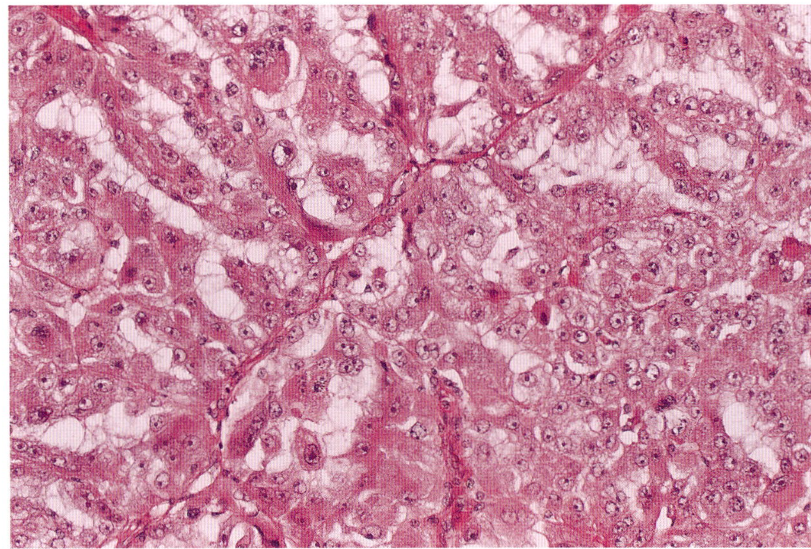

C

Figure 7–1. (*Continued*)

nal pelvis involvement and the lack of specific cellular features of the tumor make the differentiation from high-grade papillary urothelial carcinoma or adenocarcinoma of the renal pelvis or even renal cell carcinoma difficult. In fact, urinary cytology is useful for diagnosing malignancy, but the specific diagnosis usually requires tissue confirmation.

Primary sarcoma of the kidney in adults is rare, and the exfoliation of spindle malignant cells into the urine stream is even rarer. However, an occasional case report has addressed the presentation of a primary renal myxoid leiomyosarcoma and ipsilateral urothelial carcinoma of the ureter, both detected by urine cytology examination (Yokose et al., 1991). More frequently, one finds clusters of malignant small-blue-cell tumor cells with elongated nuclei arranged in tubular formations in cases of Wilms' tumor in children (Helson and Hajdu, 1972).

Differential diagnosis. Cells of renal cell clear cell carcinoma must be differenti-

Table 7–2. Differential Diagnoses of Clear Cells in Urine Cytology

	Diagnostic Clues
XGP	Histiocytes with bland nucleus and finely vacuolated cytoplasm. Multinucleated histiocytes. Cytoplasmic inclusions (PAS −). CD68+, keratin−. Urine culture: *Escherichia coli* and *Proteus* spp. Unilateral renal pelvic mass. Staghorn calculus may be present.
Clear cell RCC	Finely vacuolated, clear, or granular cytoplasm. Round anaplastic nucleus with large nucleoli. Mitoses. Atypical multinucleated giant cells. CD68−, keratin+. Subtle abnormalities in low-grade carcinoma. Clinical and radiologic findings are useful.
Tuberculosis (renal, bladder)	Foamy and epithelioid histiocytes with delicate, thin nuclear membrane. Round or elongated nucleus. Granulomas. Langhans-type multinucleated giant cells with bland nuclear features. Urine culture: *Mycobacterium* spp.
Malacoplakia (renal, bladder)	Eosinophilic histiocytes of von Hansemann. Michaelis-Gutmann cytoplasmic inclusions (PAS+, calcium+, iron+). Urine culture: *E. coli.*
Renal tubular injury (osmotic nephrosis)	Macrophage-like epithelial cells with vacuolated cytoplasm, round nucleus, thin uniform nuclear membrane, and inconspicuous nucleolus. Clean background. Clinical history of intravenous administration of hypertonic solution, i.e., sucrose.

Abbreviations: XGP = xanthogranulomatous pyelonephritis; RCC = renal cell carcinoma; PAS = periodic acid–Schiff.

ated from neoplastic and nonneoplastic lesions showing clear cells in the urine, as described in Table 7–2. Multinucleated cells must be differentiated from the benign umbrella cells and giant cells present in infectious and inflammatory processes. The distinction of carcinoma cells from multinucleated cells, macrophages, or injured renal tubular cells is based on the presence of nuclear anaplasia and nucleolar characteristics (Ho et al., 1996; Khalil et al., 2000). The size of the nucleolus in renal cell carcinoma easily exceeds 4 μm, which is more than half the diameter of a red blood cell. However, clinical and radiologic correlation should always support the presence of a renal neoplasm.

Spindle cells of a urinary tract sarcoma, most commonly of the urinary bladder, should be differentiated from nonneoplastic spindle cell proliferations and from the effects of radiotherapy or chemotherapy. In all of these processes, with the notable exception of changes associated with chemotherapy, cellular anaplasia is important in distinguishing benign from malignant cells. Low-grade urothelial carcinoma commonly shows "cercariform" cells with bland nuclear features that are easily distinguishable from those of high-grade sarcoma. The finding of keratinized and elongated cells with nuclear anaplasia favors high-grade urothelial carcinoma. In addition, cellular smears are common in urothelial malignancies, in contrast to spindle cell sarcomas or reactive spindle cell processes, which characteristically exfoliate few cells. These lesions are covered in Chapters 5 and 6. Further categorization is based on the clinical history, cystoscopic evaluation, radiologic examination, and ultimately tissue samples.

Prostatic Carcinomas

Prostatic adenocarcinoma may be of acinar or ductal origin, the former being the more frequent. The tumor can involve the bladder neck or prostatic urethra or metastasize to the urinary bladder wall. In these instances, spontaneously exfoliated malignant cells can be detected in urine specimens; however, they are more evident in urine samples obtained after prostate palpation at the time of digital rectal examination or prostatic massage (Garret and Jassie, 1976; Varma et al., 1988). Characteristically, prostate carcinomas with extraprostatic involvement are of a high stage, have a Gleason score higher than 6, and show marked nuclear anaplasia (Bardales et al., 1998; Varma et al., 1988). However, a substantial number of patients with acinar prostatic adenocarcinoma may exfoliate malignant cells without having involvement of the bladder wall or evidence by cystoscopy (Foot et al., 1958). Therefore, when malignant cells are identified in the urine specimen, the cytopathologist must diagnose the primary site correctly and exclude urinary tract adenocarcinoma and high-grade urothelial carcinoma because of the therapeutic implications.

Retrospective review has shown that the specific cytologic diagnosis of acinar prostate adenocarcinoma is made in approximately 40% of histologically proven cases (Bardales et al., 1998). However, identification of the characteristic, although not pathognomonic, cytologic features will increase the diagnostic accuracy (Bardales et al., 1998; Krishnan and Truong, 2000) (Box 7–2).

Cytology. Depending on the collection method, urine specimens are of variable cellularity, the lowest being found in voided urine. In the presence of bladder wall involvement with mucosal ulceration, the smear background shows necrosis and acute inflammation, in contrast to the clean back-

Box 7–2.

Characteristically, prostate carcinomas involving the bladder wall are of a high stage and high Gleason score.

Smears show cells in small clusters forming occasional acini.

Cells have scant vacuolated cytoplasm, a round nucleus, and prominent nucleoli.

Occasionally, the cellular features are similar to those of high-grade urothelial carcinoma.

Pleomorphic nuclei and large cytoplasmic vacuoles are seen in prostatic duct carcinoma.

ground seen in cases lacking mucosal ulceration.

Malignant cells of acinar-type adenocarcinoma are commonly seen in aggregates, with only a few cells found singly or in pairs. The cells exhibit ill-defined cytoplasmic borders, a high nuclear-cytoplasmic ratio, very scant vacuolated or thin cytoplasm, round or oval nuclear borders, and large, often multiple nucleoli (Fig. 7–2A–D). Occasionally, the cellular characteristics are indistinguishable from those present in a poorly differentiated urothelial primary. However, higher cell cohesiveness, vesicular chromatin, and nucleoli are more evident in prostatic carcinoma (Fig. 7–3A–D) (Bardales et al., 1998; Krishnan and Truong, 2000; Varma et al., 1988). Occasionally, neoplastic cells in urine speci-

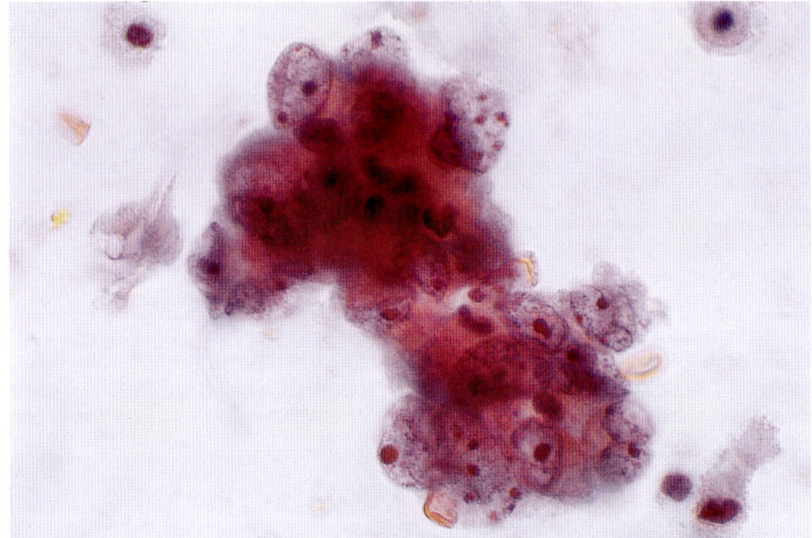

A

B

Figure 7–2. *Well-differentiated prostatic adenocarcinoma.* This type of malignancy usually exfoliates clusters of cells and a few single cells. (*A*) Malignant cells are small, with scant, finely vacuolated cytoplasm. Nuclei are commonly round and uniform (*B*), although they may have irregular nuclear contours (*C,D*). The chromatin is vesicular, and the nucleoli are prominent. All specimens are voided urine. (Papanicolaou stain, ×1000)

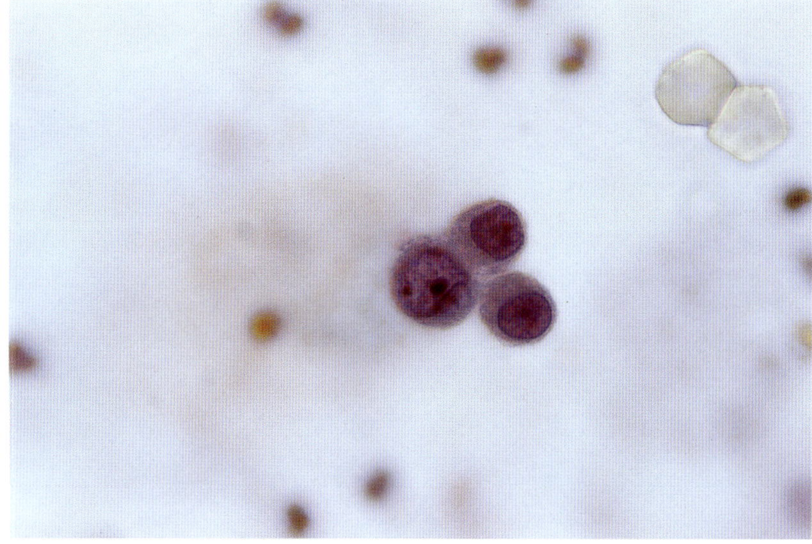
C

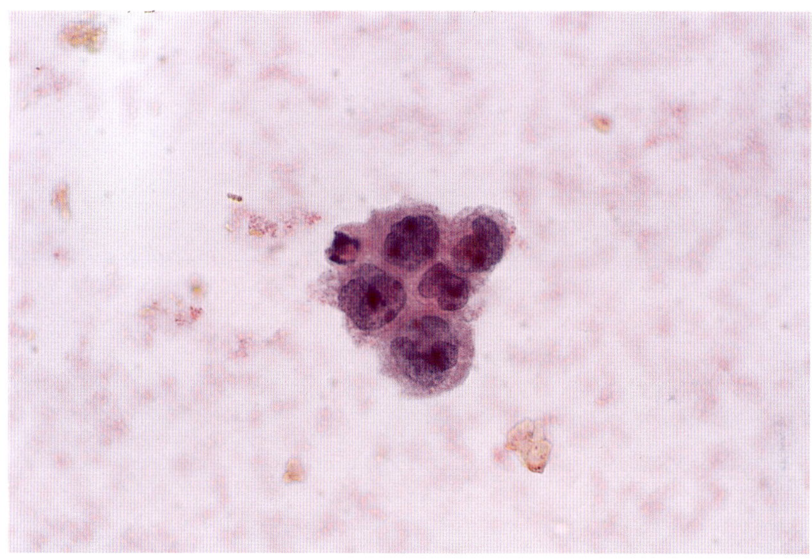
D

Figure 7–2. (*Continued*)

mens are small and well differentiated, and are easily confused with other elements of the nonneoplastic type (Rupp et al., 1994). However, in most cases, malignant features are evident.

In contrast to the cytologic findings in adenocarcinoma of the acinar type, those of the uncommon prostatic duct carcinoma include eccentric, irregular, pleomorphic nuclei and cytoplasm occupied almost entirely by large, hyperdistended secretory vacuoles (Ramzy and Larson, 1977).

The cytologic features of small-cell carcinoma of the prostate are identical to those of small-cell anaplastic carcinoma of the lung. Briefly, the cytologic features include high cellularity, loose cell aggregates with molding, and cells with scant cytoplasm and slightly pleomorphic, hyperchromatic nuclei with inconspicuous nucleoli (Nguyen-Ho et al., 1994).

Differential diagnosis. Acinar adenocarcinoma of the prostate must be differen-

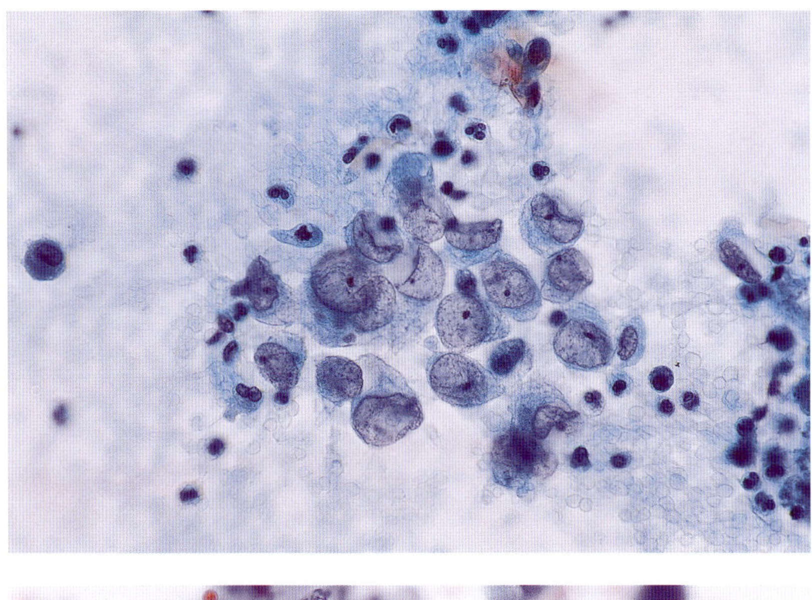

A

B

Figure 7–3. *Poorly differentiated prostatic adenocarcinoma.* This malignancy shows cells that are larger than those of well-differentiated prostatic adenocarcinoma. (*A*) Cells are less cohesive and show irregular nuclear contours, a high nuclear-cytoplasmic ratio, thin cytoplasm, vesicular chromatin, and nucleoli of variable prominence. (*B*) Cell clusters may have smaller admixed hyperchromatic "stromal" cells. (*C,D*) Tumor biopsy and immunoperoxidase stain for prostatic specific antigen may be necessary for final diagnosis. Cytology specimens are bladder washings. (A,B, Papanicolaou stain, ×600; C, hematoxylin stain, ×400; D, immunoperoxidase stain, ×400)

tiated from malignancies of urothelial origin, which are equally common in the same age group. Table 7–3 illustrates the salient features that are helpful in distinguishing among these neoplasms. Immunocytochemistry for prostatic acid phosphatase and prostate specific antigen is helpful in selected cases (Varma et al., 1988).

The characteristic large cytoplasmic vacuoles and displaced nuclei seen in prostatic duct adenocarcinoma permit its distinction from acinar prostate adenocarcinoma, high-grade urothelial carcinoma, and renal cell carcinoma. However, distinction from endometrioid carcinoma of the prostate, metastatic ovarian carcinoma, and other high-

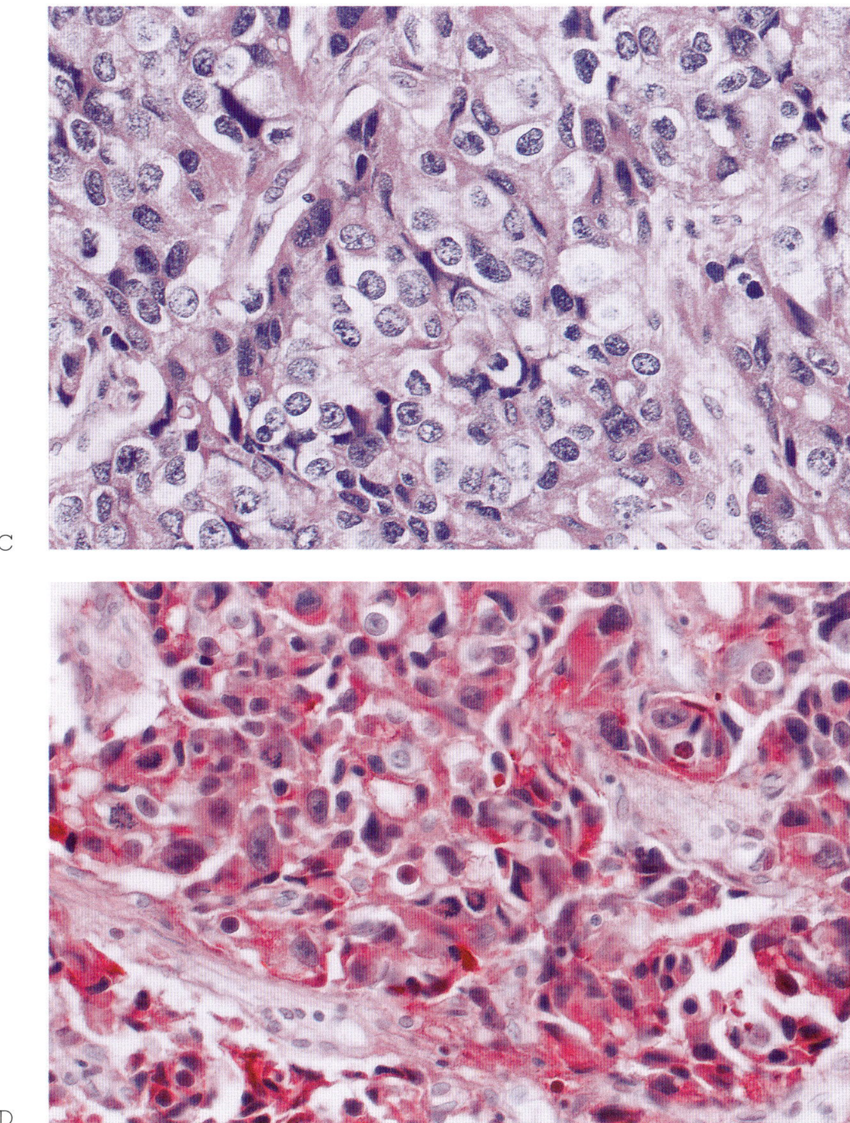

Figure 7–3. (*Continued*)

grade adenocarcinomas may be extremely difficult (Ramzy and Larson, 1977).

The differential diagnosis of small-cell prostate carcinoma includes primary and metastatic small-cell malignancies of the urinary tract. Strict clinical correlation and judicious use of special stains may be necessary.

Colonic Carcinomas

Direct involvement of the urinary bladder wall is the most common mechanism for the presence of colon cancer cells in urine cytology. Most tumors are moderately to poorly differentiated, and most patients have bladder involvement either simultaneously or up to five years after the primary diagnosis (Bardales et al., 1998).

All colonic adenocarcinomas can be diagnosed as malignant, and a high percentage have sufficient cytologic features to suggest the correct diagnosis in urine samples. Besides the cytomorphology, an important clue for the diagnosis includes an accurate clini-

Table 7–3. Comparative Cytomorphology of Acinar Adenocarcinoma of the Prostate in Urine Samples

Feature	*Prostate ACA*	*LGPUC*	*HGPUC*
Smear pattern			
Cellularity*	Low	Low	High
Single cells	Few	Few	Numerous
Acini	Present	Absent	May be present
Papillary clusters	Very rare	Present	May be present
Sheets	Present	Present	May be present
Cellular characteristics			
Shape	Round, cuboidal	Variable. "Cercariform"	Variable. "Squamoid"
NCR	Very high	Slightly increased	Very high
Cytoplasm	Clear, thin	Thin. "Ectoplasm"	Dense, vacuolated
Nucleus	Round/oval	Round/oval	Pleomorphic
Nuclear contour	Thin, irregular	Thin, regular/irregular	Thick, irregular
Chromatin	Vesicular	Slight hyperchromasia	Coarse
Nucleolus	Large	Absent	Present
Nuclear anaplasia	Present but variable	May not be identified	Marked
Background	Usually clean	Clean	Inflammation, necrosis

*Cell numbers depend on the urine collection method.

Abbreviations: ACA = adenocarcinoma; LGPUC = low-grade papillary urothelial carcinoma; HGPUC = high-grade papillary urothelial carcinoma; NCR = nuclear-cytoplasmic ratio.

cal history. In addition, the diagnosis requires the exclusion of a primary intestinal-type urinary bladder adenocarcinoma.

Cytology. Adenocarcinoma of the colon is characterized by the presence of elongated and pleomorphic cells with irregular nuclear contours, coarse chromatin, and inconspicuous nucleoli. Pleomorphic tumor cells with vacuolated cytoplasm are also found, but the presence of columnar cells and gland formation is helpful in making the diagnosis. The smears are highly cellular and show a bloody, necrotic background (Fig. 7–4A–D) (Bardales et al., 1998) (Box 7–3).

Box 7–3.

Colonic carcinoma shows variable degrees of cellular degeneration and necrosis.

The neoplastic cells are columar and pleomorphic and have coarse chromatin.

Columnar cells and partial gland formation are helpful in making the diagnosis.

Cytologic characteristics are identical to those of primary bladder adenocarcinoma of the colonic type.

Differential Diagnosis

Primary high-grade urothelial carcinoma may show pleomorphic cells with vacuolated cytoplasm and nuclear degeneration. In such cases, a diligent search for individual columnar cells or gland formation helps to identify the intestinal derivation. However, in the absence of a clinical history, it may be difficult to differentiate an invasive colonic from a primary urinary bladder intestinal-type adenocarcinoma, since the two tumor types have common immunohistochemical and ultrastructural features (Alroy et al., 1981).

Female Genital Tract Tumors

Invasive squamous cell carcinoma of the uterine cervix is the tumor that most commonly exfoliates cells into the urinary space. Other tumors include vaginal squamous carcinoma and endometrial adenocarcinoma (Creasman and Lukeman, 1972; Gupta and Taft, 1969).

In contrast to cervicovaginal cytology, urine cytology, once suggested to be a reliable screening test for gynecologic malignancies, has been shown to be of no value (Creasman and Lukeman, 1972; Graham et al., 1967; Hoskin et al., 1992). The success of routine cancer screening, particularly of cervical carcinoma, has resulted in early detection of the precursor lesions and prompt therapy. Transmural invasion of the bladder wall, the most common mechanism for the presence of cancer cells in urine, is exceptional, and metastases to the urinary tract are even more rare. Therefore, the finding of malignant cells of genital cancer in urine cytology constitutes a rarity.

Cytology. Patients with positive urine cytology have a clinically evident current or past high-stage genital tract malignancy

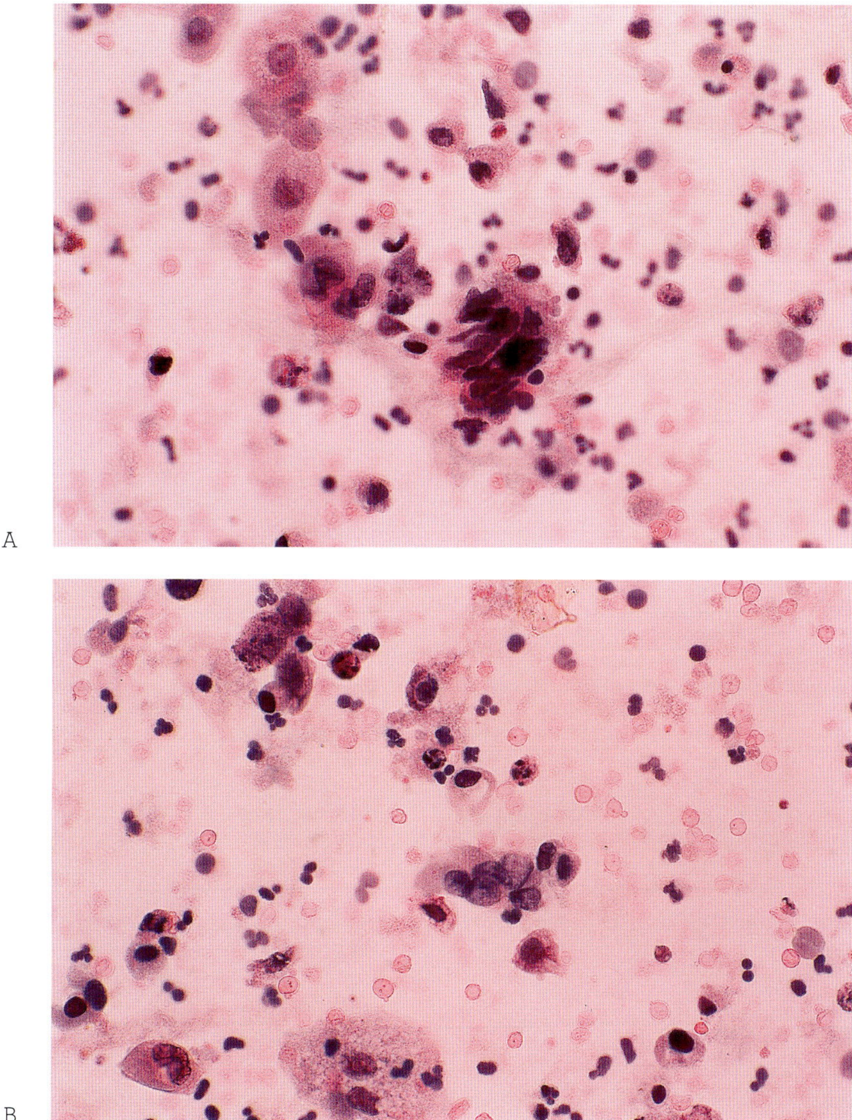

Figure 7–4. *Colonic adenocarcinoma.* (*A,B*) Marked cellular degeneration and necrosis are characteristic of these tumors. (*C*) Some single cells may be identical to those of high-grade invasive urothelial carcinoma. (*D*) Pleomorphic cells with hyperchromatic nuclei are commonly present; however, the presence of columnar cells in a gland-like arrangement is helpful for making the diagnosis. Cytologic features are identical to those of colonic-type primary adenocarcinoma of the urinary bladder. Specimens A and B are voided urine; C and D are bladder washings. (Papanicolaou stain, ×600)

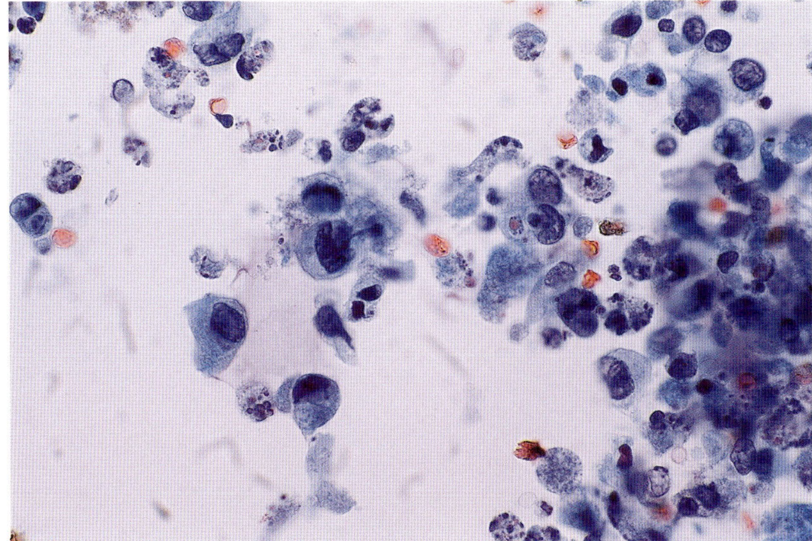

C

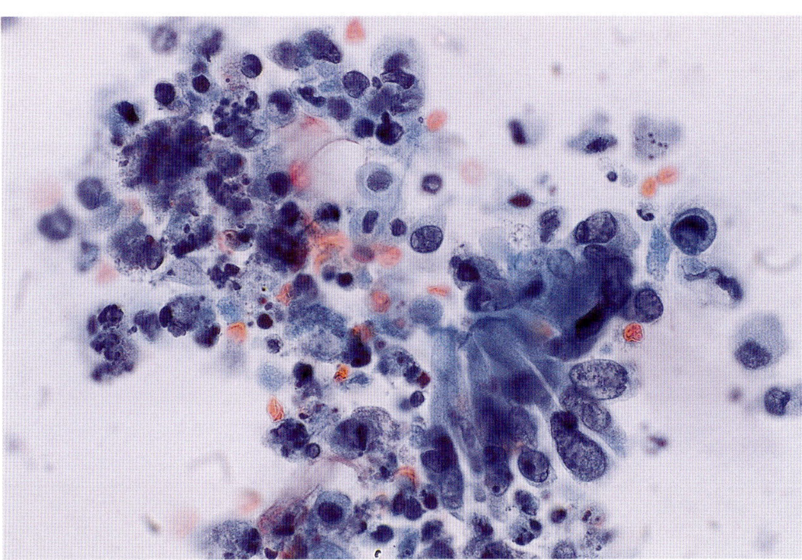

D

Figure 7–4. (*Continued*)

> **Box 7–4.**
>
> Most patients with female genital tract cancer involving the urinary bladder have a history of previous or simultaneous high-stage primary genital cancer.
>
> Squamous cell carcinoma and adenocarcinoma of extravesical origin are cytologically identical to primary bladder carcinoma of the same type. The cytologic diagnosis requires clinical correlation.

(Box 7–4). Cytologic interpretation requires clinical correlation. On a pure cytologic basis, it may be extremely difficult to differentiate between squamous cell carcinoma of genital origin and primary squamous carcinoma of the urinary bladder (Fig. 7–5A,B). Poorly differentiated urothelial carcinoma may have a predominant squamous component resembling pure bladder squamous carcinoma that may be evident in the urine smear. Large cells with an enlarged nucleus, prominent nucleoli, and vacuolated cytoplasm are seen in endometrial or endocervical adenocarcinoma and are similar to

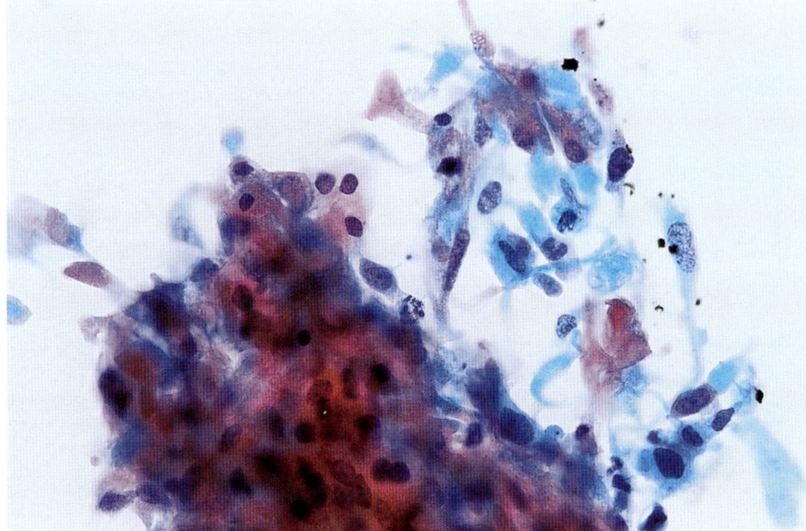

A

B

Figure 7–5. *Squamous carcinoma of the uterine cervix.* A 48-year-old woman with a history of invasive squamous cell carcinoma of the uterine cervix had hematuria and a bladder mass, the result of invasion of the primary tumor. (*A,B*) Smears show spindle cells with keratinized cytoplasm, hyperchromatic nuclei, and necrosis. Voided urine specimen prepared by the monolayer technique. (Papanicolaou stain, ×600)

those of primary urothelial adenocarcinoma or metastases from other sources. Ultimately, tissue biopsy and comparison with tissue obtained from the primary site are necessary for a final diagnosis.

TESTICULAR TUMORS

Positive exfoliative urine cytology has been reported rarely in the diagnosis of testicular neoplasms and has been limited to preparations of seminal fluid and urinary sediment (Rojewska et al., 1989). Its use is not widespread and has not been fully accepted.

With the exception of a single case of metastatic germ cell tumor to the urinary tract with positive urine cytology, no cases of primary testicular germ cell tumors involving the urothelium of the urethra or urinary bladder by direct extension have been described (Hajdu and Nolan, 1975).

Other Retroperitoneal Tumors

Primary retroperitoneal lymphomas and sarcomas may involve the urinary tract by direct extension. The cytology of lymphoid neoplasms involving the urinary tract is discussed in the later section on metastatic hematolymphoid malignancies.

Primary sarcomas of the retroperitoneum comprise approximately 20% of soft tissue sarcomas. The retroperitoneum is the second most common site of sarcomas after the extremities. Retroperitoneal sarcomas, commonly leiomyosarcomas, comprise less than 1% of all cancers in adults. In children, they are ten times more frequent, with rhabdomyosarcoma found most commonly. High-grade sarcomas usually metastasize early, and low-grade sarcomas are locally aggressive and invade adjacent organs (Suen, 1994).

Spindle and oval malignant cells of primary retroperitoneal embryonal rhabdomyosarcoma may be present in urine samples of children who have a large pelvic mass (Helson and Hajdu, 1972). Neuroblastomas arising from sympathetic nerves in the retroperitoneum and adrenal medulla are seen primarily in infants and children and can exfoliate cells into the urinary stream. Patients usually have an abdominal mass, and urine cytology accurately assesses the extent of the disease. Exfoliated neuroblastoma cells in urine are single cells with distinct nuclear details and nucleoli, forming rosettes composed of elongated cells with fibrillary cytoplasm (Farr and Hajdu, 1972). However, cytology may not be able to differentiate neuroblastoma from other small-blue-cell tumors of childhood such as Wilms' tumor, Ewing's sarcoma, medulloblastoma, retinoblastoma, or lymphoma. Poorly formed tubule formation is seen in Wilms' tumor, in contrast to the discohesive nature of malignant cells in lymphomas and leukemias, which exhibit hematolymphoid features, large nuclei, and vacuolated cytoplasm. These neoplasms cannot be classified without clinical, radiologic, and tissue biopsy correlation and the help of immunohistochemistry, molecular biology, electron microscopy, and cytogenetics (Box 7–5).

The author is not aware of reports of primary retroperitoneal sarcomas invading the urinary tract as diagnosed by urine cytology in adults. However, the cytologic differential diagnosis of spindle cells in urine includes primary or metastatic sarcoma, spindle cell carcinoma, carcinosarcoma, melanoma, and inflammatory pseudotumor. Melanoma commonly exfoliates pigmented pleomorphic, clearly malignant cells in a patient who has evidence of disseminated disease and a history of cutaneous melanoma. A high-grade sarcoma may exfoliate spindle anaplastic malignant cells in the presence of a retroperitoneal mass. A bladder pseudotumor lacks nuclear anaplasia, and the patient has a bladder mass but no retroperitoneal disease (Sonobe et al., 1999). Spindle cell carcinoma and other high-grade carcinomas yield smears of high cellularity, in contrast to the hypocellular smears of spindle cell sarcomas. The spindle cell component of carcinosarcoma is not present in urine cytology, and smears show only the poorly differentiated non–spindle cell epithelial component. The cytologic assessment of these unusual neoplasms must be supported by immunohistochemistry.

Box 7–5.

Retroperitoneal sarcomas may invade the urinary tract and exfoliate malignant cells that are visible in the urine sediment.

Spindle and oval cells may be seen in the urine of children with primary retroperitoneal embryonal rhabdomyosarcoma. The differential diagnosis includes other small-cell tumors of infancy.

METASTASES INVOLVING THE URINARY TRACT

In general, by virtue of the lymphatic and venous circulation, metastases develop initially in the wall of the urinary tract under the mucosal surface. The surface urothelium is involved subsequently by extension of the metastatic focus, and exfoliation of malignant cells into the urine stream occurs. Most metastases do not cause ulceration of the mucosa. Therefore, the submucosal nature and the tumor differentiation of some metastatic deposits influence the diagnostic accuracy of urine cytology in the detection of these deposits.

In addition, the cytomorphology of metastatic malignancies, in particular carcinomas, may be similar if not identical to that of primary carcinomas of the urinary tract, making the diagnosis more difficult (Bardales et al., 1998). However, in most instances, a metastasis to the urinary tract is a manifestation of generalized metastatic disease, and the history of primary malignancy is known. Comparison of cell morphology between the metastasis and the primary site is advised in such cases.

As in tissue biopsy examination, metastasis should be suspected in any malignant growth with unusual cytologic features.

Box 7–6.

Malignant melanoma and stomach, breast, and lung carcinomas are the most frequent malignancies involving the urinary bladder.

Lung carcinoma is the most common tumor metastatic to the kidney; metastases to the renal pelvis are rare.

Malignant lymphoma accounts for approximately 50% of malignancies involving the ureters.

The urinary bladder, prostate, colorectum, kidney, and lung account for most metastases to the urethra and penis.

Distant Primary Sites and Urologic Clinical Manifestations

Metastases to the urinary tract, a small but significant percentage of all tumors of the urinary tract, are of particular interest to the physician because of the treatment implications of such a diagnosis. Most metastatic deposits are silent, and those that are symptomatic do not necessarily have positive urine cytology. This explains the rarity of urine cytology reports in comparison to those found in the surgical pathology literature. Hematuria and urinary tract obstruction are common clinical manifestations (Box 7–6).

Urinary Bladder

The most common primary carcinomas disseminating to the urinary bladder originate in the skin (malignant melanoma), stomach, breast, and lung in order of decreasing frequency. However, metastases from diverse tumor types and sites have been described (Petersen, 1992c). Rarely, metastasis can be the first manifestation of a malignant process, or the urinary tract may be the first site of metastasis of a known primary (Silverstein et al., 1987; van Driel et al., 1987). Because of the ample size of the urinary bladder cavity, urinary obstruction is usually not a main clinical feature; the tumor may cause nonspecific irritative symptoms or hematuria; in most cases, it is silent. However, if the tumor is located in the area of the trigone, any of the ureteral or urethral orifices can be involved, causing urinary obstruction.

Kidney and Renal Pelvis

Lung carcinoma is the most common tumor that is metastatic to the renal parenchyma. Involvement of the renal interstitium together with the renal tubules and renal pelvis causes the malignant cells to gain access to the urinary stream, to be identified by urine cytology. In contrast, metastases to the renal pelvis are exceedingly rare, lymphomas and carcinomas particularly of lung, breast, and gastrointestinal tract origin being the most common primaries (Melamed and Reuter, 1993). Metastases to the renal pelvis from a contralateral renal cell carcinoma have also been reported (Lindell et al., 1991; Takashi et al., 1993). These deposits are usually asymptomatic or may cause flank pain with intermittent hematuria or obstruction, particularly when they are located in the renal pelvis.

Ureter

More than 300 cases of metastatic ureteral neoplasms from diverse primary sites have been reported. Malignant lymphoma accounts for approximately 50%, and, in decreasing order of frequency, breast, stomach, uterine cervix, kidney, colorectum, and bladder comprise most of the remainder of these malignancies (Petersen, 1992a). Ureteral and renal pelvic metastases may be asymptomatic. However, obstruction and hydronephrosis can be seen.

Urethra

Malignancies metastatic to the male or female urethra are rare and usually are characterized clinically by hematuria, urethral obstruction, and pain. The urinary bladder, prostate, colorectum, and kidney are the most common primary sites (Kotecha and Gentile, 1974; Selikowitz and Olsson, 1973). It should be stressed that metastasis to the penis usually lacks involvement of the urothelium, accounting for the absence of malignant cells in the urine of these patients. As is the case for the urethra, approximately 90% of metastases to the penis originate in the prostate, urinary bladder, colon, kidney, and lung in order of decreasing frequency. The remaining 10% originate from various other primary sites (Petersen, 1992b).

Diagnostic considerations. The clinical diagnosis of a metastasis to the urinary tract raises questions about the tumor type and site of involvement. The latter can be determined by clinical and radiologic means. However, pathologic examination must confirm the diagnosis. In this regard, urine cytology is a minimally invasive or noninvasive method of determining the malignant nature of the cells. In addition, the site of involvement, that is renal versus lower urinary tract, can also be suspected cytologically by evaluation of the nonneoplastic elements present in the smear. Findings of fragments of necrotic renal parenchyma, degenerated tubular cells, and epithelial casts strongly suggest a renal metastasis.

In general, an accurate diagnosis relies heavily upon good communication with the clinical team for correlation of their findings with tumor cell morphology. In this regard, cystoscopic examination, radiologic evaluation, and the clinical history are invaluable for the cytopathologist, whose main role is to separate a primary urinary tract malignancy accurately from a secondary deposit.

Carcinomas

Squamous Carcinoma

Positive urine cytology showing neoplastic squamous cells consistent with metastatic squamous cell carcinoma has been described in renal metastases with involvement of the renal pelvis from disseminated bronchogenic and esophageal squamous carcinoma (Chung, 1979; Satoh et al., 1989) (Box 7–7).

> **Box 7–7.**
>
> Smears of metastatic and primary squamous cell carcinoma of the urinary tract are identical. Clinical correlation is necessary to distinguish between the two.
>
> Malignant squamous cells are also present in high-grade urothelial carcinoma with squamous differentiation. A finding of high-grade urothelial carcinoma cells confirms such a diagnosis.
>
> Urolithiasis, chronic inflammation that occurs with double-J ureteral stent placement, schistosomiasis, and renal tuberculosis may cause keratinizing squamous metaplasia (leukoplakia) with marked cytologic atypia.
>
> Keratinized cells with anaplastic nuclei are important for the diagnosis.

Cytology. One report of metastatic squamous carcinoma of the lung describes poorly preserved, malignant, keratinized and nonkeratinized pleomorphic squamous cells in the absence of a necrotic or inflammatory background (Chung, 1979). The cellular characteristics are identical in both primary renal pelvis and bladder squamous carcinoma and in metastatic squamous carcinoma.

Differential diagnosis. Before the cytologic diagnosis of metastasis is made, conclusive clinical evidence of a primary squamous carcinoma, usually in the lung or esophagus, must be present, and a primary squamous cell carcinoma of the bladder or renal pelvis must be excluded. The absence of tumor diathesis favors a metastatic deposit over a primary tumor, and the presence of damaged renal tubular cells and casts favors renal involvement. However, necrosis can be seen in both instances, and the final diagnosis must be based on a clinical–cytologic correlation.

We must remember that squamous differentiation is not unusual in cases of primary high-grade urothelial carcinoma. In these cases, the cellularity is higher, cell degeneration is less apparent, and clusters and single high-grade urothelial malignant cells predominate over the squamous cells. High cellularity and necrosis are often present in invasive tumors.

Occasionally, bladder or renal pelvis calculi can cause extensive tissue damage, with cellular degeneration and reactive squamous cell changes mimicking squamous carcinoma. Keratinization may be present in both instances, but reactive cells lack nuclear anaplasia. However, we must remember that primary squamous cell carcinoma may be associated with bladder or renal pelvic lithiasis. In such cases, the presence of urolithiasis is a distractor, and the diagnosis requires careful clinical, cytologic, and at times tissue correlation. Atypical squamous metaplasia can be seen in urine specimens from patients with a permanent double-J ureteral stent placed to alleviate ureteral obstruction due to lithiasis or malignancy (see Fig. 6–34 in Chapter 6).

Box 7–8.

The breast, gastrointestinal tract, and kidney are frequent primary sites for metastatic adenocarcinoma of the urinary tract.

Large cells with vacuolated cytoplasm and large, round, eosinophilic nucleoli suggest a renal origin. Large cells with multiple cytoplasmic vacuoles and pleomorphic, convoluted nuclei suggest an ovarian origin. Signet-ring cells suggest a breast or gastric origin in addition to primary signet-ring cell adenocarcinoma of the bladder. A colonic-type adenocarcinoma may be primary or metastatic. Clinical correlation and tissue examination, including comparison with tissue from the primary site, are necessary.

A history of a primary malignancy is almost always known at the time of urine cytologic examination.

Adenocarcinoma

In our experience and after review of 46 urine specimens from 41 patients with the diagnosis of bladder adenocarcinoma, 30 patients had a history of extravesical malignancy, breast (3 patients) and kidney (2 patients) being the most common noncontiguous primary sources of involvement (Bardales et al., 1998) (Box 7–8).

Cytology. Voided urine specimens from ductal breast carcinoma metastatic to the urinary tract are characterized by the presence of large, round cells with well-defined cytoplasmic borders, ample cytoplasm with large vacuole(s), an eccentric nucleus, and often a prominent nucleolus. The cellularity is scant, and the cells show variable degrees of degeneration that do not preclude the diagnosis of malignancy. The background may be clean (Fig. 7–6A,B).

Urine specimens from metastatic lobular breast carcinoma show smaller cells with scant cytoplasm and hyperchromatic nuclei. The cells are arranged singly or in short linear arrangements or clusters (Fig. 7–7A,B).

In the ureter, metastatic breast carcinomas may present as urinoma, a collection of urine that may be aspirated for diagnostic purposes (Canossi et al., 1995; Zunarelli et al., 1995).

Instrumented urine collected from renal cell carcinoma metastatic to the bladder shows large cells with abundant cytoplasm, round nuclei, prominent round nucleoli, and a bloody background. Similar findings are present in metastases along the ureteral stump after nephrectomy for renal carcinoma (Fig. 7–8). It has been suggested that the presence of renal cell carcinoma cells in the urine may increase the risk of tumor recurrence in the ureteral stump. Thus, for some authors, such urine findings should be an indication for radical nephroureterectomy instead of radical nephrectomy alone (Leonard et al., 1996).

Papillary serous ovarian adenocarcinoma metastatic to the bladder in bladder washings shows high cellularity and predominantly tissue fragments and papillary-like clusters composed of large cells with multivacuolated cytoplasm, large pleomorphic nuclei, and prominent nucleoli. Multinucleation may be present along with rare psammoma bodies and debris (Edgerton et al., 1999).

Metastatic adenocarcinoma of colorectal origin shows extensive columnar cells and

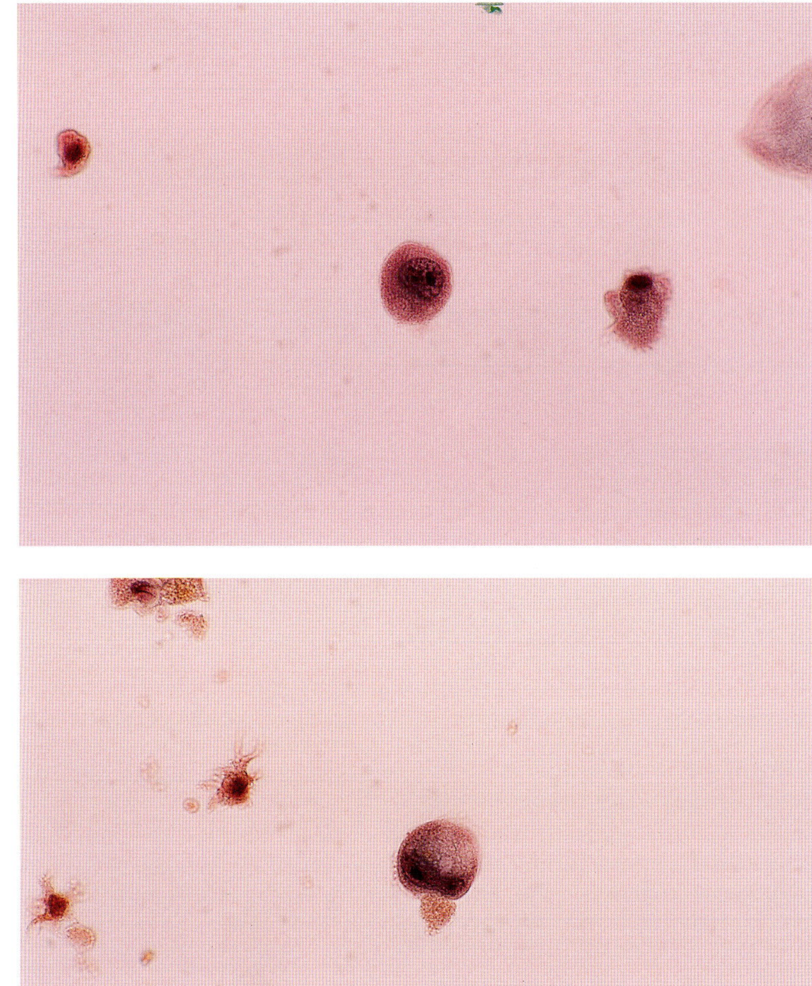

Figure 7–6. *Metastatic ductal carcinoma of the breast.* This voided urine specimen shows a few round cells with large nuclei, vesicular chromatin, and prominent nucleoli. An irregular nuclear contour (*A*) and a large cytoplasmic vacuole (*B*) are noted. Cells are identical to those of primary bladder adenocarcinoma. (Papanicolaou stain, ×600)

variable necrosis. Similarly, gastric adenocarcinoma shows malignant cells of the signet-ring cell type.

Differential diagnosis. Criteria for the differential diagnosis of clear cells identified in urine smears, that is, reactive urothelial cells, macrophages, clear cell renal cell carcinoma, and so on in urine, are presented in Chapter 5.

Before a cytologic diagnosis of metastatic adenocarcinoma is made, one must exclude a primary adenocarcinoma of the urothelial tract, a distinction that cannot be made with certainty on cytologic grounds alone. However, there are certain cytologic characteristics that may lead to the diagnosis or narrow the list of possible primary sites. Large cells with clear, thin, or vacuolated cytoplasm and large, round,

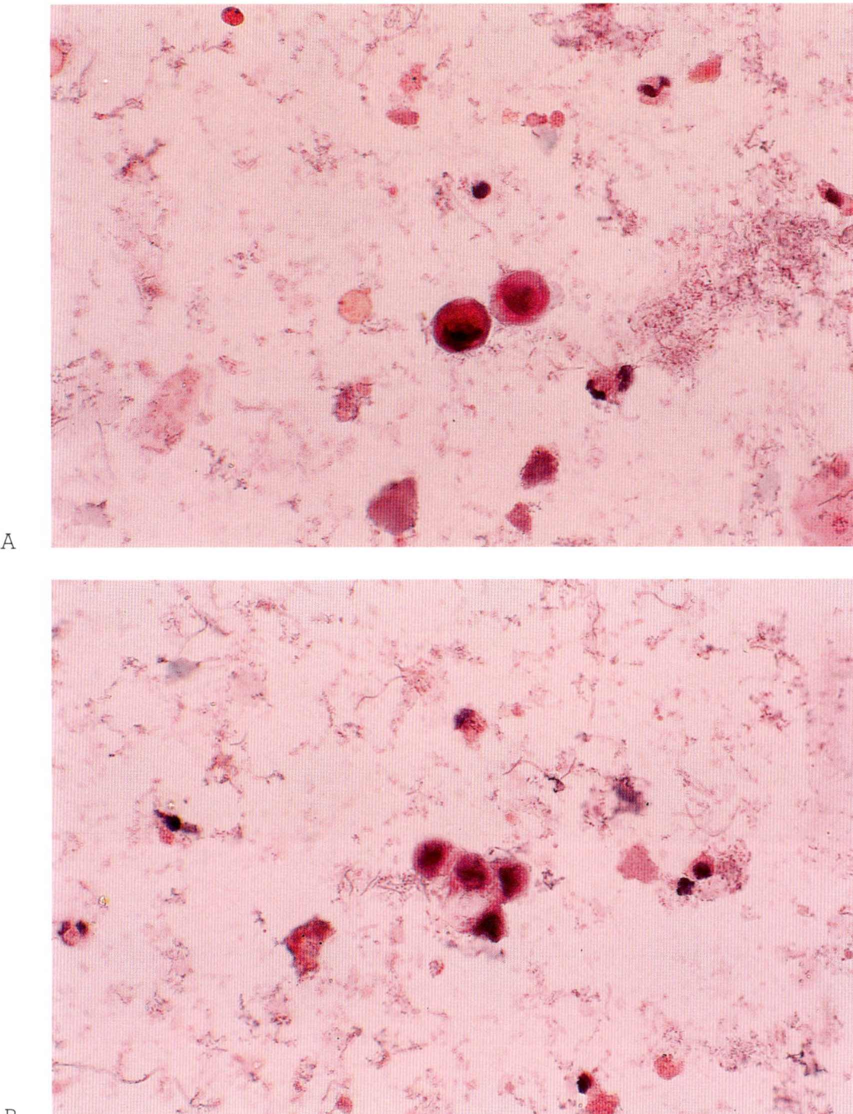

Figure 7–7. *Metastatic lobular carcinoma of the breast.* (*A,B*) Single cells and short linear arrangements with irregular nuclear contours and vacuolated cytoplasm are evident in the voided urine of a woman with a history of lobular carcinoma of the breast. Necrosis is also noted. (Papanicolaou stain, ×600)

eosinophilic nucleoli suggest a renal origin. Large cells with multiple cytoplasmic vacuoles and pleomorphic, convoluted nuclei strongly suggest an ovarian origin. Signet-ring cells suggest a breast or gastric origin in addition to primary signet-ring cell adenocarcinoma.

Unfortunately, primary adenocarcinomas of the intestinal type and metastatic gastrointestinal adenocarcinomas have similar cytologic features. Therefore, clinical and radiologic information is important, because with few exceptions, patients have a history of a distant primary tumor and multiorgan involvement at the time of positive urine cytology.

Small-Cell Carcinoma

At the time of metastasis to the urinary tract, small-cell carcinoma is usually widely metastatic, and the patient has a clinical history of a lung primary. A mediastinal mass in a

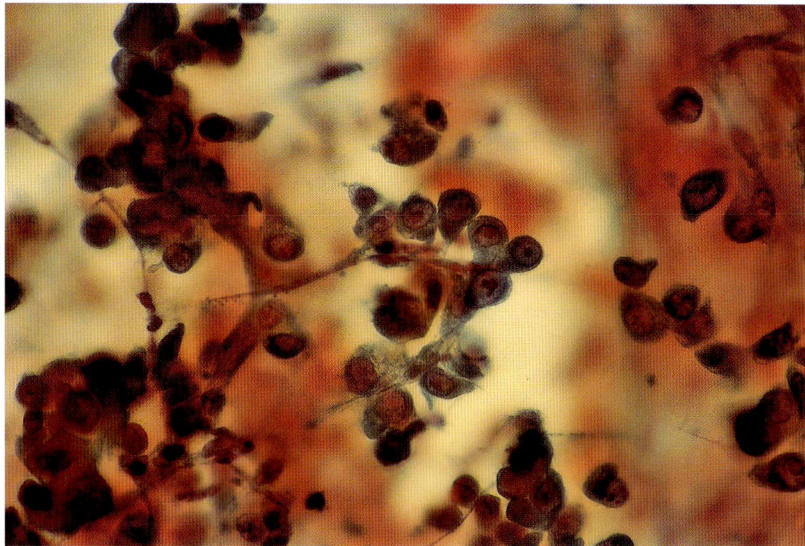

Figure 7–8. *Metastatic renal cell carcinoma.* A 63-year-old man had a history of nephrectomy for renal cell carcinoma. Bladder brushings of a mass in the ureterovesical orifice show large cells with scant cytoplasm, round vesicular nuclei, and large round nucleoli characteristic of renal cell carcinoma. (Papanicolaou stain, ×600)

patient with a history of smoking is the common accompanying clinical scenario.

Cytology. The cytologic features of small cell carcinoma include hypercellularity, with isolated single and clustered small cells showing scant cytoplasm, hyperchromatic nuclei, finely granular nuclear chromatin, rare nucleoli, and a high mitotic index. Nuclear molding is prominent, and the background is commonly bloody, necrotic, or inflamed, with numerous degenerated cells (Box 7–9).

Box 7–9.

Widely metastatic disease and a clinical history of lung cancer are commonly present in small-cell carcinoma metastatic to the urinary tract.

Smears show extensive necrosis and small cells, with scant cytoplasm, prominent nuclear molding, degenerated hyperchromatic nuclei, and absent or inconspicuous nucleoli.

Primary small-cell carcinoma of the bladder and prostate, metastatic Merkel cell carcinoma, and small-cell-type malignant lymphoma are the main differential diagnostic considerations.

Differential diagnosis. Small-cell malignant neoplasms, including primary bladder and prostatic small cell neuroendocrine carcinoma, metastatic Merkel cell carcinoma, malignant lymphoma, and high-grade urothelial carcinoma, show similar cytologic findings.

A small-cell undifferentiated tumor of the bladder and prostate shows single cells and small aggregates of numerous discohesive, round to oval, hyperchromatic cells with coarse chromatin, nuclear molding, and high nuclear-cytoplasmic ratios. It is impossible to differentiate this tumor from a metastasis (Nguyen-Ho et al., 1994; Rollins and Schumann, 1991). Merkel cell carcinoma has identical cytologic features; however, the clinical history of a primary small-cell carcinoma of the skin is essential for making the diagnosis.

In contrast to small-cell carcinoma, malignant lymphoma is characterized by the presence of discohesive, monotonous cells with scant cytoplasm and lack of cell aggregates (Andrion et al., 1993). The cytologic features are described in the next section.

The use of immunocytochemical stains for neuroendocrine (neuron-specific enolase, chromogranin, synaptophysin), epithelial (cytokeratin, epithelial membrane antigen, prostatic acid phosphatase, and prostate-specific phosphatase), and lymphoid (leukocyte common antigen) differentiation is helpful in equivocal cases. Electron microscopy may be helpful in selected cases (Nguyen-Ho et al., 1994; Rollins and Schumann, 1991).

The distinction from high-grade urothelial carcinoma may be difficult because of the extensive necrosis. However, a search for larger, well-preserved cells in clusters, with squamous features not uncommon in high-grade urothelial carcinoma, is helpful (Ali et al., 1997; Zukerberg et al., 1991).

The more cohesive nature of carcinoid tumor cells forming groups or sheets of polygonal small cells with scant cytoplasm and oval nuclei allows the differentiation of carcinoid from small-cell carcinoma (Rudrick et al., 1995).

As with other metastatic deposits, the diagnosis of metastatic small-cell carcinoma should be made after a primary small-cell carcinoma of the urinary tract, commonly in the bladder, is excluded. Positive results of clinical and radiologic investigations for a primary tumor elsewhere, including the prostate, favor metastatic small cell carcinoma (McRae and Garcia, 1997).

Hematolymphoid Malignancies

Malignant Lymphoma

The overall incidence of genitourinary tract involvement by malignant lymphoma in 1,269 autopsies has been reported to be 57%; 47% of these cases had lesions in the renal parenchyma, and less commonly in the urinary bladder and ureters (Rosenberg et al., 1961). Most cases of lymphomatous involvement of the urinary tract have been of the non-Hodgkin's type, but rare cases of Hodgkin's disease have been reported (Andrion et al., 1993; Bocian et al., 1982; Cheson et al., 1984; Givler, 1971). The metastases are usually asymptomatic, particularly in patients with renal involvement (Richmond et al., 1962).

The precise pathologic mechanisms by which lymphoma cells reach the urine remain to be elucidated. Renal involvement by disseminated lymphoma causing diffusely enlarged kidneys can directly damage the interstitium and tubules, with penetration of malignant cells to the urinary stream. A similar mechanism has been proposed to explain the presence of abnormal lymphoid cells in the urine from patients with acute cellular renal allograft rejection and multiple myeloma (Cheson et al., 1984). However, a lymphomatous mass in the renal parenchyma and, less frequently, lymph node disease with involvement of the ureter(s) or urinary bladder constitute direct pathways of urinary tract involvement.

The rarity of the antemortem diagnosis of urinary tract involvement is in part the result of a silent clinical presentation, lack of suspicion, and lack of noninvasive, sensitive diagnostic tests. In this regard, the role of urine cytology is not completely clear. In one antemortem study, 28% of 50 randomly selected patients with malignant lymphoma had malignant lymphoid cells in the urine, establishing the diagnosis of urinary tract involvement (Cheson et al., 1984). Similar studies on a larger patient population need to be conducted.

Most urine cytology diagnoses involve Hodgkin's disease, small-cell lymphoma, and large-cell lymphoma, the last being the most frequent. Most large- and small-cell lymphomas are of higher grade and have multivisceral involvement including the kidney. Aggressive lymphomas composed of small cells that may lead to renal involvement include Burkitt's lymphoma and diffuse, nodular, and mixed follicle center cell lymphomas.

The cytologic diagnosis must be correlated with the clinical history and tissue biopsy findings. Although, at times, malignant lymphoid cells in urine are the first manifestation of systemic lymphoma, most patients have evidence of systemic lymphomatous involvement at the time of urine evaluation (Cheson et al., 1984). Thus, cytologic evidence of malignancy is strongly supported by clinical findings in most cases. In equivocal cases, however, detection of light-chain restriction or other markers can be assessed by immunocytochemistry performed on cytospin smears (Tanaka et al., 1993; Yam and Janckila, 1985).

Box 7–10.

Casts, reactive renal tubular cells, and necrotic renal parenchyma in addition to malignant lymphoid cells are found in the urine sediment of patients with renal involvement by malignant lymphoma.

Malignant lymphoid cells in urine sediment that lacks evidence of renal tissue damage strongly suggest lower urinary tract involvement.

Smears show cell degeneration, necrosis, and inflammation, along with monomorphic and dissociated lymphoid cells.

Alcohol-fixed Papanicolaou-stained cytospin smears are most commonly used for cytologic interpretation of urine sediment. However, the use of air-dried, Romanovsky-stained smears facilitates the evaluation of cytoplasmic features complementing the former. In any case, the interpretation must be made on intact and well-preserved cells only (Box 7–10).

Cytology. In general, a high percentage of urine sediment abnormalities is detectable by cytology in lymphoma patients. Thus, the diagnosis requires careful evaluation of the abnormal cells, of noncellular elements, and of the background.

In cases of renal involvement, malignant cells are present along with pathologic casts, damaged renal tubular cells, or fragments of necrotic parenchyma. These findings, although not specific, are more common and more evident in patients with positive cytology (100% of cases) than in those with negative cytology (56% of cases). The presence of such changes in negative cases suggests either occult renal involvement or renal damage of other etiology (Cheson et al., 1984). Hematuria has been found in both negative and positive cases. However, in positive cases, the finding of hematuria and urothelial cells in the absence of casts and damaged tubular cells strongly suggests lower urinary tract and not renal involvement.

Depending on the urine collection method, smears are variably cellular, showing single monomorphic large (large-cell lymphoma) or medium-sized (small-cell lymphoma) mononuclear cells. Prominent nucleoli, nuclear folds, and coarse chromatin are common findings (Fig. 7–9). A variable degree of cell degeneration is always present, along with a background of necrosis and inflammation. The blastic phase of leuke-

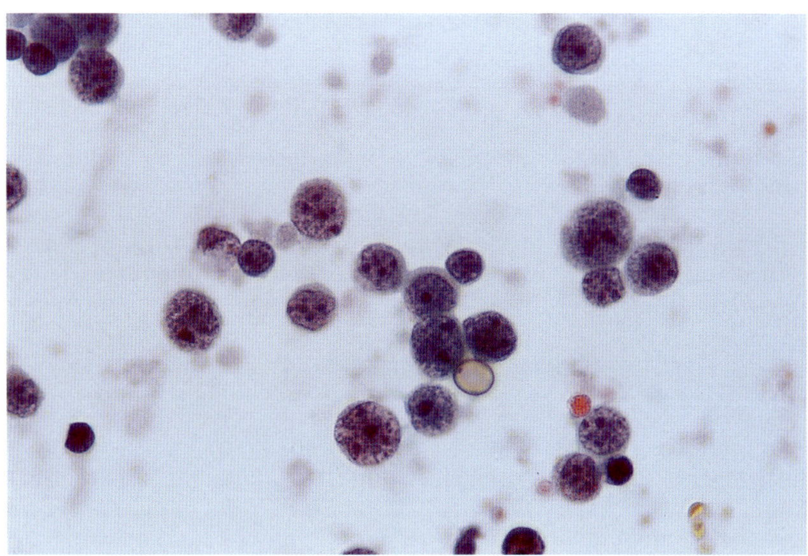

Figure 7–9. *Large cell non-Hodgkin's lymphoma.* Characteristically, these cells are single and show marked nuclear anaplasia in a background of necrosis. Bladder washing. (Papanicolaou stain, ×1000)

mias and lymphomas has been detected in urine specimens associated with hematuria (Cheson et al., 1981).

In cases of Hodgkin's disease, bladder involvement may occur many years after the primary lymph node diagnosis has been made. In these cases, the identification of Reed-Sternberg cells in the urine supports the diagnosis (Bocian et al., 1982). However, very large cells with vacuolated or dense cytoplasm, pleomorphic multilobated nuclei, and prominent nucleoli similar to those present in Hodgkin's disease can be present in Ki-1-positive anaplastic large-cell lymphoma (Tanaka et al., 1993). Immunocytochemistry showing CD15+, CD30+, and CD45− stain in the large malignant cells supports the diagnosis of Hodgkin's disease. Ki-1-positive anaplastic large-cell lymphoma shows CD15−, CD30+, CD45+/−, and EMA+ stain in the anaplastic large cells.

In summary, urine cytology enables an experienced clinician to diagnose and localize lymphomatous involvement of the urinary tract.

Differential diagnosis. Malignant lymphomas, including Hodgkin's disease involving the urinary tract, must be differentiated from primary malignancies of the same type in the urinary bladder of immunocompetent and immunosuppressed patients (Gupta et al., 1989; Mincione and Gasbarre, 1995). Poorly differentiated urothelial carcinoma or adenocarcinoma, undifferentiated carcinoma, and small-cell anaplastic carcinoma of the bladder or prostate may shed malignant cells in urine, and these are difficult to distinguish from cells of hematolymphoid malignancies. However, the presence of clusters and single cells favors a nonlymphoid origin. The judicious use of cell markers further helps in the diagnosis of difficult cases (Zukerberg et al., 1991).

Plasma Cell Myeloma

Plasma cell myeloma may involve the urinary tract as a primary or metastatic disease (Chaitin et al., 1984; Hill et al., 1983; Igel et al., 1991; Lemos et al., 2000; Roth et al., 1977; Yang et al., 1982).

Cytology. Smears show abnormal plasma cells with dense cytoplasm with focal paranuclear clearing, eccentric nuclei, coarse and clumped chromatin more prominent toward the nuclear membrane, and variably seen nucleoli (Fig. 7–10). A smaller amount of cytoplasm, loss of paranuclear clearing, and more prominent nucleoli are seen in poorly differentiated plasma cell myeloma. Cells indistinguishable from those of large-cell lymphoma are seen in myeloma with lymphomatous transformation. Large cells with fragile cytoplasm and large nucleoli are seen in the immunoblastic lymphoma type of transformation. These features are more evident when the smears are stained with Romanovsky stain.

Differential diagnosis. Plasma cells must be differentiated from urothelial cells exfoliated from urothelial carcinoma with plasmacytoid cytomorphology. In both entities, plasma cell myeloma and urothelial carcinoma with plasmacytoid features, patients may present with multiorgan involvement including multiple bone metastases (Sahin et al., 1991; Zukerberg et al., 1991). Amelanotic metastatic malignant melanoma with plasmacytoid or lymphoid features is another consideration in the differential diagnosis. Positive epithelial markers (cytokeratin), intracytoplasmic light chain (kappa or lambda) restriction, and positive melanoma markers (S-100, melan-A, and HMB-45) demonstrated by immunocytochemistry are helpful to establish the diagnosis of urothelial carcinoma, plasma cell myeloma, and malignant melanoma (Box 7–11).

Box 7–11.

Smears show abnormal plasma cells. However, urinary tract involvement by poorly differentiated plasma cell myeloma shows cells indistinguishable from those of large-cell malignant lymphoma.

Before the diagnosis of plasma cell myeloma is made, plamacytoid urothelial carcinoma and malignant melanoma must be excluded.

A clinical history of systemic plasma cell myeloma and intracytoplasmic light-chain restriction demonstrable by immunocytochemistry are essential for making the diagnosis.

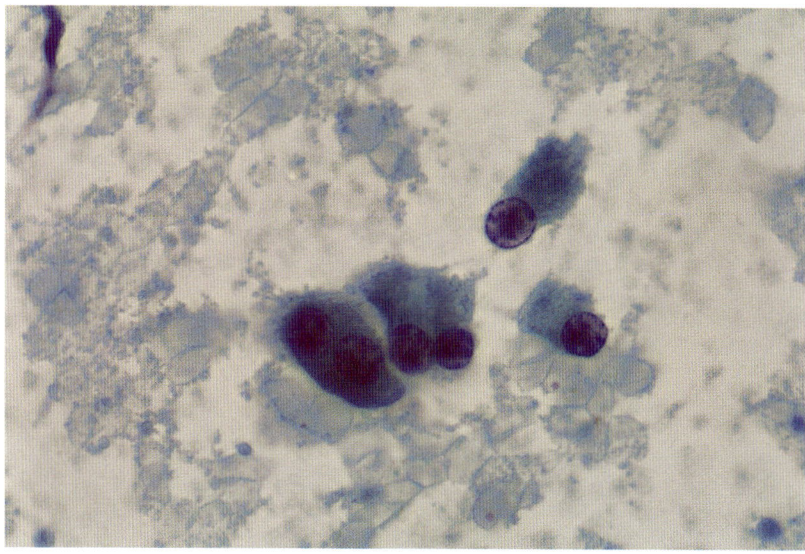

Figure 7–10. *Plasma cell myeloma.* In a background of blood and necrosis, malignant plasma cells show abundant, dense, granular cytoplasm with eccentric round nuclei and coarse chromatin. Voided urine. (Papanicolaou stain, ×1000)

Leukemias

Leukemias involve the urinary bladder in 20% of cases. However, they rarely manifest clinically (Chaitin et al., 1984; Givler, 1971).

Cytology. In chronic myelogenous leukemia, the smears show immature myeloid cells in various stages of differentiation, particularly myelocytes and metamyelocytes. Blast cells are seen in acute leukemias. Unless there is clinical information or the presence of Auer rods, differentiation of myeloid from lymphoid malignancies cannot be established with certainty by cytology alone and may require the use of immunocytochemistry. Air-dried, Romanovsky-stained smears are useful for evaluating cytoplasmic features.

Occasionally, urine specimens from patients with leukemia may be contaminated with peripheral blood involved with the hematopoietic malignancy. This is particularly important in patients with bleeding diatheses and when urine is obtained by instrumentation or through an inserted Foley catheter. In these cases, urine cytology results need to be correlated with findings in the peripheral blood before a diagnosis of leukemic involvement of the urinary tract is rendered (Box 7–12).

> **Box 7–12.**
>
> Leukemias rarely involve the urinary tract.
>
> Unless the clinical history is provided, the differentiation between myeloid and lymphoid lineage cannot be made with certainty on examination of urine sediment only.
>
> In the presence of hematuria, contamination of urine specimens with peripheral blood involved with the hematopoietic malignancy should be excluded before the diagnosis of leukemia is made.
>
> Cytologic evaluation of urine sediment should include Papanicolaou-stained and air-dried, Romanovsky-stained smears.

Malignant Melanoma

Primary malignant melanoma of the urinary tract is rare, occurring mainly in the urethra and penis (Anichkov and Nikonov, 1982; Oldbring and Mikulowski, 1987). The diagnosis of primary melanoma is made only when an exhaustive clinical history and meticulous physical examination fail to exclude a cutaneous, ocular, or visceral melanoma, and when a transition between normal and neoplastic tissue is present on histologic examination (Ainsworth et al., 1976).

Melanomas metastatic to the urinary tract are more frequent, the kidney and bladder being the most commonly affected sites (Das Gupta and Brasfield, 1964). Although urinary tract involvement is usually accompanied by systemic metastatic disease, on rare occasions the urinary tract may be the first site of visceral metastasis, usually years after regional nodal disease has developed (Piva and Koss, 1964; Zogno et al., 1997). On cystoscopy, brown-black nodules are apparent in most of these cases, facilitating the diagnosis. Commonly, degranulation of intracellular pigment and exfoliation of melanin-laden cells darken the urine (melanuria), as is evident on naked-eye examination of unstained smears (McClements et al., 1988; Woodard et al., 1978).

The diagnosis of melanomatous involvement of the urinary tract, including the prostate, by means of urine cytology is accurate and is important for assessment of the spread of metastatic disease. However, the diagnosis usually indicated a poor prognosis (Piva and Koss, 1964; Zogno et al., 1997).

Cytology. Malignant cells have variable amounts of intracellular melanin pigment, and some may be hypomelanotic or amelanotic; however, the nuclear characteristics are similar in both types. Exfoliation of melanoma cells to the urine stream may be intermittent, and negative cytologic results may be present among positive ones (Piva and Koss, 1964).

Smears are usually cellular and show large, predominantly dissociated, variably shaped cells with marked nuclear anaplasia. The cytoplasm is usually dense and cyanophilic and contains a fine, dusty to coarse, granular brown-black melanin pigment in varying proportions. Commonly, the cytoplasmic pigmentation obscures the nuclear characteristics of both malignant cells and melanophages. The nuclear-cytoplasmic ratio is usually high. The nuclei are large, vesicular, irregular, and eccentric, with red, prominent macronucleoli (Piva and Koss, 1964; Woodard et al., 1978; Zogno et al., 1997). Intranuclear cytoplasmic invaginations may be seen. Occasionally, cells have a uniform shape resembling that of small-cell malignant lymphoma and finely granular melanin pigment surrounding a bland nucleus with an inconspicuous nucleolus (Fig. 7–11A,B). In questionable cases, comparison with tissue sections from the original tumor is recommended.

Melanophages with a low nuclear-cytoplasmic ratio and benign-appearing nuclei are also present together with variable amounts of extracellular melanin pigment granules. Reactive pigmented renal tubular cells may be present, indicating renal injury. A coincidental association of melanoma cells with *Wuchereria bancrofti* microfilaria in urine sediment has been described (Kapila and Verma, 1986).

In summary, most diagnoses of melanoma in urine specimens require the presence of melanin pigment within the cytoplasm of cells exhibiting nuclear anaplasia (Piva and Koss, 1964) (Box 7–13).

Box 7–13.

"Melanuria" can be evident on gross examination of the urine sediment.

Melanoma cells have variable amounts of cytoplasmic melanin granules, nuclear anaplasia, and macronucleoli.

Melanophages have cytoplasmic melanin granules, a low nuclear-cytoplasmic ratio, and benign-appearing nuclei.

Pigmented renal tubular cells with varying degrees of atypia, but no anaplasia, may be seen in cases of melanomatous renal involvement.

The unequivocal diagnosis of malignant melanoma is based on the presence of cells with nuclear anaplasia.

Differential diagnosis. Melanin-pigmented renal tubular cells can be seen in the urine in melanosis of the kidney in the absence of detectable renal metastasis (Piva and Koss, 1964). These cells indicate renal tubular interstitial damage and show varying degrees of atypia but no features of malignancy. To avoid an erroneous interpreta-

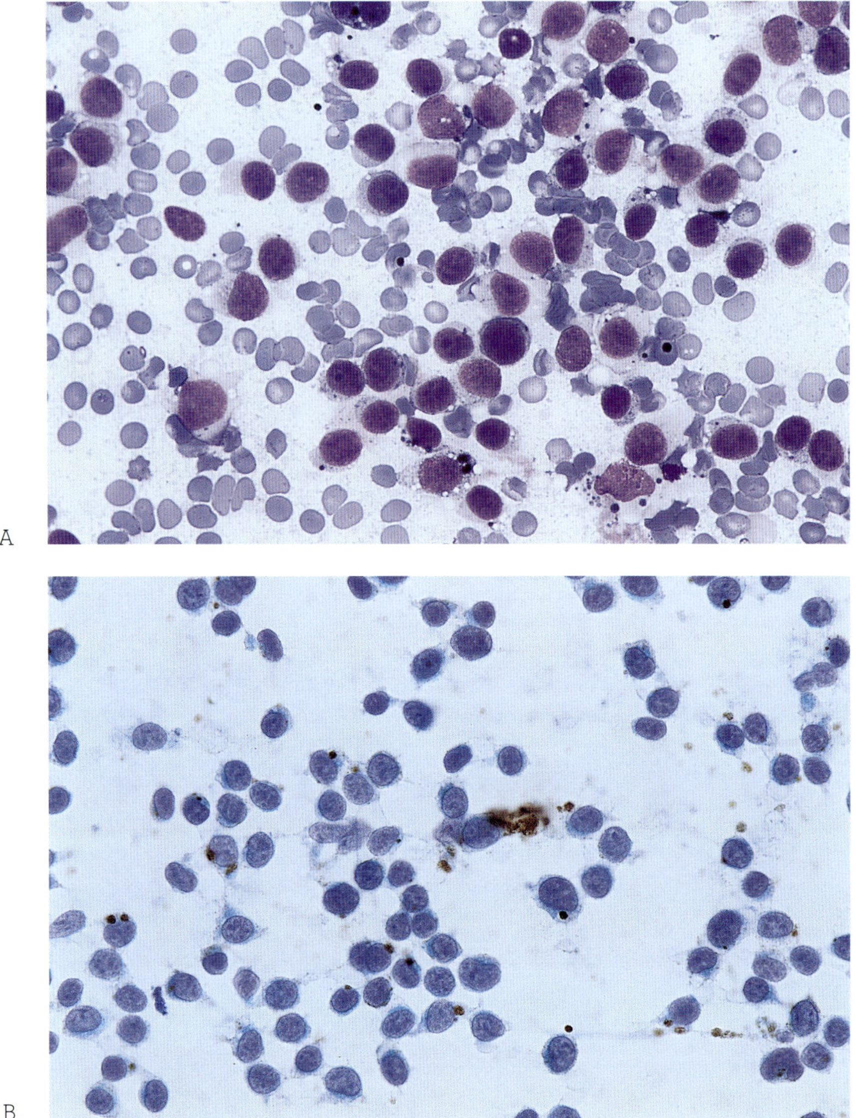

Figure 7–11. *Malignant melanoma.* (*A,B*) Brushing smears from a brown bladder nodule show round/oval cells with finely granular cytoplasmic melanin pigment and round nuclei with minimal anaplasia. The patient had a primary "lymphocyte"-type malignant melanoma of the skin. (A, Diff-Quik stain, ×600; B, Papanicolaou stain, ×600)

tion, one should make the diagnosis of melanoma based on the presence of unequivocal malignant nuclear features and not solely on the basis of "pigmented" cells.

The cytologic diagnosis of amelanotic melanoma must be supported by strong clinical evidence, and the melanoma has to be differentiated from primary or metastatic epithelial and hematolymphoid malignancies. Positive immunocytochemistry for S-100, melan-A, and HMB-45, as well as ultrastructural analysis demonstrating mature melanosomes in the tumor cells, are diagnostic of melanoma.

Germ Cell Tumors

Gonadal or extragonadal germ cell tumors can extrinsically compress or metastasize to the ureter, bladder, or urethra, although metastases to the urinary tract from germ cell tumors are extremely rare (Davies et al., 1992; Hajdu and Nolan, 1975; Viddeleer et al., 1992). Mainly the kidney is affected by metastatic disease, and the patient may have acute renal failure, hematuria, and massive proteinuria due to diffuse tumor infiltration in both kidneys (Abboud et al., 1981). Intraluminal metastasis to the ureter causes obstructive uropathy (Davies et al., 1992).

Seminoma and dysgerminoma cells exfoliated into the urinary lumen can be evident on cytologic examination of the urine sediment (Hajdu and Nolan, 1975; Viddeleer et al., 1992). Cell exfoliation may be intermittent, and the patient may be asymptomatic.

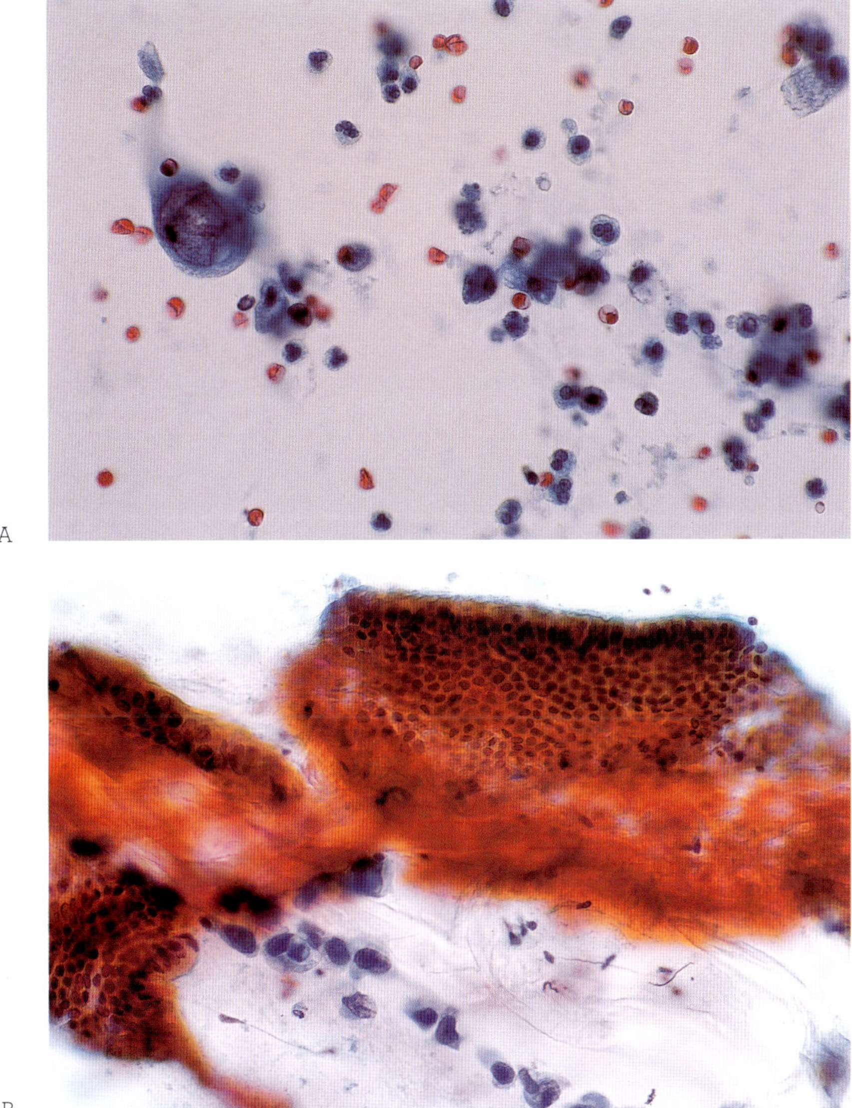

Figure 7–12. *Carcinoma in ileal-conduit urine.* Malignancies that develop in the mucosa of an ileal loop are of urothelial origin and can produce obstruction. (*A*) Urine cytology shows a few highly anaplastic malignant cells in a background of cellular degeneration with round cells that have pyknotic nuclei, characteristic of ileal-conduit urine. (*B,C*) Brushing smears show sheets of benign intestinal epithelium and both clustered and single malignant high-grade urothelial carcinoma cells. (Papanicolaou stain, ×600)

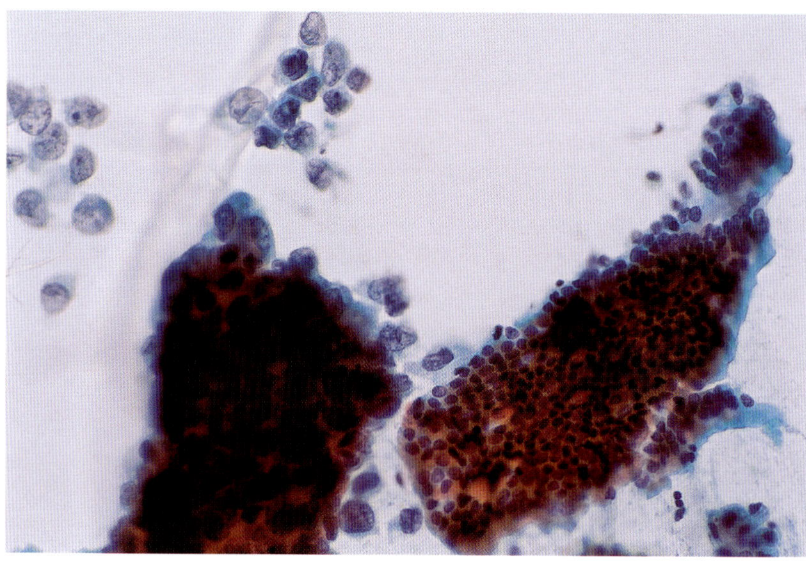

C

Figure 7–12. (*Continued*)

Cytology. In one study, urine cytology of a seminoma metastatic to the urinary bladder showed clusters of large, round cells with scant, thin cytoplasm, round nuclei, and large, round nucleoli in a background of tissue damage (Viddeleer et al., 1992).

Neoplasms Arising in Intestinal Loops

An ileal conduit remains the preferred form of urinary diversion for patients undergoing total cystectomy for invasive urothelial carcinoma of the bladder. In rare cases, neoplasms can develop in the ileal mucosa and are limited to urothelial tumors (Fig. 7–12A–C). Adenomatous polyps and carcinoid tumors have been described as causing ileal conduit stenosis and hydronephrosis, but primary ileal adenocarcinoma has not been reported (Kerfoot et al., 1999; Shokeir et al., 1995; Tomera et al., 1982).

Malignancy, usually of intestinal origin, is a late complication of nonileal intestinal segments used in urinary diversion. Adenocarcinoma developing in the intestinal loop, most commonly at the anastomotic site, has an incidence of 2% to 15% after ureterosigmoidostomy, with a much higher incidence of polyps, in the range of 40%. This translates to a 500- to 7,000-fold increase in the risk of adenocarcinoma compared to the general population (Azimuddin et al., 1999; Malone et al., 1997). Malignant changes not only occur after ureterosigmoidostomy, but are observed after other nonileal loop conduit uroenteric reconstructions as well. They

Box 7–14.

Patients with nonileal-conduit urinary diversions are at high risk of developing adenocarcinoma in the intestinal loop.

Malignant neoplasms arising in the ileal mucosa of patients with ileal conduit are limited to urothelial carcinomas. No ileal adenocarcinomas have been reported.

Urine sediment commonly shows evidence of high-grade urothelial carcinoma.

Adenocarcinoma sheds single and clustered columnar cells with pleomorphic nuclei in a background of necrosis.

Long-term surveillance with annual flexible sigmoidoscopy, random biopsies, and urine cytology is advocated for patients with nonileal-conduit urinary diversions. Urine cytology examination is recommended for patients with ileal conduit diversions.

appear to be multifactorial and occur in as many as 40% of patients undergoing uroenteric reconstructions after decades of follow-up (Azimuddin et al., 1999; Malone et al., 1997; Treiger and Marshall, 1991). An interval of 53 years is the longest reported (Harford et al., 1984). It appears that the latency period is shorter when the ureterosigmoidostomy is performed for malignant lesions, with some cases developing within two years after the procedure (Badalament et al., 1990). The patients commonly have obstructive urinary symptoms.

Sigmoid polyps, commonly found in these patients, strongly suggest that these tumors also follow the sequence of polyp-adenocarcinoma that is seen in colorectal neoplasia. Long-term surveillance with annual flexible sigmoidoscopy, random biopsies, and urine cytology is advocated in these patients so that early premalignant lesions can be diagnosed and excised. The survival rate after detection of a malignant tumor is approximately 30%, in contrast to 100% survival of patients undergoing colonoscopic excision of nonmalignant polyps (Azimuddin et al., 1999) (Box 7–14).

REFERENCES

Abboud, H. E., Galla, J. H., Lucas, B. A., Curtis, J. J., Luke, R. G. (1981). Acute renal failure due to seminoma metastatic to the kidneys. *Nephron* 27: 101–3.

Ainsworth, A. M., Clark, W. H., Mastrangelo, M., Conger, K. B. (1976). Primary malignant melanoma of the urinary bladder. *Cancer* 37: 1928–36.

Ali, S. Z., Reuter, V. E., Zakowski, M. F. (1997). Small cell neuroendocrine carcinoma of the urinary bladder. A clinicopathologic study with emphasis on cytologic features. *Cancer* 79: 356–61.

Alroy, J., Roganovic, D., Banner, B. F., Jacobs, J. B., Merk, F. B., Ucci, A. A., Kwan, P. W., Coon, J. S. T., Miller, A. W. D. (1981). Primary adenocarcinomas of the human urinary bladder: histochemical, immunological and ultrastructural studies. *Virchows Arch [Pathol Anat]* 393: 165–81.

Andrion, A., Gaglio, A., Zai, G. (1993). Bladder involvement in disseminated malignant lymphoma diagnosed by voided urine cytology. *Cytopathology* 4: 115–17.

Anichkov, N. M., Nikonov, A. A. (1982). Primary malignant melanomas of the bladder. *J Urol* 128: 813–15.

Azimuddin, K., Khubchandani, I. T., Stasik, J. J., Rosen, L., Riether, R. D. (1999). Neoplasia after ureterosigmoidostomy. *Dis Colon Rectum* 42: 1632–8.

Badalament, R. A., Cirulli, C., Zerick, W., Lucas, J. G., Drago, J. R. (1990). Colon carcinoma associated with ureterosigmoidostomy. *J Surg Oncol* 45: 207–11.

Bardales, R. H. (1998). Fine needle aspiration cytology of papillary neoplasms. *Clin Lab Med* 18: 373–99, v.

Bardales, R. H., Pitman, M. B., Stanley, M. W., Korourian, S., Suhrland, M. J. (1998). Urine cytology of primary and secondary urinary bladder adenocarcinoma. *Cancer* 84: 335–43.

Bocian, J. J., Flam, M. S., Mendoza, C. A. (1982). Hodgkin's disease involving the urinary bladder diagnosed by urinary cytology: a case report. *Cancer* 50: 2482–5.

Bonsib, S. M., Gibson, D. C., Mhoon, N., Greene, G. F. (2000). Renal sinus invasion in renal cell carcinoma. *Mod Pathol* 13: 94A.

Bunge, R. G., Kraushaar, O. F. (1950). An early renal malignancy, diagnosed preoperatively. *J. Urol* 63: 475–9.

Canossi, B., Renzi, E., Santini, D. (1995). [Ureteral metastasis from lobular breast carcinoma. Unusual cause of urinoma]. *Radiol Med* 90: 158–60.

Chaitin, B. A., Manning, J. T., Ordonez, N. G. (1984). Hematologic neoplasms with initial manifestations in lower urinary tract. *Urology* 23: 35–42.

Cheson, B. D., Johnston, J. L., del Junco, G., Kjeldsberg, C. R. (1981). Cytologic evidence for disseminated immunoblastic lymphoma. *Am J Clin Pathol* 75: 621–5.

Cheson, B. D., Schumann, J. L., Schumann, G. B. (1984). Urinary cytodiagnostic abnormalities in 50 patients with non-Hodgkin's lymphomas. *Cancer* 54: 1914–19.

Chung, H. R. (1979). Bronchogenic squamous cell carcinoma metastatic to kidney. Detection by urine sediment cytology. *Urology* 13: 561–4.

Creasman, W. T., Lukeman, J. (1972). Unreliability of urinary cytology in detecting gynecologic malignancy. *Cancer* 30: 148–9.

Das Gupta, T., Brasfield, R. (1964). Metastatic melanoma. A clinicopathological study. *Cancer* 17: 1323–39.

Davies, R. J., King, D. M., Hendry, W. F. (1992). Case report: intraureteric metastasis from testicular teratoma. *Clin Radiol* 46: 354–6.

DeMay, R. M. (1996). Secondary tumors of the urinary tract. In *The Art and Science of Cytopathology* (R. M. DeMay, ed.), Vol. I, pp. 408–9. ASCP Press, Chicago.

Edgerton, M. E., Hoda, R. S., Gupta, P. K. (1999). Cytologic diagnosis of metastatic ovarian adenocarcinoma in the urinary bladder: a case report and review of the literature. *Diagn Cytopathol* 20: 156–9.

Fallick, M. L., Hutchinson, M., Alroy, J., Long, J. P. (1997). Collecting-duct carcinoma presenting as upper tract lesion with abnormal urine cytology. *Diagn Cytopathol* 16: 258–61.

Farr, G. H., Hajdu, S. I. (1972). Exfoliative cytology of metastatic neuroblastoma. *Acta Cytol* 16: 203–6.

Foot, N. C., Papanicolaou, G. N., Holmquist, N. D., Seybolt, J. F. (1958). Exfoliative cytology of urinary sediments: a review of 2829 cases. *Cancer* 11: 127–37.

Garret, M., Jassie, M. (1976). Cytologic examination of postprostatic massage specimens as an aid in diagnosis of carcinoma of the prostate. *Acta Cytol* 20: 126–31.

Givler, R. L. (1971). Involvement of the bladder in leukemia and lymphoma. *J Urol* 105: 667–70.

Graham, R. M., Graham, J. B., Stevens, J. L. (1967). Voided urine in cytologic screening for female genital tract. *Obstet Gynecol* 30: 253–6.

Gupta, P. K., Taft, P. D. (1969). Urinary cytology in female genital cancer. *Am J Obstet Gynecol* 104: 1043–6.

Gupta, R., Belitsky, P., Stewart, L., Cohen, A. D., Handa, S. P. (1989). Urine cytology in a case of B-cell lymphoma in a transplanted kidney. *Transplant Proc* 21: 3585–7.

Hajdu, S. I., Nolan, M. A. (1975). Exfoliative cytology of malignant germ cell tumors. *Acta Cytol* 19: 255–60.

Hajdu, S. I., Savino, A., Hajdu, E. O., Koss, L. G. (1971). Cytologic diagnosis of renal cell carcinoma with the aid of fat stain. *Acta Cytol* 15: 31–3.

Harford, F. J., Fazio, V. W., Epstein, L. M., Hewitt, C. B. (1984). Rectosigmoid carcinoma occurring after ureterosigmoidostomy. *Dis Colon Rectum* 27: 321–4.

Helson, L., Hajdu, S. I. (1972). The cytology of urine of pediatric cancer patients. *J Urol* 108: 660–2.

Hill, G. S., Morel-Maroger, L., Mery, J. P., Brouet, J. C., Mignon, F. (1983). Renal lesions in multiple myeloma: their relationship to associated protein abnormalities. *Am J Kidney Dis* 2: 423–38.

Ho, C., Hguyen, G. K., Schumann, G. B. (1996). Reparative renal tubular cells in urine sediments: A potential false-positive diagnostic pitfall. *Acta Cytol* 40: 1025 (Abstract).

Hoskin, P. J., Blake, P. R., Trott, P. A. (1992). The role of routine urine and parametrial aspirate cytology in staging carcinoma of the cervix. *Clin Oncol* 4: 183–5.

Igel, T. C., Engen, D. E., Banks, P. M., Keeney, G. L. (1991). Renal plasmacytoma: Mayo Clinic experience and review of the literature. *Urology* 37: 385–9.

Kapila, K., Verma, K. (1986). Coexistent metastatic malignant melanoma cells and *Wuchereria bancrofti* microfilariae in urinary sediment [letter]. *Acta Cytol* 30: 696–7.

Kerfoot, B. P., Steele, G. S., Datta, M. W., Richie, J. P. (1999). Carcinoid tumor in an ileal conduit diversion. *J Urol* 162: 1685–6.

Khalil, M., Shin, H. J., Tan, A., DuBose, T. D., Jr., Ordonez, N., Katz, R. L. (2000). Macrophagelike vacuolated renal tubular cells in the urine of a male with osmotic nephrosis associated with intravenous immunoglobulin therapy. A case report. *Acta Cytol* 44: 86–90.

Koss, L. G. (1992). Tumors of the urinary tract and prostate in urinary sediment. In *Diagnostic Cytology and Its Histopathologic Bases* (L. G. Koss, ed.), 4th ed., Vol. I, pp. 934–1017. J. B. Lippincott, Philadelphia.

Koss, L. G. (1996). Tumors of the bladder. In *Diagnostic Cytology of the Urinary Tract with Histopathologic and Clinical Correlations* (L. G. Koss, ed.), pp. 71–139. J. B. Lippincott, Philadelphia.

Kotecha, N., Gentile, R. L. (1974). Carcinoma of prostate with urethral metastasis. *Urology* 3: 85–6.

Krishnan, B., Truong, L. D. (2000). Prostatic adenocarcinoma diagnosed by urinary cytology. *Am J Clin Pathol* 113: 29–34.

Lemos, N., Melo, C. R., Soares, I. C., Lemos, R. R., Lemos, F. R. (2000). Plasmacytoma of the urethra treated by excisional biopsy. *Scand J Urol Nephrol* 34: 75–6.

Leonard, A. M., Lefevre, A., Lagrange, W. (1996). Ureteral stump metastasis from renal adenocarcinoma: case report and literature review. *Acta Urol Belg* 64: 23–5.

Lindell, O. I., Grein, H. U., Schreiter, F. J. (1991). Opposite renal pelvic metastasis of renal adenocarcinoma. Case report. *Scand J Urol Nephrol* 25: 165–7.

Malone, M. J., Izes, J. K., Hurley, L. J. (1997). Carcinogenesis. The fate of intestinal segments used in urinary reconstruction. *Urol Clin North Am* 24: 723–8.

Marshall, F. F., Stewart, A. K., Menck, H. R. (1997). The National Cancer Data Base: report on kidney cancers. The American College of Surgeons Commission on Cancer and the American Cancer Society. *Cancer* 80: 2167–74.

Mauri, M. F., Bonzanini, M., Luciani, L., Dalla Palma, P. (1994). Renal collecting duct carcinoma. Report of a case with urinary cytologic findings. *Acta Cytol* 38: 755–8.

McClements, B. M., McDowell, I. F., McCluskey, D. R. (1988). A case of dark urine, hyperpigmentation and hepatomegaly. *Irish J Med Sci* 157: 157.

McRae, S., Garcia, B. M. (1997). Cytologic diagnosis of a primary pure oat cell carcinoma of the bladder in voided urine. A case report. *Acta Cytol* 41: 1279–83.

Meisels, A. (1963). Cytology of carcinoma of the kidney. *Acta Cytol* 7: 239–44.

Melamed, M. R., Reuter, V. E. (1993). Pathology and staging of urothelial tumors of the kidney and ureter. *Urol Clin North Am* 20: 333–47.

Milsten, R., Frable, W. J., Texter, J. H., Jr., Paxson, L. (1973). Evaluation of lipid stain in renal neoplasms as adjunct to routine exfoliative cytology. *J Urol* 110: 169–71.

Mincione, G. P., Gasbarre, M. (1995). Primary vesical malignant lymphoma of histiocytes detected by urine cytology. *Pathologica* 87: 559–62.

Nguyen, G. K., Akin, M. R. (1998). Fine needle aspiration cytology of the kidney, renal pelvis, and adrenal. *Clin Lab Med* 18: 429–59, vi.

Nguyen, G. K., Schumann, G. B. (1997). Cytopathology of renal collecting duct carcinoma in urine sediment. *Diagn Cytopathol* 16: 446–9.

Nguyen-Ho, P., Nguyen, G. K., Villanueva, R. R. (1994). Small cell anaplastic carcinoma of the prostate: report of a case with positive urine cytology. *Diagn Cytopathol* 10: 159–61.

Oldbring, J., Mikulowski, P. (1987). Malignant melanoma of the penis and male urethra. Report of nine cases and review of the literature. *Cancer* 59: 581–7.

Petersen, R. O. (1992a). Neoplasms metastatic to the ureter. In *Urologic Pathology* (R. O. Petersen, ed.), 2nd ed., pp. 244–8. J. B. Lippincott, Philadelphia.

Petersen, R. O. (1992b). Penis: metastatic neoplasms. In *Urologic Pathology* (R. O. Petersen, ed.), 2nd ed., p. 699. J. B. Lippincott, Philadelphia.

Petersen, R. O. (1992c). Urinary bladder: metastatic neoplasms. In *Urologic Pathology* (R. O. Petersen, ed.), 2nd ed., pp. 365–7. J. B. Lippincott, Philadelphia.

Piscioli, F., Detassis, C., Polla, E., Pusiol, T., Reich, A., Luciani, L. (1983). Cytologic presentation of renal adenocarcinoma in urinary sediment. *Acta Cytol* 27: 383–90.

Piscioli, F., Pusiol, T., Scappini, P., Luciani, L. (1985). Urine cytology in the detection of renal adenocarcinoma. *Cancer* 56: 2251–5.

Piva, A. E., Koss, L. G. (1964). Cytologic diagnosis of metastatic malignant melanoma in urinary sediment. *Acta Cytol* 8: 398–402.

Ramzy, I., Larson, V. (1977). Prostatic duct carcinoma: exfoliative cytology. *Acta Cytol* 21: 417–20.

Richmond, J., Sherman, R. S., Diamond, H. D., Craver, L. F. (1962). Renal lesions associated with malignant lymphomas. *Am J Med* 32: 184–207.

Rojewska, J., Pykalo, R., Czaplicki, M. (1989). Urine and semen cytomorphology in patients with testicular tumors. *Diagn Cytopathol* 5: 9–13.

Rollins, S., Schumann, G. B. (1991). Primary urinary cytodiagnosis of a bladder small-cell carcinoma. *Diagn Cytopathol* 7: 79–82.

Rosenberg, S., Diamond, H., Jasiowitz, B., Craver, L. (1961). Lymphosarcoma: a review of 1269 cases. *Medicine (Balt)* 40: 31–84.

Roth, R. M., Glovsky, M. M., Cooper, J. F., 3rd, Douglas, S. D. (1977). Gamma A myeloma with hyperviscosity and obstructive uropathy. *J Urol* 117: 527–9.

Rudrick, B., Nguyen, G. K., Lakey, W. H. (1995). Carcinoid tumor of the renal pelvis: report of a case with positive urine cytology. *Diagn Cytopathol* 12: 360–3.

Rupp, M., O'Hara, B., McCullough, L., Saxena, S., Olchiewski, J. (1994). Prostatic carcinoma cells in urine specimens. *Cytopathology* 5: 164–70.

Sahin, A. A., Myhre, M., Ro, J. Y., Sneige, N., Dekmezian, R. H., Ayala, A. G. (1991). Plasmacytoid transitional cell carcinoma. Report of a case with initial presentation mimicking multiple myeloma. *Acta Cytol* 35: 277–80.

Satoh, S., Ujiie, T., Nomura, K., Okamoto, T., Kubo, T., Abe, T. (1989). [Metastatic renal tumor of esophageal carcinoma: report of a case]. *Hinyokika Kiyo—Acta Urol Jpn* 35: 1025–9.

Schumann, G. B., Colon, V. F. (1980). Urine cytology. Part II: renal cytology. *Am Family Physician* 21: 102–6.

Selikowitz, S. M., Olsson, C. A. (1973). Metastatic urethral obstruction. *Arch Surg* 107: 906–8.

Shokeir, A. A., Shamaa, M., el-Mekresh, M. M., el-Baz, M., Ghoneim, M. A. (1995). Late malignancy in bowel segments exposed to urine without fecal stream. *Urology* 46: 657–61.

Silverstein, L. I., Plaine, L., Davis, J. E., Kabakow, B. (1987). Breast carcinoma metastatic to bladder. *Urology* 29: 544–7.

Sonobe, H., Okada, Y., Sudo, S., Iwata, J., Ohtsuki, Y. (1999). Inflammatory pseudotumor of the urinary bladder with aberrant expression of cytokeratin. Report of a case with cytologic, immunocytochemical and cytogenetic findings. *Acta Cytol* 43: 257–62.

Suen, K. C. (1994). Retroperitoneum III: soft tissue tumors. In *Guides to Clinical Aspiration Cytology: Retroperitoneum and Intestine.* 2nd ed., (T. S. Kline, ed.), pp. 85–133. Igaku-Shoin, New York.

Takashi, M., Sakata, T., Sai, Y., Shimoji, T., Miyake, K. (1993). Metastasis of renal cell carcinoma to the contralateral renal pelvis. *Urol Int* 51: 102–4.

Tanaka, T., Yoshimi, N., Sawada, K., Takami, T., Sugie, S., Etori, F., Kachi, H., Mori, H. (1993). Ki-1-positive large cell anaplastic lymphoma diagnosed by urinary cytology. A case report. *Acta Cytol* 37: 520–4.

Tomera, K. M., Unni, K. K., Utz, D. C. (1982). Adenomatous polyp in ileal conduit. *J Urol* 128: 1025–6.

Treiger, B. F., Marshall, F. F. (1991). Carcinogenesis and the use of intestinal segments in the urinary tract. *Urol Clin North Am* 18: 737–42.

van Driel, M. F., Ypma, A. F., van Gelder, B. (1987). Gastric carcinoma metastatic to the bladder. *Br J Urol* 59: 193–4.

Varma, V. A., Fekete, P. S., Franks, M. J., Walther, M. M. (1988). Cytologic features of prostatic adenocarcinoma in urine: a clinicopathologic and immunocytochemical study. *Diagn Cytopathol* 4: 300–5.

Viddeleer, A. C., Lycklama a Nijeholt, G. A., Beekhuis-Brussee, J. A. (1992). A late manifestation of testicular seminoma in the bladder in a renal transplant recipient: a case report. *J Urol* 148: 401–2.

Woodard, B. H., Ideker, R. E., Johnston, W. W. (1978). Cytologic detection of malignant melanoma in urine. A case report. *Acta Cytol* 22: 350–2.

Yam, L. T., Janckila, A. J. (1985). Immunocytochemical diagnosis of lymphoma from urine sediment. *Acta Cytol* 29: 827–32.

Yang, C., Motteram, R., Sandeman, T. F. (1982). Extramedullary plasmacytoma of the bladder: a case report and review of literature. *Cancer* 50: 146–9.

Yokose, T., Fukuda, H., Ogiwara, A., Sakai, K., Saitoh, K. (1991). Myxoid leiomyosarcoma of the kidney accompanying ipsilateral ureteral transitional cell carcinoma. A case report with cytological, immunohistochemical and ultrastructural study. *Acta Pathol Jpn* 41: 694–700.

Zogno, C., Schiaffino, E., Boeri, R., Schmid, C. (1997). Cytologic detection of metastatic malignant melanoma in urine. A report of three cases. *Acta Cytol* 41: 1332–6.

Zukerberg, L. R., Harris, N. L., Young, R. H. (1991). Carcinomas of the urinary bladder simulating malignant lymphoma. A report of five cases. *Am J Surg Pathol* 15: 569–76.

Zunarelli, E., Pollastri, C. A., Calo, M., Migaldi, M., Galizia, G., Criscuolo, M. (1995). Metastatic breast carcinoma presenting as urinoma. *Pathologica* 87: 712–4.

8

DIAGNOSTIC VALUE OF URINE CYTOLOGY EXAMINATION

LUIS EDUARDO DE LAS CASAS, M.D.
RICARDO H. BARDALES, M.D.

IN 1945, George Papanicolaou introduced urine cytology examination as a test for diagnosing cancer of the urinary tract, including the bladder, the prostate, and the kidney (Papanicolaou and Marshall, 1945). Since then, numerous studies have confirmed the value of urine cytology for diagnosis of cancer of the bladder and the upper urinary tract. However, this test is of limited value in the diagnosis of cancer of the kidney, prostate, and testicles.

Urine cytology examination should be performed so that urinary tract malignancies can be excluded in patients with hematuria and irritative symptoms (Harrison et al., 1951; Papanicolaou, 1947; Sarnacki et al., 1971). However, the usefulness of this test as a screening technique for detection of urothelial neoplasms in patients with asymptomatic hematuria is controversial.

Urine cytology examination should be used in the following circumstances:

- As a screening method for urothelial carcinoma in a population exposed to bladder carcinogens.
- For detection of high-grade urothelial neoplasms.
- For follow-up of patients with a low-grade urothelial neoplasm to detect progression to high-grade carcinoma.
- For follow-up of patients with urothelial carcinoma treated with intravesical therapy to detect recurrences.
- For postoperative follow-up of patients with urinary tract malignancies, particularly detection of urethral, ureteral, or renal pelvic neoplasms after cystectomy.
- For diagnosis of nonbacterial infections and infestations of the urinary tract.
- For detection of noninfectious inflammatory conditions of the urinary tract.

The accuracy of bladder cancer diagnosis by urine cytology examination ranges between 26% and 100%, with a mean of 71.6% (Umiker, 1964). False-negative diagnoses have been reported to be 23% to 38%, and false-positive results range from 1.3% to 11.9%, with a mean of 2.3% (Prall et al., 1972; Sarnacki et al., 1971). In contrast to the detection of high-grade carcinoma, urine cytology is not totally reliable in the detection of low-grade papillary urothelial neoplasms, where the low sensitivity of this test is notorious. When urothelial papillomas, which are not expected to be diagnosed by urine cytology, are excluded from the statistical analysis, the reliability of a positive diagnosis by urine cytology (positive predictive value) is 92% (Malik and Murphy, 1999).

VARIABLES THAT AFFECT THE URINE CYTODIAGNOSIS OF CANCER

Urine can be obtained not only as a spontaneously voided sample, but also by catheterization or by instrumentation techniques that include urethroscopy, cystoscopy, and ureteropyeloscopy with washings and brushings. Regardless of the urine procurement technique, special training is required for adequate cytologic interpretation of the smear.

Variables that affect the diagnostic accuracy of urine cytology examination include the following:

- Type of urinary specimen, that is, voided or instrumented urine, washings, brushing.
- Histologic grade of the tumor, that is, low-grade or high-grade.
- Solitary versus multifocal tumor.
- Noninvasive and invasive malignancies.
- Source of the urinary specimen, that is, urethra, bladder, ureter, renal pelvis.
- De novo or recurrent neoplasms.
- Number of urine samples examined.
- Specimen adequacy.
- Experience of the pathologist.
- Architectural pattern of the tumor, that is, papillary or flat.
- Adequate clinical information.

Type of Urinary Specimen

Cytology of bladder washings (66% overall diagnostic accuracy) provides considerably more information than does cytology of voided urine (49% overall accuracy) (Maier et al., 1995; Trott and Edwards, 1973; Zein et al., 1984). In general, voided urine specimens are frequently hypocellular and have varying degrees of cellular degeneration. The first morning voided urine and 24-hour urine collection have poorly preserved cells and should be avoided as providing inadequate samples for urine cytology examination. However, acceptable voided urine specimens include the second morning urine or any random sample if it is processed immediately.

In a recent study of 227 specimens, a positive cytohistologic correlation was found in 77% of 185 bladder washings and 71% of 36 voided urine specimens, with an overall 72% sensitivity (biopsy-proved neoplasm detected by urine cytology) and 82% specificity (absent neoplasm with negative cytology) (Malik and Murphy, 1999). When different tumor grades were considered, bladder wash cytology detected 52% of grade 1, 88% of grade 2, and 99% of grade 3 urothelial carcinomas among 361 bladder tumors. In the same study with the same number of samples, voided urine cytology detected 21% of grade 1, 76% of grade 2, and 96% of grade 3 urothelial carcinomas (Maier et al., 1995). Despite the better performance of bladder washings, it appears that the differences in sensitivity and specificity for the cytologic detection of malignant cells in different portions of voided urine or bladder washing are not statistically significant (Mungan et al., 1999).

Voided urine cytology is not reliable for the follow-up of low-grade neoplasms unless there is progression to high-grade carcinoma. However, voided urine cytology is the method of choice for follow-up of high-grade urothelial carcinoma (Koss et al., 1985). Data suggest that, in patients with initially positive voided urine cytology and negative workup, if the cytology findings subsequently become negative, the likelihood of the development of urothelial carcinoma is low. However, if the initially positive cytology continues to remain positive, there is a much higher probability of urothelial carcinoma being detected in this population (Konety et al., 1999).

Histologic Grade of the Tumor

A precise evaluation of the accuracy of urine cytology examination for diagnosis of noninvasive papillary urothelial neoplasms is limited by the histologic classification used for correlation and by the cytomorphologic criteria used for cytologic diagnosis. However, regardless of the various schemes used for classifying papillary urothelial neoplasms, the consensus is that urine cytology examination is deficient in the diagnosis of papillary urothelial neoplasms of low malignant potential (PUNLMP, WHO grade 1) and in the diagnosis of 30% to 60% of low-grade papillary urothelial carcinomas (WHO grade 2) with a poor cytohistologic correlation (Beyer-Boon et al., 1978; Brown, 2000; Esposti and Zajicek, 1972; Grossman, 1998; Rife et al., 1979; Shenoy et al., 1985). Furthermore, distinction between urothelial papilloma and nonneoplastic lesions of the bladder is not possible by urine cytology (Esposti and Zajicek, 1972).

The sensitivity and specificity for the diagnosis of low-grade papillary urothelial neoplasms range from 0% to 100%, depending on the number of low-grade papillary urothelial carcinomas included (Renshaw et al., 1996). Thus, the search for cytologic and ancillary methods for improving the diagnosis of these tumors has been the focus of research in the last decade. The use of cytologic criteria identified by logistic regression analysis has yielded a 76% diagnostic accuracy for the detection of low-grade urothelial carcinoma in bladder washings, with 85% sensitivity, 96% specificity, 87% positive predictive value, and 77% negative predictive value (Raab et al., 1994). However, when these criteria were applied prospectively, the sensitivity of diagnosis of these neoplasms decreased to 33% (Raab et al., 1996). Furthermore, these cytologic criteria may not be fully reliable in the evaluation of PUNLMP (Renshaw et al., 1996). In summary, there are no cytologic criteria for accurate diagnosis of PUNLMP and low-grade papillary urothelial carcinoma.

The diagnostic accuracy of detecting urothelial carcinoma is higher with the examination of high-grade tumors. Koss et al. reported that cytology of voided urine is highly reliable, with 94.2% sensitivity in the diagnosis of high-grade tumors and with 100% sensitivity in the diagnosis of primary flat carcinoma in situ. The method failed to recognize PUNLMP and about one-third of low-grade papillary urothelial carcinomas (Koss et al., 1985). The results from a large series of urinary cytology specimens of 1,672 patients confirmed that urinary cytology detects approximately 79% of high-grade urothelial neoplasms. A negative result is predictive of no cancer in more than 90% of cases, regardless of whether the pathologist's practice setting is referral-based or community-based (Bastacky et al., 1999). The overall diagnostic yield of urine cytology in the evaluation of low-grade urothelial neoplasms is as high as 76%, but the false-positive rate is 10% to 14% (Murphy et al., 1984).

The causes of false-positive diagnoses cannot be classified accurately. Adequate length of follow-up is important because it is known that positive cytology findings can precede histologic confirmation of urothelial carcinoma by many years (Murphy et al., 1984). Also, when extensive urothelial denudation occurs in a focus of carcinoma, malignant cells may be evident in cytology specimens only, resulting in a negative biopsy diagnosis with an apparent cytohistologic discrepancy (DeBellis and Schumann, 1986). Polyomavirus infection and intravesical or systemic therapy effects can be confused with high-grade urothelial carcinoma (Bastacky et al., 1999). Although some false-positive results may be obtained by urine cytology after intravesical chemotherapy, urine cytology is a reasonable alternative to more invasive and expensive methods during follow-up of bladder tumors (Baltaci et al., 1996). Occasionally, markedly reactive urothelial cells with prominent nucleoli resemble high-grade papillary carcinoma or adenocarcinoma. Other causes of false-positive results include instrumentation effect, urolithiasis, infection, and inflammation of the urinary tract, which sheds pseudopapillary clusters that can be mistaken for low-grade papillary urothelial neoplasms. Criteria applied for distinguishing such clusters from true papillary fragments of low-grade urothelial neoplasms have resulted in 95% sensitivity and 83% specificity in differentiating between benign and malignant processes. However, these criteria need to be tested in a large number of patients (Kannan and Bose, 1993; Kannan and Gupta, 1999). Awareness of these diagnostic difficulties and careful evaluation of the cellular detail increase the diagnostic accuracy of the urine cytology examination.

Most false-negative diagnoses are not the result of interpretive errors, but are due to the lack of tumor cells in the urine cytology specimens (Malik and Murphy, 1999). However, a study of false-negative cases showed that 65 of 180 patients (36%) with bladder cancer, including 27 patients (15%) with PUNLMP, had consistently negative urine cytology results. Approximately 40% of the false-negative cytology smears, on review, had rare degenerated malignant or suspicious cells that were missed by the pathologist. Malignant cells were not identified by cytology in the remainder of the cases (Bastacky et al., 1999).

Statistical analysis of dysplasia in urine cytology is difficult to perform because of lack of uniform cytologic and histologic criteria for making such a diagnosis. As in low-grade papillary urothelial neoplasms, the cytologic diagnosis of dysplasia is not reproducible, and a clear statistical value cannot be extracted (Sekine, 1989). In addition, because dysplasia frequently coexists with urothelial carcinoma, the cytologic distinction between the two entities may not be of practical value (Renshaw, 2000).

Solitary versus Multifocal Tumor

Positive voided urine cytology of noninvasive urothelial carcinomas appears to be related mainly to tumor grade and stage rather than to tumor size (Chow et al., 1994; Tut et al., 1998).

The relationship between positive voided urine cytology and multifocal disease is controversial. Chow et al. reported that positive voided urine cytology at diagnosis is a predictor of multifocal disease and of tumor recurrence (Chow et al., 1994). However, Tut et al., in a study of 109 patients who had newly diagnosed Ta/T1 urothelial carcinoma of the bladder, as assessed by voided urine cytology at diagnosis, found that tumor multifocality at the time of diagnosis had an impact on the subsequent tumor recurrence rate; however, they found no association between voided urine cytology and multifocal disease (Tut et al., 1998).

Noninvasive and Invasive Malignancies

The accuracy of urine cytology depends not only on the tumor grade but also on the stage of the tumor, that is, the depth of invasion (DiBonito et al., 1992). Cells exfoliated from noninvasive high-grade papillary urothelial carcinoma and invasive urothelial carcinoma are similar both morphologically and architecturally. In such cases, necrosis helps one to make the cytologic diagnosis of invasive carcinoma. However, the absence of necrosis does not exclude such a diagnosis. In high-grade cases, "cannibalism" (cytophagocytosis), a characteristic histologic finding associated with malignancy, appears to be an indicator of both the anaplastic grade and the invasiveness of urothelial carcinoma of the bladder, and it has significant predictive value for tumor progression (Kojima et al., 1998).

For practical purposes, distinction between invasive and noninvasive urothelial carcinomas may not be important, since the diagnosis of high-grade carcinoma is usually made with little difficulty by urine cytology and is followed by cystoscopic and histologic confirmation.

Source of the Urinary Specimen

The diagnosis of urothelial carcinoma of the ureter or renal pelvis is more difficult than that of the bladder because of the location of the tumor. Maier et al. found that bladder washings diagnosed 66% of 36 urothelial carcinomas of the upper urinary tract, in contrast to voided urine cytology, which diagnosed 33% of 31 urothelial carcinomas of the upper urinary tract. The diagnosis of high-grade carcinomas in this area was 81% by bladder washings, in contrast to 64% by voided urine cytology (Maier et al., 1995). It must be remembered that instrumented urine specimens of the ureters or renal pelvis show larger aggregates than those in the bladder, a perfect invitation for an erroneous diagnosis of a low-grade papillary neoplasm.

De Novo and Recurrent Neoplasms

Urine cytology is a useful method for detecting tumor recurrence. The presence of high-grade urothelial carcinoma cells in the urine cytology sample of a patient with a known low-grade urothelial lesion is an indication for immediate cystoscopy and tissue diagnosis. The cytologic diagnosis of a recurrent high-grade tumor is usually not difficult.

Among 111 patients undergoing surveillance for urothelial carcinoma, 51% of those with positive urine cytology were shown to have recurrent urothelial carcinoma, but 87% of those with negative cytology had no recurrent tumors (Loh et al., 1996). Thus, urine cytology is a good adjunct to assessing the suitability of flexible cystoscopy in the follow-up of patients with urothelial tumors.

Number of Samples Examined

It must be noted that not all patients with carcinoma affecting the urinary tract have malignant cells detectable in all urine specimens. The diagnostic value of voided urine in the detection of urothelial carcinoma can be increased when three specimens are obtained on three consecutive days. In a study of 151 positive cases, Koss et al. found that the cytologic diagnosis of urothelial carcinoma was established on the first specimen in 79%, on the second specimen in an additional 14%, and on the third specimen in an additional 7% of cases (Koss et al., 1985). Similar findings have been reported for bladder washings (Malik and Murphy, 1999).

Specimen Adequacy

Proper collection, rapid transport to the laboratory, and immediate processing are of paramount importance. Alternatively, the urine sample can be kept refrigerated at 4°C for a few hours. However, immediate preparation is strongly advised.

In general, obscuring blood and inflammation, extensive cell degeneration, and scant cellularity are causes of specimen inadequacy. However, specific guidelines for adequacy of the urine specimen at the time of cytologic interpretation have not been established. Some authors have proposed that one should have at least 15 basal and intermediate cells for a smear to be classified as adequate (Bastacky et al., 1999).

Experience of the Pathologist

Interpretive errors are inevitable and inherent in the degree of the pathologist's experience, gained after examination of numerous samples and after learning from previous mistakes. Experienced pathologists rarely agree on the cytologic diagnosis of a single low-grade case, as shown by the agreement reached by four expert cytopathologists in only 20% of ten of such tumors (van der Poel et al., 1997). However, agreement is high for the diagnosis of high-grade carcinomas.

There is considerable patient inconvenience, expense, and risk of infection with cystoscopy compared to cytology. Therefore, when an experienced pathologist is available, patients with PUNLMP or low-grade papillary noninvasive urothelial carcinoma may be monitored primarily by urinary cytology, with less frequent endoscopy (Morrison et al., 1984).

Architectural Pattern of the Tumor

A low-grade papillary urothelial neoplasm, an almost innocuous tumor of low malignant potential, is difficult to detect on cytology but is readily visible by cystoscopy. In contrast, carcinoma in situ, a flat, high-grade carcinoma, may not be visualized by cystocopy, but cytology is highly accurate (Koss et al., 1985). Awareness of these facts reinforces the concepts that urine cytology examination and endoscopy are complementary in the diagnosis of low-grade papillary neoplasms, and that cooperation should exist between the pathologist and the urologist.

Adequate Clinical Information

One of the most important variables for accurate interpretation of the urine cytology sample is the clinical information the pathologist receives. This information should be provided in the requisition form, obtained from the computer database, or transmitted directly by the treating physician. The preliminary clinical impression, the most recent patient treatment history, previous treatments that may induce cytologic changes that might lead to misdiagnosis, and availability of previous cytologic or tissue samples are pieces of information that should be known at the time of cytologic evaluation. However, in most instances, the pathologist is not aware of such information, and this increases the likelihood of an erroneous diagnosis.

Thus, establishment of a close working relationship between the urologist and the pathologist is the single most important requirement for a successful urine cytology interpretation (Huben and Gaeta, 1996). When adequate clinical information is provided, and the urinary specimen is correctly handled and processed from procurement to cytopreparation, the diagnostic accuracy of urine cytology is approximately 90%, ranging from 88% for low-grade tumors to 96% for high-grade carcinomas (Maier et al., 1995; Malik and Murphy, 1999).

The application of ancillary techniques, including evaluation of bladder tumor (basement membrane) antigen, DNA ploidy analysis, tumor-specific nuclear matrix proteins, molecular genetic markers, blood group antigen expression, and others, aimed at improving the sensitivity of urine cytology in the diagnosis of bladder cancer, is limited by the lower specificity than that of urine cytology, in particular for the diagnosis of low-grade urothelial neoplasms (Pirtskalaishvili et al., 1999). These techniques are described in Chapter 9.

VALUE OF UPPER URINARY TRACT CYTOLOGY

After a diagnosis of urothelial carcinoma of the bladder, the frequency of upper urinary tract urothelial carcinoma ranges between 1.5% and 21% (Sadek et al., 1999). However, the incidence varies according to the histologic grade of the bladder tumor. The estimated incidence of upper urinary tract carcinomas in patients with low-grade urothelial carcinoma of the bladder has been stated to be less than 5% (Sadek et al., 1999). The incidence of high-grade carcinoma ranges from 1.7% to 21% in patients followed for at least seven years (Cookson et al., 1997; Oldbring et al., 1989; Schwartz et al., 1992).

Voided urine cytology examination detects 25% to 59% (mean 38%) of renal pelvic carcinomas and 35% of ureteral carcinomas (Chow et al., 1994; Sadek et al., 1999; Sarnacki et al., 1971; Wagle et al., 1974). In a study of 42 patients with suspected tumors (34 urothelial carcinomas and 8 lithiases) of the upper urinary tract, voided urine and upper urinary tract washing detected malignancy in 76% and 97% of cases, respectively, with only one false-positive result (Raica et al., 1995).

Brush cytology is a specific and more sensitive sampling method than irrigation or catheterized urine in detecting urothelial carcinoma of the upper urinary tract. The calculated sensitivity for diagnosis of urothelial carcinoma by this technique was 72% and the specificity was 94%. However, in this study, brush cytology was successful in diagnosing dysplasia or carcinoma in situ. In contrast, the irrigation or catheterized urine specimens from these same patients yielded 48% sensitivity, and only higher-grade lesions were detected. Brush cytology was less effective in diagnosing PUNLMP or low-grade papillary urothelial carcinoma (Dodd et al., 1997).

Regardless of the urinary sample procurement method, the causes of false-positive and false-negative results for the upper urinary tract have the same variables as those of the bladder. Raica et al. reported a case of squamous metaplasia of the renal pelvis that was misinterpreted as squamous carcinoma (Raica et al., 1995). In other cases, the tumor cells visualized may not represent upper urinary tract carcinoma, but are probably malignant cells of bladder origin that were carried into the upper urinary tract by means of the instruments used (Sadek et al., 1999).

In the view of most urologists, the validity of upper urinary tract cytology in the evaluation of urothelial carcinoma, particularly after bladder cancer, is questionable. Considering that upper urinary tract sampling for cytology is an invasive procedure that may carry significant morbidity, upper urinary tract surveillance must be limited to specific cases and weighed against the expected risk (Sadek et al., 1999). However, there is no consensus regarding optimal monitoring of the urinary tract in patients with primary bladder cancer.

It seems reasonable that patients with high-grade bladder cancer who are at significantly higher risk for upper urinary tract cancer must be considered for upper endoscopy with urine sampling. However, other diagnostic modalities such as imaging techniques must be applied after treatment of primary bladder tumors and before endoscopy (Nielsen and Ostri, 1988). The series examining the value of cytology of the upper urinary tract that include a significant number of cases reported in the literature are insufficient to prove or disprove such an approach. Unfortunately, the lack of sensitivity of urine cytology in diagnosing low-grade urothelial tumors is the major drawback to considering cytologic examination as the main diagnostic modality for detection of low-grade tumors of the upper urinary tract (Chow et al., 1994).

In any individual case, the surgical decision must be based on a combination of imaging techniques, clinical evaluation, and urine cytology results.

REFERENCES

Baltaci, S., Suzer, O., Ozer, G., Beduk, Y., Gogus, O. (1996). The efficacy of urinary cytology in the detection of recurrent bladder tumours. *Int Urol Nephrol* 28: 649–53.

Bastacky, S., Ibrahim, S., Wilczynski, S. P., Murphy, W. M. (1999). The accuracy of urinary cytology in daily practice. *Cancer* 87: 118–28.

Beyer-Boon, M. E., de Voogt, H. J., van der Velde, E. A., Brussee, J. A., Schaberg, A. (1978). The efficacy of urinary cytology in the detection of urothelial tumours. Sensitivity and specificity of urinary cytology. *Urol Res* 6: 3–12.

Brown, F. M. (2000). Urine cytology. Is it still the gold standard for screening? *Urol Clin North Am* 27: 25–37.

Chow, N. H., Tzai, T. S., Cheng, H. L., Chan, S. H., Lin, J. S. (1994). Urinary cytodiagnosis: can it have a different prognostic implication than a diagnostic test? *Urol Int* 53: 18–23.

Cookson, M. S., Herr, H. W., Zhang, Z. F., Soloway, S., Sogani, P. C., Fair, W. R. (1997). The treated natural history of high risk superficial bladder cancer: 15-year outcome. *J Urol* 158: 62–7.

DeBellis, C. C., Schumann, G. B. (1986). Cystoscopic biopsy supernate. A new cytologic approach for diagnosing urothelial carcinoma in situ. *Acta Cytol* 30: 356–9.

DiBonito, L., Musse, M. M., Dudine, S., Falconieri, G. (1992). Cytology of transitional-cell carcinoma of the urinary bladder: diagnostic yield and histologic basis. *Diagn Cytopathol* 8: 124–7.

Dodd, L. G., Johnston, W. W., Robertson, C. N., Layfield, L. J. (1997). Endoscopic brush cytology of the upper urinary tract. Evaluation of its efficacy and potential limitations in diagnosis. *Acta Cytol* 41: 377–84.

Esposti, P. L., Zajicek, J. (1972). Grading of transitional cell neoplasms of the urinary bladder from smears of bladder washings. A critical review of 326 tumors. *Acta Cytol* 16: 529–37.

Grossman, H. B. (1998). New methods for detection of bladder cancer. *Semin Urol Oncol* 16: 17–22.

Harrison, J. H., Bostford, T. W., Tucker, M. R. (1951). The use of the smear of the urinary sediment in the diagnosis and management of neoplasm of the kidney and bladder. *Surg Gynecol Obstet* 92: 129–139.

Huben, R. P., Gaeta, J. (1996). Pathology and its importance in evaluating outcome in patients with superficial bladder cancer. *Semin Urol Oncol* 14: 23–9.

Kannan, V., Bose, S. (1993). Low grade transitional cell carcinoma and instrument artifact. A challenge in urinary cytology. *Acta Cytol* 37: 899–902.

Kannan, V., Gupta, D. (1999). Calculus artifact. A challenge in urinary cytology. *Acta Cytol* 43: 794–800.

Kojima, S., Sekine, H., Fukui, I., Ohshima, H. (1998). Clinical significance of "cannibalism" in urinary cytology of bladder cancer. *Acta Cytol* 42: 1365–9.

Konety, B. R., Metro, M. J., Melham, M. F., Salup, R. R. (1999). Diagnostic value of voided urine and bladder barbotage cytology in detecting transitional cell carcinoma of the urinary tract. *Urol Int* 62: 26–30.

Koss, L. G., Deitch, D., Ramanathan, R., Sherman, A. B. (1985). Diagnostic value of cytology of voided urine. *Acta Cytol* 29: 810–16.

Loh, C. S., Spedding, A. V., Ashworth, M. T., Kenyon, W. E., Desmond, A. D. (1996). The value of exfoliative urine cytology in combination with flexible cystoscopy in the diagnosis of recurrent transitional cell carcinoma of the urinary bladder. *Br J Urol* 77: 655–8.

Maier, U., Simak, R., Neuhold, N. (1995). The clinical value of urinary cytology: 12 years of experience with 615 patients. *J Clin Pathol* 48: 314–17.

Malik, S. N., Murphy, W. M. (1999). Monitoring patients for bladder neoplasms: what can be expected of urinary cytology consultations in clinical practice. *Urology* 54: 62–6.

Morrison, D. A., Murphy, W. M., Ford, K. S., Soloway, M. S. (1984). Surveillance of stage O, grade I bladder cancer by cytology alone—is it acceptable? *J Urol* 132: 672–4.

Mungan, N. A., Kulacoglu, S., Basar, M., Sahin, M., Witjes, J. A. (1999). Can sensitivity of voided urinary cytology or bladder wash cytology be improved by the use of different urinary portions? *Urol Int* 62: 209–12.

Murphy, W. M., Soloway, M. S., Jukkola, A. F., Crabtree, W. N., Ford, K. S. (1984). Urinary cytology and bladder cancer. The cellular features of transitional cell neoplasms. *Cancer* 53: 1555–65.

Nielsen, K., Ostri, P. (1988). Primary tumors of the renal pelvis: evaluation of clinical and pathological features in a consecutive series of 10 years. *J Urol* 140: 19–21.

Oldbring, J., Glifberg, I., Mikulowski, P., Hellsten, S. (1989). Carcinoma of the renal pelvis and ureter following bladder carcinoma: frequency, risk factors and clinicopathological findings. *J Urol* 141: 1311–13.

Papanicolaou, G. N. (1947). Cytology of the urine sediment in neoplasms of the urinary tract. *J Urol (Balt)* 57: 375–9.

Papanicolaou, G. N., Marshall, V. F. (1945). Urine sediment smears as a diagnostic procedure in cancers of the urinary tract. *Science* 101: 519–21.

Pirtskalaishvili, G., Getzenberg, R. H., Konety, B. R. (1999). Use of urine-based markers for detection and monitoring of bladder cancer. *Tech Urol* 5: 179–84.

Prall, R. H., Wernett, C., Mims, M. M. (1972). Diagnostic cytology in urinary tract malignancy. *Cancer* 29: 1084–9.

Raab, S. S., Lenel, J. C., Cohen, M. B. (1994). Low grade transitional cell carcinoma of the bladder. Cytologic diagnosis by key features as identified by logistic regression analysis. *Cancer* 74: 1621–6.

Raab, S. S., Slagel, D. D., Jensen, C. S., Teague, M. W., Savell, V. H., Ozkutlu, D., Lenel, J. C., Cohen, M. B. (1996). Low-grade transitional cell carcinoma of the urinary bladder: application of select cytologic criteria to improve diagnostic accuracy [corrected] [published erratum appears in Mod Pathol 1996 Jul;9(7):803]. *Mod Pathol* 9: 225–32.

Raica, M., Mederle, O., Ioiart, I. (1995). Upper urinary tract washing in the diagnosis of transitional cell carcinoma of the renal pelvis and calices. A comparison with voided urine. *Rom J Morphol Embryol* 41: 101–5.

Renshaw, A. A. (2000). Compassionate conservatism in urinary cytology [editorial]. *Diagn Cytopathol* 22: 137–8.

Renshaw, A. A., Nappi, D., Weinberg, D. S. (1996). Cytology of grade 1 papillary transitional cell carcinoma. A comparison of cytologic, architectural and morphometric criteria in cystoscopically obtained urine. *Acta Cytol* 40: 676–82.

Rife, C. C., Farrow, G. M., Utz, D. C. (1979). Urine cytology of transitional cell neoplasms. *Urol Clin North Am* 6: 599–612.

Sadek, S., Soloway, M. S., Hook, S., Civantos, F. (1999). The value of upper tract cytology after transurethral resection of bladder tumor in patients with bladder transitional cell cancer. *J Urol* 161: 77–9; discussion 79–80.

Sarnacki, C. T., McCormack, L. J., Kiser, W. S., Hazard, J. B., McLaughlin, T. C., Belovich, D. M. (1971). Urinary cytology and the clinical diagnosis of urinary tract malignancy: a clinicopathologic study of 1,400 patients. *J Urol* 106: 761–4.

Schwartz, C. B., Bekirov, H., Melman, A. (1992). Urothelial tumors of upper tract following treatment of primary bladder transitional cell carcinoma. *Urology* 40: 509–11.

Sekine, H. (1989). A study of dysplasia associated with bladder cancer—histopathological findings of bladder giant sections and related urinary cytology. *Jap J Urol* 80: 545–54.

Shenoy, U. A., Colby, T. V., Schumann, G. B. (1985). Reliability of urinary cytodiagnosis in urothelial neoplasms. *Cancer* 56: 2041–5.

Trott, P. A., Edwards, L. (1973). Comparison of bladder washings and urine cytology in the diagnosis of bladder cancer. *J Urol* 110: 664–6.

Tut, V. M., Hildreth, A. J., Kumar, M., Mellon, J. K. (1998). Does voided urine cytology have biological significance? *Br J Urol* 82: 655–9.

Umiker, W. (1964). Accuracy of cytologic diagnosis of cancer of the urinary tract. *Acta Cytol* 8: 186–93.

van der Poel, H. G., Boon, M. E., van Stratum, P., Ooms, E. C., Wiener, H., Debruyne, F. M., Witjes, J. A., Schalken, J. A., Murphy, W. M. (1997). Conventional bladder wash cytology performed by four experts versus quantitative image analysis. *Mod Pathol* 10: 976–82.

Wagle, D. G., Moore, R. H., Murphy, G. P. (1974). Primary carcinoma of the renal pelvis. *Cancer* 33: 1642–8.

Zein, T., Wajsman, Z., Englander, L. S., Gamarra, M., Lopez, C., Huben, R. P., Pontes, J. E. (1984). Evaluation of bladder washings and urine cytology in the diagnosis of bladder cancer and its correlation with selected biopsies of the bladder mucosa. *J Urol* 132: 670–1.

9

BIOMARKERS AND SPECIAL TECHNIQUES FOR DETECTION OF UROTHELIAL CANCER

PERKINS MUKUNYADZI, M.D.

UROTHELIAL carcinoma is a neoplasm arising from the urothelium of the renal pelvis, the ureters, and the bladder. In the bladder, the incidence of urothelial carcinoma has been increasing steadily over the last few decades (Burchardt et al., 2000; Landis et al., 1999). Bladder cancer accounts for more than 50,000 new cases per year and kills about 12,500 patients a year in the United States (Landis et al., 1999). The majority of the patients (75%) have noninvasive urothelial carcinoma, whereas about 20% and 5% have invasive cancer and metastatic disease, respectively, at the time of initial diagnosis. Most tumors are characterized by a high rate (60%) of recurrence. Fewer tumors show a tendency to progress to more advanced disease (40% progression over 10 years) (Burchardt et al., 2000). The prognosis is still, to a large extent, dependent on the tumor stage. However, high-grade histologic or cytologic features correlate with a poor outcome. Urothelial carcinoma in situ (CIS), a high-grade, flat lesion of the urothelium, has a bad prognosis. Table 9–1 shows the relationship between the noninvasive cancer stage and grade to recurrence and disease progression.

ROLE OF URINE CYTOLOGY

Urine cytology remains the gold standard for detection of urothelial tumors and for follow-up of treated bladder cancer patients (Burchardt, 2000). The main goal is to detect the cancer before it progresses to invasive disease. However, the diagnosis of low-grade tumors is problematic because its accuracy remains low. Low-grade tumors lack sufficient cytologic malignant features to allow for easy separation from a variety of benign or reactive conditions of the bladder, such as inflammatory changes, urolithiasis, or instrumentation effects (Kannan and Bose, 1993).

The diagnostic sensitivity of urine cytology is low; the quoted range in one study was 20% to 60%. Well-differentiated tumors do not exfoliate easily, resulting in scant cellularity, especially in voided urine specimens. Bladder washings or instrumented specimens from catheterized patients can improve the cell yield, but the sensitivity remains low in low-grade tumors. The cytologic interpretation and diagnosis of low-grade lesions are heavily dependent on the experience of the pathologist and therefore vary widely. Reactive changes due to infectious and inflammatory conditions can lead to a false-positive diagnosis (1%–12%). On the other hand, it is relatively easy to make the cytologic diagnosis of high-grade cancers, including CIS, because the malignant cellular features are usually obvious (Brown, 2000).

Cystoscopy provides direct visualization of the bladder urothelium and has had an es-

Table 9–1. Relationship Between Noninvasive Urothelial Cancer Stage and Grade to Clinical Outcome

Stage	Progression	Recurrence
T_a	4%	52%
T_1	30%	77%
Grade	Progression	Recurrence
1	2%–10%	63%
2	11%–19%	67%
3	33%–45%	71%

tablished role in the diagnosis and monitoring of bladder cancer. Cystoscopy has a sensitivity of about 70%. Although it provides valuable information, its main disadvantage is that it is an invasive procedure.

Any special techniques or use of biomarkers, therefore, that seek to replace cytology or supplement urine cytology have to perform better than cytology in terms of sensitivity and specificity, cost, and ease of performance. This chapter reviews and discusses some of the biomarkers and special techniques that are currently available and that are used in some centers in the detection of bladder cancer and in follow-up of treated patients.

BIOMARKERS AND SPECIAL TECHNIQUES FOR THE DETECTION OF BLADDER CANCER

Biomarkers and other special techniques have a potential role in the improvement of screening for and detection of bladder cancer. Several biomarkers that have received Food and Drug Administration (FDA) approval are now available for monitoring of patients with bladder cancer, and newer markers and techniques continue to be developed. Good biomarkers should have high diagnostic accuracy, should be easy to use and cost-effective, and should provide prognostic information as well. The FDA approved the use of nuclear matrix (NMP22), fibrin/fibrinogen degradation products (FDP), and bladder tumor antigen (BTA) tests for monitoring bladder cancer patients. Some of the biomarkers are listed in Table 9–2 and will be discussed briefly.

Bladder Tumor Antigen Test

The first generation of BTA tests included the original BTA test (C.R. BARD Inc., Redmond, WA), a quantitative latex agglutination test performed on a urine sample (Burchardt et al., 2000; Sarosdy, 1977; Sarosdy et al., 1997). The BTA *stat* test (Sarosdy et al., 1997) and the BTA TRAK assay (Ellis et al., 1997) are newer tests.

BTA Tests

The original BTA tests were dependent on the measurement of some incompletely defined bladder tumor basement membrane complex, consisting of specific polypeptides ranging from 16 to 165 kD in molecular mass (Conn et al., 1997; Malkowicz, 2000). The membrane complex could be degraded into its constituent parts, which were measurable by use of human IgG, latex particles, and blocking agents that formed an agglutination reaction in a colorimetric assay. The result was a visual color change seen on a test strip.

The original BTA test had a sensitivity range of 34% to 70% but lower specificity. The sensitivity was higher for high-grade cancer and lower for low-grade tumors. The BTA test was positive in 65% of patients with CIS or Ta tumors. This test was, however, limited by its low specificity, with positive results seen in several nonurothelial carcinomas (86.4% in patients with renal cell and prostate carcinomas) and in those with benign urologic diseases such as urinary tract infection, lithiasis, and benign prostatic hyperplasia (Burchardt et al., 2000; Malkowicz,

Table 9–2. Biomarkers Used for Detection of Bladder Cancer

Bladder tumor antigen (BTA)
Nuclear matrix proteins
Fibrin/fibrinogen degradation products (FDP)
Hyaluronic acid and hyaluronidase
Telomerase
Cell cycle regulators
Flow cytometry, image cytometry, and fluorescence in situ hybridization
Blood group–related antigens
Others

2000). In general, the sensitivity of the BTA test is not significantly higher than that of urine cytology, especially for low-grade lesions, and its specificity is limited.

BTA *stat* TEST and BTA TRAK Assay

Changes and modifications that were intended to improve the sensitivity of the original BTA test were introduced. The result was the BTA *stat* test and the BTA TRAK (Blow Diagnostic Sciences, Redmond, WA) test; both assays are based on the human complement factor H–related protein (hCFHrp) (Burchardt et al., 2000). The hCFHrp has a functional and structural resemblance to human complement factor H and is produced by bladder cancer cells but not by other epithelial tumor cells (Heicappell et al., 1999; Thomas et al., 1999). Human complement factor H is important in the inhibition of the alternate complement pathway, thus preventing lysis of foreign cells. Therefore, the production of hCFHrp by bladder tumor cells would facilitate the survival of the tumor cells by avoiding destruction by the immune system.

The BTA *stat* test is an immunochromatographic test that is performed as a dipstick test in the clinic or the physician's office. It is a qualitative test that can provide an answer within five minutes. The BTA *stat* test has higher sensitivity (50%) for low-grade lesions than does cytology (20%–40%). However, the sensitivity of the BTA *stat* test for high-grade tumors is lower than that of cytology (for the BTA *stat* test, the range is from 29% to 83% for grade 2 to grade 3 tumors, whereas for cytology the range is 70% to 100%) (Murphy et al., 1997; Pode et al., 1999; Sanchez-Carbazo et al., 1999). The specificity range of the BTA *stat* test is 72% to 95% (thus less than that of cytology, which is above 90%). The specificity of the BTA *stat* test was also reduced in individuals with benign prostatic hypertrophy (BPH) (88%), inflammatory conditions (70%), and bladder calculi (50%), as well as those with invasive procedures or manipulation of the lower urinary tract (33%) (Pode et al., 1999; Sarosdy, 1997).

The BTA TRAK test is a quantitative assay that measures the hCFHrp and is done in reference laboratories (Heicappell et al., 1999; Thomas et al., 1999). It is performed as a quantitative sandwich immunoassay that can be done on fresh, refrigerated, or frozen urine samples. Test samples are added to microwells that contain coated antibodies, followed by incubation and washing for removal of excess reagents. A conjugate reagent enzyme is then added to the mixture and binds to the antigen-antibody complex, followed by washing and addition of the substrate. A color change is produced during the incubation phase. The amount or intensity of the color generated is proportional to the amount of the hCFHrp in the urine specimen. Absorbance in each well is read at 405 nm, and a standard graph is used to read the actual hCFHrp concentration (0–100 U/ml). The BTA TRAK test has a sensitivity superior to that of cytology for all tumor grades (Ellis et al., 1997; Leyh et al., 1997; Thomas et al., 1999). The overall sensitivity is 68%, with a range of 54% (grade 1 tumors) to 78% (grade 3 tumors). The downside of the BTA TRAK test lies in its low discriminating power: 31%–50% of patients with nonurothelial tumors, as well as 43% and about 50% of patients with urolithiasis and urinary tract infections, respectively, show elevation of the hCFHrp (Burchardt et al., 2000; Heicappell et al., 1999; Thomas et al., 1999). The test could also give a false-positive result if done within two weeks after instrumentation procedures on the urinary tract.

Long-term studies are needed to evaluate the significance of a positive BTA TRAK test in follow-up of patients with a history of bladder cancer who have negative results on a cystoscopic examination and cytology (39% of such patients have a positive BTA TRAK test). Such results could represent either false positives or a significant finding in predicting recurrence of the tumor before gross detection of disease is possible.

NUCLEAR MATRIX PROTEIN TEST

Nuclear matrix proteins (NMPs) form a structural scaffold that holds the nucleus together. Because of their unique position, the NMPs are involved in DNA replication, RNA processing and transcription, and regulation of gene expression (Berezney, 1991). Many organ- and cancer-specific NMPs have been recognized. Nuclear matrix proteins associated with specific cancers such as urothelial, colon, breast, and bone cancers have been identified (Carpinito et al., 1996).

The NMP22 test (Matritech, Inc., Newton, MA) is a quantitative test that measures a nuclear mitotic protein, NMP22, by enzyme immunoassay, using monoclonal antibodies. The mitotic protein is associated with chromatid distribution during mitosis and is elevated significantly (25-fold) in the urine of patients with urothelial cancer. It is released and shed into the urine as the tumor cells undergo apoptosis. The results are reported in units per milliliter. In several clinical studies, significant differences in the level of NMP22 between normal individuals and patients with urothelial malignancies have been found (Carpinito et al., 1996; Keesee et al., 1996). High sensitivity rates for the NMP22 test have been reported, ranging from 68% to 100%, better than those of cytology at 20% to 40% (Grocela, 2000; Soloway et al., 1996; Zippe et al., 1999).

Different specificity values of the NMP22 test have been published in different studies and appear to result from the different cutoff values selected in reading the test results. However, the specificity ranges from 61% (Serretta et al., 1998) to 85% (Zippe et al., 1999) in patients with bladder cancer, whereas it is 86% in patients with other, nonmalignant urologic diseases (Grocela, 2000). High false-positive rates are reported in benign conditions such as benign prostatic hypertrophy (15.6%) and lithiasis (50%) (Miyanaga et al., 1997). The NMP22 test appears to have a role in monitoring treated bladder cancer patients in the search for recurrent disease (Berezney, 1991; Carpinito et al., 1996; Serretta et al., 1998). Some studies have shown that patients with values higher than 10 μg/ml, and particularly those greater than 20 U/ml, after resection have high recurrence rates. The NMP22 test appears to perform better than cytology and may provide cost savings, particularly in the detection of upper urinary tract tumors that may not be easy to find by urine cytology (Grocela, 2000).

FIBRIN/FIBRINOGEN DEGRADATION PRODUCTS TEST

Fibrin/fibrinogen degradation products (FDPs) are associated with bladder cancers and can be measured by the FDP test (Accu-Dx, Intracel Corp., Rockville, MD) in urine (McCabe et al., 1984). This test is related to the production of vascular endothelial growth factor by bladder tumors, leading to an increase in vascular permeability and subsequently to leakage of plasma proteins such as plasminogen, fibrinogen, and other clotting factors. In the extravascular space, fibrinogen is converted to fibrin, which is then degraded by plasmin (from the conversion of plasminogen by the exposed lysine residues on the fibrin) to FDPs. These degradation products enter the circulation and can be detected in urine by the FDP test (McCabe et al., 1984; Schmetter et al., 1997; Tsihlias and Grossman, 2000).

Initially, an immunoassay was used to detect the FDPs, but it was time-consuming and difficult to perform. This test was replaced by a commercial test kit, which employed the hemagglutination immunoinhibition method. In this method, formalin-treated sheep red blood cells are sensitized to human fibrinogen and therefore agglutinate in the presence of antibodies to fibrinogen. To perform the test, one adds diluted (to levels just sufficient to cause agglutination) specific antifibrinogen serum to the urine samples. The presence of FDPs in the urine samples causes the antibodies to attach to the FDPs and thus inhibits the agglutination of the sheep red blood cells.

Another method, latex agglutination, uses latex particles coated with rabbit antihuman fibrinogen serum and has higher sensitivity. Serial dilutions can be used to convert this qualitative test into a quantitative method. An enzyme-linked immunosorbent assay (ELISA) using monoclonal antibodies is also available. In this test, beads are coated with the antibody and are used to capture the FDPs from urine. The beads are then incubated with horseradish peroxidase–conjugated goat antibody to human fibrinogen. The mixture is reacted with a chromogen, resulting in a colored product whose absorbance can be measured at 405 nm. This method shows a 4-fold increase in sensitivity over the hemagglutination immunoinhibition test, but the ELISA test is time-consuming and expensive.

The newest FDP test is a commercial kit that uses a monoclonal antibody and is an easy, rapid qualitative test. Its sensitivity appears to vary in different studies and has been found to range from 27% to 100%. The overall sensitivity of about 82% (63.2% for grade 1, 88.2% for grade 2, and 95.0% for grade 3 tumors) is much higher than that for cytology, which is about 38% (Johnston et al., 1997). The specificity is about 80% (Schmetter et al., 1997) in patients who are suspected to have bladder cancer, but who have negative cystoscopic findings, and 86% in patients with other nonmalignant urologic diseases (Ewing et al., 1987; Tsihlias and Grossman, 2000).

This assay needs to be used in conjunction with urine cytology, because other, nonmalignant clotting disorders or diseases may lead to a significant presence of FDPs in urine.

Hyaluronic Acid-Hyaluronidase Test

Hyaluronic acid (HA) is a nonsulfated glycosaminoglycan that is a component of the tissue matrix and tissue fluids. Hyaluronic acid is important in maintenance of tissue integrity, water homeostasis, and regulation of the osmotic balance. Hyaluronidases (HAases) are enzymes (endoglycosidases) that have the ability to degrade the hyaluronidase. Studies have shown an elevation of HAase in several tumor types, including bladder cancer (Lokeshwar et al., 1998). Metastatic tumors also produce HAases that degrade the matrix to facilitate tumor spread (Csoka et al., 1997). These substances can be detected in urine. It has been shown that HA is elevated five to seven times (Pham et al., 1997) in bladder cancers, whereas HAase is increased only in high-grade tumors, on the order of four to seven times.

The HA test is based on an ELISA-like assay, which in turn is based on the principle of competitive binding. Varying volumes of urine are added and incubated with human umbilical cord HA that is coated on microtiter wells. Biotinylated nasal cartilage HA-binding protein (HABP) is also added to the wells, and during incubation there is competition between urine HA and HA attached to the wells to combine with biotinylated HABP. After incubation, the unbound HABP is washed off and a substrate solution is added, leading to color development. Absorbance of the product is measured at 405 nm, and a standard graph is used to read the urinary HA value in milligrams per milliliter. The sensitivity and specificity of both HA and HAase have been reported to range between 86% and 92% (Lokeshwar and Block, 2000; Lokeshwar et al., 1998; Pham et al., 1997). Increases in urinary HA levels are unrelated to tumor grade because the sensitivity is essentially the same for both low- and high-grade tumors. Therefore, the ability of this test to detect low-grade tumors makes it a good noninvasive screening test that can be used as an adjunct to cytology.

In the HAase test, one measures the HAase activity in urine by incubating the urine sample with HA coated on microtiter plates (plus an appropriate buffer). The coated HA is degraded by HAase, followed by a washing step for removal of the degraded HA. Sequential incubations with biotinylated HABP are then used to measure the remaining coated HA, as described in the HA test. The HAase value is read off from standard curves, which are plotted by absorbance (control OD_{405} − sample OD_{405}) against the *Streptomyces* HAase concentration (milliunits per milliliter). The HAase test has high sensitivity for the detection of higher-grade tumors (70% and 90% for G2 and G3 tumors, respectively) compared to 27% for low-grade tumors. The overall sensitivity for all high-grade tumors is 82.5% (Pham et al., 1997).

The combined HA-HAase test can be used to yield more information than is provided by either test used separately, because the HA test detects all grades of bladder tumors, whereas the HAase test detects only high-grade tumors. A positive result on the combined test indicates a G2 or higher-grade tumor, whereas a positive HA test with a negative HAase test implies a G1 tumor. Additionally, a positive result on both the HA and HAase tests in the combined test raises its sensitivity to 92.2% without compromising its specificity. The specificity ranges from 80% to 90% (Burchardt et al., 2000). The HA-HAase test, which is noninvasive and inexpensive, is also important in the follow-up of patients with a prior diagnosis of bladder cancer.

Telomerase Test

Telomeres are short DNA sequences that are found at the terminal ends of chromosomes (Harley et al., 1999). In somatic cells, the telomeres undergo degradation as the cells age; therefore, with each replication cycle, they become progressively shorter. Loss of a critical number of telomeres eventually leads to chromosomal instability and cell death (Blackburn, 1991; Harley et al., 1990). Germ cells are able to avoid the aging process as a result of loss of the protective telomeres by producing an enzyme, telomerase, that is responsible for maintaining the telomeres (Blackburn, 1992). This enzyme has also been found in cancer cells, including urothelial tumor cells. Telomerase activity has been found in urine from patients with all grades of bladder tumors. Telomeric repeat amplification protocols are used to measure telomerase activity (Dalbagni et al., 1997; Kinoshita et al., 1997). The sensitivity of this method is on the order of 60%–70%, and the specificity is about 80% (Dalbagni et al., 1997). Nonmalignant urologic conditions such as benign prostatic hyperplasia, urolithiasis, and inflammation lead to false-positive rates of about 23.3% (Burchardt et al., 2000; Kavaler et al., 1998).

Cell Cycle Regulators

Under normal conditions, cell cycle regulatory proteins are involved in control of proliferation and progression of the cell cycle. In cancer there is uncontrolled cell growth as a result of disturbed cell cycle regulation, which is a consequence of altered genes and proteins that are important in the cell cycle. A number of alterations of proteins and genes involved in the cell cycle have been linked to the development and progression of bladder cancer (Burchardt et al., 2000).

The *p53* tumor suppressor gene plays an important role in cell cycle regulation. It either facilitates cell-cycle arrest in G1, and therefore allows for repair after DNA damage, or it promotes apoptosis. *p53* acts in concert with other genes and regulators such as the mdm2 protein, ARF, WAF1 protein, and the retinoblastoma tumor suppressor gene (*Rb*). Overexpression of mdm2 leads to degradation of p53, whereas mutations of either *p53* or mdm2 cause stabilization of *p53*. ARF prevents the degradation of *p*53 by binding mdm2. Diminished mdm2 has the opposite effect, causing increased levels of p53 and thus cell cycle arrest followed by apoptosis.

p53 alterations are present in advanced bladder cancer but may also be seen in CIS (Habuchi et al., 1994; Sarkis et al., 1993). Increased levels of *p53* can be detected by immunohistochemistry in tissues, although it is difficult to quantify them in a reproducible fashion. *p53* mutations can be detected by use of the polymerase chain reaction on urine samples in patients with bladder cancer. Similarly, *p53* mutations can be detected in urine samples by immunoreactivity methods. Immunoreactivity has a sensitivity of about 23.5% and a specificity of about 75%. Other molecular methods have also been tried but have not been standardized.

More studies are still needed to evaluate the use of *p53* alterations, because current information is conflicting and confusing. Some reports have indicated that positive immunoreactivity for p53 carries a higher risk of recurrence, an association with cancer progression, and a poor response to chemotherapy and bacillus Calmette-Guerin (BCG) treatment, but other publications show no significant benefits of p53 as a marker for disease progression (Burchardt et al., 2000; Lowe et al., 1994; Sarkis et al., 1993). Mixed results of increased immunoreactivity for p53 in patients with aggressive noninvasive bladder cancer have been reported. These patients are usually treated by instillations of BCG after transurethral resections. It appears that positive immunoreactivity for p53 does not predict which patients will fail BCG therapy. However, this is important information, as it implies that the disease in such patients is likely to progress and therefore requires other forms of therapy (Lowe et al., 1994; Sarkis et al., 1993).

Testing for *p53* alterations in low-grade bladder tumors does not significantly improve the diagnosis of these tumors over cytology, because *p53* alterations are often absent.

Flow Cytometry, Image Cytometry, and Fluorescence in Situ Hybridization

High-grade bladder tumors, including high-grade lesions in situ, demonstrate aneuploid cell populations on flow cytometry (Burchardt et al., 2000). In contrast, low-grade tumors are diploid. Thus, although flow cytometry (FCM) can be used to discriminate between high- and low-grade tumors, its perfomance does not exceed that of cytology. The sensitivity of DNA FCM has been reported as 45%, with a specificity of 87% (Burchardt et al., 2000; Koss et al., 1989; Mora et al., 1996). Flow cytometry often requires large numbers of abnormal cells to give an accurate value.

Image analysis is better for detecting higher-grade tumors than for detecting well-differentiated tumors. Parameters such as nuclear-cytoplasmic ratio, size, shape, and chromatin quality are used for separating low-grade tumor cells from normal and reactive cells. Published studies have shown image cytometry to give better results than urine cytology (Carter et al., 1987; Katz et al., 1997). Better accuracy in characterizing individual tumor cells can be achieved by the use of laser scanning cytometry, which combines FCM and image cytometry (Kamentsky and Kamentsky, 1991).

The fluorescence in situ hybridization (FISH) technique can be used to detect abnormal changes in chromosomal numbers and DNA content in individual cells. Some studies have indicated that FISH is able to distinguish between Ta and T1 tumors (Carter et al., 1987; Pycha et al., 1998). The loss of chromosome 9, detected by FISH in bladder washings of patients who were treated with BCG, is associated with a high recurrence rate of bladder cancer (Pycha et al., 1998).

Blood Group–Related Antigens

The normal urothelium of 75%–80% of the population expresses the ABH and Lewis blood group–related antigens (Le^{a}, Le^{b}, Le^{y}). Tumorigenesis is associated with altered expression of these antigens (Hakomori, 1985; Huben, 1984), such as loss of the ABH antigens or enhanced expression of the Lewis X antigen (Cordon-Cardo et al., 1988; Huber, 1984). Immunocytochemistry can be used to measure various tumor-associated, blood group–related antigens (Cordon-Cardo et al., 1998; Golijanin et al., 1995). Lewis antigen is the only blood group–related antigen that appears to have a potential clinical application. When monoclonal antibodies to this antigen are used, the sensitivity ranges from 85% to 89% (Burchardt et al., 2000; Shenfeld et al., 1990). This test, however, has poor specificity because 51% of reactive urothelial cells also express the Lewis X antigen (Halachmi et al., 1998).

Other Biomarkers

There are several other biomarkers that have potential use in detecting of urothelial cancer, but they have not been fully investigated. These include cell cycle regulatory proteins such as pRb, p15, and p21, as well as cyclin-dependent kinases. Oncogenes (*C-ERB*, *C-RAS*, etc.), proliferating antigens, Ki-67 and proliferating cell nuclear antigen (PCNA), growth factors, and cell adhesion molecules such as cadherins and integrins all have potential use in detecting and monitoring urothelial cancer. Other urothelial carcinoma–associated tumor antigens such as 486P3/12, M344, 19 A211, and tissue polypeptide antigen (TPA) have been studied, but most lack specificity (Cohen et al., 1997; Halachmi et al., 1998).

In general, some biomarkers have better sensitivity than does urine cytology. Although most are limited by low specificity, they have a potentially important role when used in combination with urine cytology. Urine cytology still remains the gold standard for the detection of urothelial cancer despite its low sensitivity for low-grade tumors.

REFERENCES

Berezney, R. (1991). The nuclear matrix: a heuristic model for investigating genomic organiztion and function in the cell nucleus. *J Cell Biochem* 47: 109–23.

Blackburn, E. (1991). Structure and function of the telomeres. *Nature* 350: 569–73.

Blakburn, E. (1992). Telomerases. *Annu Rev Biochem* 61: 113–29.

Brown, F. (2000). Urine cytology. Is it still the gold standard for screening? *Urol Clin North Am* 27: 25–37.

Burchardt, M., Burdardt, T., Shabsigh, A., de la Taille, A., Benson, M., Sawczuk, I. (2000). Current concepts in biomarker technology for bladder cancers. *Clin Chem* 46: 595–605.

Carter, H., Amberson, J., Bander, N., Bandalament, R., Gorelick, J., Vaughan, E.J., et al. (1978). New diagnostic techniques for bladder cancer. *Urol Clin North Am* 14: 763–9.

Carpinito, G., Stadler, W., Briggman, J., Chodak, G., Church, P., Lamm, D., et al. (1996). Urinary nuclear matrix protein as a marker for transitional cell carcinoma of the urinary tract. *J Urol* 156: 1280–5.

Cohen, M., Griebling, T., Ahaghotu, C., Rokhlin, O., Ross, J. (1997). Cellular adhesion molecules in urologic malignancies. *Am J Clin Pathol* 107: 56–63.

Conn, I., Crocker, J., Wallace, D., Hughes, M., Hilton, C. (1987). Basement membranes in urothelial carcinoma. *Br J Urol* 60: 536–42.

Cordon-Cardo, C., Reuter, V., Lloyd, K., Sheifeld, J., Fair, W., Old, L., et al. (1988). Blood group-related antigens in human urothelium: enhanced expression of precursor LeX and LeY determinants in urothelial carcinoma. *Cancer Res* 48: 4113–20.

Csoka, T., Frost, G., Stern, R. (1997). Hyaluronidases in tissue invasion. *Invasion Metastases* 17: 297–311.

Dalbagni, G., Han, W., Zhang, Z., Cordon-Cardo, C., Saigo, P., Fair, W., et al. (1997). Evaluation of the telomeric repeat amplification protocol (TRAP) assay for telomerase as a diagnostic modality in recurrent bladder cancer. *Clin Cancer Res* 3: 1593–8.

Ellis, W. J., Blumentein, B., Ishak, L., Enfield, D. (1997). Clinical evaluation of the BTA TRAK assay and comparison to voided urine cytology and the Bard BTA test in patients with recurrent bladder tumors. The Multi Center Study Group. *Urology* 50: 882–7.

Ewing, R., Tate, G., Hetherington, J. (1987). Urinary fibrin/fibrinogen degradation products in transitional cell carcinoma of the bladder. *Br J Urol* 59: 53.

Golijanin, D., Sherman, Y., Shapiro, A., Pode, D. (1995). Detection of bladder tumors by immunostaining of the Lewis X antigen in cells from voided urine. *Urology* 46: 173–7.

Grocela, J. A., McDougal, W. S. (2000). Utility of nuclear matrix protein (NMP22) in the detection of recurrent bladder cancer. *Urol Clin North Am* 27: 47–51.

Habuchi, T., Kinoshita, H., Yamada, H., Kakehi, Y., Ogawa, O., Wu, W., et al. (1994). Oncogene amplification in urothelial cancers with p53 gene mutation or MDM2 amplification. *J Natl Cancer Inst* 86: 1331–5.

Hakomori, S. (1985). Aberrant glycosylation in cancer cell membranes as focused on glycolipids: overview and perspectives. *Cancer Res* 45: 2405–14.

Halachmi, S., Linn, J., Amiel, G., Moskovitz, B., Nativ, O. (1998). Urine cytology, tumor markers and bladder cancer. *Br J Urol* 82: 647–54.

Harley, C., Futcher, A., Greider, C. (1990). Telomeres shorten during aging of human fibroblasts. *Nature* 345: 458–60.

Heicappell, R., Wettig, I., Schostak, M., Muller, M., Steiner, U., Sauter, T., et al. (1999). Quantitative detection of human complement factor H-related protein in transtional cell carcinoma of the urinary bladder. *Eur Urol* 35: 81–7.

Huben, R. (1984). Tumor markers in bladder cancer. *Urol* 23(3Suppl): 10–14.

Johnston, B., Morales, A., Emerson, L., Lundie, M. (1997). Rapid detection of bladder cancer: a comparative study of point of care tests. *J Urol* 158: 2098–2101.

Kamentsky, L., Kamentsky, L. (1991). Microscope-based multiparameter laser scanning cytometer yielding data comparable to flow cytometry data. *Cytometry* 12: 381–7.

Kannan, V., Bose, S. (1993). Low grade transitional cell carcinoma and instrument artifact: a challenge in urinary cytology. *Acta Cytol* 37: 899–902.

Katz, R., Sinkre, P., Zhang, H., Kidd, L., Johnston, D. (1997). Clinical significance of negative and equivocal urinary bladder cytology alone and in combination with DNA image analysis and cystoscopy. *Cancer Cytopathol* 81: 354–64.

Kavaler, E., Landman, J., Chang, Y., Droller, M., Liu, B. (1998). Detecting human bladder carcinoma cells in voided urine samples by assaying for the presence of telomerase activity. *Cancer* 82: 708–14.

Keesee, S., Briggman, J., Thill, G., et al. (1996). Utilization of nuclear matrix proteins for cancer diagnosis. *Crit Rev Eukaryot Gen Expr* 6: 189–214.

Kinoshita, H., Ogawa, O., Kakehi, Y., Mishina, M., Mitsumori, K., Itoh, N., et al. (1997). Detection of telomerase activity in exfoliated cells in the urine of patients with bladder cancer. *J Natl Cancer Inst* 89: 724–30.

Koss, L., Czerniak, B., Herz, F., Wersto, R. (1989). Flow cytometric measurements of DNA and other cell components in human tumors: a critical appraisal. *Human Pathol* 20: 528–48.

Landis, S., Murray, T., Bolden, S., Wingo, P. (1999). Cancer statistics, 1999. *CA. Cancer J Clin* 49: 8–31.

Leyh, S., Hall, R., Mazeman, E., et al. (1997). Comparison of the Bard BTA test with voided urine and bladder wash cytology in the diagnosis and management of cancer of the bladder. *Urology* 50: 49–53.

Lokeshwar, V., Block, N. (2000). HA-HAase urine test. A sensitive and specific method for detecting bladder cancer and evaluating its grade. *Urol Clin North Am* 27: 53–61.

Lokeshwar, V., Soloway, M., Block, N. (1998). Secretion of bladder tumor-derived hyaluronidase activity by invasive bladder tumor cells. *Cancer Lett* 131: 21–7.

Lowe, S., Bodis, S., McClatchey, A., Remington, L., Ruley, H., Fisher, D., et al. (1994). p53 status and the efficay of cancer therapy in vivo. *Science* 266: 807–10.

Malkowicz, S. (2000). The application of human compliment factor H-related protein (BTA TRAK) in monitoring patients with bladder cancer. *Urol Clin North Am* 27: 63–73.

McCabe, R., Lamm, D., Haspel, M., Pomato, N., Smith, K., Thompson, E., et al. (1984). A diagnostic-prognostic test for bladder cancer using monoclonal antibody-based enzyme-linked immunoassay for detection of urinary fibrin(nogen) degradation products. *Cancer* 44: 5886–93.

Miyanaga, N., Akaza, H., Ishikawa, S., Ohtani, M., Noguchi, R., Kawai, K., et al. (1997). Clinical evaluation of nuclear matrix protein 22 in (NMP22) in urine as a novel marker of urothelial cancer. *Eur Urol* 31: 163–8.

Mora, L., Nicosia, S., Pow-Sang, J., Ku, N., Diaz, J., Lockhart, J., et al. (1996). Ancillary techniques in the followup of transitional cell carcinoma: a comparison of cytology, histology, and deoxyribonucleic acid image analysis cytometry in 91 patients. *J Urol* 156: 49–54.

Murphy, W., Rivera-Ramirez, I., Meina, C., et al. (1997). The bladder tumor antigen (BTA) test compared to urine cytology in the detection of bladder neoplasms. *J Urol* 158: 2102–6.

Pham, H., Block, N., Lokeshwar, V. (1997). Tumor-derived hayluronidase: a diagnostic urine marker for high-grade bladder cancer. *Cancer Res* 57: 778–83.

Pode, D., Shapiro, A., Wald, M., Nativ, O., Laufer, M., Kaver, I. (1999). Noninvasive detection of bladder cancer with the BTA stat test. *J Urol* 161: 443–6.

Pycha, A., Mian, C., Hofbauer, J., Haitel, A., Wiener, H., Marberger, M. (1998). Does topical instillation therapy influence chromosomal aberrations in superficial bladder cancer? *J Urol* 159: 265–9.

Sanchez-Carbayo, M., Herrero, E., Megias, J., Mira, A., Espasa, A., Chinchilla, V., et al. (1999). Initial evaluation of the diagnostic performance of the new urinary bladder cancer antigen test as a tumor marker for transitional cell carcinoma of the bladder. *J Urol* 161: 1110–15.

Sarkis, A., Dalbagni, G., Cordon-Cardo, C., Zhang, Z., Sheinfeld, J., Fair, W., et al. (1993). Nuclear overexpression of p53 protein in transitional cell bladder carcinoma: a marker of disease progression. *J Natl Cancer Inst* 85: 53–9.

Sarosdy, M. (1997). The use of the BTA test in the detection of persistent or recurrent transitional cell carcinoma of the bladder. *World J Urol* 15: 103–6.

Sarosdy, M., Hudson, M., Ellis, W., Soloway, M., de Vere White, R., Sheinfeld, J., et al. (1997). Improved detection of recurrent bladder cancer using the bard BTA stat test. *Urology* 50: 349–53.

Schmetter, B., Habicht, K., Lamm, D. L., Morales, A., Bander, N., Grossman, H., et al. (1997). A multicenter trial evaluation of the fibrin/fibrinogen degradation products test for detection and monitoring of bladder cancer. *J Urol* 158: 801–5.

Serretta, V., Lo Presti, D., Vasile, P., Gange, E., Esposito, E., Menozzi, I. (1998). Urinary NMP22 for the detection of recurrence after transurethral resection of transitional cell carcinoma of the bladder: experience on 137 patients. *Urology* 52: 793–6.

Sheinfeld, J., Reuter, V., Melamed, M., Fair, W., Morse, M., Sogani, P., et al. (1990). Enhanced bladder cancer detection with Lewis X antigen as a marker of neoplastic transformation. *J Urol* 143: 285–8.

Soloway, M., Briggman, V., Carpinito, G., Chodak, G., Church, P., Lamm, D., et al. (1996). Use of a new tumor marker, urinary NMP22, in the detection of occult or rapidly recurring transitional cell carcinoma of the urinary tract following surgical treatment. *J Urol* 156: 363–7.

Thomas, L., Leyh, H., Marberger, M., Bombardieri, E., Bassi, P., Pagano, F., et al. (1999). Multicenter trial of the quantitative BTA TRAK assay in the detection of bladder cancer. *Clin Chem* 45: 472–7.

Tsihlias, J., Grossman, B. (2000). The utility of fibrin/fibrinogen degradation products in superficial bladder cancer. *Urol Clin North Am* 27: 39–46.

Zippe, C., Pandrangi, L., Agarwal, A. (1999). NMP22 is a sensitive, cost-effective test in patients at risk for bladder cancer. *J Urol* 161: 62–5.

10

FINE-NEEDLE ASPIRATION BIOPSY OF URINARY TRACT NEOPLASMS

FINE-NEEDLE aspiration biopsy (FNAB) is safe, cost effective, and accurate, and, in many instances it eliminates the need for tissue biopsy. A full review of FNAB of cysts and nonurothelial malignancies of the kidney, extensively covered in review articles, specialized textbooks, and cytology treatises, is beyond the scope of this book. Fine-needle aspiration biopsy of the prostate, once commonly used for diagnosing adenocarcinoma, is not widely used except in some Scandinavian countries and will not be discussed in this chapter.

This chapter covers FNAB of primary and secondary masses of the urinary tract. These lesions are located mainly in the area of the renal pelvis and proximal ureter and are sampled under radiologic guidance, such as by ultrasound or computed tomography (CT). Common and uncommon lesions sampled by FNAB are listed in Table 10–1.

Primary epithelial tumors of the renal pelvis, particularly urothelial carcinoma, comprise most of the cases and are covered in detail, followed by secondary neoplasms that are mentioned briefly. Fine-needle aspiration biopsy of metastatic urothelial carcinoma and a brief review of FNAB in the evaluation of renal allografts are included at the end of this chapter.

RADIOLOGICALLY GUIDED FINE-NEEDLE ASPIRATION BIOPSY

Most renal cell carcinomas produce an expansile radiographic effect, as opposed to the infiltrative effect typical of invasive urothelial carcinoma, renal lymphoma, squamous cell carcinoma of the renal pelvis, and metastases to the kidney (Hartman et al., 1988). This parameter, which is important for radiologic diagnosis, usually refines the clinical diagnosis but does not replace the pathologic diagnosis. Percutaneous FNAB under ultrasound, fluoroscopic, or CT guidance provides a preoperative diagnosis and often the staging of these tumors.

This clinical, radiologic, and pathologic approach, which is widely accepted as the standard of care for patients with deep-seated masses, helps to provide a well-planned radical surgical procedure or another form of therapy. However, many urologists and radiologists who are afraid of needle tract tumor seeding believe that kidney masses should not be sampled prior to surgical exploration. Many centers perform FNAB of renal pelvis masses only when the possibility of lymphoma or metastatic disease is suspected or in patients with high risk for surgery.

Procedure. The radiologist performs FNAB in the radiology suite, using a 20 to 22-gauge spinal-type needle, 8 to 20 cm long, attached to a 10 or 20 cc disposable plastic syringe. Usually, one or two passes are required to obtain sufficient diagnostic material. Additional passes may be required to obtain material for special studies such as electron microscopy, immunocytochemistry, ploidy analysis, flow cytometry, and cytogenetics.

Table 10–1. Renal Pelvic Masses Amenable to Sampling by Fine Needle Aspiration Biopsy

Primary neoplasms
Carcinoma: papillary, squamous, glandular, sarcomatoid, neuroendocrine.
Lymphoid: lymphoma and plasmacytoma.
Sarcoma: leiomyosarcoma, liposarcoma.
Miscellaneous: malignant neoplasms with rhabdoid features, melanoma, carcinoid tumor, amyloidoma.
Secondary neoplasms: metastasis to the renal pelvis.

Cytologic interpretation. Immediate cytologic interpretation is made by the use of a Diff-Quik stain. The objectives are to determine whether the specimen is adequate, and to obtain a preliminary diagnosis and frequently a definitive diagnosis. Diagnostic accuracy is based largely on the cytopathologist's experience and on adequate sampling.

Complications. Life-threatening complications are rare. Perirenal hemorrhage, pneumothorax, arteriovenous fistula, urinoma, and infections have been reported, but transient gross and microscopic hematuria is the most common complication (Lang, 1977; Nguyen, 1987). Fine-needle aspiration biopsy does not produce significant tissue damage, nor does it preclude further histologic interpretation when tumor resection is needed. An occasional case of recurrent urothelial carcinoma attributed to needle tract seeding of tumor eight months after FNAB performed with a 22-gauge needle has been reported (Slywotzky and Maya, 1994).

PRIMARY NEOPLASMS OF THE RENAL PELVIS

Malignancies of the renal pelvis are 14 times less common than those of the kidney (Zajicek, 1979). Malignant epithelial neoplasms of the renal pelvis constitute approximately 5% of all urinary tract tumors. Papillary urothelial carcinoma is by far the most common (92%), followed by squamous carcinoma (6%) and adenocarcinoma (1%) (Zajicek, 1979). Metastatic deposits are less common.

Clinical manifestations include gross hematuria and, less commonly, pain and hydronephrosis. A renal pelvis filling defect is usually seen in the intravenous pyelogram. In some institutions, a case clinically and radiographically suspicious for renal pelvis carcinoma with negative urine cytology is evaluated by means of percutaneous FNAB. However, this approach is not widely used (Zajicek, 1979).

Epithelial Neoplasms

Papillary Urothelial Carcinoma

Papillary urothelial carcinoma is most common in male adults after the sixth decade, although pediatric cases have also been reported. It is associated with tobacco and phenacetin abuse, exposure to industrial carcinogens, and cyclophosphamide and radiation therapy, particularly in the presence of a genetic predisposition (Brenner and Schellhammer, 1987; Bringuier et al., 1998; Rohde and Jakse, 1998). There is a marked increase in risk associated with Balkan nephropathy, an unusual form of chronic renal failure of unknown cause common along the river Danube (Bonsib and Eble, 1997). Individuals with kidney or ureter stones appear to be at increased risk of developing renal pelvis/ureter or bladder cancer but not renal cell carcinoma (Chow et al., 1997).

A direct relationship between phenacetin consumption dose and the degree of papillary scarring, a higher histologic grade, and tumor progression has been shown in patients with urothelial tumors of the renal pelvis (Stewart et al., 1999). A high incidence of renal pelvis carcinoma in renal transplant patients (15%) seems to be related to the underlying cause of renal failure, especially analgesic nephropathy, which is most likely related to prior analgesic abuse and not to immunosuppressive therapy (Penn, 1995). In nontransplant cases, the karyotypic profile supports the presence of similar pathogenetic mechanisms in urothelial carcinomas of both the renal pelvis and bladder (Fadl-Elmula et al., 1999).

These tumors may reach a large size, with extensive necrosis, venous invasion, calcification, lymphadenopathy, and spontaneous hemorrhage (Bree et al., 1990). They are similar histologically to tumors of the bladder. Papillary neoplasms of low malignant potential (WHO grade 1) and low-grade papillary carcinomas (WHO grade 2) have a prominent papillary pattern. In contrast, high-grade papillary carcinomas (WHO grade 3) are poorly differentiated, with squamous, glandular, and occasionally sarcomatoid components. The tumor arises from the surface urothelium, although an occasional tumor arising in a preexisting pyelocaliceal cyst has been described (Friedman et al., 1999).

Approximately 60% of carcinomas of the renal pelvis can be diagnosed by urine cytology (Grabstald et al., 1971). However, higher accuracy can be obtained by means of endocopic retrograde tumor brushings or FNAB of large renal pelvis tumors.

Cytology. Fine needle aspiration biopsy diagnosis of papillary urothelial carcinomas of the renal pelvis can be made with accuracy. Smears from adequately sampled tumors are highly cellular. The cells are arranged in thick clusters, sheets, and papillary structures, particularly in papillary neoplasms of low malignant potential and low-grade papillary carcinomas. Variable numbers of small aggregates and single cells are seen in such cases. A marked lack of cohesiveness, as well as cell pleomorphism and variable necrosis, are seen in high-grade papillary carcinomas (Table 10–2).

Nguyen and Schumann reported FNAB characteristics in five cases of papillary neoplasms of low malignant potential. Benign-appearing urothelial cells that have a low nuclear-cytoplasmic ratio, dense and homogeneous cytoplasm, well-defined cytoplasmic borders, oval and eccentric nuclei, powdery chromatin, smooth nuclear contours, and inconspicuous nucleoli comprised the predominant cell population. Many cells exhibited a cytoplasmic extension or "tail," and others had a more columnar shape and slightly pleomorphic nuclei (Nguyen and Schumann, 1997). Zajicek stated that these cytoplasmic elongations are characteristic of urothelial tumors (Fig. 10–1A–C) (Zajicek, 1979).

Santamaria et al. reported on the FNAB findings in one papillary neoplasm of low malignant potential and three low-grade papillary carcinomas. The latter show architectural features similar to those present in papillary neoplasms of low malignant potential; however, their lack of cohesiveness is more evident, and they show slight nuclear enlargement, nuclear hyperchromasia with granular chromatin, and moderate pleomorphism. Cells with a cytoplasmic tail are also present (Fig. 10–2A,B) (Santamaria et al., 1995).

The architectural pattern and cell characteristics present in FNAB of the above-mentioned tumors are lost in high-grade papillary carcinomas (Fig. 10–3A–D). Lack of cellular cohesiveness predominates, and the smear background may be necrotic. Marked variation in the size and shape of cells with a high nuclear-cytoplasmic ratio, coarse chromatin, irregular nuclear contours, and prominent nucleoli make the distinction from other malignancies difficult. It is not unusual to find evidence of squamous differentiation and keratinization (Fig. 10–4A–C).

Table 10–2. Features of Papillary Urothelial Carcinoma on Fine Needle Aspiration Biopsy

High cellularity
Papillary structures, sheets, clusters*
Variable lack of cell cohesiveness
Urothelial cells with cytoplasmic "tails"*
Pleomorphic clearly malignant cells†
Squamous, glandular, sarcomatoid features†

*Predominantly in papillary urothelial neoplasms of low malignant potential (WHO, grade 1) and low-grade papillary urothelial carcinomas (WHO grade 2).

†Predominantly in high-grade papillary urothelial carcinomas (WHO grade 3).

Differential diagnosis. The diagnosis of renal pelvis papillary carcinoma is not refutable in most cases. However, on occasion, primary renal tumors with papillary architecture and poorly differentiated metastatic malignancies need to be considered in the differential diagnosis of high-grade papillary carcinoma. Close clinical–radiologic correlation is important for diagnostic accuracy.

When the FNAB pattern shows lack of cellular cohesion and evidence of a high-grade malignancy, high-grade urothelial carcinoma, lymphoma, and metastasis must be considered in the differential diagnosis. An exhaustive search for cell clusters reveals the

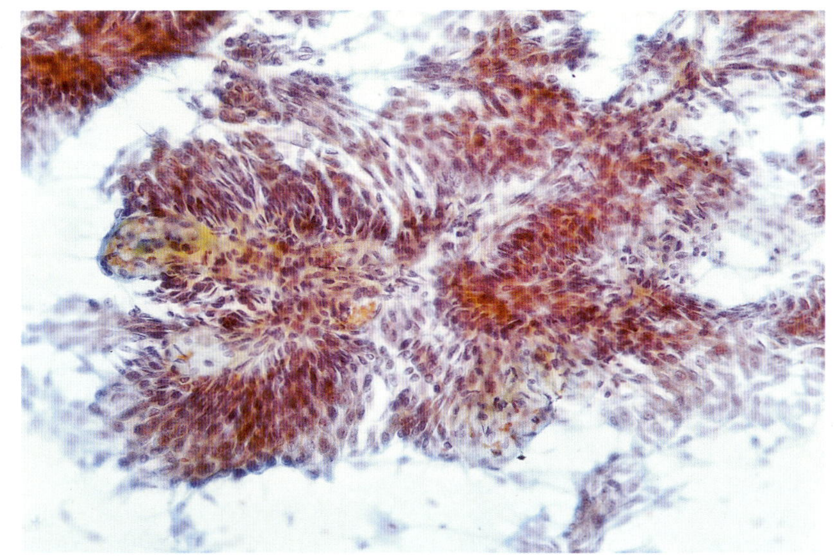
A

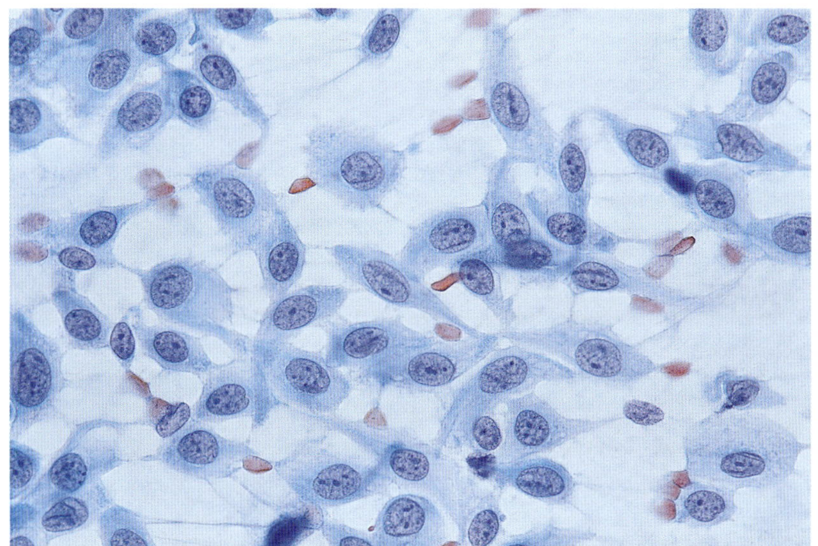
B

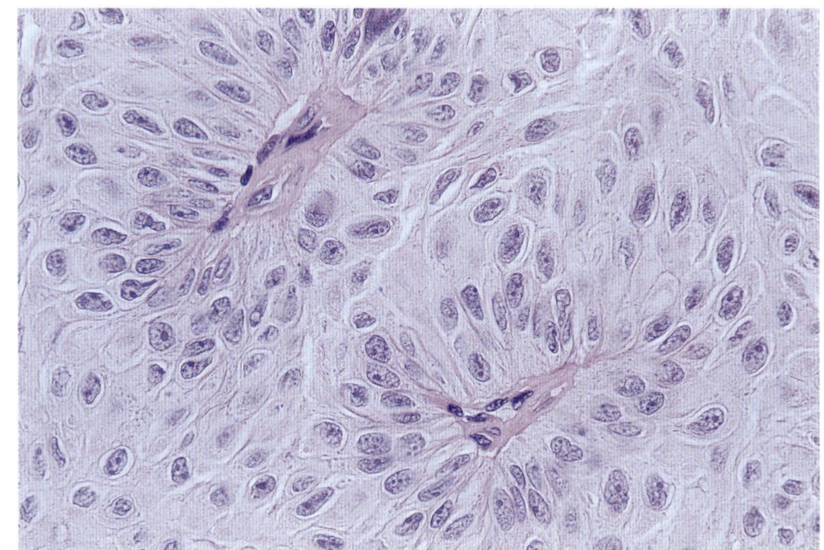
C

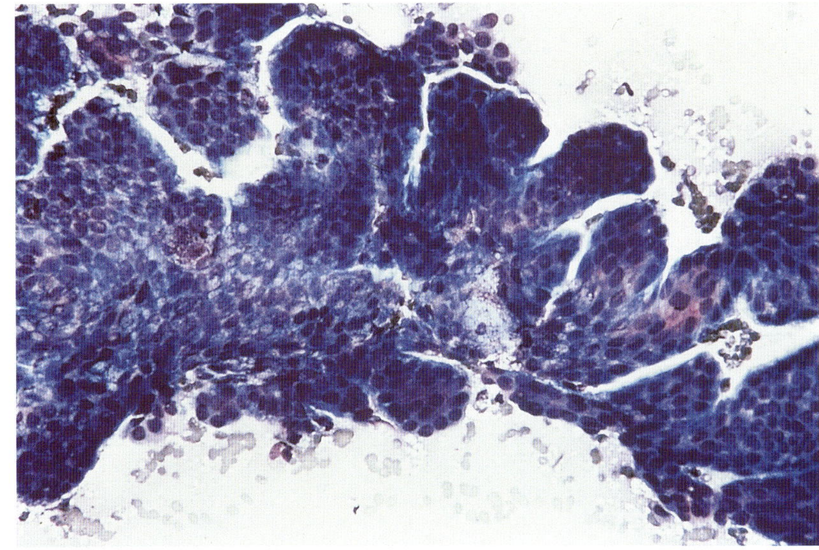

A

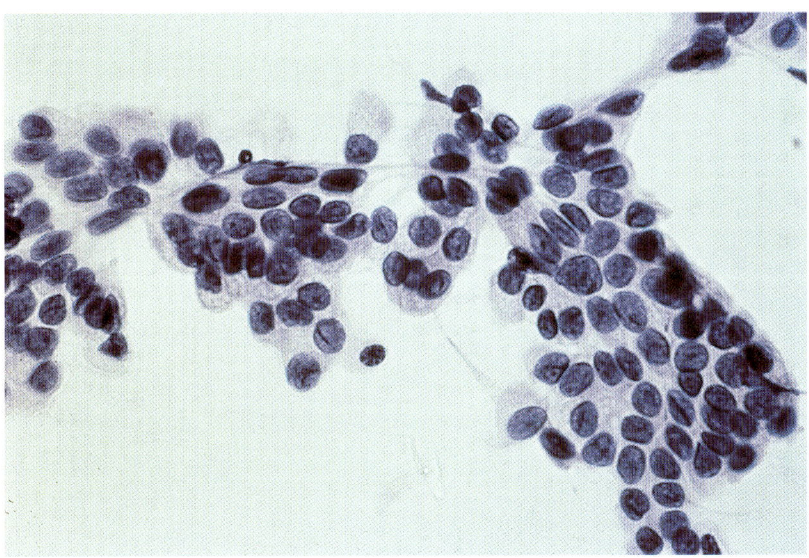

B

Figure 10–2. *Fine-needle aspiration biopsy of low-grade papillary urothelial carcinoma (grade 2 papillary urothelial carcinoma).* (*A,B*) Tri-dimensional papillary clusters and cohesive sheets of cuboidal neoplastic cells with round/oval hyperchromatic nuclei, uniform granular chromatin, smooth nuclear contours, and nuclear folds are present. Cells with cytoplasmic elongations were identified in other fields of the smears. (Diff-Quik stain, (A) ×200; Papanicolau stain, (B) ×400). Courtesy of Dr. Michael W. Stanley, Chair of Pathology, Hennepin County Medical Center, Minneapolis, MN.

Figure 10–1. *Fine-needle aspiration biopsy of papillary urothelial neoplasm of low malignant potential (grade 1 papillary urothelial carcinoma).* (*A*) Smears prepared from a 2 cm tumor of the proximal ureter show prominent papillary fragments supported by fibrovascular cores. (*B*) Cells are loosely cohesive and exhibit delicate cytoplasmic elongations and an eccentric, round/oval, bland, uniform nucleus. (*C*) No mitosis, necrosis, or pleomorphism is present in the tissue sections. (Papanicolaou stain, (A) ×200, (B) ×600; hematoxylin and eosin stain, (C) ×600)

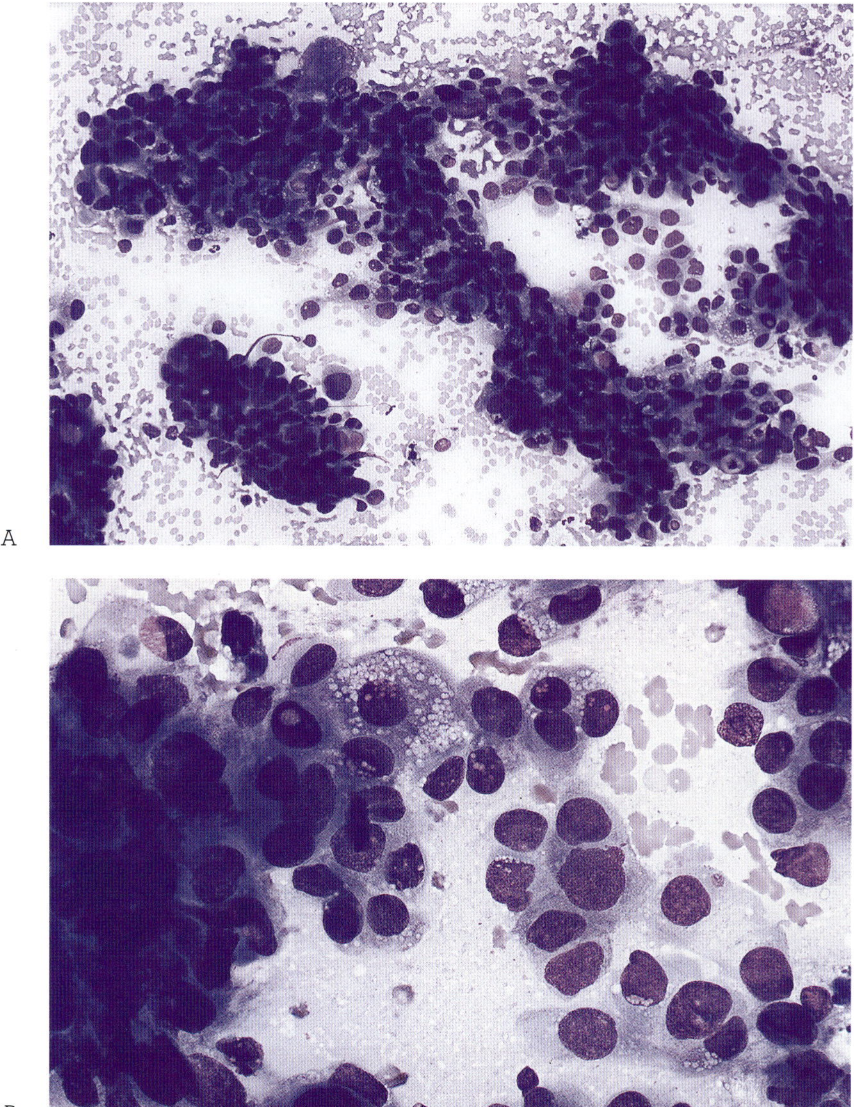

Figure 10–3. *Fine-needle aspiration biopsy of high-grade papillary urothelial carcinoma (grade 3 papillary urothelial carcinoma).* (*A,B*) Clusters and poorly formed papillary aggregates exhibiting large, moderately pleomorphic cells with dense, homogeneous cytoplasm and a large, irregular nucleus are present in the aspirate smears of a large renal pelvis tumor. (*C*) Cellular degeneration with cytoplasmic vacuolization is present, resembling renal cell clear cell carcinoma. (*D*) Large pleomorphic cells and necrotic debris are present in the cell block preparation. (Diff-Quik stain, (A) ×200, (B,C) ×600; hematoxylin and eosin stain, (C) ×400)

epithelial nature of high-grade urothelial carcinoma, and the presence of papillary structures supports this diagnosis. A history of a primary malignancy in a distant organ and clinical evidence of tumor in other body sites strongly suggest metastasis. In questionable cases, the use of immunocytochemistry for workup of a poorly differentiated large-cell neoplasm is helpful. The presence of squamous differentiation, which is frequent in high-grade urothelial carcinoma, is not indicative of squamous cell carcinoma.

When papillary architecture predominates, primary renal neoplasms should be in-

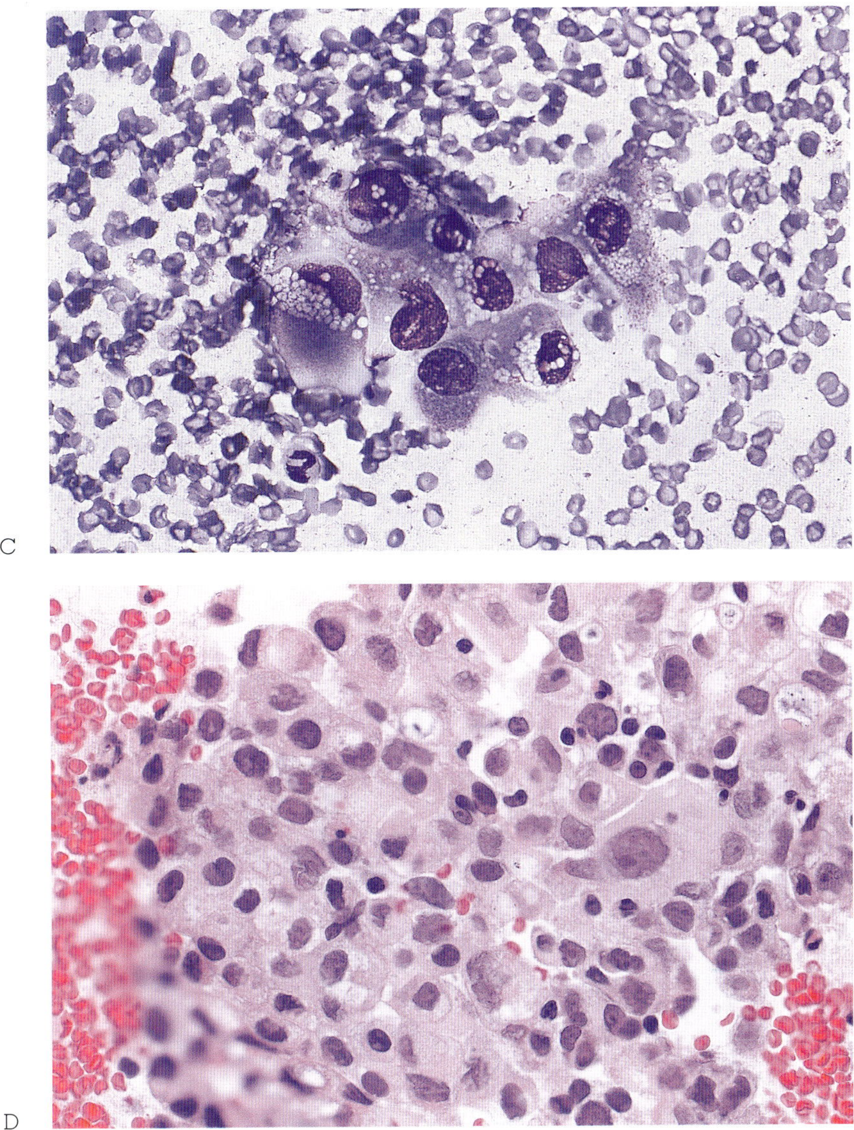

Figure 10–3. (*Continued*)

cluded in the differential diagnosis. Renal cell carcinoma of the clear-cell type with papillary features can be differentiated from renal pelvis urothelial carcinoma by its abundant cytoplasm with variable vacuolization, coarse chromatin, and a large, eosinophilic nucleolus. Papillary renal cell carcinoma (chromophil tumors), with its characteristic cytogenetic profile, shows papillary fronds covered by large columnar to cuboidal cells with granular, dense, eosinophilic or basophilic cytoplasm, a high nuclear-cytoplasmic ratio, irregular nuclear contours, granular chromatin, and variable nucleoli. Fine-needle aspiration biopsy of collecting duct carcinoma, an aggressive malignancy occurring in a young age group, shows papillary fragments and cell groups with vacuolated cytoplasm and considerable nuclear atypia (Bardales, 1998; Caraway et al., 1995; Dekmezian et al., 1991).

In summary, the degree of cohesiveness, papillary arrangement, cytoplasmic characteristics, nuclear-cytoplasmic ratio, and nuclear features are important in the cytologic differential diagnosis.

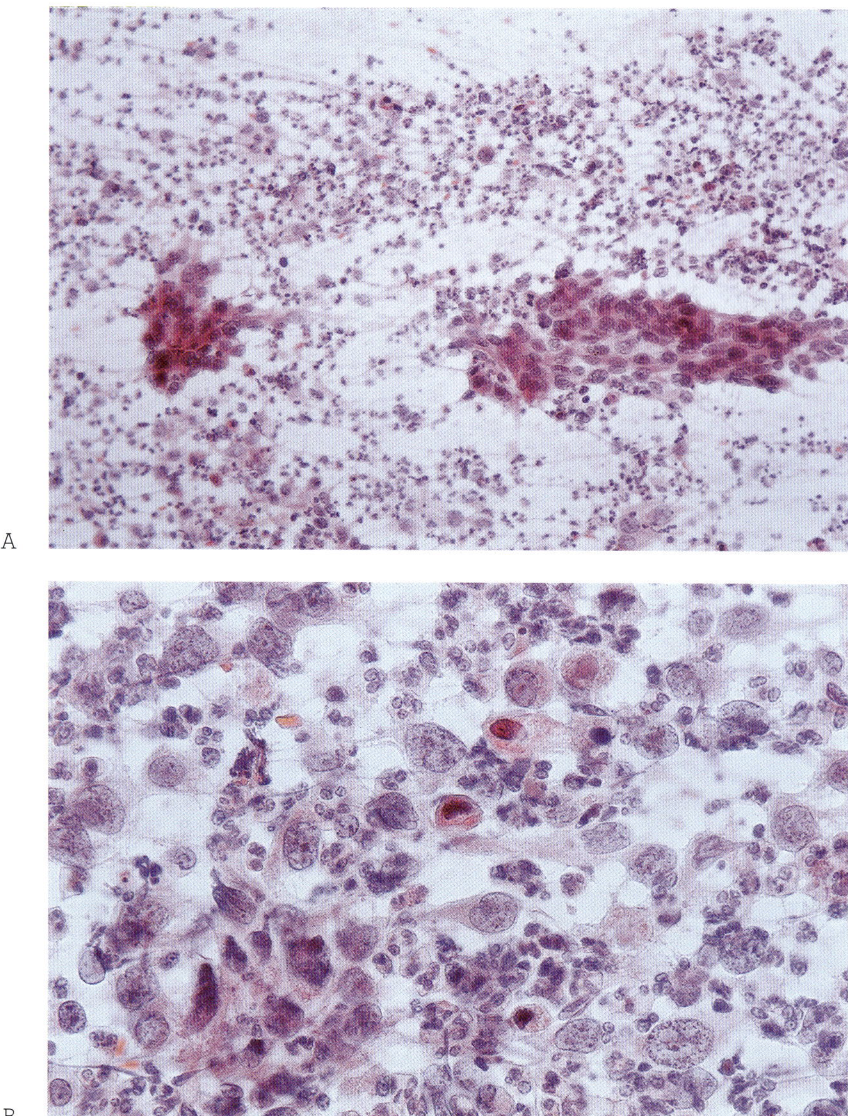

Figure 10–4. *Fine-needle aspiration biopsy of high-grade urothelial carcinoma with squamous features.* (*A*) Poorly differentiated urothelial carcinoma showing lack of papillary architecture, extensive necrosis, and cell pleomorphism. (*B*) Cells with squamous features, large irregular nuclei, and coarse chromatin are noted. (*C*) Tissue section shows numerous mitoses and nests of poorly differentiated carcinoma. (Papanicolaou stain, (A) ×100, (B) ×600; hematoxylin and eosin stain, (C) ×400)

Squamous Cell Carcinoma

Squamous cell carcinoma is associated with renal calculi, chronic inflammation, and keratinizing squamous metaplasia or leukoplakia. Association with the last characteristic in the absence of renal calculi or inflammation is debatable, although squamous metaplasia is present in the mucosa adjacent to the squamous carcinoma in one-third of cases (Bonsib and Eble, 1997; Melamed and Reuter, 1993).

Squamous cell carcinoma is nonpapillary and infiltrating, and usually is of high grade and high stage, with extensive renal involvement at the time of diagnosis (Melamed and Reuter, 1993). The histopathology is similar to that of squamous cell carcinoma of the bladder. Keratinization and necrosis are common.

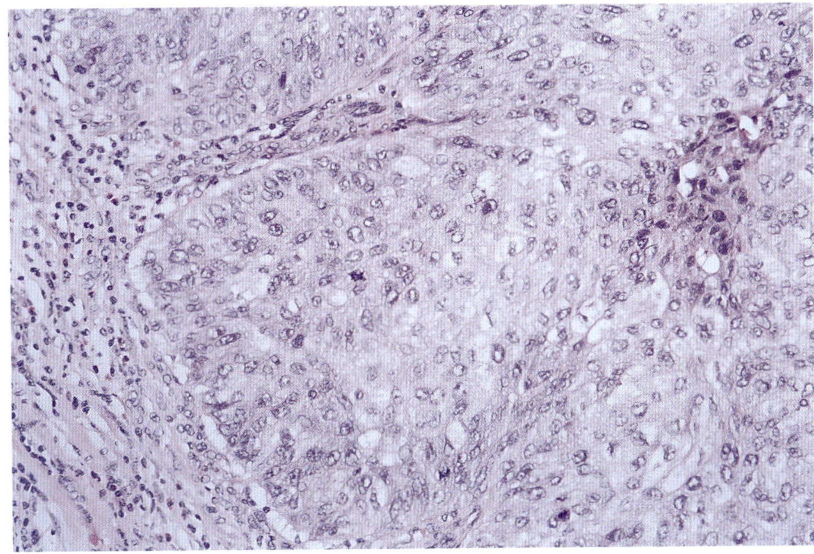
C

Figure 10–4. (*Continued*)

One should reserve the term *squamous cell carcinoma* for those cases in which, after meticulous histologic examination, the squamous elements are found to comprise 90% or more of the tumor (Box 10–1).

Cytology. Fine needle aspiration biopsy of squamous carcinoma from the renal pelvis does not differ from samples obtained from other sources. Cells are pleomorphic, with dense, orangeophilic cytoplasm and well-defined cytoplasmic borders. The nuclei are hyperchromatic and may be pyknotic, with irregular contours and coarse chromatin. The presence of large, irregular, eosinophilic nucleoli is variable. Keratinization is evident and extensive, with a necrotic background.

> **Box 10–1.**
>
> Squamous cell carcinoma of the renal pelvis is associated with chronic inflammation and keratinizing squamous metaplasia or leukoplakia.
>
> Smears show necrosis and keratinization. Cells are pleomorphic, with frank nuclear anaplasia.
>
> The diagnosis of squamous cell carcinoma is made histologically after careful sampling of the resected specimen.

Differential diagnosis. Squamous differentiation is not unusual in high-grade urothelial carcinoma. Thus, a diagnosis of squamous cell carcinoma should be reserved for cases in which the diagnostic criteria are fulfilled and extensive keratinization is present. Differentiation from abscess is based on careful evaluation of the background and cellular detail, particularly when the sample exhibits extensive necrosis and scant cellularity. The results of FNAB of primary and metastatic squamous cell carcinoma are indistinguishable, and clinical–radiologic correlation is necessary for differentiating between the two.

Adenocarcinoma

Adenocarcinoma is a neoplasm of high grade and high stage at the time of diagnosis. As in squamous cell carcinoma, lithiasis, infection, and inflammation with glandular metaplasia appear to be predisposing conditions (Melamed and Reuter, 1993; Rao et al., 1996).

Adenocarcinoma of the renal pelvis occurs equally often in men and women. Affected patients are most commonly in the fifth decade of life but have a broad age

> **Box 10–2.**
>
> Infection, lithiasis, and chronic inflammation with glandular metaplasia seem to be predisposing factors for the development of adenocarcinoma of the renal pelvis.
>
> Histologically, renal pelvis adenocarcinoma can be of the mucinous, tubulovillous, or papillary type.
>
> Fine needle aspiration biopsy smears show malignant cells with ill-defined cytoplasmic borders, clear or vacuolated cytoplasm, vesicular chromatin, and conspicuous nucleoli in a background of necrosis.

range, from 12 to 87 years (Spires et al., 1993). Histologically, the majority of adenocarcinomas exhibit an intestinal pattern, either tubulovillous or mucinous, but a small percentage are papillary and nonintestinal (Bonsib and Eble, 1997; Melamed and Reuter, 1993).

Mucinous intestinal-type tumors occur in older patients, are associated with infection, hydronephrosis, or calculous disease, and have a better prognosis than do their morphologically similar tubulovillous counterparts. Tubulovillous tumors are associated with a bad prognosis (Spires et al., 1993).

The histologic association of malignant squamous and glandular epithelium, such as in adenosquamous carcinoma, suggests glandular and squamous metaplasia of the urothelium with subsequent adenosquamous neoplasia (Howat et al., 1983) (Box 10–2).

Cytology. Fine needle aspiration biopsy findings of adenocarcinoma of the renal pelvis do not differ from those at other sites, and as in cases of squamous carcinoma, the diagnosis is made only when metastasis has been excluded (Suen, 1994a). Numerous single and clustered mucus-secreting malignant cells of variable shape showing anaplastic nuclei are present in the FNAB smears of intestinal-type adenocarcinomas; glandular fragments composed of cells of variable columnarity may be identified. Smears of nonintestinal-type tumors are highly cellular and show cells with ill-defined cytoplasmic borders, thin or vacuolated cytoplasm, nuclei of variable pleomorphism, vesicular chromatin, and conspicuous nucleoli. A background of necrosis is invariably present in the smear background (Nguyen, 1987). Malignant mucus-secreting glandular and squamous cells have been described in association with a case of recurrent mixed adenosquamous carcinoma of the renal pelvis (Wahl, 1985).

Small-Cell Carcinoma

In the urinary tract, primary small-cell carcinoma, a locally aggressive and rapidly fatal neoplasm of elderly people, occurs most commonly in the bladder. It is exceedingly rare in the renal pelvis. Heavy smoking and radiation therapy have been suggested as predisposing factors (Guillou et al., 1993; Mazzucchelli et al., 1995).

Association with synchronous or metachronous urothelial carcinoma of the ipsi- or contralateral renal pelvis may be present, as well as squamous, glandular, or sarcomatous components (Weiss, 1993).

The morphologic spectrum of the tumor is similar to that seen in neuroendocrine tumors of the lung. Neuroendocrine differentiation is supported by the presence of neurosecretory granules at the ultrastructural level and by a positive immunoreaction to chromogranin A, argyrophilic (Grimelius) silver stain, neuron-specific enolase, and synaptophysin (Essenfeld et al., 1990; Guillou et al., 1993; Kitamura et al., 1997) (Box 10–3).

Cytology. Fine needle aspiration biopsy smears of small-cell carcinoma of the renal pelvis show aggregates of small malignant cells with scant cytoplasm, a high nuclear-cytoplasmic ratio, and hyperchromatic nu-

> **Box 10–3.**
>
> Fine needle aspiration biopsy cytology of small-cell carcinoma of the renal pelvis is similar to that of the lung.
>
> Smears show numerous clusters and single small cells with prominent nuclear molding, variable degrees of "crushing" artifact, and necrosis.

clei. Prominent cellular molding, markedly damaged single cells, and a background exhibiting nuclear debris, "crushing" artifact, and apoptotic bodies are features similar to those of small-cell carcinoma of the lung (Nguyen, 1987).

Sarcomatoid Carcinoma and Carcinosarcoma

As with adenocarcinoma and squamous cell carcinoma, the term *sarcomatoid carcinoma* should be reserved for those cases in which the sarcomatoid pattern comprises 90% or more of the tumor (Reuter, 1993).

Sarcomatoid carcinomas are rare tumors composed of mixed epithelial and sarcomatoid elements. The sarcomatoid component of these tumors commonly expresses cytokeratin and EMA positivity (Piscioli et al., 1984). In equivocal cases, ultrastructural studies showing true desmosomes and cytoplasmic intermediate filaments suggestive of keratin are conclusive in showing the epithelial lineage (Reuter, 1993).

Similar to sarcomatoid carcinoma, carcinosarcoma is also a high-grade, high-stage malignancy that occurs more frequently in the urinary bladder than in the renal pelvis (Chen et al., 1983; Orsatti et al., 1993; Rao et al., 1986). Its malignant mesenchymal component may be chondrosarcomatous, osteosarcomatous, or myosarcomatous (Reuter, 1993). No specific etiologic association has been found for these tumors.

The FNAB cytomorphology of similar tumors in other organs has been described (de la Torre and Larsson, 1995; Finley et al., 1988; Gorczyca et al., 1992). The author is not aware of FNAB findings for these tumors in the renal pelvis.

Undifferentiated Carcinoma

Primary undifferentiated (lymphoepithelial-like) carcinoma rarely occurs outside the nasopharynx. In the urinary tract, the bladder and ureter are affected more than the renal pelvis. Two cases have been reported in the latter organ (Cohen et al., 1999; Fukunaga and Ushigome, 1998).

The histomorphology of undifferentiated carcinoma is similar to that of its nasopharyngeal counterpart, including nests and syncytial sheets of round/elongated, large, pale epithelial cells surrounded by nonneoplastic lymphohistiocytic stroma. The epithelial cells in one of the reported cases were cytokeratin 7+, cytokeratin 20+, EMA+, and vimentin−. The lymphocytes were predominantly UCHL-1+ T cells. No Epstein-Barr viral genomic sequences were detected by in situ hybridization, and the tumor had an aneuploid DNA content (Fukunaga and Ushigome, 1998).

This tumor should be distinguished from primary and metastatic malignant neoplasms with spindle cell features, including sarcoma, melanoma, sarcomatoid carcinoma, and inflammatory pseudosarcomatous lesions. The use of a panel of immunostains and of electron microscopy is helpful.

Fine needle aspiration biopsy cytomorphology has not been reported.

Lymphoid Neoplasms

Lymphomas of the urinary tract are rare. Those affecting the upper urinary tract manifest as multiple nodules infiltrating each kidney, a single large renal mass, diffuse infiltration of both kidneys, or a lesion of the renal hilus encasing the renal pelvis and proximal ureter (Rubin, 1979).

Lymphomas of the urinary tract should be suspected in patients with systemic lymphoma and immunosuppression (Begara Morillas et al., 1996; Maeda et al., 1986) (Box 10–4).

Renal plasmacytoma, a rare extramedullary tumoral form of plasma cell myeloma, may be difficult to distinguish preoperatively from renal cell carcinoma,

Box 10–4.

Malignant lymphomas of the urinary tract are usually associated with systemic disease.

Fine needle aspiration biopsy smears are highly cellular and show "lymphoglandular bodies," with variable numbers of naked nuclei in the background. Cellular features are similar to those seen in FNAB of lymph nodes involved with malignant lymphoma.

Immunocytochemistry may be necessary to exclude a metastatic, poorly differentiated adenocarcinoma or amelanotic melanoma.

urothelial carcinoma, and other renal pelvis tumors (Igel et al., 1991; Silver et al., 1977).

Fine needle aspiration biopsy is a useful diagnostic approach, because the management and surgical approach are different in lymphoid and nonlymphoid neoplasms. Patients having a high surgical risk are likely to benefit from FNAB, which provides a diagnosis and material for special studies including cytogenetics, flow cytometry, proliferation markers, immunocytochemistry, and research protocols.

Cytology. Fine needle aspiration biopsy cytomorphology of these tumors is similar to that in lymph nodes and is best evaluated in Romanovsky-stained smears.

A

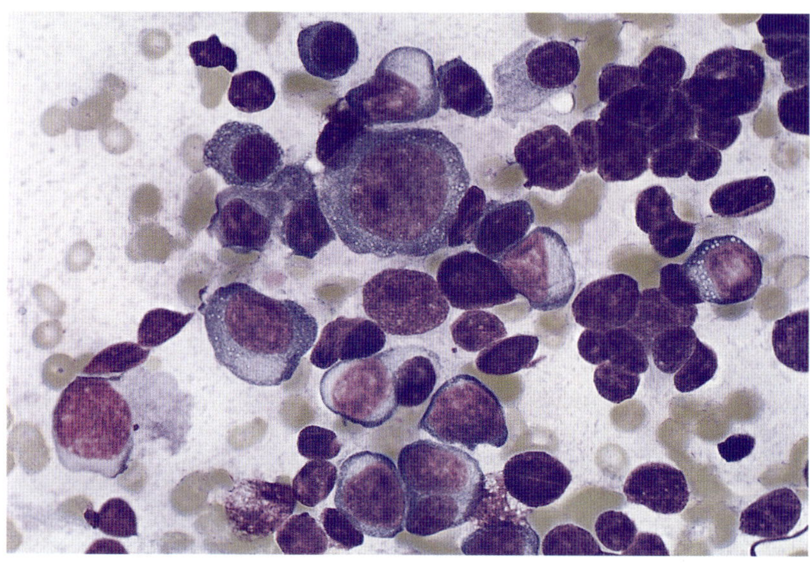

B

Figure 10–5. *Fine-needle aspiration biopsy of plasma cell myeloma involving the renal pelvis.* (*A*) Large cells with eccentric nuclei and ill-defined nuclear clearing are features that facilitate the diagnosis of plasma cell myeloma. (*B*) Nuclear pleomorphism and prominent nucleoli are features that indicate immunoblastic transformation. These patients had a history of plasma cell myeloma and had hematuria and a retroperitoneal mass involving the kidney and renal pelvis. (Diff-Quik stain, (A) ×1000; Papanicolaou stain, (B) ×1000)

Briefly, smears show a monomorphic population of dishesive malignant lymphoid cells. In large-cell lymphoma, smears may show cells with poorly preserved cytoplasm that result in numerous "lymphoglandular bodies" and free-lying nuclei in the smear background. Well-preserved cells show irregular nuclear contours, coarse chromatin, and nuceloli of variable size that are occasionally located peripherally toward the nuclear membrane. Prominent nuclear clefts are present in follicle center cell malignant lymphomas. Abnormal plasma cells with varying degrees of dedifferentiation are present in the smears from patients with plasma cell myeloma (Fig.10–5A,B).

Differential diagnosis. In most cases, malignant lymphoma and plasmacytoma can be accurately diagnosed and distinguished from renal cell carcinoma and metastatic malignancies by means of FNAB (Murphy et al., 1985). Occasionally, metastatic amelanotic melanoma may exhibit cytomorphology almost identical to that of large-cell malignant lymphoma or plasma cell myeloma. The use of immunocytochemistry for cytokeratin, CD45, HMB-45, S100, B- and T-cell markers, and light chain monoclonality may be necessary in selected cases.

Mesenchymal Neoplasms

Sarcomas of the renal pelvis are exceedingly rare. Their clinical picture and radiologic characteristics are similar to those of epithelial neoplasms and include hematuria, flank pain, and a renal mass.

The histologic characteristics of mesenchymal neoplasms are similar to those in other body sites. Leiomyosarcomas are the most common, although other types such as liposarcoma, malignant fibrous histiocytoma, rhabdomyosarcoma, and extragonadal germ-cell tumor exist (Melamed and Reuter, 1993; Vahlensieck et al., 1991). When they attain a large size, it is difficult to determine whether they are primary tumors from the renal pelvis or retroperitoneal involving the renal pelvis (Box 10–5).

Box 10–5.

Sarcomas of the renal pelvis are rare and may be the result of secondary involvement from a primary retroperitoneal sarcoma.

Leiomyosarcoma shows spindle cells with ill-defined cytoplasm and elongated nucei with blunt ends.

Myxoid liposarcoma shows spindle cells embedded in a myxoid stroma containing scattered lipoblasts and supported by a delicate capillary network.

Cytology

Fine needle aspiration biopsy features of soft-tissue tumors of the renal pelvis are similar to those of tumors found in the retroperitoneum (Suen, 1994b). The FNAB diagnosis of a mesenchymal neoplasm is followed by surgical resection for histologic classification in most cases (Zajicek, 1979).

Fine needle aspiration biopsy of smooth muscle neoplasms shows spindle cells with scant, ill-defined cytoplasm, nuclei with blunt ends, and granular chromatin. The degree of cellular cohesion, cellular atypia, and nuclear pleomorphism correlates with the histologic grade of the tumor. Pleomorphism, mitoses, and necrosis indicate high-grade leiomyosarcoma. However, in some instances, low-grade leiomyosarcomas cannot be distinguished from leiomyomas on FNAB cytology (Fig. 10–6A,B) (Tao and Davidson, 1993).

Low-grade myxoid liposarcoma exhibits spindle, oval, and round cells admixed in a myxoid stroma supported by a delicate capillary network. The diagnostic lipoblasts are scattered, show variable degrees of nuclear atypia, and contain cytoplasmic vacuoles that indent the nuclear membrane. It must be emphasized that the myxoid stroma can be seen in other spindle cell neoplasms, that is, peripheral nerve sheath tumor, myxoma, and malignant fibrous histiocytoma (Suen, 1994b).

In high-grade sarcomas, immunocytochemical stains performed in the cell block material can be helpful in the diagnosis, particularly when preoperative radiation therapy is planned (Chow et al., 1994).

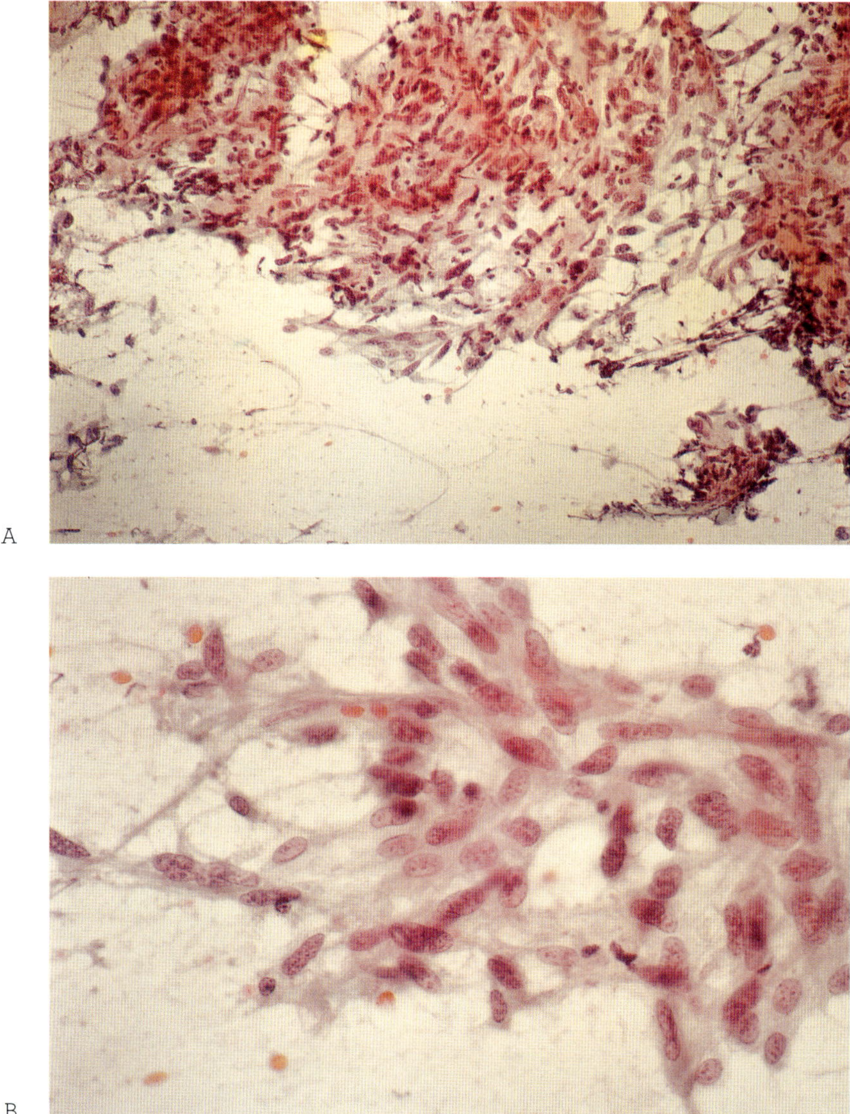

Figure 10–6. *Fine-needle aspiration biopsy of leiomyosarcoma of the renal pelvis.* (*A*) Aspirate from a leiomyosarcoma of the renal pelvis shows loosely cohesive spindle cells embedded in a fibrillary stroma. (*B*) Nuclei are elongated and mildly pleomorphic and have blunt ends. Necrosis is present. (Papanicolaou stain, (A) ×100, (B) ×600)

Differential diagnosis. The differential diagnosis includes other primary and metastatic spindle cell neoplasms of the renal pelvis and kidney, including sarcomatoid carcinoma, melanoma, sarcoma, and carcinosarcoma. Judicious use of immunostains and electron microscopy is helpful in the diagnosis. HMB-45, a melanoma marker, may be positive in the cells of some smooth-muscle tumors, particularly those that appear to arise from the renal capsule (Bonsib, 1996).

MISCELLANEOUS NEOPLASMS

Malignant Neoplasms with Rhabdoid Features

Malignant rhabdoid tumor is a distinct, highly malignant renal neoplasm in chil-

Box 10–6.

Some of the miscellaneous neoplasms that may be sampled by FNAB include malignant rhabdoid tumors, collision tumors, primary malignant melanoma, carcinoid tumor, and amyloidoma.

Fine needle aspiration biopsy of only few of these tumors has been described.

dren. Tumors with similar histomorphology in extrarenal sites and in adults have been described (Parham et al., 1994). Few cases have been reported in the renal pelvis of adults, but some are observed in association with papillary urothelial carcinoma, mucin-producing papillary adenocarcinoma, and glandular metaplasia (Kumar et al., 1992; Shintaku et al., 2000; Yusuf et al., 1996). Speculation remains as to whether these tumors of the renal pelvis in adults represent unusual morphologic variants of a urothelial carcinoma or a distinct clinicopathologic entity, as in childhood renal rhabdoid tumors. These tumors most likely are examples of primary renal pelvis malignancies with rhabdoid features (Kumar et al., 1992; Parham et al., 1994) (Box 10–6).

Cytology. Fine needle aspiration biopsy in a single case report (Yusuf et al., 1996) of recurrent urothelial carcinoma with rhabdoid features of the kidney in an adult showed discohesive, large to medium-sized polygonal cells with eccentric nuclei, prominent nucleoli, abundant pink cytoplasm, and paranuclear eosinophilic density. The last corresponds to the cytoplasmic eosinophilic globules seen in the histologic sections and cytoplasmic filamentous inclusions seen ultrastructurally.

Differential Diagnosis. Accurate elimination of metastatic, systemic, or local reactive processes will prevent unnecessary radical surgery. Large-cell malignant neoplasms such as melanoma, immunoblastic lymphoma, anaplastic large-cell lymphoma, sarcoma, and adenocarcinoma need to be excluded with the use of immunostains and accurate clinical–radiologic correlation. Similarly, extra-adrenal paraganglioma, ganglioneuroma, and proliferative myositis (a benign reactive process) should be ruled out based on the same parameters. Electron microscopy may be helpful in selected cases.

Collision Tumor

Histologic features of a collision tumor (papillary urothelial carcinoma and peripheral primitive neuroectodermal tumor) of the renal pelvis have been described (Phan et al., 1997), but no FNAB findings have been reported.

Malignant Melanoma

Rare cases of malignant melanoma primary to the renal pelvis have been described. One of these occurred in a three-year old girl (Ehara et al., 1997; Frasier et al., 1988). There are no reports on the FNAB of these tumors, however, most likely, FNAB findings would be identical to those of metastatic melanoma in other organs. A definitive diagnosis is reached only after a metastatic deposit from a skin or ocular primary is excluded.

Carcinoid Tumor

The cytomorphologic features of a large carcinoid tumor of the renal pelvis in fresh voided urine have been described (Ji and Li, 1994; Rudrick et al., 1995). No description of FNAB cytomorphology is available, but it is expected to be similar to that of other organs.

Amyloidoma

Localized amyloidosis (amyloid tumor) of the genitourinary tract is rare. When it occurs in the renal pelvis with hematuria and flank pain or in the urinary bladder, the clinical exclusion of malignancy is difficult (Agarwal et al., 1995; Murphy et al., 1986). Localized amyloidosis in the renal pelvis may be associated with amyloidosis of the bladder and ureter (Dias et al., 1979; Fox et al., 1984; Sparwasser et al., 1991).

Characteristic histologic findings consist of deposits of amorphous eosinophilic material that produce an apple-green birefringence with Congo red stain. Electron microscopy showing aggregates of linear, nonbranching fibrils supports the diagnosis. Calcification of the renal pelvis wall may be seen.

Cytology. No cytologic description of amyloid tumor of the renal pelvis has been reported. However, the cytomorphology is expected to be identical to that of similar tumors in the lung. On Diff-Quik stain, aspirates are acellular, showing clumps of amorphous, irregular, and waxy basophilic to metachromatic material mimicking "puffy clouds." Papanicolaou staining shows "brittle" eosinophilic or cyanophilic clumps of amorphous material with occasional prominent fissures.

As in amyloid tumors of the lung, FNAB is useful in distinguishing these tumors from other primary or metastatic malignancies of the renal pelvis (Halliday et al., 1998).

SECONDARY NEOPLASMS OF THE RENAL PELVIS

Metastases to the renal pelvis and ureter are exceedingly rare, and they may be of any histologic type. Lymphomas and carcinomas, particularly from the lung, breast, and gastrointestinal tract, are the most common (Melamed and Reuter, 1993). Metastases from contralateral renal cell carcinoma have been reported as well (Lindell et al., 1991; Takashi et al., 1993).

The therapeutic approach to patients with disease metastatic to the renal pelvis or ureter is different from that of patients with primary tumors of this region. Thus, FNAB offers advantages in diagnosing these tumors. Most patients with metastases to the renal pelvis have a history of or a synchronous primary tumor. Thus, clinical correlation and comparison with the original malignancy, when available, is helpful.

Fine needle aspiration biopsy of such cases has not been reported, although they are expected to have features similar to those of other metastatic tumors.

METASTATIC UROTHELIAL CARCINOMA

Tumor stage and grade, localization, and residual tumor after surgery are the most important prognostic parameters for evaluating patients who develop recurrent and/or metastatic urothelial carcinoma (Ozsahin et al., 1999). Patients younger than 60 years and those with cancer of the renal pelvis have distant metastases more often than older patients with bladder cancer (Sengelov et al., 1996).

It has been found that patients with recurrent invasive urothelial carcinoma of the bladder are likely to have metastases elsewhere, and that patterns of recurrence and metastasis are not dependent on the features of the primary tumor, for example, the histology, grade, and location (Sengelov et al., 1996; Wallmeroth et al., 1999). It appears that the frequency of metastases from invasive bladder cancer is slightly higher in patients treated with cystectomy, strongly suggesting the lack of a full understanding of the natural history of the disease (Wallmeroth et al., 1999).

Approximately two-thirds of patients with muscle-invasive bladder carcinoma, and 88% of those with renal pelvis or ureter carcinoma, have metastases to regional lymph nodes (Saitoh et al., 1982; Wallmeroth et al., 1999). The liver, lung, bone, peritoneum, pleura, kidney, adrenal gland, and intestine are also affected in order of decreasing frequency (Wallmeroth et al., 1999). Metastases to the lungs, peritoneum, and ipsilateral adrenal are more frequent in tumors of the renal pelvis than in those of the ureter (Saitoh et al., 1982).

Metastases to the oral cavity, skin, eyes, colon, penis, brain, and small bones from urothelial carcinomas of the bladder and renal pelvis have been reported (Atta, 1983; Bucin et al., 1982; Chitale et al., 1997; Girgis et al., 1999; Guo et al., 1998; Hayes et al., 1992; Kiff and Hulton, 1983; Skoog et al., 1998; Weithman et al., 1988). Exceptionally, a solitary metastatic deposit is the first clinical evidence of disseminated urothelial carcinoma (Saito, 1998).

Fine needle aspiration biopsy has proved invaluable in the diagnosis of metastatic urothelial carcinoma. It is inexpensive and safe, and can have a high accuracy in diagnosing local tumor spread and regional or distant metastases (Wajsman et al., 1982b).

Cytology. Fine needle aspiration biopsy cytology of urothelial carcinoma metastatic to different body sites has been described sporadically in small series and case reports (Epstein, 1977; Hayes et al., 1992; Weithman et al., 1988).

Fine needle aspiration biopsy of metastatic papillary urothelial carcinoma of low nuclear grade shows high cellularity, with numerous single and clustered cells. Occasional mitoses may be observed. However, characteristically, all cases show "cercariform cells" (Table 10–3). This cell type, although not pathognomonic, provides a significant clue to the urothelial origin of the metastatic deposit (Powers and Elbadawi, 1995). Similar cells with a cytoplasmic tail have been described in FNAB of primary renal pelvis papillary urothelial neoplasms of low malignant potential (WHO, grade 1) and low- and high-grade papillary urothelial carcinomas (WHO, grades 2 and 3) (Fig. 10–7A,B) (Nguyen and Schumann, 1997; Santamaria et al., 1995; Zajicek, 1979).

The "cercariform cell" has distinct characteristics that include a nucleated globular body and a cytoplasmic process with a nontapering, flattened, and bulbous or fishtail-like end. Cercariform cells correspond to intermediate urothelial cells in histologic and ultrastructural preparations of normal urothelium (Table 10–4) (Powers and Elbadawi, 1995). However, cells with shorter, broader tails and a flattened end appear to be more common (Renshaw and Madge, 1997). Similar urothelial cells, which are spindled, pyramidal, and/or racquet-shaped, with eccentric nuclei and cytoplasmic features of both squamous and glandular differentiation, have been described in metastases from high-grade urothelial carcinomas (Figs. 10–8A–C) (Johnson and Kini, 1993).

Table 10–3. Fine Needle Aspiration Biopsy Features of Metastatic Papillary Urothelial Carcinoma

Cellular smears
Clusters and occasional papillary structures*
Cercariform cells*
Spindle, pyramidal, or racket-shaped cells†
Moderate to marked pleomorphism†
Squamous and glandular features†

*More numerous in papillary neoplasms of low malignant potential (WHO grade 1) and low-grade papillary carcinomas (WHO grade 2).

†Predominantly in high-grade papillary carcinomas (WHO grade 3).

Data from two series of metastatic papillary carcinoma show that 14 of 25 (56%) cases had more than ten cercariform cells per case (Hida and Gupta, 1999; Renshaw and Madge, 1997). Malignant cells with squamous features can be seen admixed with cercariform cells. However, metastatic high-grade urothelial carcinoma with extensive squamous differentiation lacks cercariform cells (Renshaw and Madge, 1997). Thus, it appears that cells with cercariform morphology are more evident in metastases from papillary urothelial carcinoma of low nuclear grade, and less numerous and less apparent in metastases from urothelial carcinoma of high nuclear grade.

Fine needle aspiration biopsy of urothelial carcinoma of high nuclear grade also shows cells with moderate to marked pleomorphism and a higher degree of incohesiveness. Individual cells are large and obviously malignant, with dense cytoplasm and well-defined cytoplasmic borders, hyperchromatic nuclei, and prominent nucleoli. Cytoplasmic vacuolization and squamous features may be present (Fig. 10–9A–C). Mucoid material in cytoplasmic vacuoles has been described in metastatic, poorly differentiated urothelial carcinoma. The mucoid inclusion is identical to that present in cells of adenocarcinoma and usually parallels the grade of malignancy. Therefore, the presence of mucin-positive cells in a patient with a history of high-grade urothelial carcinoma does not exclude the possibility of a metastatic urothelial carcinoma. This may be a potential source of misdiagnosis (Guo et al., 1998).

A clinical history of primary invasive urothelial carcinoma is almost always present and is very useful in the diagnosis of metastatic disease. However, in the absence of a known primary, the above-mentioned FNAB cytology features may be strongly suggestive of metastatic urothelial carcinoma.

Metastases from squamous carcinoma are similar to those of lung or skin origin. Likewise, metastatic adenocarcinoma of urinary tract origin are identical to those of gastrointestinal or breast origin (Fig. 10–10A,B). Comparison with tissue from the primary site is necessary, along with clinical correlation.

Metastatic small-cell carcinoma from the urinary tract cannot be differentiated from

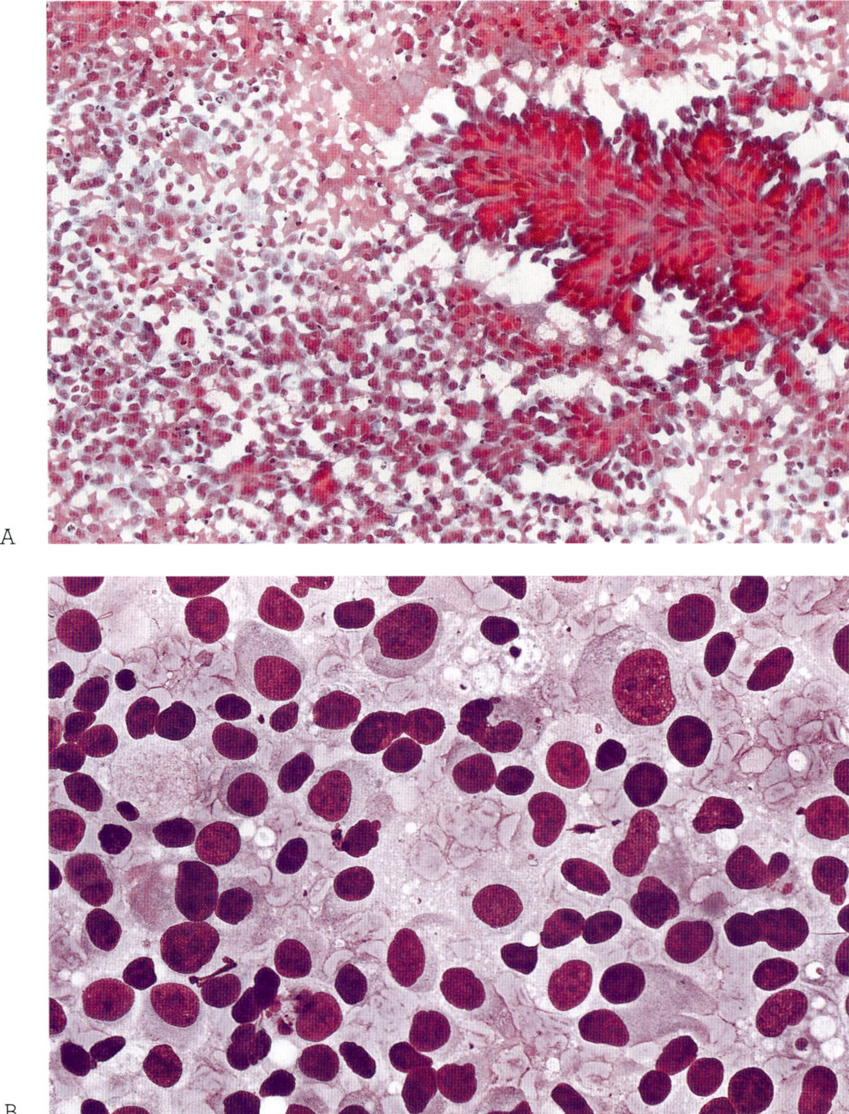

Figure 10–7. *Fine-needle aspiration biopsy of metastatic low-grade papillary urothelial carcinoma.* (*A*) A papillary cluster, necrosis, and numerous cells are present in the aspirate smears of a lung metastasis from a urothelial carcinoma of the bladder. (*B*) Mild to moderate pleomorphic cells with homogeneous cytoplasm and occasional cytoplasmic elongations are identified. (Papanicolaou stain, (A) ×100; Diff-Quik stain, (B) ×600)

Table 10–4. Characteristics of Cercariform Cells*

Nucleated globular body
Variable nuclear hyperchromasia but no pyknosis
Unipolar, nontapering cytoplasmic process
Flat, bulbous, fish-tail-like cytoplasmic end

*Absent in cases of carcinoma with extensive squamous differentiation.

Figure 10–8. *Cercariform cells in FNAB of metastatic low-grade papillary urothelial carcinoma.* (*A–C*) Aspirate smears from a subcutaneous mass metastastic from a bladder urothelial carcinoma show numerous cells with cytoplasmic processes conferring a spindled, pyramidal, globoid, and cercariform shape. The presence of these cells strongly suggests a urothelial origin for the metastasis. (Diff-Quik stain, (A) ×100, (B) ×600; Papanicolaou stain, (C) ×600)

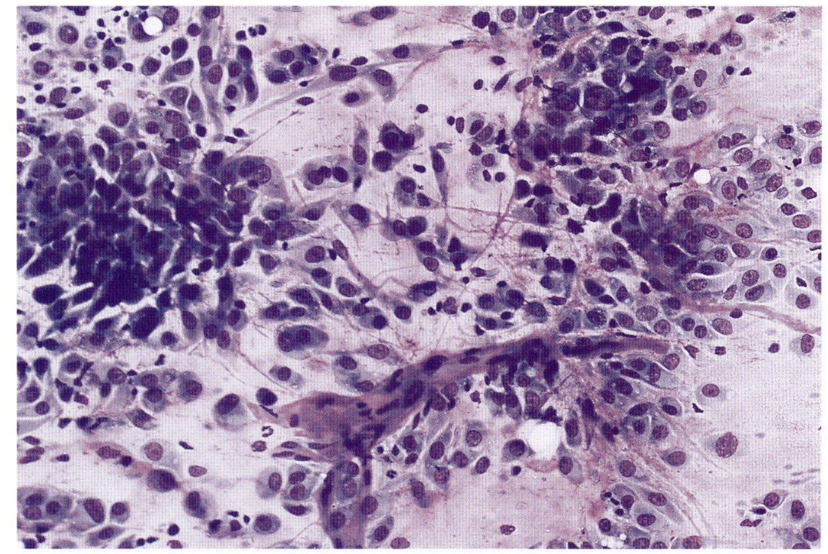
A

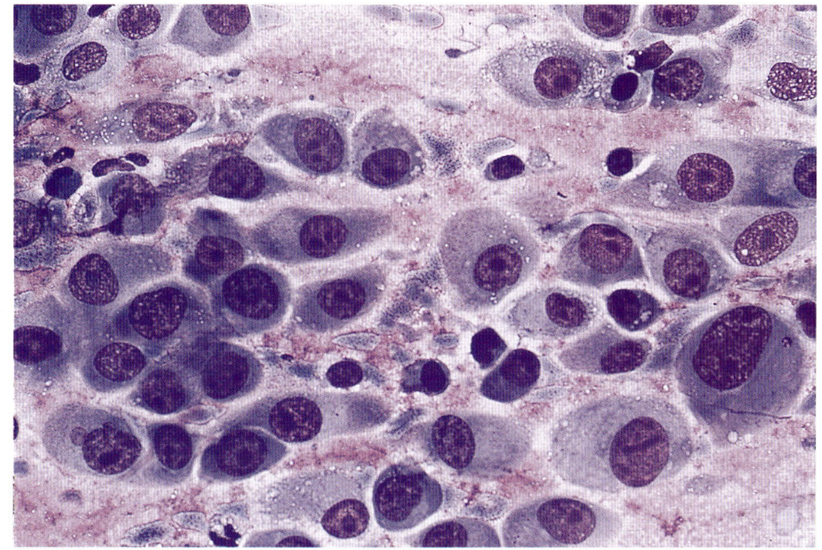
B

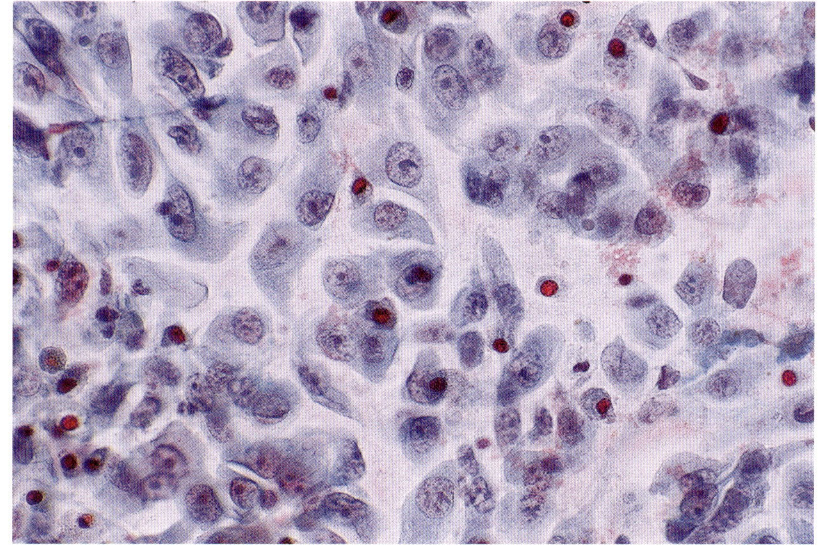
C

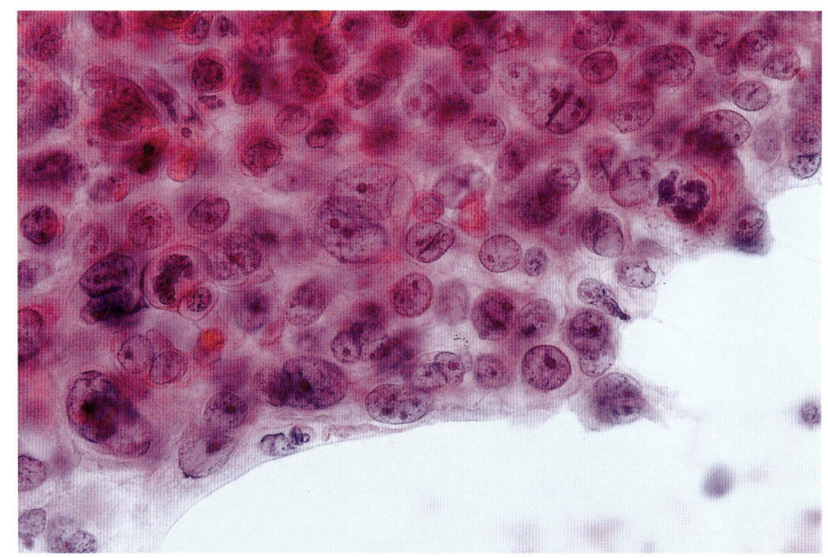

A

B

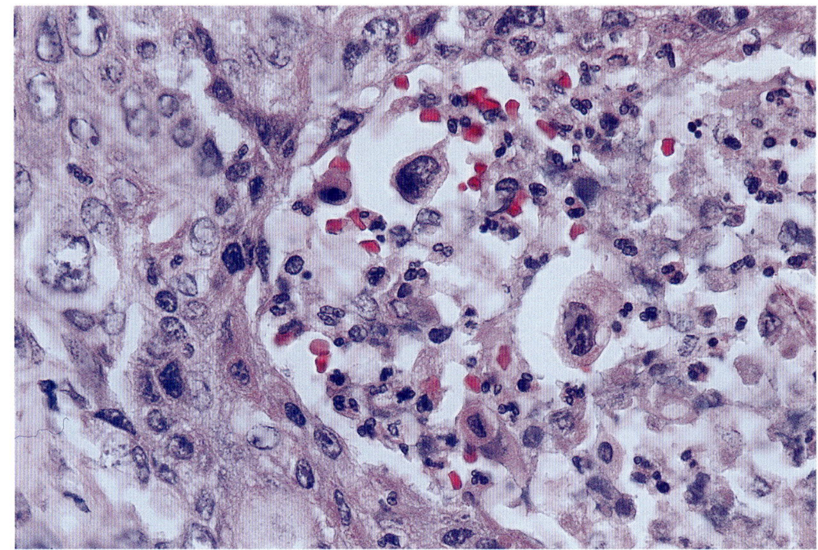

C

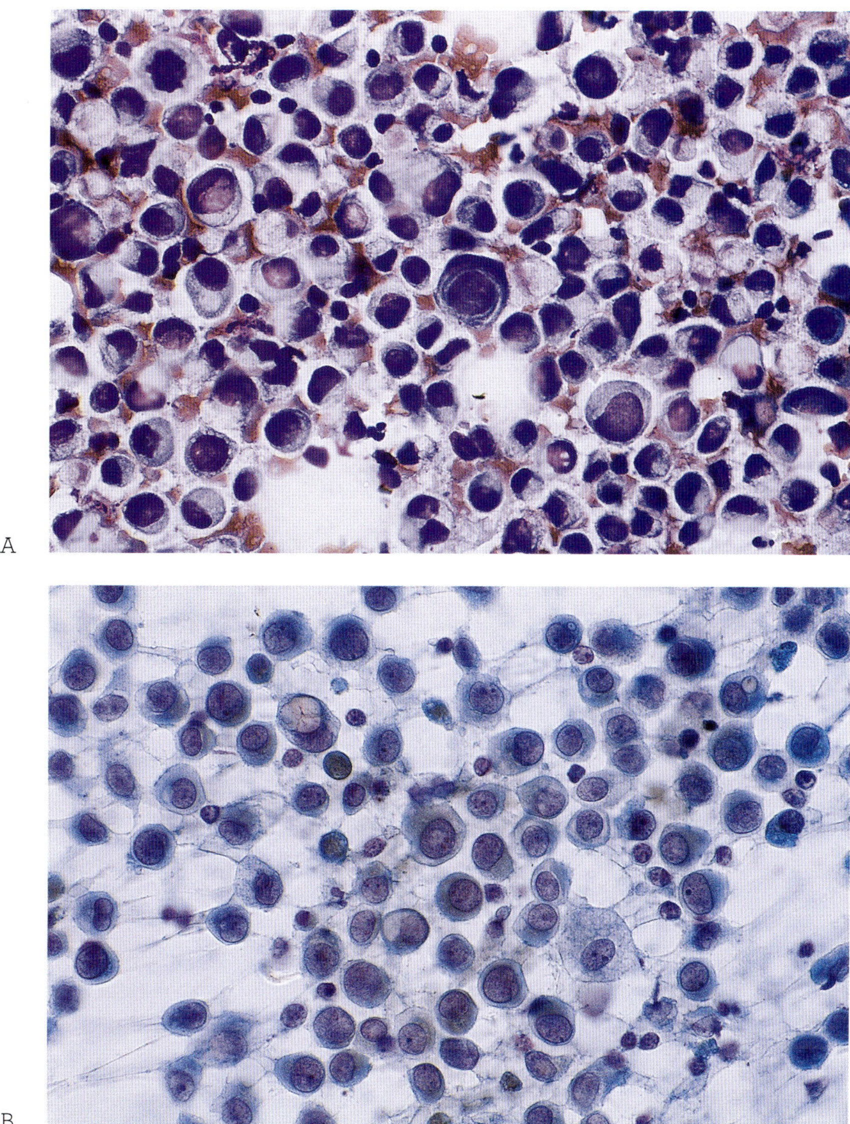

Figure 10–10. *Fine-needle aspiration biopsy of metastatic adenocarcinoma of urinary tract origin.* (*A,B*) Aspirate smears of a lymph node metastasis from a primary urinary bladder adenocarcinoma show numerous single round cells with thin cytoplasm, occasional large cytoplasmic vacuoles, round nuclei, vesicular uniform chromatin, inconspicuous nucleoli, and a background of necrosis. These features are diagnostic of adenocarcinoma but do not help to establish the primary site. (Diff-Quik stain, (A) ×600; Papanicolaou stain, (B) ×600)

◄

Figure 10–9. *Fine-needle aspiration biopsy of metastatic high-grade urothelial carcinoma.* (*A,B*) Aspirate smears from a mesenteric mass in a patient with a history of invasive bladder carcinoma exhibit a high nuclear grade and squamous features. These findings are not uncommon in primary and metastatic, poorly differentiated urothelial carcinomas. (*C*) Similar features are identified in the tissue sections obtained from the resected primary bladder tumor. (Papanicolaou stain, (A,B) ×600; hematoxylin and eosin stain, (C), ×400)

that of other sites. Its distinction requires clinical–radiographic correlation (Shin and Caraway, 1998). Smears show a marked lack of cohesion, occasional clustering, and numerous mitoses. Cells are small, with scant cytoplasm, nuclear molding, coarse chromatin, and inconspicuous nucleoli. Fine needle aspiration biopsy is even more important in these cases because the diagnosis may alter the clinical management.

Differential diagnosis. Cercariform cells must be differentiated from spindle cells of squamous and mesenchymal origin, particularly in FNAB of masses in deep-seated organs, to exclude non-small-cell lung carcinoma and a low-grade spindle-cell neoplasm. Powers and Elbadawi state that cells of low-grade spindle neoplasms often have a mixture of unipolar and bipolar cells with frayed and often tapered cytoplasm, in contrast to the unipolar and flattened ends of cercariform urothelial cells. Squamous cells may have cytoplasmic tails ("tadpole" and "strap" cells), but their nucleus is usually pyknotic, their cytoplasm is keratinized, and their tail (often bipolar) is tapered into a frayed, irregular end rather than into a flat end (Powers and Elbadawi, 1995). The presence of numerous cercariform cells in an FNAB specimen favors a diagnosis of urothelial rather than nonsmall cell lung carcinoma (Renshaw and Madge, 1997).

Cercariform cells may be identified in the FNAB material of approximately one-third of cases of poorly differentiated epithelial malignancies metastatic to different organs and from different primary sites. More than ten cercariform cells have been found in metastases from pancreatic adenocarcinoma metastatic to the liver, squamous carcinoma of the cervix to lymph nodes, and primary lung adeno- and adenosquamous carcinomas (Hida and Gupta, 1999). However cercariform cells are more numerous in urothelial carcinomas (Hida and Gupta, 1999; Renshaw and Madge, 1997).

In cases of small-cell carcinoma, the diagnosis of a primary tumor usually antecedes the FNAB, and a panel of immunocytochemical studies may be useful for differentiating other small-cell neoplasms, including neuroblastoma, Ewing's sarcoma, melanoma, and lymphoma. These studies, although useful for the diagnosis, have limited value for distinguishing the primary site (Shin and Caraway, 1998).

DIAGNOSTIC ACCURACY

Fine Needle Aspiration Biopsy of Primary Renal Pelvis Masses

Most series of FNABs of renal pelvis masses are composed of a few cases described in conjunction with FNAB of renal masses (Murphy et al., 1985; Nguyen, 1987). However, fluoroscopic percutaneous FNAB of renal pelvis masses has a success rate of 85% (Nguyen, 1987).

In a compilation of two series, 12 cases of primary urothelial carcinoma of the renal pelvis were accurately diagnosed by FNAB, with no false-positive or false-negative diagnoses (Nguyen and Schumann, 1997; Santamaria et al., 1995).

Fine Needle Aspiration Biopsy of Metastatic Urothelial Carcinoma

Fine needle aspiration biopsy is particularly useful for evaluating metastatic disease in patients with urologic cancer. Noninvasive techniques for assessing the lymphatic spread of urologic neoplasms have proved to be of limited value, and a positive FNAB result may prevent a radical operation in patients with clinically localized disease. Percutaneous FNAB of retroperitoneal pelvic and abdominal lymph nodes detected metastases in 20% of patients with clinically localized disease (Wajsman et al., 1982a).

The method is inexpensive, with low morbidity and no mortality, and can have high accuracy in diagnosing local tumor spread or distant metastases (Wajsman et al., 1982b). However, for a negative result to be used in clinical management, the cytologic interpretation must be made with samples that have adequate cellularity.

Percutaneous FNAB of regional nodal chains in patients with urologic cancer has shown an overall diagnostic accuracy of 93% to 100%; thus, it is more accurate and specific than lymphography (Luciani and Piscioli, 1985, 1986; Piscioli et al., 1985a). Others reported a false-negative rate of 5% to 6% and no false-positive diagnoses

(Cochand-Priollet et al., 1987; Piscioli et al., 1985b).

Thus, in the management of urologic cancer, a positive aspiration biopsy should be considered diagnostically definitive in determining the tumor stage. Negative aspirations may also be accepted as definitive of diagnostic staging, because the sensitivity of the method is more than 93% (Luciani et al., 1983).

FINE NEEDLE ASPIRATION BIOPSY IN RENAL ALLOGRAFT REJECTION

Needle core biopsy remains the gold standard for diagnosis of renal allograft rejection and other intragraft complications. Noninvasive methods such as urine and blood examination have proved unsuccessful. However, FNAB has been reported to be equally cost effective, rapid, simple, and accurate, with high intra- and interobserver reproducibility and with higher than 90% needle core biopsy correlation (Belitsky et al., 1985; Delaney et al., 1993; Manfro et al., 1995; von Willebrand and Hayry, 1984). The major advantages of FNAB in renal allograft rejection are listed in Table 10–5.

However, despite these apparent advantages, the use of FNAB in the diagnosis of renal allograft rejection or complications has declined in the last two decades, and currently it is not widely used. It appears that the drawbacks, although fewer, are more limiting than the advantages for the acceptance of FNAB.

When a sample is not adequate, or when the results of FNAB are inconclusive or do not agree with the clinical diagnosis, a needle core biopsy must be performed. This delay, in the face of a difficult clinical problem, is a major argument against the use of FNAB in the management of renal transplant patients. Needless to say, information derived from FNAB is more limited than that obtained from a needle core biopsy, and interpreting smears accurately requires strong clinical correlation and great expertise on the part of the cytopathologist. Furthermore, improvements in radiologic techniques, the use of thinner needles to procure adequate material for histologic examination, and the reduced number of complications have made needle core biopsy the preferred method in most transplant centers (Gecim et al., 1995).

It has been proposed that FNAB may be a useful adjunct in the diagnosis and monitoring of acute cellular rejection, especially in the first three months (Hayry and Lautenschlager, 1992). However, no data demonstrating improved graft survival with FNAB monitoring are available (Bogman, 1995). At present, FNAB is not a replacement for needle core biopsy and does not allow comprehensive evaluation of the transplant. However, for those who use FNAB to monitor the outcome of a renal transplant, the low frequency of never-functioning grafts is due at least in part to the frequent use of

Table 10–5. Use of Fine Needle Aspiration Biopsy in Renal Allograft Rejection

Advantages	*Disadvantages*
Less traumatic	Results may be inconclusive
Very low risk of complications	Multiparametric diagnosis of rejection
Frequent samples, even daily	Diagnose the presence, but not the type or degree of acute cellular rejection
Feasible in patients with bleeding	Does not diagnose acute vascular rejection
Immediate MGG stain in cytospin smears	Does not diagnose chronic rejection
Rapid diagnosis (two to four hours)	
Provides a specimen for special studies	

Abbreviation: MGG = May-Grunwald-Giemsa.

Table 10–6. Features of Renal Cellular Allograft Rejection on Fine Needle Aspiration Biopsy

Tubular cells: large and swollen to small and necrotic
Mononuclear cell inflammation: blastic cells, macrophages
Activated endothelial cells
Platelet clumps

FNAB to monitor the allograft, allowing prompt and appropriate treatment of rejection episodes or cyclosporine toxicity (Hayry, 1989).

Technical aspects, diagnostic criteria, and criteria for specimen adequacy have been described. The correct interpretation of inflammatory cell counts, that is, of lymphoblasts, plasmablasts, plasma cells, monoblasts, and macrophages, requires comparison with the peripheral blood count and application of a correction factor. The sum of the corrected values represents the intensity of inflammation. Abnormalities of parenchymal cells are graded only when epithelial and inflammatory cells are available in sufficient numbers (Belitsky et al., 1985; Hayry, 1989; Hayry et al., 1990). Glomeruli may be present, but they can be evaluated only by special methods (Hayry, 1989).

Special studies such as flow cytometry, immunocytochemistry, and molecular analysis, applied for further cellular characterization and evaluation of inflammatory mediators in the aspirate material, could significantly improve the diagnostic accuracy of FNAB and the management of kidney transplant patients (Chandraker, 1999; Kim and Nast, 1993; Nishioka et al., 1992; Oliveira et al., 1997a; 1998).

Cytology. Allograft FNAB cytology may be useful during the first weeks after renal transplantation, when the main clinical considerations include acute cellular rejection, acute tubular necrosis, cyclosporine toxicity, and viral infection, particularly CMV and human polyomavirus–associated nephropathy. The last is a recently recognized symptomatic infection caused by the BK strain of the polyomavirus family (Nickeleit et al., 2000).

In FNAB smears from patients with acute cellular rejection (Table 10–6), the tubular cells range from small and degenerated to large and swollen and lack cytoplasmic vacuolization, although occasional cytoplasmic vacuoles may be seen (Fig. 10–11). Mononu-

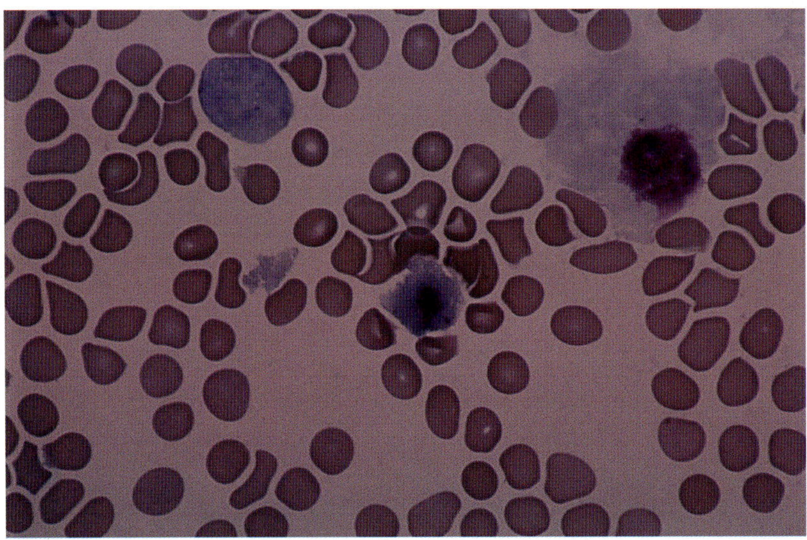

Figure 10–11. *Renal tubular cells in acute cellular rejection.* During episodes of acute rejection, tubular cells may be considerably smaller than intact tubular cells, with pyknotic nuclei and condensed cytoplasm that stains dark blue with May-Grunwald-Giemsa stain (×600). Courtesy of Dr. Cynthia C. Nast, Department of Pathology and Laboratory Medicine, Cedars-Sinai Medical Center and UCLA School of Medicine, Los Angeles, CA.

Table 10–7. Features of Acute Tubular Necrosis on Fine Needle Aspiration Biopsy

Swollen tubular cells
"Puffy" vacuolization
Mononuclear cell inflammation: minimal or absent

clear cell inflammation is prominent, with the presence of immature mononuclear cells, macrophages, and platelet clumps (Egidi, 1990). Fine needle aspiration biopsy cannot diagnose acute vascular rejection. However, lymphocyte and macrophage predominance with an absence of blastic cells, and an increased yield of activated endothelial cells with prominent nucleoli, are suggestive of acute vascular rejection (Egidi, 1990; Hayry, 1989). Few macrophages may be present during the first days after transplantation in stable patients; however, numerous macrophages are characteristic of irreversible rejection (Oliveira et al., 1997b). Necrotic tubular cells are seen in late and irreversible rejection.

In samples from transplanted patients and acute tubular necrosis (Table 10–7), tubular cells appear swollen and often vacuolated in a characteristic "puffy" pattern, containing large, nonisometric vacuoles (Fig. 10–12). No or only minimal inflammation may be present. These changes are reversible with improving graft function (Egidi, 1990; Hayry et al., 1990).

In samples exhibiting cyclosporine toxicity (Table 10–8), the tubular cells are also swollen and vacuolated. In contrast to acute tubular necrosis, the vacuoles are small and uniform, with an isometric pattern (Fig. 10–13A,B). Endothelial cell vacuolization and erythrophagocytosis can be seen. Mononuclear cell inflammation may be minimal or absent (Egidi, 1990). This diagnosis can be confirmed by immunocytochemistry with the use of antibodies to demonstrate cyclosporine deposits in the tubular cells (Hayry et al., 1990). It has been suggested that isometric vacuolization is more readily seen in FNAB specimens than in needle core biopsy specimens (Hayry, 1989). Similar cellular changes have been described in tacrolimus toxicity and with administration of hyperosmotic fluids such as sucrose ("osmotic nephrosis") (Stahl et al., 1998). These changes are reversible when the dose of the drug is lowered.

The microscopic characteristics of the cytomegalovirus cytopathic effect of CMV are

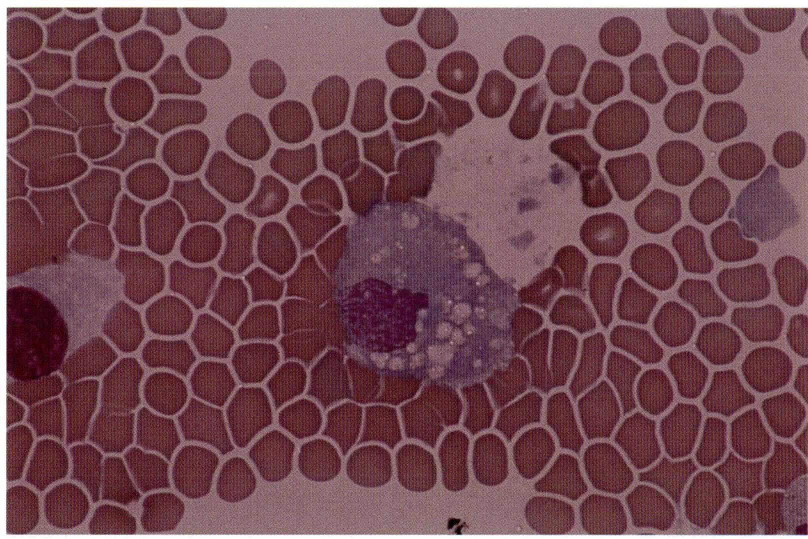

Figure 10–12. *Renal tubular cells in acute tubular necrosis.* Necrotic tubular cells display cytoplasmic swelling and irregularly shaped vacuoles. (May-Grunwald-Giemsa stain, ×600). Courtesy of Dr. Cynthia C. Nast, Department of Pathology and Laboratory Medicine, Cedars-Sinai Medical Center and UCLA School of Medicine, Los Angeles, CA.

Table 10–8. Features of Cyclosporine Toxicity on Fine Needle Aspiration Biopsy

Swollen tubular cells
Isometric cytoplasmic vacuolization*
Endothelial cytoplasmic vacuolization may be present
Erythrophagocytosis may be present
Mononuclear cell inflammation: minimal or absent

*More evident in FNAB biopsy specimens than in core biopsy specimens.

well known. The use of molecular biology techniques in FNAB material has improved the accuracy of the diagnosis of this infection in renal allografts (Murer et al., 1992; Nast et al., 1991).

Polyomavirus produces a large intranuclear inclusion within the slightly enlarged renal tubular cells. The inclusion is homogeneous and basophilic and occupies almost the entire nucleus, with a thin rim of clear-

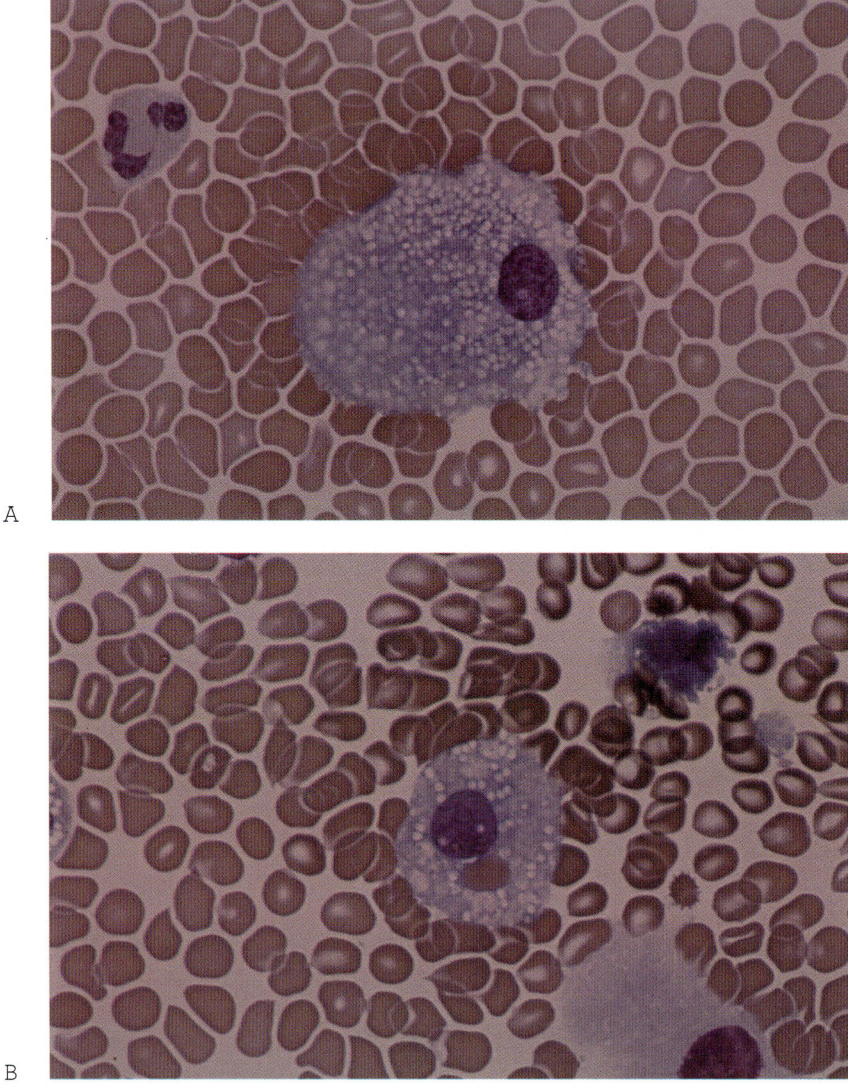

Figure 10–13. *Renal tubular cells in cyclosporine/FK506 nephrotoxicity.* (*A*) The tubular cells have swollen cytoplasm that contains small isometric vacuoles. (*B*) Occasionally, tubular cell phagocytosis of erythrocytes is evident, a process that occurs in vitro while cells are in the medium before aspirate processing. (May-Grunwald-Giemsa stain, ×600). Courtesy of Dr. Cynthia C. Nast, Department of Pathology and Laboratory Medicine, Cedars-Sinai Medical Center and UCLA School of Medicine, Los Angeles, CA.

ing around the inclusion. Detection of polyomavirus DNA in the plasma and FNAB material by polymerase chain reaction was reported to be closely associated with nephropathy (Nickeleit et al., 2000). Large numbers of cells with a viral cytopathic effect in urine cytology are present in these patients and are used as an adjunctive diagnostic tool (Drachenberg et al., 1999; Nickeleit et al., 1999).

In addition, the differential diagnosis of intranuclear viral inclusion bodies in a renal allograft includes infections with adenovirus and herpes simplex virus, which have not yet been reported in renal transplant FNAB.

The search for less invasive procedures for diagnosing acute and chronic renal allograft rejection continues. I agree that, with the application of molecular biology techniques to the aspirated material, FNAB offers a major advantage to the clinician in understanding transplant biology, and that the final aim is to provide the patient with more accurate therapy based on a more accurate diagnosis (Helderman et al., 1988).

REFERENCES

Agarwal, S. K., Walmsley, B. H., Marley, N. J. (1995). Primary amyloidosis of urinary bladder. *J R Soc Med* 88: 171P–172P.

Atta, H. R. (1983). Presumed metastatic transitional cell carcinoma of the choroids. *Br J Ophthalmol* 67: 830–3.

Bardales, R. H. (1998). Fine needle aspiration cytology of papillary neoplasms. *Clin Lab Med* 18: 373–99, v.

Begara Morillas, F., Silmi Moyano, A., Hermida Gutierrez, J., Chicharro Almarza, J., Fernandez Acenero, M. J., Martin Rodilla, C., Ramirez Fernandez, J. C., Rapariz Gonzalez, M., Salinas Casado, J., Resel Estevez, L. (1996). [Lymphoproliferative pathology of the genitourinary tract. Report of 6 cases and review of the literature]. *Arch Esp Urol* 49: 562–70.

Belitsky, P., Campbell, J., Gupta, R. (1985). Serial biopsy controlled evaluation of fine needle aspiration in renal allograft rejection. *Lab Invest* 53: 580–5.

Bogman, M. J. (1995). Fine-needle aspiration biopsy of the renal transplant—is it worthwhile or a waste of time? [editorial]. *Nephrol, Dialy, Transplant* 10: 2182–4.

Bonsib, S. M. (1996). HMB-45 reactivity in renal leiomyomas and leiomyosarcomas. *Mod Pathol* 9: 664–9.

Bonsib, S. M., Eble, J. N. (1997). Renal pelvis and ureter. In *Urologic Surgical Pathology* (D. G. Bostwick, J. N. Eble, eds.), pp. 150–65. Mosby, St. Louis.

Bree, R. L., Schultz, S. R., Hayes, R. (1990). Large infiltrating renal transitional cell carcinomas: CT and ultrasound features. *J Comput Assist Tomogr* 14: 381–5.

Brenner, D. W., Schellhammer, P. F. (1987). Upper tract urothelial malignancy after cyclophosphamide therapy: a case report and literature review. *J Urol* 137: 1226–7.

Bringuier, P. P., McCredie, M., Sauter, G., Bilous, M., Stewart, J., Mihatsch, M. J., Kleihues, P., Ohgaki, H. (1998). Carcinomas of the renal pelvis associated with smoking and phenacetin abuse: p53 mutations and polymorphism of carcinogen-metabolising enzymes. *Int J Cancer* 79: 531–6.

Bucin, E., Andreasson, L., Bjorlin, G. (1982). Metastases in the oral cavity. Case reports. *Int J Oral Surg* 11: 321–5.

Caraway, N. P., Wojcik, E. M., Katz, R. L., Ro, J. Y., Ordonez, N. G. (1995). Cytologic findings of collecting duct carcinoma of the kidney. *Diagn Cytopathol* 13: 304–9.

Chandraker, A. (1999). Diagnostic techniques in the work-up of renal allograft dysfunction—an update. *Curr Opin Nephrol Hypertens* 8: 723–8.

Chen, K. T., Workman, R. D., Flam, M. S., DeKlotz, R. J. (1983). Carcinosarcoma of renal pelvis. *Urology* 22: 429–31.

Chitale, S. V., Morrow, D. R., Patel, R., Gaches, C. G., Ball, R. Y. (1997). Cutaneous metastases from transitional cell carcinomas of the bladder and renal pelvis. *Br J Urol* 79: 292–3.

Chow, L. T., Chan, S. K., Chow, W. H. (1994). Fine needle aspiration cytodiagnosis of leiomyosarcoma of the renal pelvis. A case report with immunohistochemical study. *Acta Cytol* 38: 759–63.

Chow, W. H., Lindblad, P., Gridley, G., Nyren, O., McLaughlin, J. K., Linet, M. S., Pennello, G. A., Adami, H. O., Fraumeni, J. F., Jr. (1997). Risk of urinary tract cancers following kidney or ureter stones [see comments]. *J Natl Cancer Inst* 89: 1453–7.

Cochand-Priollet, B., Roger, B., Boccon-Gibod, I., Ferrand, J., Faure, B., Blery, M. (1987). Retroperitoneal lymph node aspiration biopsy in staging of pelvic cancer: a cytological study of 228 consecutive cases. *Diagn Cytopathol* 3: 102–7.

Cohen, R. J., Stanley, J. C., Dawkins, H. J. (1999). Lymphoepithelioma-like carcinoma of the renal pelvis. *Pathology* 31: 434–5.

de la Torre, M., Larsson, E. (1995). Fine-needle aspiration cytology of carcinosarcoma of the parotid gland: cytohistological and immunohistochemical findings. *Diagn Cytopathol* 12: 350–3.

Dekmezian, R., Sneige, N., Shabb, N. (1991). Papillary renal-cell carcinoma: fine-needle aspiration of 15 cases. *Diagn Cytopathol* 7: 198–203.

Delaney, V., Ling, B. N., Campbell, W. G., Bourke, J. E., Fekete, P. S., O'Brien, D. P. d., Taylor, A. T., Whelchel, J. D. (1993). Comparison of fine-needle aspiration biopsy, Doppler ultrasound, and radionuclide scintigraphy in the diagnosis of acute allograft dysfunction in renal transplant recipients: sensitivity, specificity, and cost analysis. *Nephron* 63: 263–72.

Dias, R., Fernandes, M., Patel, R. C., de Shadarevian, J. J., Lavengood, R. W. (1979). Amyloidosis of renal pelvis and urinary bladder. *Urology* 14: 401–4.

Drachenberg, C. B., Beskow, C. O., Cangro, C. B., Bourquin, P. M., Simsir, A., Fink, J., Weir, M. R., Klassen, D. K., Bartlett, S. T., Papadimitriou, J. C. (1999). Human polyoma virus in renal allograft biopsies: morphological findings and correlation with urine cytology. *Hum Pathol* 30: 970–7.

Egidi, M. F. (1990). Fine-needle aspiration biopsy in renal transplantation: a review of cytologic features. *Diagn Cytopathol* 6: 330–5.

Ehara, H., Takahashi, Y., Saitoh, A., Kawada, Y., Shimokawa, K., Kanemura, T. (1997). Clear cell melanoma of the renal pelvis presenting as a primary tumor. *J Urol* 157: 634.

Epstein, N. A. (1977). The cytologic appearance of metastatic transitional cell carcinoma. *Acta Cytol* 21: 723–5.

Essenfeld, H., Manivel, J. C., Benedetto, P., Albores-Saavedra, J. (1990). Small cell carcinoma of the renal pelvis: a clinicopathological, morphological and immunohistochemical study of 2 cases. *J Urol* 144: 344–7.

Fadl-Elmula, I., Gorunova, L., Mandahl, N., Elfving, P., Lundgren, R., Rademark, C., Heim, S. (1999). Cytogenetic analysis of upper urinary tract transitional cell carcinomas. *Cancer Genet Cytogenet* 115: 123–7.

Finley, J. L., Silverman, J. F., Dabbs, D. J. (1988). Fine-needle aspiration cytology of pulmonary carcinosarcoma with immunocytochemical and ultrastructural observations. *Diagn Cytopathol* 4: 239–43.

Fox, M., Hammond, J. C., Knox, R., Underwood, J. C. (1984). Localised primary amyloidosis of the renal pelvis. *Br J Urol* 56: 223–4.

Frasier, B. L., Wachs, B. H., Watson, L. R., Tomasulo, J. P. (1988). Malignant melanoma of the renal pelvis presenting as a primary tumor. *J Urol* 140: 812–14.

Friedman, H. D., Nsouli, I. S., Krauss, D. J., Vohra, S., Powers, C. N. (1999). Transitional cell carcinoma arising in a pyelocaliceal cyst. An unusual cystic renal lesion with cytologic and imaging findings. *Virchows Arch* 434: 459–62.

Fukunaga, M., Ushigome, S. (1998). Lymphoepithelioma-like carcinoma of the renal pelvis: a case report with immunohistochemical analysis and in situ hybridization for the Epstein-Barr viral genome. *Mod Pathol* 11: 1252–6.

Gecim, I. E., Rowlands, P., McDicken, I., Bakran, A., Sells, R. A., Gladman, M., Gillies, J. (1995). Core needle biopsy in renal transplantation. *Int Urol Nephrol* 27: 357–63.

Girgis, S., Ramzy, I., Baer, S. C., Schwartz, M. R. (1999). Fine needle aspiration diagnosis of transitional cell carcinoma metastatic to the brain. A case report. *Acta Cytol* 43: 235–8.

Gorczyca, W., Olszewski, W., Tuziak, T., Kram, A., Woyke, S., Ucinski, M. (1992). Fine needle aspiration cytology of rare malignant tumors of the breast. *Acta Cytol* 36: 918–26.

Grabstald, H., Whitmore, W. F., Melamed, M. R. (1971). Renal pelvic tumors. *JAMA* 218: 845–54.

Guillou, L., Duvoisin, B., Chobaz, C., Chapuis, G., Costa, J. (1993). Combined small-cell and transitional cell carcinoma of the renal pelvis. A light microscopic, immunohistochemical, and ultrastructural study of a case with literature review [see comments]. *Arch Pathol Lab Med* 117: 239–43.

Guo, M., Lemos, L. B., Baliga, M., Fowler, J. E., Jr. (1998). Fine needle aspiration biopsy of the penis: transitional cell carcinoma of the urinary bladder with mucinous differentiation. *Urology* 51: 1040–2.

Halliday, B. E., Silverman, J. F., Finley, J. L. (1998). Fine-needle aspiration cytology of amyloid associated with nonneoplastic and malignant lesions. *Diagn Cytopathol* 18: 270–5.

Hartman, D. S., Davidson, A. J., Davis, C. J., Jr., Goldman, S. M. (1988). Infiltrative renal lesions: CT-sonographic-pathologic correlation. *Am J Roentgenol* 150: 1061–4.

Hayes, M. M., Jones, E. C., Verma, A. K., Lim, C. H., Milne, G., Tse, E. (1992). Transitional cell carcinoma of the renal pelvis metastatic to the metacarpal. A case report correlating cytologic and histologic findings. *Acta Cytol* 36: 946–50.

Hayry, P. J. (1989). Fine-needle aspiration biopsy in renal transplantation [clinical conference]. *Kidney Int* 36: 130–41.

Hayry, P. J., Lautenschlager, I. (1992). Fine-needle aspiration biopsy in transplantation pathology. *Semin Diagn Pathol* 9: 232–7.

Hayry, P. J., von Willebrand, E., Lautenschlager, I., Taskinen, E., Krogerus, L. (1990). Diagnosis of rejection: role of fine-needle aspiration biopsy. *Transplant Proc* 22: 2597–6000.

Helderman, J. H., Hernandez, J., Sagalowsky, A., Dawidson, I., Glennie, J., Womble, D., Toto, R. D., Brinker, K., Hull, A. R. (1988). Confirmation of the utility of fine needle aspiration biopsy of the renal allograft. *Kidney Int* 34: 376–81.

Hida, C. A., Gupta, P. K. (1999). Cercariform cells: are they specific for transitional cell carcinoma? *Cancer* 87: 69–74.

Howat, A. J., Scott, E., Mackie, D. B., Pinkerton, J. R. (1983). Adenosquamous carcinoma of the renal pelvis. *Am J Clin Pathol* 79: 731–3.

Igel, T. C., Engen, D. E., Banks, P. M., Keeney, G. L. (1991). Renal plasmacytoma: Mayo Clinic experience and review of the literature. *Urology* 37: 385–9.

Ji, X., Li, W. (1994). Primary carcinoid of the renal pelvis. *J Environ Pathol Toxicol Oncol* 13: 269–71.

Johnson, T. L., Kini, S. R. (1993). Cytologic features of metastatic transitional cell carcinoma. *Diagn Cytopathol* 9: 270–8.

Kiff, E. S., Hulton, N. (1983). A case of colonic metastasis from a primary renal pelvic carcinoma. *Clin Oncol* 9: 353–4.

Kim, Y. J., Nast, C. C. (1993). Identification of tubular cell nephron segment origin in renal fine-needle aspirates. *Mod Pathol* 6: 150–4.

Kitamura, M., Miyanaga, T., Hamada, M., Nakata, Y., Satoh, Y., Terakawa, T. (1997). Small cell carcinoma of the kidney: case report. *Int J Urol* 4: 422–4.

Kumar, S., Kumar, D., Cowan, D. F. (1992). Transitional cell carcinoma with rhabdoid features. *Am J Surg Pathol* 16: 515–21.

Lang, E. K. (1977). Renal cyst puncture and aspiration: a survey of complications. *Am J Roentgenol* 128: 723–7.

Lindell, O. I., Grein, H. U., Schreiter, F. J. (1991). Opposite renal pelvic metastasis of renal adenocarcinoma. Case report. *Scand J Urol Nephrol* 25: 165–7.

Luciani, L., Piscioli, F. (1985). Experience with modified Chiba needle in staging of genitourinary tumors. *Urology* 25: 568–72.

Luciani, L., Piscioli, F., Menichelli, E., Pusiol, T. (1983). The value and role of percutaneous pelvic lymph node aspiration biopsy in definitive staging of prostatic and bladder carcinoma. *Eur Urol* 9: 216–20.

Luciani, L., Piscioli, F., Pusiol, T., Scappini, P. (1986). The value of aspiration cytology in the definitive staging of bladder carcinoma. *Br J Urol* 58: 26–30.

Maeda, K., Hawkins, E. T., Oh, H. K., Kini, S. R., Van Dyke, D. L. (1986). Malignant lymphoma in transplanted renal pelvis. *Arch Pathol Lab Med* 110: 626–9.

Manfro, R. C., Goncalves, L. F., de Moura, L. A. (1995). Reproducibility of fine-needle aspiration biopsy in the diagnosis of acute rejection of renal allografts. *Nephrol Dial Transplant* 10: 2306–9.

Mazzucchelli, L., Studer, U. E., Kraft, R. (1995). Small-cell undifferentiated carcinoma of the renal pelvis 26 years after subdiaphragmatic irradiation for non-Hodgkin's lymphoma. *Br J Urol* 76: 403–4.

Melamed, M. R., Reuter, V. E. (1993). Pathology and staging of urothelial tumors of the kidney and ureter. *Urol Clin North Am* 20: 333–47.

Murer, L., Zacchello, G., Basso, G., Palu, G., Barbato, A., Zanon, G. F., Zacchello, F. (1992). Early and rapid diagnosis of CMV infection by nonradioactive in situ hybridization in pediatric kidney transplant recipients [see comments]. *Nephron* 60: 25–9.

Murphy, M. N., Alguacil-Garcia, A., MacDonald, R. G. (1986). Primary amyloidosis of renal pelvis with duplicate collecting system. *Urology* 27: 470–3.

Murphy, W. M., Zambroni, B. R., Emerson, L. D., Moinuddin, S., Lee, L. H. (1985). Aspiration biopsy of the kidney. Simultaneous collection of cytologic and histologic specimens. *Cancer* 56: 200–5.

Nast, C. C., Wilkinson, A., Rosenthal, J. T., Blifeld, C., Ettenger, R. B., Beaumont, P., Danovitch, G. M. (1991). Diagnosis of cytomegalovirus infection in renal allografts using fine needle aspiration biopsy. *Transplant Proc* 23: 1354.

Nguyen, G. K. (1987). Percutaneous fine-needle aspiration biopsy cytology of the kidney and adrenal. *Pathol Annu* 22: 163–91.

Nguyen, G. K., Schumann, G. B. (1997). Needle aspiration cytology of low-grade transitional cell carcinoma of the renal pelvis. *Diagn Cytopathol* 16: 437–41.

Nickeleit, V., Hirsch, H. H., Binet, I. F., Gudat, F., Prince, O., Dalquen, P., Thiel, G., Mihatsch, M. J. (1999). Polyomavirus infection of renal allograft recipients: from latent infection to manifest disease. *J Am Soc Nephrol* 10: 1080–9.

Nickeleit, V., Hirsch, H. H., Zeiler, M., Gudat, F., Prince, O., Thiel, G., Mihatsch, M. J. (2000). BK-virus nephropathy in renal transplants—tubular necrosis, MHC-class II expression and rejection in a puzzling game. *Nephrol Dial Transplant* 15: 324–32.

Nishioka, T., Negita, M., Ikegami, M., Imanishi, M., Ishii, T., Uemura, T., Kunikata, S., Kanda, H., Matsuura, T., Akiyama, T., et al. (1992). Fine needle aspiration biopsy of human renal transplants. *Transplant Proc* 24: 1551–3.

Oliveira, G., Xavier, P., Mendes, A., Guerra, L. E. (1997a). Cultures of aspiration biopsy specimens in the immunological monitoring of renal transplants. *Nephron* 76: 310–14.

Oliveira, G., Xavier, P., Murphy, B., Neto, S., Mendes, A., Sayegh, M. H., Guerra, L. E. (1998). Cytokine analysis of human renal allograft aspiration biopsy cultures supernatants predicts acute rejection. *Nephrol Dial Transplant* 13: 417–22.

Oliveira, J. G., Xavier, P., Neto, S., Mendes, A. A., Guerra, L. E. (1997b). Monocytes-macrophages and cytokines/chemokines in fine-needle aspiration biopsy cultures: enhanced interleukin-1 receptor antagonist synthesis in rejection-free kidney transplant patients. *Transplantation* 63: 1751–6.

Orsatti, G., Corgan, F. J., Goldberg, S. A. (1993). Carcinosarcoma of urothelial organs: sequential involvement of urinary bladder, ureter, and renal pelvis. *Urology* 41: 289–91.

Ozsahin, M., Zouhair, A., Villa, S., Storme, G., Chauvet, B., Taussky, D., Gouders, D., Ries, G., Bontemps, P., Coucke, P. A., Mirimanoff, R. O. (1999). Prognostic factors in urothelial renal pelvis and ureter tumours: a multicentre Rare Cancer Network study. *Eur J Cancer* 35: 738–43.

Parham, D. M., Weeks, D. A., Beckwith, J. B. (1994). The clinicopathologic spectrum of putative extrarenal rhabdoid tumors. An analysis of 42 cases studied with immunohistochemistry or electron microscopy. *Am J Surg Pathol* 18: 1010–29.

Penn, I. (1995). Primary kidney tumors before and after renal transplantation. *Transplantation* 59: 480–5.

Phan, S. T., Meng, M., Weidner, N. (1997). Collision tumor: a peripheral neuroepithelioma and a transitional-cell carcinoma occurring simultaneously in the renal pelvis. *Ann Diagn Pathol* 1: 91–8.

Piscioli, F., Bondi, A., Scappini, P., Luciani, L. (1984). "True" sarcomatoid carcinoma of the renal pelvis. First case report with immunocytochemical study. *Eur Urol* 10: 350–5.

Piscioli, F., Pusiol, T., Leonardi, E., Luciani, L. (1985a). Role of percutaneous pelvic node aspiration cytology in the management of bladder carcinoma. *Acta Cytol* 29: 37–43.

Piscioli, F., Scappini, P., Luciani, L. (1985b). Aspiration cytology in the staging of urologic cancer. *Cancer* 56: 1173–80.

Powers, C. N., Elbadawi, A. (1995). "Cercariform" cells: a clue to the cytodiagnosis of transitional cell origin of metastatic neoplasms? *Diagn Cytopathol* 13: 15–21.

Rao, D. S., Krigman, H. R., Walther, P. J. (1996). Enteric type adenocarcinoma of the upper tract urothelium associated with ectopic ureter and renal dysplasia: an oncological rationale for complete extirpation of this aberrant developmental anomaly. *J Urol* 156: 1272–4.

Rao, S. S., Rao, N. N., Venkataratnam, G. (1986). Carcinosarcoma of renal pelvis in a child. A case report. *Ind J Pathol Microbiol* 29: 313–16.

Renshaw, A. A., Madge, R. (1997). Cercariform cells for helping distinguish transitional cell carcinoma from non–small cell lung carcinoma in fine needle aspirates. *Acta Cytol* 41: 999–1007.

Reuter, V. E. (1993). Sarcomatoid lesions of the urogenital tract. *Semin Diagn Pathol* 10: 188–201.

Rohde, D., Jakse, G. (1998). Involvement of the urogenital tract in patients with five or more separate malignant neoplasms. Case and review. *Eur Urol* 34: 512–17.

Rubin, B. E. (1979). Computed tomography in the evaluation of renal lymphoma. *J Comput Assist Tomogr* 3: 759–64.

Rudrick, B., Nguyen, G. K., Lakey, W. H. (1995). Carcinoid tumor of the renal pelvis: report of a case with positive urine cytology. *Diagn Cytopathol* 12: 360–3.

Saito, S. (1998). Solitary cutaneous metastasis of superficial bladder cancer. *Urol Int* 61: 126–7.

Saitoh, H., Hida, M., Nakamura, K., Satoh, T. (1982). Distant metastasis of urothelial tumors of the renal pelvis and ureter. *Tokai J Exp Clin Med* 7: 355–64.

Santamaria, M., Jauregui, I., Urtasun, F., Bertol, A. (1995). Fine needle aspiration biopsy in urothelial carcinoma of the renal pelvis. *Acta Cytol* 39: 443–8.

Sengelov, L., Kamby, C., von der Maase, H. (1996). Pattern of metastases in relation to characteristics of primary tumor and treatment in patients with disseminated urothelial carcinoma [see comments]. *J Urol* 155: 111–14.

Shin, H. J., Caraway, N. P. (1998). Fine-needle aspiration biopsy of metastatic small cell carcinoma from extrapulmonary sites. *Diagn Cytopathol* 19: 177–81.

Shintaku, M., Megumi, Y., Maekura, S. (2000). Adenocarcinoma of the renal pelvis with vimentin-positive intracytoplasmic inclusions. *Pathol Int* 50: 48–51.

Silver, T. M., Thornbury, J. R., Teears, R. J. (1977). Renal peripelvic plasmacytoma: unusual radiographic findings. *Am J Roentgenol* 128: 313–15.

Skoog, L., Collins, B. T., Tani, E., Ramos, R. R. (1998). Fine needle aspiration cytology of penile tumors. *Acta Cytol* 42: 1336–40.

Slywotzky, C., Maya, M. (1994). Needle tract seeding of transitional cell carcinoma following fine-needle aspiration of a renal mass. *Abdominal Imaging* 19: 174–6.

Sparwasser, C., Gilbert, P., Mohr, W., Linke, R. P. (1991). Unilateral extended amyloidosis of the renal pelvis and ureter: a case report. *Urol Int* 46: 208–10.

Spires, S. E., Banks, E. R., Cibull, M. L., Munch, L., Delworth, M., Alexander, N. J. (1993). Adenocarcinoma of renal pelvis [published erratum appears in Arch Pathol Lab Med 1994 Sep;118(9):872]. *Arch Pathol Lab Med* 117: 1156–60.

Stahl, M., Vonwiller, H. M., Mihatsch, M. J., Schifferli, J. A. (1998). Prolonged acute renal failure after i.v. immunoglobulin therapy in the refeeding phase of anorexia nervosa. *Nephrol Dial Transplant* 13: 2364–7.

Stewart, J. H., Hobbs, J. B., McCredie, M. R. (1999). Morphologic evidence that analgesic-induced kidney pathology contributes to the progression of tumors of the renal pelvis. *Cancer* 86: 1576–82.

Suen, K. C. (1994a). Kidneys and urinary tract. In *Guides to Clinical Aspiration Biopsy. Retroperitoneum and Intestine*, 2nd ed., (T. S. Kline, ed.), pp. 247–286. Igaku-Shoin, New York.

Suen, K. C. (1994b). Retroperitoneum III: soft tissue tumors. In *Guides to Clinical Aspiration Cytology: Retroperitoneum and Intestine*, 2nd ed., (T. S. Kline, ed.), pp. 85–133. Igaku-Shoin, New York.

Takashi, M., Sakata, T., Sai, Y., Shimoji, T., Miyake, K. (1993). Metastasis of renal cell carcinoma to the contralateral renal pelvis. *Urol Int* 51: 102–4.

Tao, L. C., Davidson, D. D. (1993). Aspiration biopsy cytology of smooth muscle tumors. A cytologic approach to the differentiation between leiomyosarcoma and leiomyoma. *Acta Cytol* 37: 300–8.

Vahlensieck, W., Jr., Riede, U., Wimmer, B., Ihling, C. (1991). Beta-human chorionic gonadotropin-positive extragonadal germ cell neoplasia of the renal pelvis. *Cancer* 67: 3146–9.

von Willebrand, E., Hayry, P. (1984). Reproducibility of the fine-needle aspiration biopsy. Analysis of 93 double biopsies. *Transplantation* 38: 314–16.

Wahl, R. W. (1985). Fine needle aspiration study of metastatic mixed adenosquamous carcinoma of the renal pelvis. A case report. *Acta Cytol* 29: 580–3.

Wajsman, Z., Gamarra, M., Park, J. J., Beckley, S., Pontes, J. E. (1982a). Transabdominal fine needle aspiration of retroperitoneal lymph nodes in staging of genitourinary tract cancer (correlation with lymphography and lymph node dissection findings). *J Urol* 128: 1238–40.

Wajsman, Z., Gamarra, M., Park, J. J., Beckley, S. A., Pontes, J. E., Murphy, G. P. (1982b). Fine-needle aspiration of metastatic lesions and regional lymph nodes in genitourinary cancer. *Urology* 19: 356–60.

Wallmeroth, A., Wagner, U., Moch, H., Gasser, T. C., Sauter, G., Mihatsch, M. J. (1999). Patterns of metastasis in muscle-invasive bladder cancer (pT2–4): An autopsy study on 367 patients. *Urol Int* 62: 69–75.

Weiss, M. A. (1993). Small-cell carcinoma of the urinary tract. Important prognostic and therapeutic implications [editorial; comment]. *Arch Pathol Lab Med* 117: 237–8.

Weithman, A. M., Morrison, G., Hurt, M. A. (1988). Metastatic transitional-cell carcinoma to the mandible: a case report. *Diagn Cytopathol* 4: 156–8.

Yusuf, Y., Belmonte, A. H., Tchertkoff, V. (1996). Fine needle aspiration cytology of a recurrent malignant tumor of the kidney with rhabdoid features in an adult. A case report. *Acta Cytol* 40: 1313–16.

Zajicek, J. (1979). Cytology of infradiaphragmatic organs. 1. Kidney and renal pelvis. *Monogr Clin Cytol* 7: 1–37.

INDEX

Page numbers followed by "b" indicate boxed content; numbers followed by "f" indicate figures; numbers followed by "t" indicate tables.